Plant Responses to Biotic and Abiotic Stresses: Crosstalk between Biochemistry and Ecophysiology

Plant Responses to Biotic and Abiotic Stresses: Crosstalk between Biochemistry and Ecophysiology

Editors

M. Iftikhar Hussain
Adele Muscolo
Mukhtar Ahmed

MDPI • Basel • Beijing • Wuhan • Barcelona • Belgrade • Manchester • Tokyo • Cluj • Tianjin

MDPI

Editors

M. Iftikhar Hussain
Department of Plant Biology & Soil Science
Universidad de Vigo
Vigo
Spain

Adele Muscolo
Department of Agriculture
Mediterranea University
Reggio Calabria
Italy

Mukhtar Ahmed
Department of Agronomy
PMAS Arid Agriculture University
Rawalpindi
Pakistan

Editorial Office
MDPI
St. Alban-Anlage 66
4052 Basel, Switzerland

This is a reprint of articles from the Special Issue published online in the open access journal *Plants* (ISSN 2223-7747) (available at: www.mdpi.com/journal/plants/special_issues/plant_Crosstalk).

For citation purposes, cite each article independently as indicated on the article page online and as indicated below:

LastName, A.A.; LastName, B.B.; LastName, C.C. Article Title. *Journal Name* **Year**, *Volume Number*, Page Range.

ISBN 978-3-0365-8401-0 (Hbk)
ISBN 978-3-0365-8400-3 (PDF)

Cover image courtesy of Iftikhar Hussain

Contents

About the Editors

M. Iftikhar Hussain

Dr. M. Iftikhar Hussain has worked as a research scientist at the Department of Plant Biology and Soil Science, Universidad de Vigo, Spain. His current research interests lie in the agronomical, physiological, biochemical, and stable isotopic responses of allelochemicals, heavy metals, and salinity in the model plant Arabidopsis, cereals, oil seed crops, halophytes, forages, weeds, and fruit trees. It also involves screening, selection, and evaluation of different crop genotypes that can be better adapted to the marginal environment. He graduated in Plant Sciences at the University of Agriculture, Faisalabad, Pakistan, and obtained his PhD with honors in Plant Biology at the Universidad de Vigo, Spain, in 2011. Following his PhD, he served the International Center for Biosaline Agriculture, Dubai, United Arab Emirates, and the University of Sharjah, UAE, in the Applied Biology Department as a postdoctoral researcher. He was appointed as a research scientist in a European Union-funded project (ECOBREED) at the UVIGO (Spain) in August 2019. Among his recognitions, he was awarded the "European Doctorate" and Outstanding Ph.D. Thesis award (Medal, certificate) for his PhD work by the UVIGO (Spain). He is the holder of a "Hosted Scientist Program Award" from the Global Forum for Innovations in Agriculture, Abu Dhabi, UAE, for the years 2016 and 2018. He is the author of more than 142 peer-reviewed publications. He has delivered several invited lectures at different international conferences and serves as a reviewer and guest editor for numerous international journals. He serves as Guest Editor for different special issues in MDPI journals, including *Plants*, *Sustainability*, and *Horticulturae*.

Adele Muscolo

Professor Adele Muscolo is a soil scientist specializing in the areas of soil fertility and soil quality. She graduated in Biological Sciences at Messina University, Italy, and completed her PhD in Food Science at Naples University, Italy. She started her professional career as a researcher at the Mediterranea University of Reggio Calabria, where she is still working as a full professor in soil chemistry and ecology. Since 1988, she has investigated the connections between soil chemistry, biochemistry, and function in regulating ecosystem-level processes in forests, managed ecosystems (e.g., tree nurseries), grasslands, and urban environments. Her main research areas include the characterization and dynamics of soil's organic matter in relation to soil fertility, soil erosion, soil degradation, soil salinization, the relationship between soil organic matter (labile and recalcitrant C fractions) and soil biodiversity in driving soil ecosystem functioning, and environmental issues such as biotic and abiotic pressure on soil. Her research goal is to provide innovative solutions for soil management that will increase productivity while ensuring soil and environmental quality. She is a reviewer for international scientific journals and an associate editor for the Journal of *Forestry Research and Forests*. She is an evaluator of research projects for the European Community, International Funding Research Agencies, and Italian and Foreign Research Ministries. She is the the author/co-author of over 200 papers published in international journals with an impact factor. She is part of the list of "World's Top 2% of scientist in their main subfield discipline for 2020 and 2021" (Ioannidis JPA, Boyack KW, Baas J; 2020-Updated science-wide author databases of standardized citation indicators. PLoS Biol 18(10): e3000918). She was appointed as a member of the Intergovernmental Technical Panel on Soils (ITPS) (FAO-Global Soil Partnership 2022–2025).

Mukhtar Ahmed

Mukhtar Ahmed's research focuses on the impact of climate change/climate variability on crop ecology, crop physiology, cropping systems, and rain-fed ecosystem management through remote sensing and modeling. He has been involved in teaching and research since 2005 and started his career at PMAS Arid Agriculture University in Rawalpindi. During his PhD and visit to Sydney University, Australia, he worked on the application of APSIM as a decision support tool and rainfall forecasting using generalized additive models. He was awarded a young scientist fellowship by the APCC in South Korea. He also won a research productivity award from the Pakistan Council of Science and Technology (PCST), a best research award from the PMAS Arid Agriculture University, and a Publons reviewer award. He was part of the Regional Approaches for Climate Change (REACCH) project in the USA at Washington State University Pullman, which developed multi-model ensemble approaches to minimize the uncertainties in modeling climate extremes. Dr. Ahmed was part of the Swedish University of Agricultural Sciences from 2019–2021, where his research focus was modeling the impact of climate extremes on Nordic agriculture using modern tools. He is involved in the use of statistical and dynamic models as risk management tools to mitigate the challenges of climate change. His current research includes agroecosystems modeling, precision agriculture, modeling the nutrient use efficiency of legume-based cropping systems, forage agronomy, and physiological responses to climate variability and its modeling. He is a Project co-leader in the Model Calibration Group of the Agricultural Model Intercomparison and Improvement Project (AGMIP) for Wheat and Maize Evapotranspiration. He is the editor of well-known peer-reviewed journals.

Preface to "Plant Responses to Biotic and Abiotic Stresses: Crosstalk between Biochemistry and Ecophysiology"

The growth and productivity of crops, fruit trees, legumes, and halophytes are severely impacted by the detrimental effects of biotic and abiotic stress. Biotic stress, caused by living organisms, disrupts the nutrient supply to host plants, leading to reduced vigor and, in some cases, the death of the plant. Biotic stress is a significant contributor to pre- and post-harvest losses in agriculture. On the other hand, abiotic stress, encompassing factors like drought, salinity, cold, heat, and heavy metals, is the primary cause of global crop yield reduction, causing more than a 50 percent decrease in yields of major food, oil-seed, and cash crops. These stresses induce a cascade of morphological, physiological, biochemical, and molecular changes that unfavorably affect plant growth and productivity, interfering with stress tolerance and adaptation.

Drought, salinity, extreme temperatures, and oxidative stress are examples of biotic and abiotic stresses that often occur in combination, compounding the damage to plant cells. The complexity of these stress stimuli results in equally intricate responses from plants. For instance, severe stresses during critical growth phases can directly cause mechanical damage and alter the synthesis of essential macromolecules in cells. Furthermore, all these stresses lead to oxidative damage and the generation of reactive oxygen species (ROS) within plant cells. Plants possess mechanisms to mitigate oxidative damage, such as activating antioxidant enzymes and accumulating compatible solutes that scavenge ROS. However, if the production of activated oxygen surpasses the plant's detoxification capacity, deleterious reactions occur, leading to symptoms like loss of osmotic responsiveness, wilting, and necrosis.

This Special Issue of *Plants* aims to provide a multi-perspective analysis, encompassing gas exchange, metabolomics, proteomics, isotopic, and genomic approaches, to investigate the drivers and specific strategies employed by plants for their enhanced adaptation to stressful growth conditions. By delving into the intricacies of trait selection, phenotypic plasticity, and other factors, this issue seeks to shed light on the physiological and molecular mechanisms that underlie plants' ability to thrive in challenging environments.

The collection of research presented in this Special Issue represents a collaborative effort by experts in the field, bringing together diverse insights and methodologies to deepen our understanding of plant stress responses. It is our hope that this collection will not only contribute to the body of knowledge in plant science but also inspire further research and innovative strategies to enhance plant resilience and productivity in the face of increasing stressors in our changing world.

M. Iftikhar Hussain, Adele Muscolo, and Mukhtar Ahmed
Editors

plants

MDPI

Editorial

Plant Responses to Biotic and Abiotic Stresses: Crosstalk between Biochemistry and Ecophysiology

Muhammad Iftikhar Hussain [1,*], Adele Muscolo [2] and Mukhtar Ahmed [3,4]

1 Department of Plant Biology & Soil Science, Universidad de Vigo, Campus Lagoas Marcosende, 36310 Vigo, Spain
2 Department of Agriculture, Mediterranea University, Feo di Vito, 89122 Reggio Calabria, Italy
3 Department of Agricultural Research for Northern Sweden, Swedish University of Agricultural Sciences, 90183 Umeå, Sweden
4 Department of Agronomy, Pir Mehr Ali Shah Arid Agriculture University, Murree Road, Rawalpindi 46300, Pakistan
* Correspondence: iftikhar@uvigo.es

Citation: Hussain, M.I.; Muscolo, A.; Ahmed, M. Plant Responses to Biotic and Abiotic Stresses: Crosstalk between Biochemistry and Ecophysiology. *Plants* **2022**, *11*, 3294. https://doi.org/10.3390/plants11233294

Received: 9 October 2022
Accepted: 23 November 2022
Published: 29 November 2022

Biotic and abiotic stresses, such as drought, salinity, extreme temperatures (cold and heat) and oxidative stress, are often interrelated; these conditions singularly or in combination induce cellular damage. For example, severe stresses during critical growth phases may directly result in mechanical damage and changes in the synthesis of macromolecules in cellular settings. In addition, these stresses often lead to oxidative damage and involve the formation of reactive oxygen species (ROS) in plant cells. Usually, plants have mechanisms to reduce their oxidative damage via the activation of antioxidant enzymes and the accumulation of compatible solutes that effectively scavenge ROS. However, if the production of activated oxygen exceeds the plant's capacity to detoxify it, deleterious degenerative reactions do occur, the typical symptoms being loss of osmotic responsiveness, wilting and necrosis. Given that plants face stressful conditions, imposed by changing environmental conditions that affect their growth and development during their whole life cycle, plants have to be able to perceive, process and translate different stimuli into adaptive responses.

The current Special Issue in *Plants* aims to analyze, from a multi-perspective approach (ranging from gas exchange, metabolomics, proteomics, isotopic and genomics, etc.), drivers (e.g., trait selection, phenotypic plasticity) and specific strategies used by the plants at physiological and molecular levels for their better adaptations to stressful growth conditions.

In total, 20 manuscripts (research and review) are included in this Special Issue. Furthermore, this Special Issue presents research findings in various experimental models (crops, fruit trees, legumes and halophytes) and areas ranging from cellular to ecophysiological and biochemical aspects.

Abideen et al. [1] used *Phragmites karka* to investigate the potential effects of salinity (control, 100 and 300 mM NaCl in a nutritional solution) and drought (at 50 percent water-holding capacity) and studied the correlations between stress tolerance, photosynthetic processes, biomass and ethanol output. They further discuss that plants exhibit an efficient photosynthetic system to grow in salty and dry environments, making it a viable crop for biofuel production.

Badar et al. [2] studied how to reduce the dangers of pharmaceutical pollution in the environment and investigated the bioremediation capability of edible crops and their associated microbial communities to successfully remove these pollutants. They tested paracetamol, which is also known as acetaminophen, at three concentrations (50, 100 and 200 mg/L) in terms of absorption, transport, accumulation and degradation in various organs of spinach (*Spinacia oleracea*) under controlled laboratory settings. Growth and photosynthetic machinery of the plants was negatively impacted by rising paracetamol stress levels. LC-MS data showed the drug absorption and translocation from roots to aerial parts and drug breakdown after eight days. Several bacterial strains (Burkhulderia, Sphingomonas, Pseudomonas, Staphylococcus, Stenotrophomonas and Kocuria) were isolated from spinach shoots and roots.

Hussain et al. [3] studied the rehabilitation of salt-degraded marginal soils through the selection and assessment of crop cultivars that can withstand salt stress. They evaluated the effects of different salinity levels (0, 7 and 14 dS m^{-1}) on six barley genotypes (113/1B, 59/3A, N1-10, N1-29, Barjouj and Alanda01) and evaluated different physiological, biochemical and stable isotopic responses. All measured plant traits responded to salt in a genotype-specific way. They showed that the genotypes (Barjouj and Alanda01) proved their suitability under the sandy desert soils of Dubai, UAE, as they exhibited higher grain yield, while 113/1B and Barjouj have greater grain protein content. The present research demonstrated that saline marginal sandy desert soils could support the cultivation of salt-tolerant barley genotypes for food and nutrition security as well as for the rehabilitation of marginal lands.

Riaz et al. [4] assessed the phytochemical potential of *Ziziphus* species, i.e., *Z. jujuba*, *Z. mauritiana*, *Z. spina-christi* and *Z. nummularia*, from desert environments. Leaf length, leaf width, leaf area and leaf petiole length were higher in *Z. jujube*, while *Z. mauritiana* exhibited higher dry biomass. *Z. jujube* had the largest fruit length, fruit stalk length, fruit diameter, fruit width, fruit area, seed length and seed diameter, while *Z. nummularia* had the highest fruit dry weight and widest seeds. Secondary metabolites were found in the fruits and leaves of *Ziziphus* species, including phenol, flavonoids and antioxidant activity. The highest levels of fruit phenols (304.4 mg GAE/100 g), leaf phenols (314.2 mg GAE/100 g), fruit flavonoids (123.7 mg QE/100 g) and leaf flavonoids (113.4 mg QE/100 g) were all accumulated by *Z. nummularia*. Moreover, irrigated and drought plantations led to a significant variation in morphological, fruit characteristics and phytochemical constituents that might be useful for future production technologies for this medicinal plant.

Hussain et al. [5] evaluated lowland rice genotypes under well-watered (WW) and terminal water stress (TWS) for improving drought stress and yield stability. Genotypes Look Pla and Lep Nok were found to be stress tolerant, whilst genotypes Chor Lung, Hom Nang Kaew and Hom Chan were found to be moderately tolerant genotypes. Genotypes Hom Pathum, Sang Yod, Dum Ja and Pathum Thani-1 were found to be extremely stress tolerant and relatively high yielding. Different stress-tolerance metrics, such as the stress-tolerance index (STI), the geometric mean productivity (GMP), the mean productivity index (MPRO) and the harmonic mean index (MHAR), showed significant and favorable correlations with GY during WW.

Ndiate et al. [6] demonstrated, in a greenhouse study, the impact of biochar (5%), arbuscular mycorrhizal fungus (20 g/pot, AMF) and biochar + AMF on maize (*Zea mays* L.) plants grown under salt stress (0, 50, 100, and 150 mM NaCl), to describe the mitigating technique against salinity. Plant height and fresh weight were decreased by 17.84% and 39.28%, respectively, compared to the control after 90 days of treatment with 100 mM NaCl. The growth parameters rose by 22.04%, 26.97%, 30.92% (height) and 24.79%, 62.36% and 107.7% (fresh weight) when the saline-treated soil (100 mM NaCl) was supplemented with AMF, biochar and biochar + AMF, as compared to control. The biochar + AMF enhanced plant nutrient uptake, (ii) improving soil nutrient content, (iii) increasing antioxidant enzyme activity and (iv) improving the contents of palmitoleic acid (C16:1), oleic acid (C18:1), linoleic acid (C18:2) and linolenic acid (C18:3). They concluded that biochar and AMF addition to saline and alkaline soils can successfully reduce abiotic stress and enhance plant development.

Elkelish et al. [7] investigated cysteine (Cys) (25 and 50 ppm as a foliar application) and lipoic acid (ALA) (0.02 mM, grain dipping pre-cultivation treatment) under water deficit and well-watered irrigation (100% and 70% of the required dose). The deficit irrigation increased cellular oxidative damage via increased malondialdehyde (MDA) level and hydrogen peroxide (H_2O_2). Superoxide dismutase (SOD), ascorbate peroxidase (APX), catalase (CAT) and peroxidase (POX), osmolytes and chlorophyll (Chl) were among the enzymatic antioxidants that benefited from Cys administration. The ability of the plant to scavenge reactive oxygen species (ROS), leaf relative water content (RWC), grain number, total grain yield, weight of 1000 kernels and gluten index was improved. Additionally, heatmap plot analysis uncovered numerous significant connections between the various characteristics that were examined, which may be explored.

El-Serafy et al. [8] showed that hydro-priming and halo-priming with silicon (Si) and silicon nanoparticles (SiNPs) can improve salinity tolerance of the ornamental plant *Lathyrus odoratus*. They highlighted that halo-priming with Si or SiNPs increased Lathyrus seedling salt-stress tolerance. This effect was confirmed using seawater treatments, which improved the germination percentage, seedling growth and activation of the antioxidant machinery, which detoxifies reactive oxygen species (ROS).

Alnusairi et al. [9] investigated the effect of exogenously applied nitric oxide (NO) (50 μM and 100 μM) in protecting wheat plants from NaCl-induced oxidative damage by modulating protective mechanisms. They showed that the exogenous-sourced NO at both concentrations up-regulated the antioxidant system for averting the NaCl-mediated oxidative damage on membranes. Enhancing salt tolerance by NO was concomitant with an obvious down-regulation in the relative expression of SOS1, NHX1, AQP and OSM-34, while D2 protein was up-regulated.

Farooq et al. [10] evaluated total carbon (TC), total nitrogen (TN) and isotopic natural abundance of C ($\delta^{13}C$) and N ($\delta^{15}N$) in soil and foliage of coniferous plantation (CPF), natural broadleaved forest (NBF) and mixed-forest stands at three different soil depths (i.e., 0–10, 10–20 and 20–40 cm) and how soil-available nutrients are affected by different forest types. Results showed that soil nutrient availability was higher in mixed forests. The findings provided evidence that forest type and soil depth alter TC, TN and soil $\delta^{15}N$, whereas $\delta^{13}C$ was only driven by soil depth. Moreover, plantations led to a decline in soil-available nutrient content compared with NBF and mixed-forest stands.

Farooq et al. [11] evaluated the effects of intercropping of peanut (*Arachis hypogaea* L.) with tea plants (*Camellia oleifera*), in comparison with the mono-cropping of tea and peanut. Soil health and fertility were examined. The results showed that the intercropping enhanced soil nutrient status and positively impacted soil conservation, increasing soil organic carbon (SOC), soil nutrient availability and soil enzymatic responses at different soil depths.

Rehman et al. [12] evaluated the morpho-physiological traits of two spring wheat cultivars (Millet-11, Punjab-11) and two advanced lines (V-07096, V-10110) exposed to terminal heat stress under late sowing. Results showed that improved grain yield was associated with the highest chlorophyll contents, showing stay green characteristics with maintenance of high photosynthetic rates and cooler canopies under late sowing and revealed that advanced lines and Punjab-11 with heat-adaptive traits could be promising sources for further use in the selection of heat-tolerant wheat genotypes.

Umnajkitikorn et al. [13] evaluated the potential of using elevated nitrogen priming prior to water shortage to mitigate plant stress through nitric oxide accumulation. Results indicated that plants primed with nitrogen possessed a higher photosynthetic rate, relative water content, electrolyte leakage and lipid peroxidation under water-deficit conditions, compared to control plants. The induction of water-deficit tolerance was supported by the activation of the antioxidant-defense system, induced by the accumulation of nitric oxide in leaves and roots of rice plants.

Hussain et al. [14] investigated the salinity-tolerance mechanisms of six contrasting quinoa cultivars belonging to the coastal region of Chile using agro-physiological parameters. Results suggested that all measured plant traits, except for C:N ratio, responded to salt in a genotype-specific way. Results indicated that the genotypes (Q21 and AMES13761) proved their suitability under sandy desert soils of Dubai, UAE, as they exhibited higher seed yield, while NSL106398 showed a higher seed protein content. The present research highlights the need to preserve quinoa biodiversity for a better seedling establishment, survival and stable yield in the sandy desert UAE environment.

Hassan et al. [15] evaluated the impact of various Cd concentrations (0, 5, 25, 50 and 100 M) on physiological and biochemical parameters in two sorghum (*Sorghum bicolor* L.) cultivars—JS-2002 and Chakwal Sorghum. The Cd absorption was increased in both cultivars while Cd uptake in JS-2002's leaf, stem and root was greater than that of Chakwal Sorghum. The superoxide dismutase (SOD), peroxidase (POD) and catalase activities were also lowered by Cd stress at higher levels (50 and 100 M). Results showed that JS-2002 had a higher Cd tolerance.

Hussain et al. [16] investigated the phenolic compound and flavonoid composition and allelopathic effects of an aqueous extract of aerial parts from *Acacia melanoxylon* R. Br. on seedling growth and plant biomass of the general biotest species, lettuce (*Lactuca sativa*). The acacia flower aqueous extract (AFE) and phyllodes aqueous extract (APE) reduced the leaf fresh weight, leaf dry weight, root fresh weight and root dry weight in lettuce. The mean root length decreased by 37.7% and 29.20%, following treatment with Acacia flower extract (AFE) at a concentrations of 75% and 100%, respectively. In total, 13 compounds of gallic acid, protocatechuic acid, p-hydroxybenzoic acid, p-hydroxybenzaldehyde, vanillic acid, syringic acid, p-coumaric acid and ferulic acid were among the phytochemical substances found from both parts. Rutin, luteolin, apigenin and catechin are some of the principal flavonoid chemicals found and are potential causes of the allelopathic effects of floral and phyllodes extracts from *A. melanoxylon*.

Hussain et al. [17] studied allelopathy, which is an ecological phenomenon that occurs when biomolecules from various crops, cultivated plants and bacteria or fungi are produced and released into the soil rhizosphere and have an impact on nearby species. Sorghum allelopathy has been utilized in relation to green manure, crop rotations, cover crops, intercropping and mulching, plant aqueous extracts or powder. From various plant tissues of sorghum and root exudates, a variety of allelochemicals, including benzoic acid, p-hydroxybenzoic acid, vanillic acid, ferulic acid, chlorogenic acid, m-coumaric acid, p-coumaric acid, gallic acid, caffeic acid, p-hydroxibenzaldehyde, dhurrin and sorgoleone, was identified. Among them, sorgoleone has been studied for its mode(s) of action, specific activity, selectivity, release in the rhizosphere, uptake in vulnerable species and translocation. The importance of sorghum allelopathy as an ecological tool for managing weeds is discussed in this review, which also highlights the most recent discoveries regarding the allelochemicals found in sorghum, their mechanisms of action and their place in the environment.

Khan et al. [18] reported that the global temperature has been steadily rising at a pace of 0.15–0.17 °C every decade. Therefore, measures for thermotolerance are required to maintain crop output under increased temperatures. This review was carried out with the goal of providing information on the wheat reaction in three research fields, including physiology, breeding and genetic advancements. At the heading, anthesis and grain-filling stages of wheat growth, the ideal temperatures are 16 ± 2.3 °C, 23 ± 1.75 °C and 26 ± 1.53 °C, respectively. The high temperature has a negative impact on the phenology, growth and development of the crop. The pollen viability, seed germination and embryo development are all slowed down by the high temperature before anthesis. The accumulation of starch granules, stem-reserve carbohydrates and photosynthate translocation into grains is reduced by the high post-anthesis temperature. The reviewed work showed that the genotypes with higher levels of proline, glycine betaine, heat-shock protein expression, stay green and antioxidant enzyme activity, specifically catalase, peroxidase, super oxide dismutase and glutathione reductase, can tolerate high temperature effectively by supporting cellular physiology.

Raza et al. [19] demonstrated that the maize–soybean intercropping system has many persistent constraints, including lodging, which pose severe limitations to the development and sustainability of this cropping system. The lodging phenomenon is influenced by a variety of morphological and anatomical traits. Because of the shading from maize, soybean stems develop a shade-avoidance response, which causes stem elongation and significant lodging. The primary agro-techniques needed to investigate lodging in the maize–soybean intercropping system for sustainable agriculture, however, have not yet been clearly defined. The present review suggests that controlling lodging requires a variety of strategies, including agronomic, chemical and genetic ones, that could be useful in lowering lodging hazards in the maize–soybean intercropping system. Therefore, further research is needed from agronomists, physiologists, molecular biologists and breeders to address this difficult issue.

Finally, we encourage readers to view the articles published in this Special Issue of "Plant Responses to Biotic and Abiotic Stresses: Crosstalk between Biochemistry and Ecophysiology".

Funding: This research received no external funding.

Conflicts of Interest: The authors declare no conflict of interest.

References

1. Abideen, Z.; Koyro, H.W.; Hussain, T.; Rasheed, A.; Alwahibi, M.S.; Elshikh, M.S.; Hussain, M.I.; Zulfiqar, F.; Mansoor, S.; Abbas, Z. Biomass Production and Predicted Ethanol Yield Are Linked with Optimum Photosynthesis in *Phragmites karka* under Salinity and Drought Conditions. *Plants* **2022**, *11*, 1657. [CrossRef] [PubMed]
2. Badar, Z.; Shanableh, A.; El-Keblawy, A.; Mosa, K.A.; Semerjian, L.; Mutery, A.A.; Hussain, M.I.; Bhattacharjee, S.; Tsombou, F.M.; Ayyaril, S.S.; et al. Assessment of Uptake, Accumulation and Degradation of Paracetamol in Spinach (*Spinacia oleracea* L.) under Controlled Laboratory Conditions. *Plants* **2022**, *11*, 1626. [CrossRef] [PubMed]
3. Hussain, M.I.; Khan, Z.I.; Farooq, T.H.; Al Farraj, D.A.; Elshikh, M.S. Comparative Plasticity Responses of Stable Isotopes of Carbon (δ^{13}C) and Nitrogen (δ^{15}N), Ion Homeostasis and Yield Attributes in Barley Exposed to Saline Environment. *Plants* **2022**, *11*, 1516. [CrossRef] [PubMed]
4. Riaz, M.U.; Raza, M.A.; Saeed, A.; Ahmed, M.; Hussain, T. Variations in Morphological Characters and Antioxidant Potential of Different Plant Parts of Four *Ziziphus* Mill. Species from the Cholistan. *Plants* **2021**, *10*, 2734. [CrossRef] [PubMed]
5. Hussain, T.; Hussain, N.; Ahmed, M.; Nualsri, C.; Duangpan, S. Responses of Lowland Rice Genotypes under Terminal Water Stress and Identification of Drought Tolerance to Stabilize Rice Productivity in Southern Thailand. *Plants* **2021**, *10*, 2565. [CrossRef] [PubMed]
6. Ndiate, N.I.; Saeed, Q.; Haider, F.U.; Liqun, C.; Nkoh, J.N.; Mustafa, A. Co-Application of Biochar and *Arbuscular mycorrhizal* Fungi Improves Salinity Tolerance, Growth and Lipid Metabolism of Maize (*Zea mays* L.) in an Alkaline Soil. *Plants* **2021**, *10*, 2490. [CrossRef] [PubMed]
7. Elkelish, A.; El-Mogy, M.M.; Niedbała, G.; Piekutowska, M.; Atia, M.A.M.; Hamada, M.M.A.; Shahin, M.; Mukherjee, S.; El-Yazied, A.A.; Shebl, M.; et al. Roles of Exogenous α-Lipoic Acid and Cysteine in Mitigation of Drought Stress and Restoration of Grain Quality in Wheat. *Plants* **2021**, *10*, 2318. [CrossRef] [PubMed]
8. El-Serafy, R.S.; El-Sheshtawy, A.-N.A.; Atteya, A.K.G.; Al-Hashimi, A.; Abbasi, A.M.; Al-Ashkar, I. Seed Priming with Silicon as a Potential to Increase Salt Stress Tolerance in *Lathyrus odoratus*. *Plants* **2021**, *10*, 2140. [CrossRef] [PubMed]
9. Alnusairi, G.S.H.; Mazrou, Y.S.A.; Qari, S.H.; Elkelish, A.A.; Soliman, M.H.; Eweis, M.; Abdelaal, K.; El-Samad, G.A.; Ibrahim, M.F.M.; ElNahhas, N. Exogenous Nitric Oxide Reinforces Photosynthetic Efficiency, Osmolyte, Mineral Uptake, Antioxidant, Expression of Stress-Responsive Genes and Ameliorates the Effects of Salinity Stress in Wheat. *Plants* **2021**, *10*, 1693, Correction in *Plants* **2022**, *11*, 576. [CrossRef] [CrossRef] [PubMed]
10. Farooq, T.H.; Chen, X.; Shakoor, A.; Li, Y.; Wang, J.; Rashid, M.H.U.; Kumar, U.; Yan, W. Unraveling the Influence of Land-Use Change on δ^{13}C, δ^{15}N, and Soil Nutritional Status in Coniferous, Broadleaved, and Mixed Forests in Southern China: A Field Investigation. *Plants* **2021**, *10*, 1499. [CrossRef] [PubMed]
11. Farooq, T.H.; Kumar, U.; Mo, J.; Shakoor, A.; Wang, J.; Rashid, M.H.U.; Tufail, M.A.; Chen, X.; Yan, W. Intercropping of Peanut–Tea Enhances Soil Enzymatic Activity and Soil Nutrient Status at Different Soil Profiles in Subtropical Southern China. *Plants* **2021**, *10*, 881. [CrossRef] [PubMed]
12. Rehman, H.U.; Tariq, A.; Ashraf, I.; Ahmed, M.; Muscolo, A.; Basra, S.M.A.; Reynolds, M. Evaluation of Physiological and Morphological Traits for Improving Spring Wheat Adaptation to Terminal Heat Stress. *Plants* **2021**, *10*, 455. [CrossRef] [PubMed]
13. Umnajkitikorn, K.; Fukudome, M.; Uchiumi, T.; Teaumroong, N. Elevated Nitrogen Priming Induced Oxinitro-Responses and Water Deficit Tolerance in Rice. *Plants* **2021**, *10*, 381. [CrossRef] [PubMed]
14. Hussain, M.I.; Muscolo, A.; Ahmed, M.; Asghar, M.A.; Al-Dakheel, A.J. Agro-Morphological, Yield and Quality Traits and Interrelationship with Yield Stability in Quinoa (*Chenopodium quinoa* Willd.) Genotypes under Saline Marginal Environment. *Plants* **2020**, *9*, 1763. [CrossRef] [PubMed]
15. Hassan, M.J.; Raza, M.A.; Rehman, S.U.; Ansar, M.; Gitari, H.; Khan, I.; Wajid, M.; Ahmed, M.; Shah, G.A.; Peng, Y.; et al. Effect of Cadmium Toxicity on Growth, Oxidative Damage, Antioxidant Defense System and Cadmium Accumulation in Two Sorghum Cultivars. *Plants* **2020**, *9*, 1575. [CrossRef] [PubMed]
16. Hussain, M.I.; El-Sheikh, M.A.; Reigosa, M.J. Allelopathic Potential of Aqueous Extract from *Acacia melanoxylon* R. Br. on *Lactuca sativa*. *Plants* **2020**, *9*, 1228. [CrossRef] [PubMed]

17. Hussain, M.I.; Danish, S.; Sánchez-Moreiras, A.M.; Vicente, Ó.; Jabran, K.; Chaudhry, U.K.; Branca, F.; Reigosa, M.J. Unraveling Sorghum Allelopathy in Agriculture: Concepts and Implications. *Plants* **2021**, *10*, 1795. [CrossRef] [PubMed]
18. Khan, A.; Ahmad, M.; Ahmed, M.; Hussain, M.I. Rising Atmospheric Temperature Impact on Wheat and Thermotolerance Strategies. *Plants* **2021**, *10*, 43. [CrossRef] [PubMed]
19. Raza, A.; Asghar, M.A.; Ahmad, B.; Bin, C.; Hussain, M.I.; Li, W.; Iqbal, T.; Yaseen, M.; Shafiq, I.; Yi, Z.; et al. Agro-Techniques for Lodging Stress Management in Maize-Soybean Intercropping System—A Review. *Plants* **2020**, *9*, 1592. [CrossRef] [PubMed]

plants

MDPI

Article

Biomass Production and Predicted Ethanol Yield Are Linked with Optimum Photosynthesis in *Phragmites karka* under Salinity and Drought Conditions

Zainul Abideen [1,*], Hans Werner Koyro [2], Tabassum Hussain [1], Aysha Rasheed [1], Mona S. Alwahibi [3], Mohamed S. Elshikh [3], Muhammad Iftikhar Hussain [4,*], Faisal Zulfiqar [5], Simeen Mansoor [6] and Zaheer Abbas [7]

1 Dr. Mouhammed Ajmal Khan Institute of Sustainable Halophyte Utilization, University of Karachi, Karachi 75270, Pakistan; thussain@uok.edu.pk (T.H.); halophyte_aysha@yahoo.com (A.R.)
2 Institute of Plant Ecology, Justus-Liebig-University Giessen, D-35392 Giessen, Germany; hans-werner.koyro@bot2.bio.uni-giessen.de
3 Department of Botany and Microbiology, College of Science, King Saud University, Riyadh 11451, Saudi Arabia; malwhibi@ksu.edu.sa (M.S.A.); melshikh@ksu.edu.sa (M.S.E.)
4 Department of Plant Biology & Soil Science, Universidad de Vigo, Campus Lagoas Marcosende, 36310 Vigo, Spain
5 Department of Horticultural Sciences, Faculty of Agriculture and Environment, Islamia University of Bahawalpur, Bahawalpur 63100, Pakistan; ch.faisal.zulfiqar@gmail.com
6 Department of Genetics, University of Karachi, Karachi 75270, Pakistan; simeenm@uok.edu.pk
7 Department of Botany, Division of Science and Technology, University of Education Lahore 54770, Pakistan; fzaheerbot@gmail.com
* Correspondence: zuabideen@uok.edu.pk (Z.A.); iftikhar@uvigo.es (M.I.H.)

Citation: Abideen, Z.; Koyro, H.W.; Hussain, T.; Rasheed, A.; Alwahibi, M.S.; Elshikh, M.S.; Hussain, M.I.; Zulfiqar, F.; Mansoor, S.; Abbas, Z. Biomass Production and Predicted Ethanol Yield Are Linked with Optimum Photosynthesis in *Phragmites karka* under Salinity and Drought Conditions. *Plants* **2022**, *11*, 1657. https://doi.org/10.3390/plants11131657

Academic Editor: Kent Burkey

Received: 3 April 2022
Accepted: 18 June 2022
Published: 23 June 2022

Abstract: Plant photosynthesis and biomass production are closely associated traits but critical to unfavorable environmental constraints such as salinity and drought. The relationships among stress tolerance, photosynthetic mechanisms, biomass and ethanol yield were assessed in *Phragmites karka*. The growth parameters, leaf gas exchange and chlorophyll fluorescence of *P. karka* were studied when irrigated with the control and 100 and 300 mM NaCl in a nutrient solution and water deficit conditions (drought, at 50% water holding capacity). The plant shoot fresh biomass was increased in the low NaCl concentration; however, it significantly declined in high salinity and drought. Interestingly the addition of low salinity increased the shoot biomass and ethanol yield. The number of tillers was increased at 100 mM NaCl in comparison to the control treatment. High salinity increased the photosynthetic performance, but there were no significant changes in drought-treated plants. The saturated irradiance (Is) for photosynthesis increased significantly in low salinity, but it declined (about 50%) in high salt-stressed and (about 20%) in drought-treated plants compared to the control. The rates of dark respiration (Rd) and compensation irradiance (Ic) were decreased significantly under all treatments of salinity and drought, with the exception of unchanged Rd values in the control and drought treatments. A-Ci curve analyses revealed a significant improvement in the Jmax, Vc, max, and triose-phosphate utilization (TPU) at lower salinity levels but decreased at 300 mM NaCl and drought treatments compared to the control. In the chlorophyll fluorescence parameters (Fv/Fm, maximum photochemical quantum yield of PSII, and Y(NO)), the non-photochemical yields were not affected under the salt and drought treatments, although an effective photochemical quantum yield (YII) and electron transport rate (ETR) were significantly enhanced in water deficit compared to control plants. *P. karka* regulates an efficient photosynthesis mechanism to grow in saline and arid areas and can therefore be used as a sustainable biofuel crop.

Keywords: bioethanol; salt tolerance; water deficit conditions; chlorophyll fluorescence; photosynthetic efficiency

1. Introduction

Plants of arid and semi-arid regions display severely subdued growth and even death in the presence of either drought or saline conditions [1]. Species belonging to these conditions gradually exhibit lesser vegetation cover and can lead to desertification in the region [2]. NaCl stress and water deficit are common abiotic stress factors on a global scale and cause deleterious effects on plant biomass and stability [3–5]. Functionally, plants can reduce the harmful effects of water limitation and ion toxicity (due to soil salinity) by altering their growth, water relations, and photosynthesis [6–8]. Growth inhibition and leaf shedding under such conditions also help plants to maintain their water status and survive [9]. Most likely, biomass production in halophytes is related to photosynthesis and their protective photosystem (PS I and II) performances under salt stress [3,10]. Applying eco-physiological tools to assess the functional contribution of photosynthesis and their associated adjustments is important for biomass production [11–13].

The beneficial effects of low NaCl concentrations (100 mM) on growth and photosynthesis have been frequently observed in many studies [14,15]. It was reported that sodium ion acts as a cheap osmoticum for leaf turgor maintenance [14]. For instance, members of Chenopodiaceae attain benefits from sodium [16,17]. In C_4 species, it was assumed that Na^+ facilitate pyruvate conversion into phosphoenolpyruvate, found in mesophyll, before being added to the Calvin cycle. In addition, two halophytic species: *Kochia childsii* and *Atriplex tricolor* were cultivated in a sodium-deficient medium that declined photosystem II activities in mesophyll chloroplasts [18]. However, higher concentrations of Na^+ have deleterious effects on the photosynthetic apparatus [15].

The declined carbon fixation in salt and drought-stressed plants is also linked to lower stomatal conductance, and therefore, disturbance in the flow of electrons to Photosystem II can be possible [13,14]. The deficiency of electron and proton acceptors causes excessive light to release a surplus amount of energy as heat and chlorophyll fluorescence in plants to prevent the production of reactive oxygen species (ROS) [19,20]. Stress-tolerant plants regulate the photosynthetic rate and photoprotective mechanism to reduce the deleterious effects of ROS, which are linked with the optimum ATP synthesis, and NADP formation [21,22]. The above-said parameters are very informative in assessing the photosynthesis and physiological performance of plants [11]. The understanding of carbon assimilation and energy conversion phenomenon are linked to the production of all types of bio-compounds (e.g., ethanol), and therefore, the full potential of plants can be utilized in stressed conditions. In particular, halophytes are much-suited candidates due to their natural distributions in extreme conditions (e.g., salinity and water deficit).

It was recently demonstrated that *Phragmites karka* exhibits an efficient mechanism to tolerate salt and drought stresses, but a detailed analysis of their photochemistry and bioethanol potential is still poorly known [3]. In this study, *P. karka* plants coordinated changes involving the rate of photosynthesis and efficient photosystem II activity under saline and water deficit conditions. This plant accumulates a high amount of soluble sugar and lignocellulosic biomass [21,22]. This paper unlocks the potential of this accumulated sugar and cellulose and subsequent hemicellulose conversion into the ethanol yield on those areas that seem not suitable for agriculture. The effects of salt and drought stress on photosynthesis and their relationship with the biomass and ethanol yield was evaluated. The establishment of a suitable growing condition of the selected biofuel crop and subsequent ethanol potential can be helpful in the remediation of the increasing saline lands of Pakistan and other arid regions of the world.

2. Material and Methods

2.1. Plant Growth Conditions

Seeds of *Phragmites karka* were collected from the population located at the University of Karachi, Pakistan. A growth experiment was carried out under controlled growth chamber conditions in a growth chamber in Giessen, Germany: optimum temperature of 25 ± 2 °C, relative humidity around 50%, and photoperiod 16–8 h day–night, while the light intensity was 200–250 μmol m^{-2} s^{-1} photosynthetically active radiations. In the beginning, plant seeds were germinated in the plastic tray with wet clay soil for seedling emergence. Wuxal Super (Aglukon, Düsseldorf, Germany) was used as the nutrient for a further seven weeks. Subsequently, seedlings were transplanted into the soil (composed of 50% sand, 30% clay, and 20% gravels) in plastic pots (35 cm in height, 11 cm in diameter, and three plants per pot). The plants were irrigated periodically (12 h, from 8 am to 8 pm) with a basic nutrient solution (1/2 strength Hoagland modified after Epstein, 1972 [23]) in a quick check system [24]. The pots were divided into four groups (at the age of thirty-five days (35) after seed germination): control, at 100% water holding capacity (WHC), low and high salinity (100 and 300 mM NaCl in a nutrient solution with 100% WHC), and drought (reduced water supply at 50% WHC). Each pot contains three plants, while there are eight pots for each treatment. Plants of the control treatment were irrigated only with a nutrient solution, whereas the salt concentration was stepwise raised by adding 50 mM NaCl per day until the final concentration of desired NaCl was reached in the growth medium. In parallel to the salinity experiment, a drought treatment was started by gradually (5% per day) reducing the soil water saturation from 100 to 50%. The water holding capacity was measured as described by Veihmeyer and Hendrickson (1931) [25]. The treatments were designed based on the preliminary growth trials. The plants of all four treatments were maintained for a further five weeks. At the end of the experimental period, the soil water potential was −1.5 MPa in salinity and −0.5 MPa in drought. After 10 weeks, the plants were harvested for an eco-physiological analysis (total age) under these conditions.

2.2. Plant Harvest and Growth Parameters

Before the plant harvest, nondestructive growth parameters such as the predawn leaf water potential and midday gas exchange of leaves were recorded. The fresh weight (FW) of leaves, stems, and roots was noted. Plants were harvested and dried at 80 °C for 24 h for calculating the dry weight (DW). The leaf relative water content (RWC) was calculated separately using the equation:

$$\text{LRWC (\%)} = [(\text{FW} - \text{DW})/(\text{TW} - \text{DW})] \times 100 \text{ where TW is turgid weight}$$

A leaf is converted into a small disc and immersed for three hours to attain full turgidity at room temperature. Turgid small discs are taken out from the water and dried immediately with the help of tissue paper. The dried leaf discs are quickly weighed to determine the turgid weight (TW).

2.3. Soil Water Potential

The water potential in the soil was assessed using Wescor soil in a psychrometer (attached to a data logger) when the soil dried had an initial moisture concentration of 50% WHC; all this took about 4 weeks. Weight was assessed on a daily basis during this course of time. Soil samples were dried in the oven at the end of the experiment; the dry samples were weighed to determine the constant weight and dry weight. Water loss (up to 50% WHC for dry soil) was detected as a weight loss, which relates to the soil's ability to hold water. This function is used to calculate potential water based on the known water content.

2.4. Leaf Gas Exchange and Chlorophyll Content

LI-COR 6400 (LI-COR, Lincoln, NE, USA) was used to determine the gas exchange parameter, with 400 μmol m^{-2} s^{-1} CO_2 and a 500 μmol m^{-2} s^{-1} flow rate. Different PAR values ranged 0–2000 μmol photon m^{-2} s^{-1} to calculate the dark respiration (Rd), compensation irradiance (Ic), saturation irradiance (Is), and photosynthetic efficiency (Φc), as described by [26]. In contrast, different CO_2 concentrations were plotted to calculate the maximum Rubisco carboxylase activity (Vc, max) and maximum rate of electron transport to regenerate RuBP (Jmax) and triose-phosphate utilization (TPU) [27]. The relative chlorophyll content was measured using a SPAD 502 densitometer (Konica Minolta, Ramsey, NJ, USA).

2.5. Leaf Chlorophyll Fluorescence

The chlorophyll fluorescence was measured on similar leaf sections to those selected for the gas exchange (Pulse-controlled Junior PAM, Walz, Effeltrich, Germany). The leaves were kept in complete dark for 30 min to determine the following parameters, as described in Abideen et al. (2020) [28]: minimal fluorescence (Fo) with modulated light (<0.1 μmol photon m^{-2} s^{-1}) and maximal fluorescence (Fm) with saturating pulse (10,000 μmol photons m^{-2} s^{-1} for 0.6 s) determined the maximum photochemical quantum yield of PSII.

Maximum photochemical quantum yield of photosystem II (Fv/Fm) = Fm − Fo/Fm) [29].

Effective photochemical quantum yield YII = Fm′ − Fs/Fm′ and NPQ = Fm/Fm′ − 1 [30]. NPQ = Fm/Fm′ − 1 [30].

Non-photochemical quenching Y (NO) = F/FmY (NPQ) = F/Fm′ − F/Fm [31].

The coefficient of photochemical quenching (qP) = Fm′ − Fs)/(Fm′ − Fo′) [32].

Electron transport rate ETR = PSII × PPFD × 0.5 × 0.84 [33].

2.6. Lignocellulosic Analysis and Soluble Sugar Content

The lignocellulosic content was analyzed in dry shoots by the neutral detergent fiber (NDF) determination. The acid detergent fiber (ADF) was determined by using the residue left from the NDF analysis. Hemicellulose was determined by subtracting the ADF from NDF [34]. The ADF and NDF-treated shoot biomass were then hydrolyzed with 72% H_2SO_4 to determine the cellulose levels. Dry plant leaves were brought to a powdered form and shaken for an hour at 100 °C with deionized water, and the filtrate was obtained to treat with Anthrone's reagent to calculate the soluble sugar. The mixture was heated in a boiling water bath for 11 min, followed by cooling at room temperature. The optical density of green to dark green color was observed at 630 nm on a spectrophotometer (DU530 UV–Vis) [35].

2.7. Statistical Analysis and Calculation

Data ($n \geq 4$) were analyzed by one-way analysis of variance (ANOVA, SPSS, ver. 11), and significant differences among means ($p < 0.05$) were determined by the Bonferroni test.

The conversion of dry matter per pots into tons of biomass/hectare was performed by calculating the plant yield per pot [36]. Firstly, the surface area of the pot in cm^2 is calculated. Then, the unknown yield per hectare (x) is calculated in relation to the area that is 10,000 m^2, as shown by Zhao et al. [36]. The theoretical yield of ethanol cellulose and total soluble sugar data per hectare levels was determined by the following equation:

Ethanol yield from soluble sugar (L ha^{-1}) = total soluble sugar content (%) in dry matte (t ha^{-1}) × 0.51 (conversion factor of ethanol from sugar) × 0.85 (process efficiency of ethanol from sugar) × 1000/0.79 (specific gravity of ethanol, g mL^{-1}) [36].

The ethanol yield from cellulose and hemicellulose (L ha^{-1}) = cellulose and hemicellulose content (% DW) in dry matter × dry biomass (t ha^{-1}) × 1.11 (conversion factor of sugar from cellulose and hemicellulose) × 0.85 (process efficiency of sugar from cellulose and hemicellulose) × 0.51 (conversion factor of ethanol from sugar) × 0.85 (process efficiency of ethanol from sugar) × 1000/0.79 (specific gravity of ethanol, g mL^{-1}) [36].

3. Results

Shoot fresh biomass was stimulated in the control at 100 mM NaCl, and it significantly decreased with an increase in the NaCl concentrations, as well as in drought treatment, while the relative water content was unchanged in all treatments (Figure 1). The number of tillers was increased only at 100 mM NaCl and decreased substantially in the other treatments compared to the control plants. The number of nodes was decreased at 300 mM NaCl and drought as compared to the other treatments.

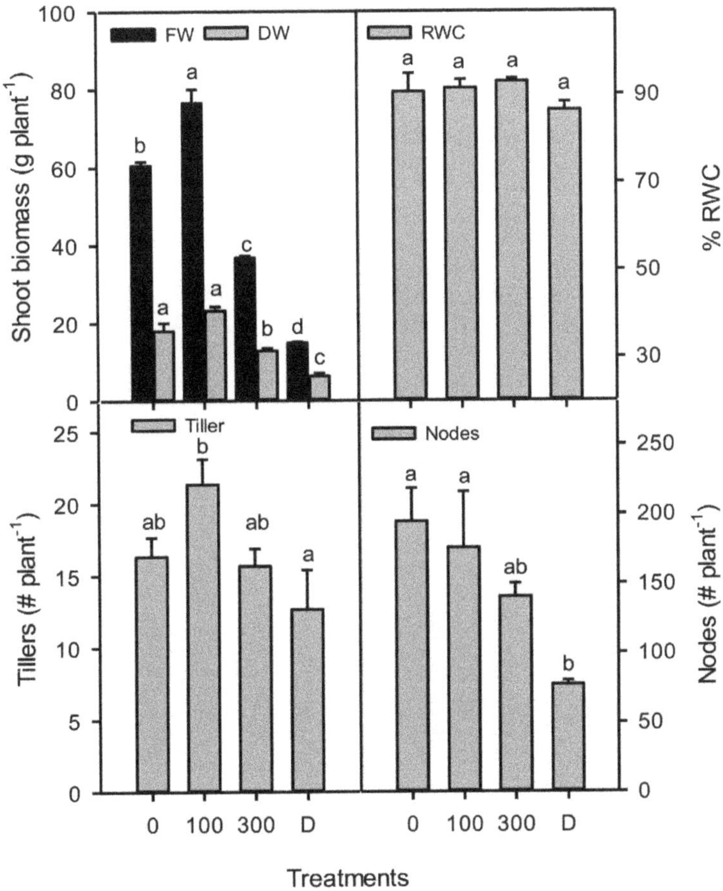

Figure 1. Plant fresh and dry shoot biomass, number of tillers, number of nodes of *Phragmites karka* grown at 0, 100, and 300 mM NaCl and drought. Different lower-case letters indicate significant differences due to salt treatments, according to Bonferroni's test ($p < 0.05$).

An analysis of the light curves (Figure 2) showed significant changes in various treatments of salinity and drought. High salinity treatment caused an increase in the photosynthetic efficiency, but there was no significant change in the drought treatment when compared to the control plants. The saturated irradiance (I_s) for photosynthesis was found to increase significantly ($p < 0.001$) between plants of the control and low salinity, but the I_s decreased significantly (about 50%) in high salt-stressed plants, with a lesser decrease (about 20%) in drought treatment with respect to control treatment. The rates of dark respiration (R_d) and compensation irradiance (I_c) were decreased significantly under all treatments of salinity and drought, but R_d was similar in the control and drought-treated plants (Figure 3) and changes in the net photosynthesis with increased CO_2 concentration and carbon assimilation under salinity and drought conditions (Figure 4). Analyses of the A-C_i curve revealed a significant improvement in the V_{cmax}, J_{max}, and TPU at low salinity but decreased at 300 mM NaCl and drought treatment, as compared to the control (Figure 5). The chlorophyll fluorescence parameters were not affected under salt and drought treatments although YII and ETR were significantly increased in drought-treated plants as compared to the NaCl treatments and non-saline control plants (Table 1).

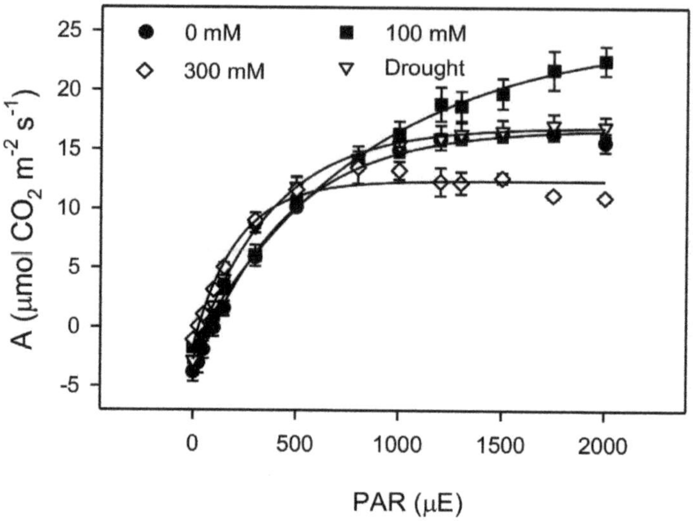

Figure 2. Light response curve between the net photosynthesis (A) and light intensities (PAR; 0–2500 μmol photon m^{-2} s^{-1}) on leaves of *Phragmites karka* under 0, 100, and 300 mM NaCl and drought.

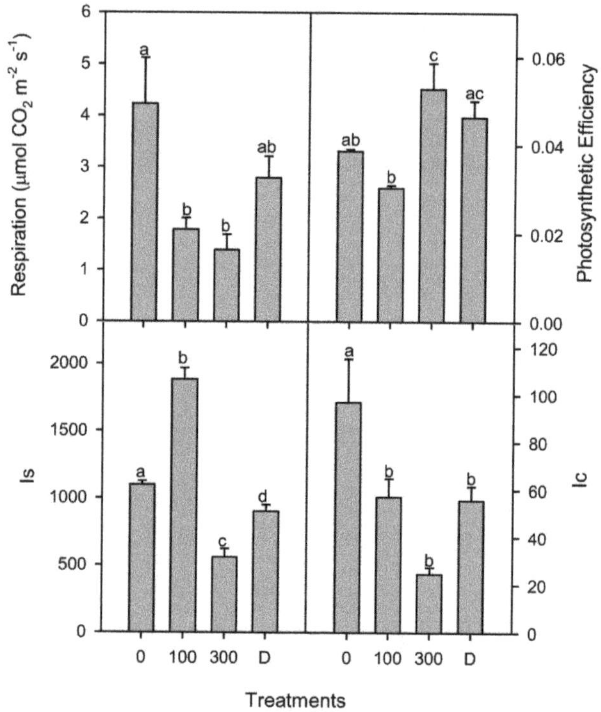

Figure 3. Dark respiration (Rd), compensation irradiance (Ic), saturation irradiance (Is), and photosynthetic efficiency (Φc) of *Phragmites karka* under 0, 100, and 300 mM NaCl and drought. Different lower-case letters indicate significant differences due to salt treatments, according to Bonferroni's test ($p < 0.05$).

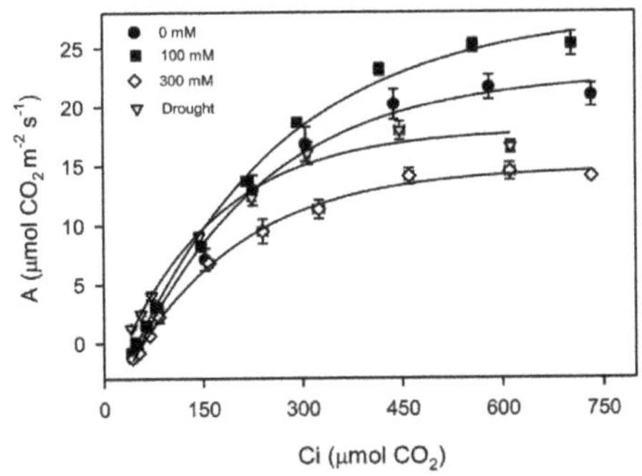

Figure 4. CO_2 response curve between net photosynthesis (A) and variable intercellular CO_2 concentrations on leaves of *Phragmites karka* under the 0, 100, and 300 mM NaCl and drought.

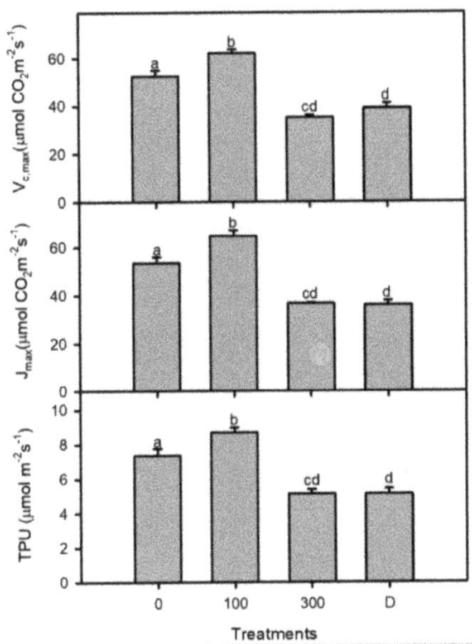

Figure 5. A-Ci curve was used to determine the following parameters: maximum rate of Rubisco carboxylase activity ($V_{c,max}$), maximum rate of electron transport (J_{max}), and utilization of triose phosphates (TPU) under 0, 100, and 300 mM NaCl. Different lower-case letters indicate significant differences due to salt treatments, according to Bonferroni's test ($p < 0.05$).

Table 1. Chlorophyll fluorescence parameters (Fv/Fm, maximum photochemical quantum yield of PSII; Y(II), effective photochemical quantum yield of PSII; coefficient of photochemical quenching (qP); Non-photochemical quenching (NPQ) Y(NPQ), yield for heat dissipation; Y(NO), and yield of non-photochemical; and ETR, electron transport rate under saline and drought conditions. Different lower-case letters indicate significant differences due to salt treatments, according to Bonferroni's test ($p < 0.05$).

Treatments	Fv/Fm	Y(II)	qP	NPQ	Y(NO)	Y(NPQ)	ETR
0	0.81 ± 0.005a	0.51 ± 0.012a	0.70 ± 0.015abc	0.59 ± 0.026b	0.30 ± 0.006a	0.18 ± 0.008abc	40.77 ± 0.91a
100	0.82 ± 0.006a	0.50 ± 0.020a	0.68 ± 0.033b	0.66 ± 0.093b	0.29 ± 0.018a	0.19 ± 0.020b	40.22 ± 1.63a
300	0.82 ± 0.005a	0.51 ± 0.030a	0.68 ± 0.032b	0.63 ± 0.057b	0.29 ± 0.013a	0.18 ± 0.020b	41.14 ± 2.45a
Drought	0.81 ± 0.004a	0.59 ± 0.014b	0.78 ± 0.017ac	0.47 ± 0.092a	0.27 ± 0.015a	0.13 ± 0.019ac	47.12 ± 1.12b

The cellulose content was enhanced with 100 mM NaCl in *Phragmites karka*, but it was reduced substantially under drought conditions. Plants treated with the 300 mM NaCl reduced the hemicellulose content as compared to control and 100 mM NaCl. The shoot total sugar was enhanced in each stress treatments as compared to the control. Plants treated with 300 mM NaCl improved the chlorophyll (SPAD) levels as compared to the control treatments (Table 2).

Table 2. Shoot cellulose (%), hemicellulose (%), total sugar (mg/g DW), and leaf chlorophyll (SPAD arbitrary values) of *Phragmites karka* under 0, 100, and 300 mM NaCl and drought. Values represent the mean ± S.E. of three replicates (n = 4). Different lower-case letters indicate significant differences due to salt treatments, according to Bonferroni's test ($p < 0.05$).

Treatments	Cellulose	Hemicellulose	Soluble Sugar	Chlorophyll
Control	29.17 ± 1.14b	22.31 ± 1.11b	51.91 ± 4.21a	40.51 ± 0.56a
100 mM NaCl	34.56 ± 1.20c	20.77 ± 2.00a	78.61 ± 2.73c	43.91 ± 0.36a
300 mM NaCl	26.67 ± 1.49b	17.72 ± 1.56a	79.40 ± 2.97c	47.47 ± 0.45b
Drought	20.06 ± 0.63a	14.82 ± 0.44a	69.50 ± 8.27b	39.69 ± 1.99a

The dry biomass per hectare was improved substantially at 100 mM NaCl compared to the other stress treatments. The ethanol yield was estimated from the total sugar and cellulose and hemicellulose data (Figure 6). Interestingly, the addition of 100 mM NaCl enhanced the ethanol yield per hectare by using the total sugar, cellulose, and hemicellulose in plants. The ethanol yield per hectare declined substantially under the higher salinity and drought conditions.

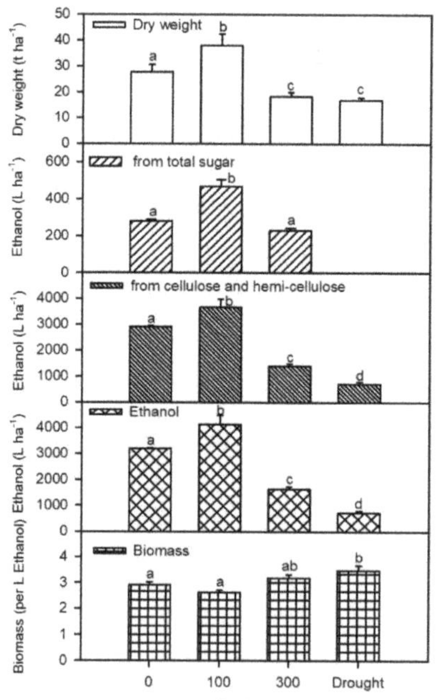

Figure 6. Plant dry biomass per hectare; ethanol yield per hectare from sugars; cellulose and hemicellulose; and the total ethanol yield of *Phragmites karka* under 0, 100, and 300 mM NaCl and drought. Values represent the mean ± S.E. of four replicates (n = 4). Different lower-case letters indicate significant differences due to salt treatments, according to Bonferroni's test ($p < 0.05$).

4. Discussions

Halophyte grasses are abundantly distributed in coastal and inland saline habitats of semi-arid regions and could be a good source of lignocellulosic biomass [37]. These plants are adapted to grow under saline conditions because of their salt resistance, high water use efficiency, and fast growth rates [38]. The cultivation of these plants is highly cost-efficient, because they utilize saline water and wastelands not fit for conventional agriculture [37–39]. The growth of different halophyte grasses has been optimized in low and/or moderately saline conditions, such as *Phragmites australis* [40], *Phragmites communis*, *Pennisetum clandestinum* [15,41], *Panicum antidotale*, and *Spartina maritima* [42]. The optimum shoot growth of *P. karka* was observed in low salinity (100 mM NaCl), and the growth decreased in higher salinity (300 mM NaCl) and drought treatments. Our results are also in agreement with several subtropical halophyte grasses, such as *Aeluropus lagopoides*, *Sporobolus ioclados*, *Urochondra setulosa*, and *Halopyrum mucronatum* that showed the optimum growth under non-saline conditions [43]. Besides *P. karka*, other species belonging to the genus *Phragmites* showed dose-dependent growth responses under saline conditions [40,41]. Therefore, we could suggest that our test species is one of the best candidates for using the sustainable utilization of saline land, particularly in arid and semi-arid regions of the world.

The rate of the leaf gas exchange varies with the duration and levels of salinity and drought conditions [44]. The ability of a plant to maintain its chlorophyll level, stomatal conductance, and rate of efficient CO_2 assimilation under saline conditions are closely related to the salt tolerance ability of the plant [45]. The photosynthetic efficiency of *P. karka* was decreased with an increase in the salinity; however, it remained comparable in drought treatment with the non-saline control (0 mM NaCl).

The survival of plants under drought and salinity without compromising the biomass is difficult; however, the salt-resistant plant maintains an optimum water use efficiency and rate of photosynthesis and fast growth rate. Under high salinity, plants improve their water use efficiency by decreasing their transpiration rate [46]; however, this reduces the CO_2 uptake, and therefore, photosynthesis is inhibited. *Phragmites karka* optimized net photosynthesis with a minimum water loss and favored higher photosynthetic rates (A) at 100 mM NaCl. However, at a higher salinity and under drought treatment, plants ensured their survival but with a growth reduction. A similar strategic reduction in photosynthetic efficiency and growth was reported for many halophytes under various abiotic stresses, such as *Desmostachya bipinnata* [47], *Aeluropus lagopoides*, and *Sporobolus tremulus* [48], under various abiotic stresses [45,47]. An effective CO_2 and water exchange is necessary for the survival of plants under stress conditions [47,48]. *Phragmites karka* exhibited higher energy requirements with increasing concentrations of NaCl, as indicated by an increase in compensation irradiance (Ic). High salt concentrations (300 mM NaCl) caused a reduction in photosynthetic machinery, which leads to a decrease in the level of Is. However, unutilized light by the photosystem may trigger photochemical damage [49]. Our data is in agreement with several published reports [44,49,50].

The data extracted from A-Ci curves showed a significant decrease in the maximum rate of Rubisco carboxylase activity (Vc, max), maximum rate of photosynthetic electron transport to regenerate Ribulose-1,5-bisphosphate (Jmax), and utilization of triose phosphate (TPU) under higher salinity and drought conditions. The reduction of TPU in *P. karka* indicates that the synthesis of sucrose/starch might be inhibited due to the reduced regeneration of phosphate (Pi) under stress conditions [15]. In addition, it may also cause growth inhibition under stress conditions, which is also evident in the lower values of the cellulose and hemicellulose contents in *P. karka* plants growing at 300 mM NaCl and drought [51,52]. Any alteration in the electron transport (ETR) disturbs the availability of the electron acceptors (like NADP+) and utilization of ADP that ultimately limits the regeneration of ribulose-1,5-bisphosphate [52]. Hence, it can be suggested that, under high salinity and drought conditions, the biochemical efficiency of the photosynthetic apparatus in *P. karka* plants decreased due to the colimitation of Vc, max, Jmax, and TPU.

The chlorophyll fluorescence data provides detailed insights into the integrity and efficient functioning of photosystem II (PSII). The maximum quantum yield of photosystem II (Fv/Fm) indicates the level of photoinhibition [53]. In the present study, unaffected Fv/Fm in all treatments suggested that there was no sign of photoinhibition, and it indicated the resilient ability of *P. karaka* in response to salt and drought stress. Our findings are also in agreement with other salt-resistant plants such as *Urochondra setulosa* and other halophytes [54,55]. It is also supported by the higher values verified for the maximum electron transport rates (ETR) in this experiment, where higher electron transport rates (ETR) were found in all treatments, especially under drought conditions. Non-Photochemical Quenching (NPQ) is an indicator of dissipating nonradiative energy from the light-harvesting complex (LHC II) of PSII that prevents the overreduction of ETC and therefore avoids damage to the photosynthetic process. Growth inhibition under water stress is caused by lower leaf expansion (due to less turgid cells, *P. karka* buffered the loss of the photosynthetic active leaf surface area by maintaining a high electron transport rate and Φ PSII under drought [3]). A higher NPQ was observed in *P. karka* at a higher salinity, indicating the efficient heat dissipation mechanism under a saline condition so NPQ serves as an index of stress for the plant [56,57]. Under severe stress situations, *P. karka* used a regulated and effective Y (NPQ) in this study to release absorbed light energy as heat that ultimately caused no change in the nonregulated process Y(NO). A similar strategy of heat dissipation has been documented in *Paspalum paspalodes* and *Paspalidium geminatum* [48]. The upregulation of the xanthophyll cycle and synthesis of photoprotective compounds such as carotenoids and the activity of photorespiration also support plant heat dissipation, which is critical to avoiding photosystem II damage under suboptimum situations [48,58,59].

The cell wall composition in grasses mostly consists of cellulose microfibrils interlinked with glucuronoarabinoxylans and polyphenolic depositions [60]. The synthesis of higher cellulose and hemicellulose in *P. karka* under saline conditions protects and supports the plant from lodging and higher light gaining, which promote growth and seedling vigor under saline and drought stress [61]. Generally, plants can reduce the cellulose synthesis and influence lignin accumulation under stress [62,63]. However, in salt-tolerant plants, the crude fiber, cellulose, and hemicellulose contents increased under salt stress [63]. Higher cellulose, hemicellulose, and total sugars in *P. karka* at a low salinity (100 mM NaCl) treatment suggest it could be a source of lignocellulose for bioethanol production in salt-affected lands. The cellulosic and hemicellulosic contents of *P. karka* are also comparable with the other bioenergy crops, such as *Cynodon dactylon* (35.7% cellulose, 25% hemicellulose) and *Panicum virgatum* (16.8% cellulose, 27.8% hemicellulose) [34].

Plants have been known as promising energy feedstock for ages and used for bioenergy production due to their lower cultivation cost, lower carbon dioxide emissions, and it is abundant in nature [64,65]. The per hectare dry biomass of *P. karka* was improved substantially at 100 mM NaCl. Higher per hectare aboveground dry biomass is reported in different feedstock crops for bioethanol [64], such as sweet sorghum [65]. The ethanol yield from the total sugars, cellulose, and hemicellulose contents were also estimated for *P. karka* in this study. Interestingly, the addition of 100 mM NaCl in the growth medium enhanced the yield of the ethanol per hectare by using the total sugar, cellulose, and hemicellulose in plants. Hence, the prospect of *P. karka* as feedstock for ethanol is probably very high, which could be helpful in utilizing the saline wastelands, as well as minimizing the energy crises and land competition for food and fuels.

5. Conclusions

This study reflects the contributions of different photochemical, stomatal, and biochemical factors on the growth performance, dry biomass, and predicted bioethanol production of *Phragmites karka* under dry, arid saline conditions. This study shows that the higher saturated irradiance (Is) of light, maximum rate of Rubisco carboxylase activity (Vc, max), maximum rate of electron transport (Jmax), and utilization of triose phosphates (TPU) are responsible for the change in the growth of *P. karka* under suboptimum conditions.

An increase in the respiratory rates exerts positive effects on the plant performance and metabolism by providing more energy to invest in the biomass and ethanol production. Growth inhibition under higher salinity and drought could be attributed to limited stomatal closure and decreased CO_2 assimilation. *P. karka* can be grown and produce a higher dry biomass and ethanol yield per hectare in saline and arid areas and could therefore be used as a sustainable biofuel crop. An increase in the maximum quantum yield, effective quantum yield, and lower photochemical quenching parameters are important in protecting plants by dissipating excessive energy, especially in drought conditions. These results clearly postulate that *P. karka* can be cultivated in areas of low salinity with the optimal photosynthetic performance. The production of higher ethanol and lignocellulosic contents in salinity can be useful in reducing the energy crises, land competition, and environmental protection.

Author Contributions: Z.A. (Zainul Abideen): Conceptualization, Formal analysis, Methodology, Data curation, Writing—original draft, and Writing—review and editing. H.W.K.: Conceptualization, Project administration, Resources, and Supervision. T.H.: Writing—original draft, Writing—review and editing, and Software. A.R.: Writing—review and editing. M.S.A. and M.S.E.: Funding acquisition and Writing—review and editing. M.I.H.: Writing—review and editing. F.Z.: Formal analysis and Writing—review and editing. S.M.: Writing—review and editing and Funding acquisition Z.A. (Zaheer Abbas): Formal analysis and Data curation. All authors have read and agreed to the published version of the manuscript.

Funding: The authors extend their appreciation to the Researchers Supporting Project number (RSP-2021/173) King Saud University, Riyadh, Saudi Arabia".

Institutional Review Board Statement: Not applicable.

Informed Consent Statement: Not applicable.

Data Availability Statement: Not applicable.

Acknowledgments: This study was conducted under the auspices of Pakistan and German Collaborative Project #: 1-308/ILPU/HEC/2009 funded Higher Education Commission (HEC) Pakistan". The authors extend their appreciation to the Researchers Supporting Project number (RSP-2021/173) King Saud University, Riyadh, Saudi Arabia.

Conflicts of Interest: The authors of the manuscript declare no conflict of interest.

References

1. Hossain, A.; Pamanick, B.; Venugopalan, V.K.; Ibrahimova, U.; Rahman, M.A.; Siyal, A.L.; Maitra, S.; Chatterjee, S.; Aftab, T. Emerging roles of plant growth regulators for plants adaptation to abiotic stress–induced oxidative stress. In *Emerging Plant Growth Regulators in Agriculture*; Academic Press: Cambridge, MA, USA, 2022; pp. 1–72.
2. Otie, V.; Udo, I.; Shao, Y.; Itam, M.O.; Okamoto, H.; An, P.; Eneji, E.A. Salinity effects on morpho-physiological and yield traits of soybean (*Glycine max* L.) as mediated by foliar spray with brassinolide. *Plants* **2021**, *10*, 541. [CrossRef]
3. Abideen, Z.; Koyro, H.W.; Huchzermeyer, B.; Ahmed, M.; Zulfiqar, F.; Egan, T.; Khan, M.A. Phragmites karka plants adopt different strategies to regulate photosynthesis and ion flux in saline and water deficit conditions. *Plant Biosyst.-Int. J. Deal. All Asp. Plant Biol.* **2021**, *155*, 524–534. [CrossRef]
4. Ziogas, V.; Tanou, G.; Morianou, G.; Kourgialas, N. Drought and salinity in citriculture: Optimal practices to alleviate salinity and water stress. *Agronomy* **2021**, *11*, 1283. [CrossRef]
5. Kumar, A.; Sharma, S.K.; Lata, C.; Devi, R.; Kulshrestha, N.; Krishnamurthy, S.L.; Singh, K.; Yadav, R.K. Impact of water deficit (salt and drought) stress on physiological, biochemical and yield attributes on wheat (*Triticum aestivum*) varieties. *Indian J. Agric. Sci.* **2018**, *88*, 1624–1632.
6. Álvarez, S.; Rodríguez, P.; Broetto, F.; Sánchez-Blanco, M.J. Long term responses and adaptive strategies of Pistacia lentiscus under moderate and severe deficit irrigation and salinity: Osmotic and elastic adjustment, growth, ion uptake and photosynthetic activity. *Agric. Water Manag.* **2018**, *202*, 253–262. [CrossRef]
7. Shahzad, B.; Fahad, S.; Tanveer, M.; Saud, S.; Khan, I.A. Plant responses and tolerance to salt stress. In *Approaches for Enhancing Abiotic Stress Tolerance in Plants*; CRC Press: Boca Raton, FL, USA, 2019; pp. 61–78.
8. Tang, L.; Zhou, Q.S.; Gao, Y.; Li, P. Biomass allocation in response to salinity and competition in native and invasive species. *Ecosphere* **2022**, *13*, e3900. [CrossRef]
9. Ma, Y.; Wei, Z.; Liu, J.; Liu, X.; Liu, F. Growth and physiological responses of cotton plants to salt stress. *J. Agron. Crop Sci.* **2021**, *207*, 565–576. [CrossRef]

10. Belghith, I.; Senkler, J.; Abdelly, C.; Braun, H.P.; Debez, A.; Turkan, I. Changes in leaf ecophysiological traits and proteome profile provide new insights into variability of salt response in the succulent halophyte Cakile maritima. *Funct. Plant Biol.* **2022**, *49*, 613–624. [CrossRef]
11. Munir, N.; Hasnain, M.; Roessner, U.; Abideen, Z. Strategies in improving plant salinity resistance and use of salinity resistant plants for economic sustainability. *Crit. Rev. Environ. Sci. Technol.* **2021**, *52*, 2150–2196. [CrossRef]
12. Ali, M.; Mustafa, A.; Abideen, Z.; Gul, B. Bioenergy Production from Halophytes Crops for Sustainable Development. In *Energy and Environmental Security in Developing Countries*; Springer: Cham, Switzerland, 2021; pp. 571–586.
13. Derbali, W.; Manaa, A.; Spengler, B.; Goussi, R.; Abideen, Z.; Ghezellou, P.; Abdelly, C.; Forreiter, C.; Koyro, H.W. Comparative proteomic approach to study the salinity effect on the growth of two contrasting quinoa genotypes. *Plant Physiol. Biochem.* **2021**, *163*, 215–229. [CrossRef]
14. Kronzucker, H.J.; Coskun, D.; Schulze, L.M.; Wong, J.R.; Britto, D.T. Sodium as nutrient and toxicant. *Plant Soil* **2013**, *369*, 1–23. [CrossRef]
15. Hussain, T.; Koyro, H.W.; Zhang, W.; Liu, X.; Gul, B.; Liu, X. Low salinity improves photosynthetic performance in Panicum antidotale under drought stress. *Front. Plant Sci.* **2020**, *11*, 481. [CrossRef]
16. Marshner, W.H. Aquinas on the evaluation of human actions. *Thomist A Specul. Q. Rev.* **1995**, *59*, 347–370. [CrossRef]
17. Subbarao, G.V.; Ito, O.; Berry, W.L.; Wheeler, R.M. Sodium—A functional plant nutrient. *Crit. Rev. Plant Sci.* **2003**, *22*, 391–416.
18. Grof, C.P.; Johnston, M.; Brownell, P.F. Effect of sodium nutrition on the ultrastructure of chloroplasts of C4 plants. *Plant Physiol.* **1989**, *89*, 539–543. [CrossRef]
19. Martínez-Peñalver, A.; Reigosa, M.J.; Sánchez-Moreiras, A.M. Imaging chlorophyll a fluorescence reveals specific spatial distributions under different stress conditions. *Flora-Morphol. Distrib. Funct. Ecol. Plants* **2011**, *206*, 836–844. [CrossRef]
20. Ghorbanzadeh, P.; Aliniaeifard, S.; Esmaeili, M.; Mashal, M.; Azadegan, B.; Seif, M. Dependency of growth, water use efficiency, chlorophyll fluorescence, and stomatal characteristics of lettuce plants to light intensity. *J. Plant Growth Regul.* **2021**, *40*, 2191–2207. [CrossRef]
21. Abideen, Z.; Ansari, R.; Khan, M.A. Halophytes: Potential source of ligno-cellulosic biomass for ethanol production. *Biomass Bioenergy* **2011**, *35*, 1818–1822. [CrossRef]
22. Abideen, Z.; Qasim, M.; Hussain, T.; Rasheed, A.; Gul, B.; Koyro, H.W.; Ansari, R.; Khan, M.A. Salinity improves growth, photosynthesis and bioenergy characteristics of Phragmites karka. *Crop Pasture Sci.* **2018**, *69*, 944–953. [CrossRef]
23. Epstein, E. *Mineral Nutrition of Plants: Principles and Perspectives*; John Wiley and Sons, Inc.: New York, NY, USA, 1972.
24. Koyro, H.W. Study of potential cash crop halophytes by a quick check system: Determination of the threshold of salinity tolerance and the ecophysiological demands. In *Cash Crop Halophytes: Recent Studies*; Springer: Dordrecht, The Netherlands, 2003; pp. 5–17.
25. Veihmeyer, F.J.; Hendrickson, A.H. The moisture equivalent as a measure of the field capacity of soils. *Soil Sci.* **1931**, *32*, 181–194. [CrossRef]
26. Schulte, M.; Offer, C.; Hansen, U. Induction of CO_2-gas exchange and electron transport: Comparison of dynamic and steady-state responses in Fagus sylvatica leaves. *Trees* **2003**, *17*, 153–163. [CrossRef]
27. Long, S.P.; Bernacchi, C.J. Gas exchange measurements, what can they tell us about the underlying limitations to photosynthesis? Procedures and sources of error. *J. Exp. Bot.* **2003**, *54*, 2393–2401. [CrossRef] [PubMed]
28. Abideen, Z.; Koyro, H.W.; Huchzermeyer, B.; Ansari, R.; Zulfiqar, F.; Gul, B.J.P.B. Ameliorating effects of biochar on photosynthetic efficiency and antioxidant defence of Phragmites karka under drought stress. *Plant Biol.* **2020**, *22*, 259–266. [CrossRef] [PubMed]
29. Kitajima, M.B.W.L.; Butler, W.L. Quenching of chlorophyll fluorescence and primary photochemistry in chloroplasts by dibromothymoquinone. *Biochim. Et Biophys. Acta (BBA)-Bioenerg.* **1975**, *376*, 105–115. [CrossRef]
30. Genty, B.; Briantais, J.M.; Baker, N.R. The relationship between the quantum yield of photosynthetic electron transport and quenching of chlorophyll fluorescence. *Biochim. Et Biophys. Acta (BBA)-Gen. Subj.* **1989**, *990*, 87–92. [CrossRef]
31. Bilger, W.; Björkman, O. Role of the xanthophyll cycle in photoprotection elucidated by measurements of light-induced absorbance changes, fluorescence and photosynthesis in leaves of Hedera canariensis. *Photosynth. Res.* **1990**, *25*, 173–185. [CrossRef]
32. Van Kooten, O.; Snel, J.F. The use of chlorophyll fluorescence nomenclature in plant stress physiology. *Photosynth. Res.* **1990**, *25*, 147–150. [CrossRef]
33. Krall, J.P.; Edwards, G.E. Relationship between photosystem II activity and CO_2 fixation in leaves. *Physiol. Plant.* **1992**, *86*, 180–187. [CrossRef]
34. Sun, Y.; Cheng, J. Hydrolysis of lignocellulosic materials for ethanol production: A review. *Bioresour. Technol.* **2002**, *83*, 1–11. [CrossRef]
35. Ludwig, T.G.; Goldberg, H.J. The anthrone method for the determination of carbohydrates in foods and in oral rinsing. *J. Dent. Res.* **1956**, *35*, 90–94. [CrossRef]
36. Zhao, Y.L.; Dolat, A.; Steinberger, Y.; Wang, X.; Osman, A.; Xie, G.H. Biomass yield and changes in chemical composition of sweet sorghum cultivars grown for biofuel. *Field Crops Res.* **2009**, *111*, 55–64. [CrossRef]
37. Abideen, Z.; Ansari, R.; Gul, B.; Khan, M.A. The place of halophytes in Pakistan's biofuel industry. *Biofuels* **2012**, *3*, 211–220. [CrossRef]
38. Gul, B.; Abideen, Z.; Ansari, R.; Khan, M.A. Halophytic biofuels revisited. *Biofuels* **2013**, *4*, 575–577. [CrossRef]
39. Khan, M.A.; Ansari, R.; Ali, H.; Gul, B.; Nielsen, B.L. Panicum turgidum a potentially sustainable cattle feed alternative to maize for saline areas. *Agric. Ecosyst. Environ.* **2009**, *129*, 542–546. [CrossRef]

40. Abideen, Z.; Koyro, H.W.; Huchzermeyer, B.; Ahmed, M.Z.; Gul, B.; Khan, M.A. Moderate salinity stimulates growth and photosynthesis of Phragmites karka by water relations and tissue specific ion regulation. *Environ. Exp. Bot.* **2014**, *105*, 70–76. [CrossRef]
41. Gorai, M.; Vadel, A.M.; Neffati, M.; Khemira, H. The effect of sodium chloride salinity on the growth, water status, and ion content of Phragmites communis. *Pak. J. Biol. Sci.* **2007**, *13*, 2225–2230. [CrossRef]
42. Gorai, M.; Ennajeh, M.; Khemira, H.; Neffati, M. Combined effect of NaCl-salinity and hypoxia on growth, photosynthesis, water relations and solute accumulation in Phragmites australis plants. *Flora* **2010**, *205*, 462–470. [CrossRef]
43. Naidoo, G.; Naidoo, Y.; Achar, P. Ecophysiological responses of the salt marsh grass Spartina maritima to salinity. *Afr. J. Aquat. Sci.* **2012**, *37*, 81–88. [CrossRef]
44. Gulzar, S.; Khan, M.A.; Liu, X. Seed germination strategies of Desmostachya bipinnata: A fodder crop for saline soils. *Rangel. Ecol. Manag.* **2007**, *60*, 401–407. [CrossRef]
45. Geissler, N.; Hussin, S.; Koyro, H.-W. Elevated atmospheric CO_2 concentration ameliorates effects of NaCl salinity on photosynthesis and leaf structure of *Aster tripolium* L. *J. Exp. Bot.* **2009**, *60*, 137–151. [CrossRef]
46. Redondo-Gómez, S.; Mateos-Naranjo, E.; Figueroa, M.E.; Davy, A.J. Salt stimulation of growth and photosynthesis in an extreme halophyte Arthrocnemum macrostachyum. *Plant Biol.* **2010**, *12*, 79–87. [CrossRef]
47. Asrar, H.; Hussain, T.; Hadi, S.M.S.; Gul, B.; Nielsen, B.L.; Khan, M.A. Salinity induced changes in light harvesting and carbon assimilating complexes of *Desmostachya bipinnata* (L.) Staph. *Environ. Exp. Bot.* **2017**, *135*, 86–95. [CrossRef]
48. Moinuddin, M.; Gulzar, S.; Hameed, A.; Gul, B.; Ajmal Khan, M.; Edwards, G.E. Differences in photosynthetic syndromes of four halophytic marsh grasses in Pakistan. *Photosynth. Res.* **2017**, *131*, 51–64. [CrossRef]
49. Pagter, M.; Bragato, C.; Malagol, M.; Brix, H. Osmotic and ionic effects of NaCl and Na_2SO_4 salinity on Phragmites australis. *Aquat. Bot.* **2009**, *90*, 43–51. [CrossRef]
50. Koyro, H.W. Effect of salinity on growth, photosynthesis, water relations and solute composition of the potential cash crop halophyte *Plantago coronopus* (L.). *Environ. Exp. Bot.* **2006**, *56*, 136–146. [CrossRef]
51. Pieters, A.J.; Paul, M.J.; Lawlor, D.W. Low sink demand limits photosynthesis under Pi deficiency. *J. Exp. Bot.* **2001**, *52*, 1083–1091. [CrossRef]
52. Lin, Z.H.; Chen, L.S.; Chen, R.B.; Zhang, F.Z.; Jiang, H.X.; Tang, N. CO_2assimilation, ribulose-1, 5-bisphosphate carboxylase/oxygenase, carbohydrates and photosynthetic electron transport probed by the JIP-test, of tea leaves in response to phosphorus supply. *BMC Plant Biol.* **2009**, *9*, 43. [CrossRef]
53. Athar, H.u.R.; Ashraf, M. Photosynthesis under Drought Stress. In *Handbook of Photosynthesis*, 2nd ed.; Pessarakli, M., Ed.; CRC Press: Boca Raton, FL, USA, 2005; pp. 793–809.
54. Bano, S.; Ahmed, M.Z.; Abideen, Z.; Qasim, M.; Gul, B.; Khan, N.U. Humic acid overcomes salinity barriers and stimulates growth of *Urochondra setulosa* by altering ion-flux and photochemistry. *Acta Physiol. Plant.* **2022**, *44*, 39. [CrossRef]
55. Ehsen, S.; Abideen, Z.; Rizvi, R.F.; Gulzar, S.; Aziz, I.; Gul, B.; Khan, M.A.; Ansari, R. Ecophysiological adaptations and anti-nutritive status of sustainable cattle feed *Haloxylon stocksii* under saline conditions. *Flora* **2019**, *257*, 151425. [CrossRef]
56. Maxwell, K.; Johnson, G.N. Chlorophyll fluorescence—A practical guide. *J. Exp. Bot.* **2000**, *51*, 659–668. [CrossRef]
57. Wei, Y.; Xu, X.; Tao, H.; Wang, P. Growth performance and physiological response in the halophyte Lycium barbarum grown at salt-affected soil. *Ann. Appl. Biol.* **2006**, *149*, 263–269. [CrossRef]
58. Qiu, N.; Lu, Q.; Lu, C. Photosynthesis, photosystem II efficiency and the xanthophyll cycle in the salt-adapted halophyte *Atriplex centralasiatica*. *New Phytol.* **2003**, *159*, 479–486. [CrossRef] [PubMed]
59. Pompeiano, A.; Landi, M.; Meloni, G.; Vita, F.; Guglielminetti, L.; Guidi, L. Allocation pattern, ion partitioning, and chlorophyll a fluorescence in *Arundo donax* L. in responses to salinity stress. *Plant Biosyst.-Int. J. Deal. All Asp. Plant Biol.* **2017**, *151*, 613–622. [CrossRef]
60. Zhang, Y.; Yu, J.; Wang, X.; Durachko, D.M.; Zhang, S.; Cosgrove, D.J. Molecular insights into the complex mechanics of plant epidermal cell walls. *Science* **2021**, *372*, 706–711. [CrossRef] [PubMed]
61. Cosgrove, D.J. Growth of the plant cell wall. *Nat. Rev. Mol. Cell Biol.* **2005**, *6*, 850–861. [CrossRef]
62. Wang, Y.; Zhang, X.; Yang, S.; Wang, C.; Lu, G.; Wang, R.; Yang, Y.; Li, D. Heterogenous expression of *Pyrus pyrifolia* PpCAD2 and PpEXP2 in tobacco impacts lignin accumulation in transgenic plants. *Gene* **2017**, *637*, 181–189. [CrossRef]
63. Maksup, S.; Sengsai, S.; Laosuntisuk, K.; Asayot, J.; Pongprayoon, W. Physiological responses and the expression of cellulose and lignin associated genes in Napier grass hybrids exposed to salt stress. *Acta Physiol. Plant.* **2020**, *42*, 109. [CrossRef]
64. Berndes, G.; Hoogwijk, M.; Broek, R. The contribution of biomass in the future global energy supply: A review of 17 studies. *Biomass Bioenergy* **2003**, *25*, 1–28. [CrossRef]
65. Dolciotti, I.; Mambelli, S.; Grandi, S.; Venturi, G. Comparison of two sorghum genotypes for sugar and fiber production. *Ind. Crops Prod.* **1998**, *7*, 265–272. [CrossRef]

plants

MDPI

Article

Assessment of Uptake, Accumulation and Degradation of Paracetamol in Spinach (*Spinacia oleracea* L.) under Controlled Laboratory Conditions

Zarreen Badar [1,*], Abdallah Shanableh [1,2,*], Ali El-Keblawy [1,3,4], Kareem A. Mosa [3,5], Lucy Semerjian [6], Abdullah Al Mutery [3,7,8], Muhammad Iftikhar Hussain [9], Sourjya Bhattacharjee [2], François Mitterand Tsombou [3,10], Sefeera Sadik Ayyaril [1], Islam M. Ahmady [3], Attiat Elnaggar [1,11], Muath Mousa [1] and Mohammad H. Semreen [12,13]

1 Research Institute of Science and Engineering (RISE), University of Sharjah, Sharjah P.O. Box 27272, United Arab Emirates; akeblawy@sharjah.ac.ae (A.E.-K.); safeeraayyaril@sharjah.ac.ae (S.S.A.); atyatnajar@yahoo.com (A.E.); mmousa2@sharjah.ac.ae (M.M.)
2 Department of Civil and Environmental Engineering, University of Sharjah, Sharjah P.O. Box 27272, United Arab Emirates; sbhattacharjee@sharjah.ac.ae
3 Department of Applied Biology, College of Sciences, University of Sharjah, Sharjah P.O. Box 27272, United Arab Emirates; kmosa@sharjah.ac.ae (K.A.M.); aalmutery@sharjah.ac.ae (A.A.M.); tsombou@yahoo.fr (F.M.T.); ialtayeb@sharjah.ac.ae (I.M.A.)
4 Department of Biology, Faculty of Science, Al-Arish University, Al-Arish 45511, Egypt
5 Department of Biotechnology, Faculty of Agriculture, Al-Azhar University, Cairo 11751, Egypt
6 Department of Environmental Health Sciences, College of Health Sciences, University of Sharjah, Sharjah P.O. Box 27272, United Arab Emirates; lsemerjian@sharjah.ac.ae
7 Human Genetics and Stem Cells Research Group, Research Institute of Sciences and Engineering, University of Sharjah, Sharjah P.O. Box 27272, United Arab Emirates
8 Molecular Genetics Lab, Biotechnology Lab, Research Institute of Sciences and Engineering, University of Sharjah, Sharjah P.O. Box 27272, United Arab Emirates
9 Department of Plant Biology and Soil Science, Campus Lagoas Marcosende, Universidad de Vigo, 36310 Vigo, Spain; mih786@gmail.com
10 Departmento de Biología Vegetal, Universidad de Málaga, 29016 Málaga, Spain
11 Department of Botany and Microbiology, Faculty of Science, Alexandria University, Alexandria 21568, Egypt
12 Department of Medicinal Chemistry, College of Pharmacy, University of Sharjah, Sharjah P.O. Box 27272, United Arab Emirates; msemreen@sharjah.ac.ae
13 Sharjah Institute for Medical Research, University of Sharjah, Sharjah P.O. Box 27272, United Arab Emirates
* Correspondence: zbadar@sharjah.ac.ae (Z.B.); shanableh@sharjah.ac.ae (A.S.)

Citation: Badar, Z.; Shanableh, A.; El-Keblawy, A.; Mosa, K.A.; Semerjian, L.; Mutery, A.A.; Hussain, M.I.; Bhattacharjee, S.; Tsombou, F.M.; Ayyaril, S.S.; et al. Assessment of Uptake, Accumulation and Degradation of Paracetamol in Spinach (*Spinacia oleracea* L.) under Controlled Laboratory Conditions. *Plants* **2022**, *11*, 1626. https://doi.org/10.3390/plants11131626

Academic Editor: Carmen Arena

Received: 17 May 2022
Accepted: 16 June 2022
Published: 21 June 2022

Abstract: The occurrence and persistence of pharmaceuticals in the food chain, particularly edible crops, can adversely affect human and environmental health. In this study, the impacts of the absorption, translocation, accumulation, and degradation of paracetamol in different organs of the leafy vegetable crop spinach (*Spinacia oleracea*) were assessed under controlled laboratory conditions. Spinach plants were exposed to 50 mg/L, 100 mg/L, and 200 mg/L paracetamol in 20% Hoagland solution at the vegetative phase in a hydroponic system. Exposed plants exhibited pronounced phytotoxic effects during the eight days trial period, with highly significant reductions seen in the plants' morphological parameters. The increasing paracetamol stress levels adversely affected the plants' photosynthetic machinery, altering the chlorophyll fluorescence parameters (F_v/F_m and PSII), photosynthetic pigments (Chl a, Chl b and carotenoid contents), and composition of essential nutrients and elements. The LC-MS results indicated that the spinach organs receiving various paracetamol levels on day four exhibited significant uptake and translocation of the drug from roots to aerial parts, while degradation of the drug was observed after eight days. The VITEK® 2 system identified several bacterial strains (e.g., members of *Burkhulderia*, *Sphingomonas*, *Pseudomonas*, *Staphylococcus*, *Stenotrophomonas* and *Kocuria*) isolated from spinach shoots and roots. These microbes have the potential to biodegrade paracetamol and other organic micro-pollutants. Our findings provide novel insights to mitigate the risks associated with pharmaceutical pollution in the environment and explore the bioremediation potential of edible crops and their associated microbial consortium to remove these pollutants effectively.

Keywords: abiotic stress; spinach; paracetamol; degradation; growth parameters; chlorophyll florescence; photosynthetic pigments; elements; microbes

1. Introduction

The United Arab Emirates (UAE) is situated in the Arabian desert, facing harsh climatic conditions and water scarcity. Such a situation poses a severe constraint to agricultural development. Therefore, alternative irrigation water sources are essential for achieving sustainable agriculture. The UAE is highly dependent on desalination for water supply and ranks among the highest per capita water consumers in the world [1]. Therefore, the reuse of treated wastewater for irrigation is a highly attractive alternative to reduce reliance on desalination and save the limited freshwater resources in the UAE. However, significant barriers exist to the widespread use of treated wastewater for irrigation due to the presence of emerging contaminants, especially pharmaceuticals, that may adversely affect soil quality and/or public health. Although these pollutants have been detected in very low concentrations (ng/L and µg/L), they pose a substantial risk to the environment due to their daily entry into terrestrial ecosystems [2].

Emerging contaminants such as pharmaceuticals have raised concerns in recent years because of the potential for chronic toxicity and the development of resistance to microbial pathogens in humans and ecosystems [3]. The presence of pharmaceuticals, their toxic degradation products, and different microscopic contaminants in reclaimed water poses a new challenge for wastewater professionals and the pharmaceutical enterprise, as wastewater treatment plants are not designed to eliminate them effectively [4]. Despite the fact that many drugs are naturally bioactive compounds, their toxicity to flora is not widely known. Thus, it is important to assess pharmaceutical contaminations and their phytotoxicity in plants.

Studies report that various drugs can enter plant organs via the roots and leaves and induce phytotoxicity, depending on the composition and concentration of the compound and plant species [5]. In addition, deadly doses vary between studies; however, the range of plant toxicity of antibiotics is from 4 to more than 10,000 µg [6]. These pollutants can be detoxified, disrupted, and separated once taken up by roots and translocated to leaves [7,8]. Moreover, the physicochemical properties of pharmaceuticals, physiological characteristics of plants, and plant–pharmaceutical interactions (e.g., water solubility, half-life, sorption affinity) all collectively influence the transport and accumulation of pharmaceuticals in plants [6,9].

Paracetamol, also named acetaminophen, is one of the most widely consumed analgesic drugs among over-the-counter (OCT) and prescribed medicines [10]. The global paracetamol market is expected to be valued at 126.2 USD million in 2022 due to the COVID-19 pandemic and is forecast to adjust to 121.7 USD million by 2028 [11]. Paracetamol is considered one of the most rapidly growing contaminants on the planet and has previously been reported in concentrations of up to 200 µg/L in wastewater effluents and 28.70 µg/L in surface waters [12], despite having high removal efficiency and a 90% elimination rate in wastewater treatment plants within 15 days [13–15].

Researchers have described the strong tendency of paracetamol and other pharmaceuticals towards sorption, accumulation, and persistence in plants and soil sediments [6,16,17], and reported a variety of phytotoxic effects in different organs of plants in soil and hydroponic systems, all of which were influenced by the applied doses [18,19]. According to recent findings, leafy vegetables (such as lettuce, spinach, cabbage, and celery) have the greatest ability to receive and accumulate emerging contaminants in their edible tissues [20,21]. Spinach and lettuce are the most common plants that uptake pharmaceuticals and other emerging contaminants from treated wastewater effluents used for irrigation [22]. In cucumber, a 5 mg/L paracetamol dose inhibited Cytochrome P450 [23], whereas beans treated for three weeks with 1–4 g/L paracetamol promoted phytotoxicity by reducing stomatal conductance and β carotene concentration [24]. Moreover, paracetamol tested at lethal doses of 151 mg/L for 168 h in *Brassica juncea* (Indian mustard) resulted in bleaching and dot-like lesions on the adaxial side of the leaf after 72 h, and necrosis was seen after seven days [25]. However, when paracetamol was applied at the concentration of 151–181 mg/L, it did not display any symptoms of toxicity in *Hordeum vulgare* (barley), *Lupinus luteolus* (pale yellow lupine), and *Phragmites autralis* [26]. Similarly, relatively lower toxic effects and only 12–18% inhibition were reported in *Linum usitatissimum* (flax) and *Aromoratia rusticana* (horseradish) [26].

Recent studies conducted at the Sharjah sewage treatment plant have reported the presence of pharmaceutical pollutants in treated wastewater in urban areas with paracetamol; a concentration of 5235 ng/L paracetamol was reported in effluent wastewaters [27,28]. Despite being classified as a pseudo-persistent drug, paracetamol is discharged into the environment on a regular basis, presenting major issues for both human health and the ecosystem [6,29,30]. Moreover, there is scarce information on the effects of the accumulation of and long-term exposure to minute concentrations of this drug on both human and plant growth and development.

Thus, spinach (*Spinacia oleracea*), a green vegetable, is the subject of our study, which aims to evaluate the absorption, translocation, accumulation, and degradation of paracetamol in various organs in a controlled laboratory setting. The study also assessed paracetamol's phytotoxicity on several morphological and physiological traits, and identified several bacterial strains in spinach roots and shoots. We hypothesized that the regular use of wastewater with low concentrations of emerging contaminants (e.g., paracetamol) in crop irrigation would concentrate them in the soil and other environmental matrixes, and result in their uptake into, translocation to and accumulation in different organs of the plants. It was also hypothesized that the accumulated paracetamol and its degradation products in the tissues of the plants would affect essential element absorption and biochemical processes, affecting pigment synthesis and photosynthesis efficiency.

2. Results

2.1. Paracetamol Uptake and Metabolism in Spinach

The results showed that paracetamol treatments, exposure periods, and their interactions significantly affected paracetamol concentrations in spinach shoots and roots (Table 1, Figure 1, Supplementary Figures S2–S4). Spinach plants treated with 50 mg/L, 100 mg/L, and 200 mg/L paracetamol accumulated 75 µg/g, 136 µg/g, and 1012 µg/g paracetamol, respectively, after four days in shoot tissues. After eight days, the drug concentrations significantly declined to 56 µg/g, 73 µg/g, and 396 µg/g in the 50 mg/L, 100 mg/L, and 200 mg/L treatments, respectively, indicating the biodegradation of paracetamol after day 4. Paracetamol concentrations in plants treated with 200 mg/L paracetamol paracetamol were significantly greater than in plants treated with 50 mg/L and 100 mg/L over the two treatment periods (Table 1, Figure 1a, Supplementary Figure S2).

The accumulated levels of paracetamol were very low in the roots compared to in the shoots. The concentrations in plants' roots treated with 50 mg/L, 100 mg/L, and 200 mg/L paracetamol were 6.5 µg/g, 7.8 µg/g, and 10.3 µg/g, respectively, after four days, with significant reductions observed in the concentrations (to 2.3 µg/g, 2.6 µg/g, and 3.6 µg/g, respectively) after eight days (Table 1, Figure 1b, Supplementary Figure S3). The results indicate high translocation of paracetamol after its uptake from the roots to the shoots. The translocation factor in plants treated with 50 mg/L, 100 mg/L, and 200 mg/L paracetamol significantly increased from 11.54, 17.46, and 98.74, respectively, after four days to 24.12, 27.63, and 112.15, respectively, after eight days (Table 1, Figure 1c). However, there were insignificant effects of the interaction between paracetamol treatments and exposure period on the translocation factor ($p \geq 0.05$, Table 1, Figure 1c).

Table 1. Results of two-way ANOVAs showing the effects of different paracetamol treatments, exposure periods, and their interactions on the concentration of paracetamol (µg/g DW) in spinach shoots, roots, and translocation factor.

Variable	Factor	df	F-Ratio	*p*-Value
Shoots	Period (P)	1	488.82	<0.001
	Treatments (T)	2	1538.01	<0.001
	P × T	2	332.65	<0.001
Roots	Period (P)	1	587.76	<0.001
	Treatments (T)	2	45.42	<0.001
	P × T	2	11.49	<0.01
Translocation Factor	Period (P)	1	26.52	<0.001
	Treatments (T)	2	591.35	<0.001
	P × T	2	0.17	ns

ns: non-significant at $p \leq 0.05$.

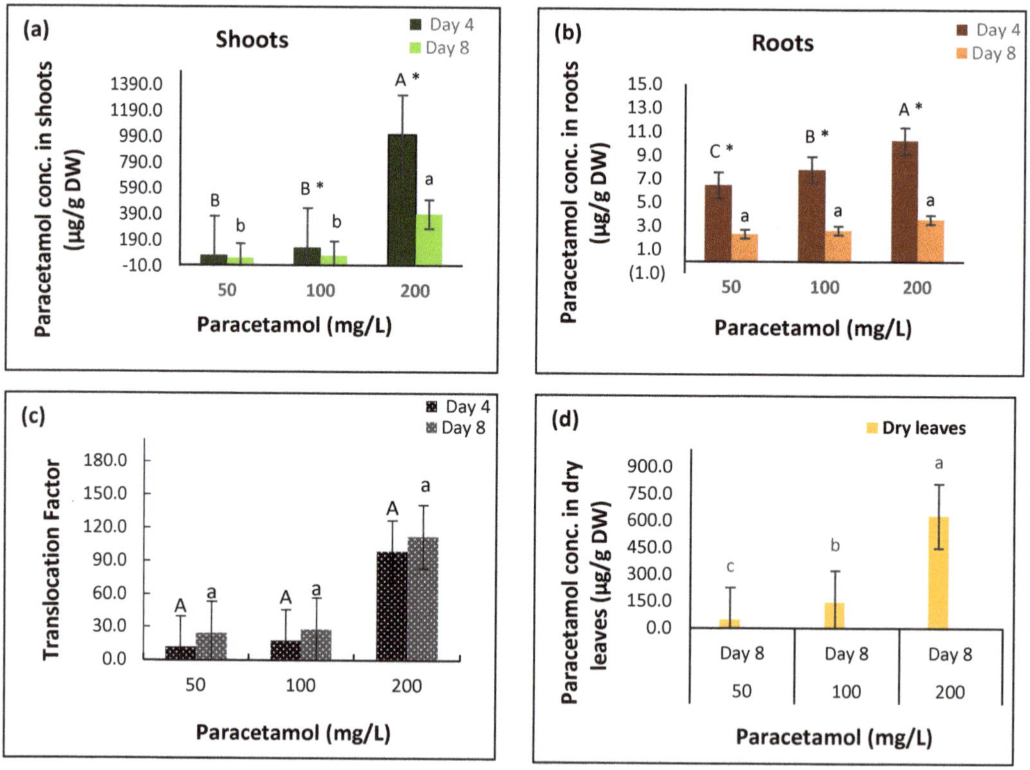

Figure 1. Effect of different paracetamol treatments at different time points on the concentration of paracetamol (µg/g DW) in spinach (**a**) shoots, (**b**) roots, (**c**) translocation factor, and (**d**) dry leaves. Error bars represent means ± S.E. of three biological replicates. Means with different upper-case and lower-case letters indicate significant differences ($p \leq 0.05$) between the different paracetamol treatments at four and eight days, respectively. Asterisks (*) indicates significant differences between four and eight days at a certain concentration.

The dried leaves from plants treated with different paracetamol concentrations were collected and assessed after eight days. On day eight, the concentration of paracetamol in dried leaves treated with 100 mg/L and 200 mg/L paracetamol was greater (143 µg/g and 628 μg/g, respectively) than in fresh leaves (73 μg/g and 396 μg/g, respectively). The accumulation of paracetamol increased in dried leaves with an increase in the applied doses of the pharmaceutical drug (Figure 1d, Supplementary Figure S4).

2.2. Influence of Paracetamol on Plant Growth Parameters

The effect of different paracetamol levels was evident on the leaves, shoot and root lengths, number of leaves, and shoot growth tolerance index (%) (Supplementary Table S1, Figure 2). Spinach root length, shoot length and number of leaves were significantly reduced in the 200 mg/L paracetamol group compared to the controls. There was no significant difference between the lower two paracetamol concentrations (50 mg/L and 100 mg/L) in all growth traits (Supplementary Table S1, Figure 2a–d). The sensitivity of the growth to paracetamol was organ-specific. For example, the reductions at 50 mg/L, 100 mg/L, and 200 mg/L, compared to controls, were 17.4%, 13.0%, and 34.8%, respectively, in leaf number, 31.0%, 41.4%, and 58.6%, respectively, in shoot length, and 20.7%, 62.3%, and 66.0%, respectively, in root length, indicating that leaf number was least affected and root length was most affected. Moreover, shoot tolerance was significantly lower ($p \leq 0.05$) at 200 mg/L than root tolerance (Supplementary Table S1, Figure 2d). In addition, plants exposed to higher paracetamol concentrations showed leaf margin necrosis and increased numbers of withered leaves and browning in the root tissues (Supplementary Figure S5).

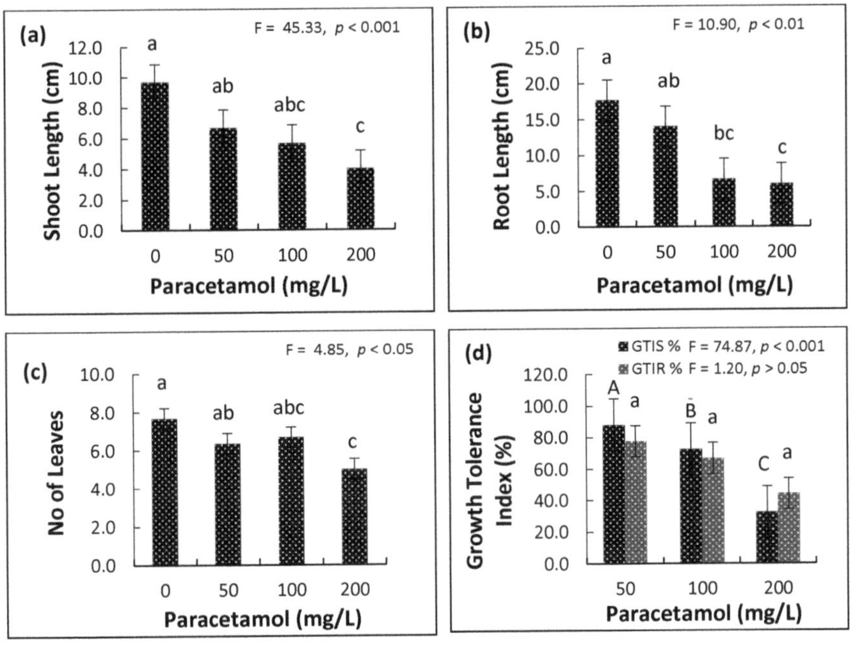

Figure 2. Effects of different paracetamol treatments on spinach (**a**) shoot length, (**b**) root length and (**c**) number of leaves, and (**d**) growth tolerance index (%) of spinach shoots (GTIS) and roots (GTIR) after eight days. Error bars represent means ± S.E. of three biological replicates. For subfigures (**a**–**c**), means with the different letters are significantly different at $p \leq 0.05$. For subfigure (**d**), different upper-case and similar lower-case letters indicate significant differences between GTIS and non-significant differences between GTIR, respectively, at the different paracetamol treatments.

2.3. Impact of Paracetamol on Photosynthetic Pigments and Chlorophyll Fluorescence

Besides detecting possible damage in the photosynthetic apparatus of plants, fluorescence measurements allow for qualitative and quantitative analysis based on the absorption and utilization of light energy [31]. Two-way ANOVA showed significant effects of paracetamol and treatment period and their interaction on the maximum quantum efficiency of PSII and F_v/F_m (Table 2). There was no significant difference in F_v/F_m between four and eight days in the non-treated plants (control). The F_v/F_m values decreased with the increase in paracetamol concentration, but the reduction was more pronounced after eight days than four days. The F_v/F_m of plants treated with 50 mg/L, 100 mg/L and 200 mg/L decreased below the control by 0.1%, 2.2% and 2.6%, respectively, after four days and 9.8%, 9.05%, and 10.3%, respectively, after eight days (Figure 3a). A significant reduction in the quantum yield efficiency of photosystem II was also seen, of 7.87%, 7.25%, and 11.84% after eight days in plants treated with 50 mg/L, 100 mg/L, and 200 mg/L paracetamol, respectively, as compared to four days with 0.13%, 1.67% and 4.25% reductions, respectively (Figure 3b).

Table 2. Results of two-way ANOVAs showing the effects of different paracetamol treatments, exposure periods, and their interactions on chlorophyll fluorescence (F_v/F_m and ΦII), Chl a, Chl b, carotenoids, and total chlorophyll content (mg/g FW) in spinach.

Variable	Factor	df	F-Ratio	*p*-Value
F_v/F_m	Period (P)	1	29.45	<0.001
	Treatment (T)	3	7.07	<0.01
	P × T	3	3.66	<0.05
ΦII	Period (P)	1	21.07	<0.01
	Treatment (T)	3	10.93	<0.01
	P × T	3	3.43	<0.05
Chl a	Period (P)	1	6.19	<0.05
	Treatment (T)	3	1.77	ns
	P × T	3	1.52	ns
Chl b	Period (P)	1	10.02	<0.01
	Treatment (T)	3	2.32	ns
	P × T	3	2.31	ns
Carotenoids	Period (P)	1	4.34	ns
	Treatment (T)	3	1.70	ns
	P × T	3	1.68	ns
Total Chlorophyll	Period (P)	1	7.11	<0.05
	Treatment (T)	3	1.89	ns
	P × T	3	1.71	ns

ns: non-significant at $p \leq 0.05$.

The two-way ANOVA results showed significant effects of paracetamol treatment period on Chl a, Chl b, and total chlorophyll (Table 2). Interestingly, Chl a, Chl b, and total chlorophyll significantly increased after eight days compared to after four days. Such differences were more obvious at higher paracetamol concentrations (100 mg/L and 200 mg/L). The results indicate that more prolonged exposure at higher paracetamol concentrations enhanced the chlorophyll synthesis. After eight days, Chl a significantly increased by 25.34% with 200 mg/L paracetamol treatment and Chl b content increased by 4.05% and 33.23% with 100 mg/L and 200 mg/L paracetamol treatments, respectively, compared to the control. However, there were insignificant effects on all the pigments for the different paracetamol treatments and the interactions between periods and treatments ($p \geq 0.05$, Table 2, Figure 3).

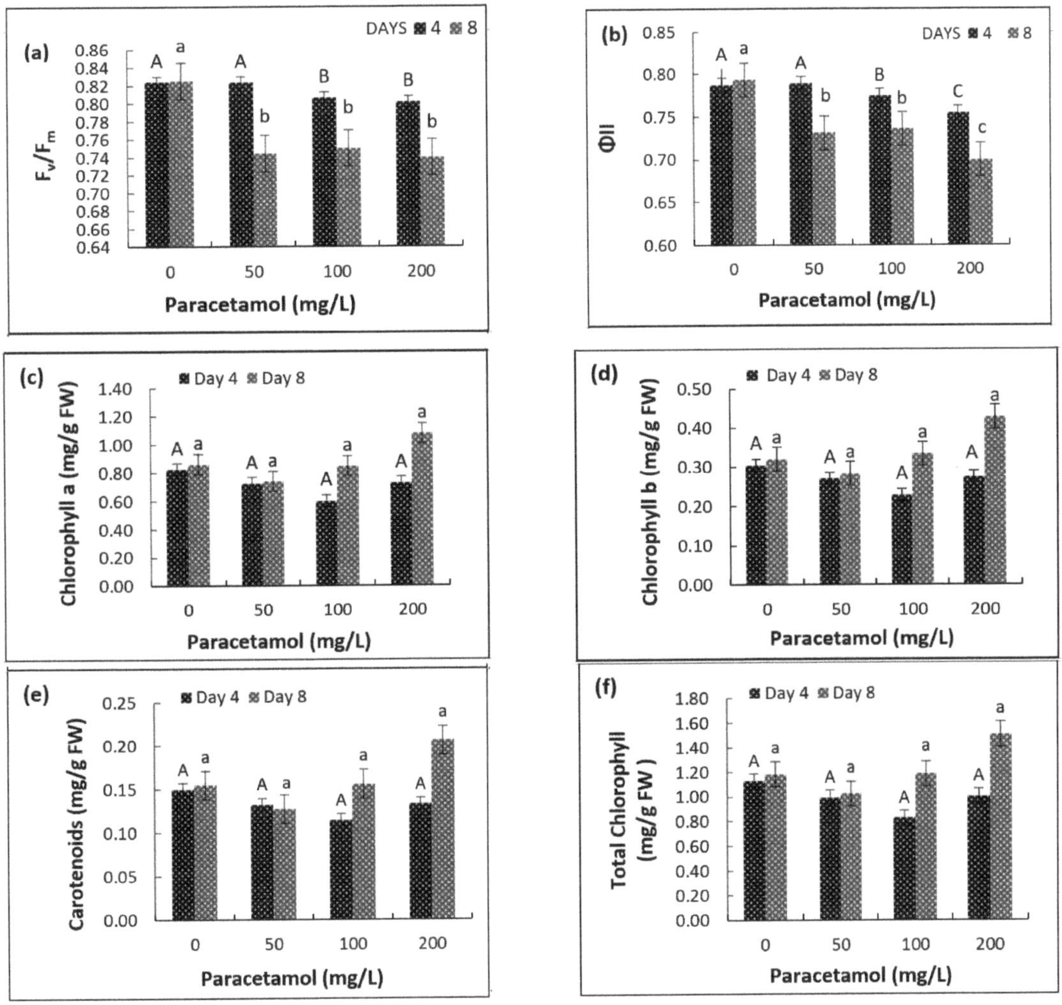

Figure 3. Effects of different paracetamol treatments and exposure periods on (**a**) F_v/F_m, (**b**) quantum yield of electron transport in photosystem II (ΦII), (**c**) Chl a, (**d**) Chl b, (**e**) carotenoids and (**f**) total chlorophyll (mg/g FW). Error bars represent ± S.E. of three biological replicates. Means with different upper-case and lower-case letters indicate significant differences ($p \leq 0.05$) at days four and eight, respectively.

2.4. Plant Nutrients and Elements

2.4.1. Macronutrients

There were significant effects of different paracetamol treatments on the concentration of nutrients in spinach shoots and roots (Supplementary Table S2, Figure 4). In the roots, there were gradual decreases in the concentrations of two macronutrients (Ca, K) with the increase in paracetamol treatments (Figure 4a,b). In the 50 mg/L, 100 mg/L, and 200 mg/L paracetamol treatments, Ca decreased by 33.1%, 39.9% and 49.8%, respectively, and K decreased by 65.1%, 69.1%, and 76.1%, respectively. For Mg, the concentration decreased by 40.2% and 46.3% in the 50 mg/L and 100 mg/L paracetamol treatments, respectively, but only 13% in the 200 mg/L treatment (Figure 4c). However, the lower paracetamol treatments (50 mg/L and 100 mg/L) significantly reduced the concentrations of the two macronutrients. Additionally, the high paracetamol treatment (200 mg/L) significantly increased the concentrations of Ca and K in the shoots (Figure 4a,b) and reduced the concentrations of Mg in the shoots compared to the control. Still, the reduction in the concentrations of Mg caused by the 200 mg/L treatment was less than that in the 50 mg/L and 100 mg/L treatments (Figure 4c).

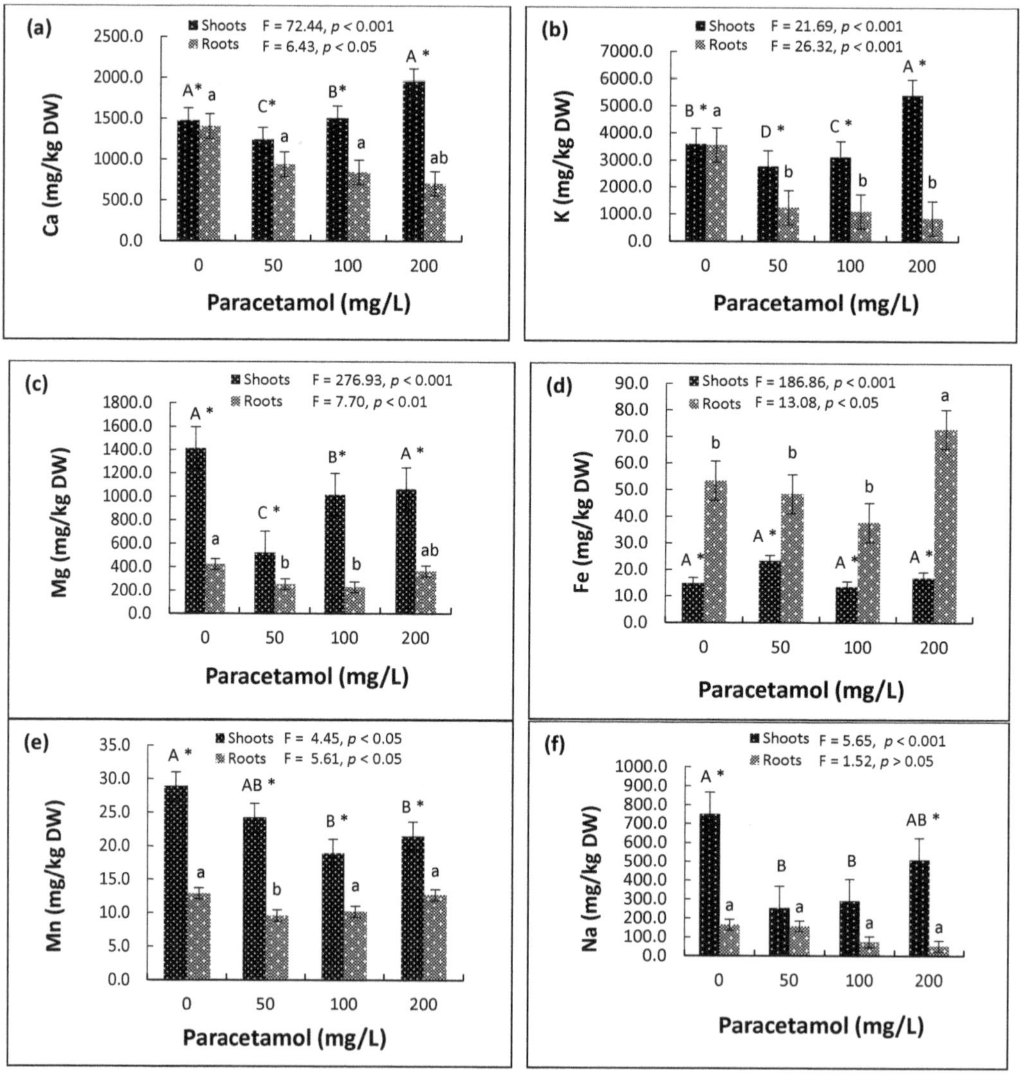

Figure 4. Effects of different paracetamol treatments on concentrations (mg/kg DW) of (**a**–**c**) macronutrients, (**d**,**e**) micronutrients and (**f**) sodium in shoot and root systems of spinach. Error bars represent means ± S.E. of three biological replicates. Means with different upper-case and lower-case letters indicate significant differences ($p \leq 0.05$) in nutrient and sodium concentrations between shoots and roots, respectively, at the different paracetamol treatments. Asterisks (*) indicate significant differences ($p \leq 0.05$) in the concentration of a nutrient or sodium between shoots and roots at a certain paracetamol level.

2.4.2. Micronutrients

The effects of the paracetamol treatments on the two micronutrients (Fe and Mn) in both roots and shoots were significant (Supplementary Table S2). Unlike most of the other elements, the concentration of Fe was significantly greater in roots than in shoots. The concentration was considerably greater in the roots of plants treated with 200 mg/L paracetamol. There was no significant difference in Fe concentration in the shoots of plants across the different paracetamol treatments (Figure 4d).

The Mn concentration was significantly reduced in shoots of plants treated with 100 mg/L and 200 mg/L paracetamol than in the control and 50 mg/L. The concentration of Mn was significantly greater in the shoots than in the roots. In the latter, Mn was significantly lower in plants treated with 50 mg/L than in the control and the higher concentrations (Figure 4e).

2.4.3. Sodium

Sodium (Na) decreased in the roots by 4.9%, 5.4%, and 6.9% in the 50 mg/L, 100 mg/L and 200 mg/L paracetamol treatments, respectively. Moreover, Na also decreased in the shoots of the treated plants as compared to the control plants (Figure 4f).

2.4.4. CHNS Analysis

Elemental analysis of spinach shoots using the CHNS analyzer revealed significant effects of paracetamol on N and S percentages in the spinach shoots (Supplementary Table S3). There was a considerable increase in the weight percentage of nitrogen in spinach shoots treated with 200 mg/L of paracetamol (3.84%), which was similar to control shoots (3.88%), and lower in those treated with 50 mg/L and 100 mg/L (2.94% and 2.78%, respectively) as shown in Figure 5. C was higher in shoot samples treated with 50 mg/L (42.01%) and 100 mg/L paracetamol (42.09%) compared to the control (41.03%), which was close to plants treated with 200 mg/L (40.74%). The percentages of H and S detected in all paracetamol-treated shoot samples were similar to control shoots (Figure 5). Overall, the C/N ratios in control and 200 mg/L paracetamol treated shoots samples were similar and lower than those in 50 mg/L and 100 mg/L treated plants (Figure 6).

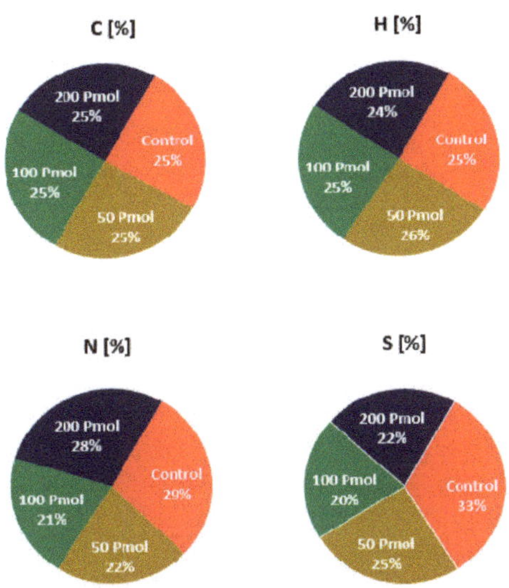

Figure 5. Variation in weight percentages of C, H, N, and S in spinach shoots after eight days of paracetamol treatment.

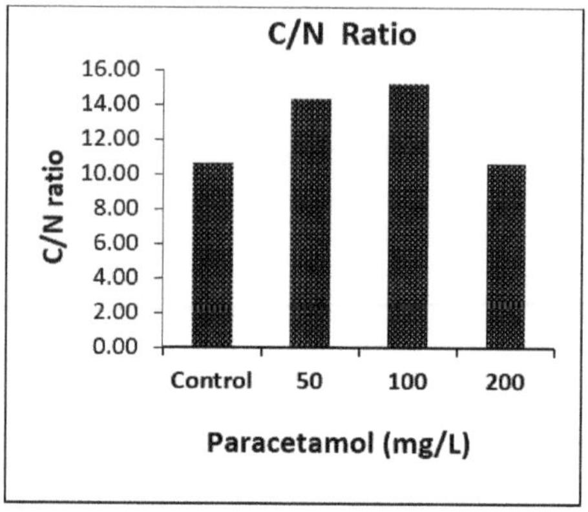

Figure 6. Variation in C/N ratio in spinach shoots after eight days of paracetamol treatments.

2.5. Microbial Analysis

VITEK® 2 microbial identification system provided rapid identification of the bacterial flora in spinach roots and shoots (Tables 3 and 4). Among the identified bacterial strains, members of *Burkhulderia, Sphingomonas, Pseudomonas, Staphylococcus, Stenotrophomonas, and Kocuria* were prominent in spinach root and shoot cultures.

Table 3. Microbial consortium of spinach roots treated with paracetamol for eight days.

Treatment	Isolate and Colony Description	Organism	Probability (%)	Confidence	Role in Plants
Control spinach roots	Gram +ve micrococcus, yellowish moist, moderate size	*Kocuria kristinae*	98	Excellent Identification	Antagonist bacteria in spinach [32].
	Gram −ve, white, large flat, and dry	*Pasteurella pneumotropica*	90	Good Identification	Endophytic bacterium capable of fixing nitrogen and solubilizing phosphate [33].
	Gram −ve, moderate moist	*Comamonas testosteroni*	99	Excellent Identification	Spinach microbiota [34]. Involved in the degradation of aromatic compounds [35].
	Gram −ve, small moist	*Acinetobacter Iwoffii*	99	Excellent Identification	Spinach microbiota [34,36]. Involved in the degradation of paracetamol and aromatic compounds [35].
50 mg/L paracetamol-treated spinach roots	Gram −ve, yellow, large, and dry	*Burkhulderia cepacia* group	95	Very Good Identification	Spinach microbiota [36]. Increased growth parameters in *Zea mays* [37].

Table 3. *Cont.*

Treatment	Isolate and Colony Description	Organism	Probability (%)	Confidence	Role in Plants
	Gram −ve, pale yellow, small, moist	*Pseudomonas florescens*	98	Excellent Identification	Involved in the degradation of paracetamol and aromatic compounds [35], Spinach microbiota [36]. Phosphorus-solubilizing, protease production in *Phragimates australis* [38]. Salt tolerant in groundnut (*Arachishypogaea*) [39]. Antifungal, plant growth promotion. Increased growth parameters, increased Pb uptake, root elongation in *Brassica napus* and *Solanum nigrum* [40].
	Gram +ve, white, large and moist	*Staphylococcus haemolyticus*	95	Very Good Identification	Spinach microbiota [36]. Increased growth parameters in *Triticum aestivum* [41].
	Gram −ve rods, pale yellow	*Stenotrophomonas maltophilia*	95	Very Good Identification	Spinach microbiota [36]. Increased growth parameters in *Triticum aestivum* [41].
100 mg/L paracetamol-treated spinach roots	Gram +ve, pale yellow, moderate, moist	*Kocuria kristinae*	87	Acceptable Identification	Antagonist bacteria in spinach [32].
	Gram +ve, white, small, round	*Kocuria rosea*	98	Excellent Identification	Capable of growing on naphthalene, phenanthrene and fluoranthene, on all three polycyclic aromatic hydrocarbons (PAHs) [42].
200 mg/L paracetamol-treated spinach roots	Gram −ve, coccobacillus, yellow, large and dry colonies	*Brucella melitensis*- Highly pathogenic	91	Good Identification	Foodborne pathogen. Contaminant of leafy vegetables and causes human brucellosis [43,44].

Table 3. *Cont.*

Treatment	Isolate and Colony Description	Organism	Probability (%)	Confidence	Role in Plants
	Gram −ve bacilli, yellow, large and moist colonies	*Sphingomonas paucimobilis*	92	Good Identification	Spinach microbiota [35]. Increase in growth parameters in *Triticum aestivum* [41]. Involved in the degradation of paracetamol and aromatic compounds [35]. Spinach microbiota, involved in PPCP biodegradation, produces lignin-degrading enzymes [36,45].
	Gram +ve, white, moderate size	*Staphylococcus auricularis*	98	Excellent Identification	Spinach microbiota, increases growth parameters in *Triticum aestivum* [36,41].

Table 4. Microbial consortium of spinach shoots treated with paracetamol for eight days.

Treatment	Isolate and Colony Description	Organism	Probability (%)	Confidence	Role in Plants
Control spinach shoots	Gram +ve, white moist, moderate size colonies	*Staphylococcus hominis ssp hominis*	N.A.	Low Discrimination Organism	Epiphytic bacteria from fruits and leafy greens, spinach microbiota [36], potential biocontrol agents, able to reduce the proliferation of *E. coli* O157:H7 and *S. enterica* in fruits and vegetables [46].
	Gram +ve, white moist, moderate size colonies	*Aerococcus viridans*	N.A.	Low Discrimination Organism	Epiphytic bacteria on leafy greens, capable of fixing nitrogen and solubilizing phosphate [33].
	Gram −ve, moderate yellow moist colonies	*Oligella ureolytica*	N.A.	Low Discrimination Organism	Pathogenic bacteria [47].
	Gram −ve, moderate yellow moist colonies	*Aeromonas salmonicida*	NA	Low Discrimination Organism	Plant growth promoting rhizobacteria, involved in biodegradation of xenobiotic compounds from contaminated water/soil environment [38,48].

Table 4. *Cont.*

Treatment	Isolate and Colony Description	Organism	Probability (%)	Confidence	Role in Plants
50 mg/L paracetamol-treated spinach shoots	Gram −ve bacilli, yellow, large and moist colonies	*Sphingomonas paucimobilis*	89	Good Identification	Spinach microbiota, involved in the degradation of paracetamol and aromatic compounds [34–36].
	Gram +ve, white, moderate size	*Staphylococcus hominis ssp hominis*	95	Very Good Identification	Spinach microbiota [36]. Epiphytic bacteria from fruits and leafy greens are potential biocontrol agents, able to reduce the proliferation of *E. coli* O157:H7 and *S. enterica* in fruits and vegetables [46].
100 mg/L paracetamol-treated spinach shoots	Gram +ve, pale yellow, moderate, moist	*Kocuria rosea*	98	Excellent Identification	Capable of growing on naphthalene, phenanthrene and fluoranthene, on all three polycyclic aromatic hydrocarbons (PAHs) [42].
	Gram −ve, white, moderate size, moist	*Escherichia coli*	93	Very Good Identification	Foodborne pathogen, spinach microbiota [34].
200 mg/L paracetamol treated spinach shoots	Gram +ve, white, moderate, moist	*Staphylococcus vitulinus*	98	Excellent Identification	Spinach microbiota [34,36].
	Gram +ve, white, moderate, moist	*Gamella bergeri*	90	Good Identification	Epiphytic bacteria from fruits and leafy greens are potential biocontrol agents, able to reduce the proliferation of *E. coli* 157:H7 and *S. enterica* in fruits and vegetables [46].

3. Discussion

Results showed that the effects of paracetamol on spinach uptake, accumulation, degradation, and phytotoxicity were concentration- and organ-dependent. There were obvious deleterious effects of increasing paracetamol stress on spinach growth, morphological and physiological parameters. Leaves displayed symptoms of withering, burning, and necrosis on the margins at higher (100–200 mg/L) paracetamol levels. This result is aligned with observations of a significant decrease in the shoot and root elongation of wheat after 21 days of application of 1.4–22.4 mg/L paracetamol treatments [49]. A similar toxic effect of higher concentrations (>10 mg) of paracetamol was also seen in cucumber plants after seven days of exposure, with a significant reduction in the biomass of leaves and roots [50]. Growth reduction is observed to be an adaptive survival mechanism for plants to endure the oxidative stress damage caused to the cellular components [51].

The photosynthesis machinery of spinach was highly susceptible to paracetamol and was impaired due to the xenobiotic stress. A significant decline occurred in the chlorophyll fluorescence parameters with time. Our results are consistent with a previous finding [24], demonstrating that PSII, PSI, and electron transport in the bean plants were inhibited by paracetamol stress, causing a potential decrease in the photosynthetic activity. A similar decreasing trend was observed in chlorophyll fluorescence for the model macrophyte *Lemna minor* used for environmental risk assessment; when treated with paracetamol, its photosynthetic activity significantly decreased by up to 37% in comparison with the control [16]. Moreover, a 40% reduction in the quantum yield of photosystem II was also reported in maize treated with 10 mg/L paracetamol [52].

Our study indicates an inconsistency in the response of chlorophyll content and photosynthetic efficiency of spinach plants treated with paracetamol. Although paracetamol significantly affected photosynthetic pigments at higher concentrations (>100 mg/L) with the exposure period, it overall decreased the photosynthesis efficiency (Figure 3, Table 2). Similarly, in the macrophyte *Lemna minor*, paracetamol exposure resulted in chlorophyll levels similar to those found in control plants. The endpoint of Chl b was more sensitive to paracetamol than that of Chl a and raised the Chl b content of *L. minor*. However, the F_v/F_m (maximal quantum yield of PSII) was not significantly affected [53]. A recent study [54] also reported that increasing paracetamol concentration and exposure time resulted in non-significant increases in total chlorophyll contents, photosynthetic capacity and photosynthetic rates, but significantly increased the photochemical reflectance index in lettuce. In addition, it has been reported that a lower concentration of Panadol (1.0 mg/L), another paracetamol, enhanced chlorophyll content and photosynthetic efficiency of *Vigna radiate* plants [55]. The authors indicated the possibility of using a lower concentration of Panadol as a plant growth regulator. Similarly, a significant increase in the chlorophyll (27.46%) and carotenoids (41.8%) was found with 500 µM chronic treatment with paracetamol in lettuce. With increasing concentrations of paracetamol, the chlorophyll content also increased. However, high concentrations of paracetamol had an inhibitory effect on the photosynthetic activity of plants and reduced the rate of photosynthesis in the experimental variants compared to the control plants [56].

In nature, plants employ various stress-tolerance strategies to combat biotic and abiotic stresses in their surrounding environment by altering gene expression, protein synthesis, and post-translational modifications that aid in the re-establishment of cellular homeostasis for their survival [51]. Our results showed a significant reduction in the growth tolerance index of spinach shoots. However, the photosynthetic pigments in paracetamol-treated plants increased with the increase in paracetamol concentration and exposure period. Interestingly, the leaves became noticeably greener at the end of the exposure time relative to the control plants. It has been reported that chlorophyll biosynthesis might offset the reduction in electron transport efficiency and Calvin cycle activities under prolonged paracetamol stress [56,57]. The chlorophyll content of cells can act as a protective mechanism against induced oxidative stress, scavenging the accumulated ROS [58]. In addition, the significant increase in the carotenoid content could help deactivate excited chlorophyll and safeguard the photosynthetic system through modulation of lipid peroxidation products, thereby attenuating and/or preventing ROS-induced damage to the photosynthetic system [59]. Still, the increases in levels of chlorophyll and carotenoids were not enough to reduce the ROS-induced damage in spinach plants, especially shoots, treated with 200 mg/L for eight days [60,61]. At 200 mg/L, spinach plants showed lower shoot tolerance with the appearance of several morphological anomalies, such as leaf withering, burning, and necrosis (Supplementary Figure S5).

Organic xenobiotics must pass via the stomata or traverse the epidermis, which is covered by the cuticle, to permeate into a leaf. Stomata are found on the leaf's lower (abaxial) side, whereas the thicker cuticle layer is located on the upper (adaxial) side. The stomatal system regulates the penetration of organics into the leaves by having multiple openings that can be enlarged as needed. The plant regulates the intake of chemicals of various molecular masses through the movement of two guard cells that control the opening and closing of the stomata by adjusting the diameter of the opening. K plays a crucial role in regulating the movement of these guard cells, cell elongation, and other vital physiological activities. A rise in K^+ concentration opens the stoma, facilitating the movement of the xenobiotic into the leaves [62]. Spinach is categorized as a nitrate accumulator, and vegetables grown on sewage sludge-contaminated soil may accumulate nitrate in edible plant portions due to extremely high nitrogen levels in the soil created by the sludge [63]. Additionally, increased potassium uptake is also reported to facilitate the uptake and transport of nitrate to the plant's aerial parts, improve nitrate metabolism and utilization, and have an indirect effect on chlorophyll synthesis [64,65].

Our results indicated significant increases in the major essential nutrients and elements, such as Ca, K, Fe, and N, in spinach plants treated with higher concentrations (100-200 mg/L) of paracetamol compared to the controls. These essential ions and elements regulate the cellular processes, are involved in chlorophyll production, and possibly contribute to photosynthesis, respiration, oxygen transport, and gene regulation, affecting plant growth and development [66]. In addition, increases in micronutrients, such as Fe and Zn, significantly increased antioxidant activities and reduced oxidative stress, enhancing chlorophyll content and gaseous exchange attributes in spinach plants irrigated with tannery wastewater with chromium [60]. A progressive decline was observed in the Na levels of both shoot and root tissues of plants treated with increasing levels of paracetamol (Figure 4f), likely supporting plants' adaptation to abiotic stress and affecting the absorption and/or distribution of organic solutes and vital nutrients in the plants [67].

Plant detoxification of xenobiotics is similar to mammalian detoxification; following a phase I activation event, a phase II conjugation process with hydrophilic molecules such as glutathione or glucose occurs. The phase III reactions of xenobiotic detoxification include xenobiotic conjugate storage, degradation, and transfer [9,64]. Conjugation with sugar is one of the main xenobiotic detoxification processes in plants. Huber et al. (2009) pioneered the discovery of paracetamol conjugates in plants, involving a dual detoxification mechanism. They discovered two metabolic pathways in the hairy root culture of horseradish, which culminates in the synthesis of glutathione and a glucose conjugate [68]. However, the fate of drugs in spinach is yet unknown.

Our results showed that the highest accumulations of paracetamol were observed after four days in the shoots and roots of spinach plants treated with 200 mg/L (Figure 1a,b and Supplementary Figures S2 and S3), and in dry leaves assessed after eight days (Figure 1d, Supplementary Figure S4). Additionally, after eight days, significant reductions of 59.06% and 60.93% in paracetamol concentrations were observed in spinach shoots after 100 mg/L and 200 mg/L paracetamol treatments, respectively (Figure 1a, Supplementary Figure S2). A similar trend of degradation with time was observed in the roots, but paracetamol accumulation was low (Figure 1b, Supplementary Figure S3). The higher translocation factor of paracetamol in spinach shoots (Figure 1c) indicates its high mobility into the aerial parts of spinach. In addition, the significant reduction in paracetamol levels after eight days indicates its conjugation ability.

Our findings coincide with a previous investigation carried out on *Typha latifolia* plants, reporting the uptake and accumulation of paracetamol to be significantly higher (0.077 μg/g FW) in leaf tissues at the end of the first week of 1 mM Paracetamol treatment, indicating a 78% transfer of the total amount. Then, after 30 days, a gradual decrease in the concentration was seen, and only 30% of the drug was present in the shoots [69]. Other studies reported a reduction in the paracetamol concentration immediately after its uptake in cucumber [50], *Solanum nigrum* [70], and *Lemna minor* [16]. The cucumber plants took up paracetamol and conjugated it quickly with glutathione. The paracetamol–glutathione conjugates in cucumber plants exposed to 5 mg/L for 144 h were 15.2 nmol/g and 1.2 nmol/g in cucumber roots and leaves, respectively [50].

Endophytes are nonpathogenic bacteria or fungi that naturally inhabit almost all plant species. Endophytes have a great ability to break down xenobiotics in plants, allowing them to withstand stress under unsuitable soil conditions [71,72]. The inoculation of endophytic bacteria enhances the phytoremediation ability of several plants for remediation of polluted soil and water [73,74]. For example, the inoculation of *Leptochloa fusca* with endophytic bacteria enhanced the biodegradation of organic and inorganic pollutants, decreasing pollutants' toxicity [74] and promoting plant growth. Although the microbial removal of paracetamol seems to be an effective remediation technique, few bacterial strains, including *Pseudomonas, Burkholderia, Stenotrophomonas, Pseudomonas, Rhodococcus, Bacillus, Delftia, Kocuria, Staphylococcus, Acinetobacter* and *Sphingomonas*, are capable of degrading paracetamol and other organic pollutants [35,75–78]. These microorganisms can use paracetamol as their sole carbon, nitrogen, and energy source.

Moreover, inoculated endophytic bacteria can also regulate the metabolic activities of organic pollutants and horizontal gene transfer from native endophytes [79]. The metabolic pathways of biodegradation, however, remain poorly characterized. Two key metabolites identified during microbial degradation of paracetamol, hydrolytic phenolic dead-end metabolite hydroquinone and 4- aminophenol, were described as carcinogenic and highly toxic compounds. Furthermore, paracetamol phenolic derivatives can cause DNA cleavage and mutagenesis in animal cell lines [76,78].

VITEK 2 provided quick and accurate identification of the microbes contributing to paracetamol degradation. We also identified bacterial strains in paracetamol-treated spinach shoots and roots capable of removing paracetamol (Tables 3 and 4). The isolated bacteria belong to the genera *Pseudomonas, Sphingomonas, Burkholderia, Staphylococcus, Stenotrophomonas,* and *Kocuria* [35,76,78]. It has been reported that a consortium of *Stenotrophomonas* and *Pseudomonas* microbial strains could degrade paracetamol up to 4 g/L. In contrast, pure cultures of the strains degraded paracetamol completely at 0.4 g/L, 2.5 g/L, and 2 g/L, respectively [77]. The consortium also had significantly higher degradation rates and greatly improved tolerance to paracetamol with a shorter lag time. This also raises serious concerns regarding the development of drug-resistant strains and promotion of the growth of opportunistic pathogens [43,44]. Hence, apart from the beneficial role of bacteria in bioremediation, their antimicrobial resistance and pathogenic potential cannot be underestimated.

4. Material and Methods

Pure analytical grade paracetamol (>98% purity) was purchased from Sigma-Aldrich. All solvents and chemical reagents used were of analytical grades.

4.1. Seed Germination and Seedling Development

Spinach (*Spinacia oleracea* L. cv. Matador) seeds were purchased from a local agricultural shop. Seeds were soaked in water for one hour to increase germination potential and then surface sterilized using 5% commercial bleach for 5–7 min, rinsed thrice with distilled water, air-dried, and soaked in pots (9 cm diameter and 7 cm height) containing approx. 350 g of garden organic potting soil. The plants were grown in a CONVIRON plant growth chamber (model E-15) adjusted to a 25/15 °C day/night regime. The lighting in the chamber was white light (1400 μmol/m^2/s of 167 photosynthetically active radiation) provided by five (400 W) metal halide and five (400 W) high-pressure sodium lamps.

4.2. Hydroponic Cultivation and Paracetamol Exposure

A hydroponic system setup, as described previously [79] was used for the experiments. Briefly, four seedlings with two true leaves were removed from the soil pots, thoroughly rinsed with distilled water, and transferred into glass jars containing 20% Hoagland nutrient solution (Hoagland's No. 2 basal salt mixture). The jars were covered with aluminum foil to prevent roots from exposure to light and then kept in the growth chambers. The culture jars were aerated frequently, and the nutrient solution was replenished as needed to maintain nutrient solution balance. All experimental material and distilled water used to prepare treatment solutions was autoclaved at 121 °C for 30 min before use to minimize the risk of contamination.

After two weeks of adaptation and upon reaching the 4-6 true leaf stage, plants were exposed to three levels of paracetamol (50 mg/L, 100 mg/L, and 200 mg/L) in 20%

Hoagland nutrient solution for an experimental period of 8 days. All treatments were performed in triplicate; each replicate was a mix of four plants in a jar. The treatment solutions were replaced every four days to maintain the same stress levels. Additionally, +ve controls (20% Hoagland nutrient solution in jars with plants) and –ve controls (20% Hoagland nutrient solution without plants) were also included in the experiments in triplicates to determine the abiotic losses of the pharmaceutical. The jars were organized in a randomized block design, with treatment as the main block. In addition, the jars of the different blocks were randomized every two days. Spinach plants were harvested at two time points, i.e., after four days and eight days of paracetamol treatments, to determine the fate of the xenobiotic. Plants were thoroughly washed with distilled water, dried, separated into shoots and roots, weighed, and stored at −80 °C until further analysis.

4.3. Assessment of Plant Growth Parameters

Spinach shoot and root lengths were measured using a standard scale, and the number of leaves was counted manually. Fresh weight (FW) of plants (shoots and roots separated) was assessed at harvesting, and dry weight (DW) was calculated after freeze-drying the samples at −84 °C in Freeze Dryer (Labconco Freezone- 6L).

The plant samples were lyophilized, and the growth tolerance indexes for shoots (GTIS %) and roots (GTIR %) were determined based on their respective dry weights [80], as follows:

$$\text{GTIS\%} = \frac{\text{Average dry weight of shoots grown in media containing paracetamol}}{\text{Average dry weight of shoots grown on media without paracetamol}} * 100 \quad (1)$$

$$\text{GTIR\%} = \frac{\text{Average dry weight of roots grown on media containing paracetamol}}{\text{Average dry weight of roots grown on medium without paracetamol}} * 100. \quad (2)$$

Chlorophyll Fluorescence Analysis: The initial fluorescence (F_o), variable fluorescence (F_v), maximum fluorescence (F_m), potential quantum efficiency of photosystem II (ΦII), (F_v/F_m), and (F_v/F_o) were recorded from the third middle section of fully expanded leaves exposed to light, using a pulse-modulated fluorescence monitoring system (FMS-2, Hansatech Instruments Ltd., Norfolk, UK). Measuring setup and calculation of selected parameters were adopted from Hussain and Reigosa (2021) [81]. After the measurements, the same leaves were collected to analyze photosynthetic pigments.

Quantification of the Photosynthetic Pigments: Three replicates of 50 mg leaves, each from four plants from a jar, were immersed in 15 mL falcon tubes containing 5 mL of methanol. The tubes were wrapped with aluminum foil and kept at 4 °C for 48 h until complete bleaching of the leaf occurred to extract the photosynthetic pigments. After complete extraction, 200 μL of supernatant was pipetted in triplicate into 96 well plates and blank wells with methanol. The absorbance of the extracts was measured using an absorbance microplate reader (E.L. x 808- Biotek) at wavelengths of 666 for chlorophyll a, 653 for chlorophyll b, and 470 for carotenoid (Cx+c). According to Lichtenthaler (1987), the concentrations of the pigments were calculated using the following absorption coefficients [82], with V being the extract volume (mL) and W the leaves fresh weight (g):

$$\text{Chlorophyll a (mg/g)} = [15.65 \times (\text{A666}) - 7.34 \times (\text{A653})] \times \text{V}/(1000 \times \text{W})$$

$$\text{Chlorophyll b (mg/g)} = [27.05 \times (\text{A653}) - 11.21 \times (\text{A666})] \times \text{V}/(1000 \times \text{W})$$

$$\text{Total Chlorophyll (mg/g)} = \text{Chlorophyll a} + \text{Chlorophyll b}$$

$$\text{Cx} + \text{c (mg/g)} = [1000 \times (\text{A470}) - 2.86 \times (\text{Chla}) - 129.2 \times (\text{Chlb})]/245 \times \text{V}/(1000 \times \text{W})$$

4.4. The Fate of Paracetamol in Spinach

We followed a previously reported [83] extraction method for determination of paracetamol concentration from plant shoot and root tissues. Briefly, 100 mg of lyophilized and ground spinach shoots and roots samples were extracted in 4 ml EDTA (150 mg/L) and vortexed for 1 min. Then, 5 mL of a mixture of ACN: MeOH (65:35) was added and vortexed for 2 min following the addition of 3 gm Na2SO4 and 0.5 gm NaCl with a 1.5 min vortex. The mixture was kept overnight at 4 °C, then centrifuged the next day at 10,000× *g* for 20 min. The supernatants were filtered through 0.22 μm sterile PES membrane syringe-driven filters (Jet-Biofil, Guangzhou, China) for LC/MS analysis.

4.4.1. Liquid Chromatography–Mass Spectrometry (LC–MS) Analysis

LC–MS analysis was performed using Waters Acquity UPLC H-Class, Xevo TQD system (Waters Corporation, Milford, M.A, USA) equipped with electrospray ionization operated in the positive ionization mode. The sample injection volume was 10 μL, and chromatographic separation of analytes was carried out on an Acquity UPLC BEH C18 1.7 μm (2.1 × 100 mm) column. Mobile phases consisted of 0.1% formic acid in water (solvent A) and acetonitrile with 0.1% formic acid (solvent B). The run time was 8 minutes, with a retention time of 3.2 min for paracetamol (Supplementary Figure S1). The gradient separation method is shown in Supplementary Table S4.

Calibration standards in the range of 0.5–200 μg/L were prepared with paracetamol in methanol in triplicate, and used to determine the paracetamol concentration in all samples from their respective peak areas. The limit of detection (LOD) and limit of quantification (LOQ) were calculated using the standard deviation (SD) of the responses and the slope (S) of the calibration curve, via the formulas LOD = 3.3 × (SD/S) and LOQ = 10 × (SD/S), yielding the values of LOD = 2.220791163 and LOQ = 6.729670192. The SD of the response was determined based on the standard deviation of y-intercepts of the calibration curve, where LOD= 0.819811731 and LOQ= 2.484277972.

4.4.2. Translocation Factor

The translocation of paracetamol from roots to shoots in different treatments was determined by calculating the translocation factor (TF) [80].

$$TF = \frac{\text{Paracetamol concentration in shoots}}{\text{Paracetamol concentration in roots}}$$

4.5. Plant Elemental Analysis

A closed-vessel microwave-assisted digestion procedure assessed the essential nutrients (macro- and micronutrients), and elements in spinach organs [84]. Briefly, 100 mg of lyophilized samples was digested with 5 mL concentrated nitric acid (70% Sigma-Aldrich, Merck, Darmstadt, Germany) and 1 mL H_2O_2 (≥30%, Sigma- Aldrich, Merck, Darmstadt, Germany) in PFA Teflon® vessels. The volume was made up to 10 mL with distilled water. The mixture was kept at room temperature for 15–20 min for cold digestion. The vessels were sealed, and samples were heated using an Anton Paar microwave digestion system using the following time and temperature program: ramping to 200 °C for 20 min, and maintaining the temperature at 200 °C for 20 min.

After cooling down to room temperature, the digested solutions were filtered through 0.22 μm sterile PES membrane syringe-driven filters (Jet-Biofil, Guangzhou, China); their volume was made up to 30 mL using Milli-Q water and they were refrigerated at 4 °C. Elemental analysis was carried out via Inductively Coupled Plasma-Optical Emission Spectroscopy (ICP-OES; ICAP 7000, Thermo Fisher Scientific, Cambridge, UK). Paracetamol with 20% Hoagland growth media solution samples (having different paracetamol concentrations) were filtered through 0.22 μm sterile PES membrane syringe-driven filters (Jetbiofil) and diluted with distilled water before ICP-OES analysis.

For simultaneous analysis of carbon (C), hydrogen (H), nitrogen (N), and sulfur (S); a known amount of plant material was oven-dried at 70 °C, finely ground and packed in tin foil capsules and analyzed by an automated Vario MACRO cube CHNS analyzer (PerkinElmer, Waltham, USA). Plant total C, N, H, and S was expressed as the percentage of elements in dried plant material.

4.6. Microbial Analysis

Our study employed the VITEK® 2 (BioMerieux, Marcy L'Etoile, France) microbial identification system to identify paracetamol-degrading bacteria in spinach roots and shoots.

Spinach shoot and root samples were weighed under aseptic conditions under a Class II Biosafety Cabinet, surface sterilized with 70% ethanol for 7 min, and washed three times with sterile distilled water. Then, 2.5% bleach was added to the samples for 10 min and they were washed thrice with sterile distilled water. After sterilization, samples were crushed in 10 mL of sterile distilled water. One milliliter of the extract was taken from each sample and serially diluted to 1×10^{-5}. Then, from each diluted sample extract, 100 μL was spread on Luria Bertani (LB) agar plates, wrapped with parafilm, and incubated at 28 °C, and bacterial growth was observed for 72 h. Gram's staining was used for preliminary identification of the isolated bacterial strains. Bacterial suspensions were prepared in sterile saline, and the density was adjusted using VITEK 2 DensiCheck (BioMerieux, Marcy L'Etoile, France) to a McFarland standard of 0.5–0.63. Gram-positive and Gram-negative bacteria were identified using G.P. and G.N. cards, respectively, via VITEK 2 G.N. (21341 BioMerieux, Marcy L'Etoile, France) and G.P. (21342 BioMerieux, Marcy L'Etoile, France) Identification Kits.

4.7. Statistical Analysis

Two-way ANOVAs were used to assess the impacts of paracetamol treatments (0 mg/L, 50 mg/L, 100 mg/L, and 200 mg/L) and exposure time (four and eight days) and their interaction on pigments, chlorophyll fluorescence traits and paracetamol concentrations in spinach organs. In addition, ANOVAs were used to assess the impacts of paracetamol concentrations on evaluated parameters, including shoot length, root length, leaf number, growth tolerance index of shoots and roots, plant nutrients and elements. Pairwise comparisons of the means were performed using the post-hoc Tukey's Honest Significant Differences (HSD) test to identify which pairs of treatments were significantly different. Three biological replicates (each replicate was a mix of four plants in a jar) were used for the assessments. Data were statistically analyzed using SYSTAT (version 13).

5. Conclusions

The results demonstrated that the edible leafy spinach plant can uptake, accumulate, and metabolize paracetamol, which is one of the most commonly used over-the-counter drugs. Despite the bioremediation process (in planta and microbial), higher doses (100 mg/L and 200 mg/L) of paracetamol and its formed metabolites induced phytotoxicity, causing oxidative stress and irreversible damage to spinach roots and edible shoots. Accumulation of these pseudo-persistent pharmaceutical pollutants in the water, soil, and plant matrix over time may increase the population of emerging opportunistic pathogens in biofilms, have a high risk for horizontal gene transfer, generate new multiple-drug-resistant strains, and thus pose a direct threat to animal and human health via the food chain.

These findings could facilitate the development of guidelines for improving wastewater treatment and utilization methods. Additionally, our research also provides a new perspective exploring the synergistic roles of plants and associated beneficial microbes in promoting plant growth and enhancing the bioremediation process of contaminants of emerging concern (CEC) in the environment. The fast growth of spinach and its fast uptake and degradation of paracetamol recommend it as a potential plant for the phytoremediation of polluted lands. The harvested plants could be used as a source of biofuel.

Our extended research plan will offer a detailed evaluation and deeper understanding of the biochemical and molecular repercussions of these pollutants and contribute to addressing the fears arising in the flora and microbial community for endorsing the application of treated wastewater to promote sustainable agricultural development in the United Arab Emirates.

Supplementary Materials: The supporting information can be downloaded at: https://www.mdpi.com/article/10.3390/plants11131626/s1, Table S1. Results of one-way ANOVAs showing the effect of different paracetamol treatments on the morphological attributes of spinach after eight days. Table S2. Results of one-way ANOVAs showing the effect of different paracetamol treatments on the concentration (mg/kg DW) of several macronutrients, micronutrients and sodium in spinach shoots and roots after eight days. Table S3. Results of one-way ANOVAs showing the effect of different paracetamol treatments on C, H, N, and S percentage in spinach. Table S4. Gradient separation method for LC–MS analysis of paracetamol, Figure S1. LC–MS chromatogram of Paracetamol standard, Figure S2. LC–MS chromatograms showing the concentration of paracetamol (Retention time: 3.2 min) in spinach shoots for (A) 50 mg/L, (B) 100 mg/L and (C) 200 mg/L paracetamol treatments after four days, and for (D) 50 mg/L, (E) 100 mg/L and (F) 200 mg/L paracetamol treatments after eight days, Figure S3. LC–MS chromatograms showing the concentration of paracetamol (Retention time: 3.2 min) in spinach roots for (A) 50 mg/L, (C) 100 mg/L and (B) 200 mg/L paracetamol treatments after four days, and for (D) 50 mg/L, (E) 100 mg/L and (F) 200 mg/L paracetamol treatments after eight days, Figure S4. LC–MS chromatograms showing the concentration of paracetamol (Retention time: 3.2 min) in spinach dry leaves for (A) 50 mg/L, (B) 100 mg/L, and (C) 200 mg/L paracetamol treatments after eight days, Figure S5. Images of untreated (control) and treated spinach plants with 50 mg/L, 100 mg/L, and 200 mg/L paracetamol (a,b).

Author Contributions: Conceptualization, A.S., A.E.-K., K.A.M., M.H.S., L.S., A.A.M., Z.B. and S.B.; methodology, A.S., A.E.-K., K.A.M., M.H.S., L.S., A.A.M., M.M., Z.B., F.M.T., A.E., I.M.A. and M.I.H.; validation, all authors; formal analysis, all authors; investigation, A.S., A.E.-K., K.A.M., M.H.S., M.M., Z.B., F.M.T., I.M.A., A.E. and S.S.A.; resources, A.S., A.E.-K., K.A.M., M.H.S., L.S., A.A.M., Z.B. and S.B.; data curation, all authors; writing, review and editing, all authors; supervision, A.S., A.E.-K., K.A.M., M.H.S. and A.A.M.; funding acquisition, A.S., A.E.-K., K.A.M., M.H.S., L.S., A.A.M. and Z.B. All authors have read and agreed to the published version of the manuscript.

Funding: Funding for this research was generously granted by the Research Institute of Sciences and Engineering, University of Sharjah (Grant Number: 21020401164-P, PI: A.S.).

Institutional Review Board Statement: Not applicable.

Informed Consent Statement: Not applicable.

Data Availability Statement: The data are available within the article and Supplementary Materials.

Conflicts of Interest: The authors declare no conflict of interest.

References

1. Albannay, S.; Kazama, S.; Oguma, K.; Hashimoto, T.; Takizawa, S.; Lo Porto, A. Water Demand Management Based on Water Consumption Data Analysis in the Emirate of Abu Dhabi. *Water* **2021**, *13*, 2827. [CrossRef]
2. Daughton, C.G. Non-Regulated Water Contaminants: Emerging Research. *Environ. Impact Assess. Rev.* **2004**, *24*, 711–732. [CrossRef]
3. Cătălina Vasilachi, I.; Asiminicesei, D.M.; Fertu, D.I.; Gavrilescu, M. Occurrence and Fate of Emerging Pollutants in Water Environment and Options for Their Removal. *Water* **2021**, *13*, 181. [CrossRef]
4. Bui, X.T.; Vo, T.P.T.; Ngo, H.H.; Guo, W.S.; Nguyen, T.T. Multicriteria Assessment of Advanced Treatment Technologies for Micropollutants Removal at Large-Scale Applications. *Sci. Total Environ.* **2016**, *563–564*, 1050–1067. [CrossRef] [PubMed]
5. Minden, V.; Deloy, A.; Volkert, A.M.; Leonhardt, S.D.; Pufal, G. Antibiotics Impact Plant Traits, Even at Small Concentrations. *AoB Plants* **2017**, *9*, plx010. [CrossRef]
6. Carvalho, P.N.; Basto, M.C.P.; Almeida, C.M.R.; Brix, H. A Review of Plant–Pharmaceutical Interactions: From Uptake and Effects in Crop Plants to Phytoremediation in Constructed Wetlands. *Environ. Sci. Pollut. Res.* **2014**, *21*, 11729–11763. [CrossRef]
7. Coleman, J.O.D.; Blake-Kalff, M.M.A.; Davies, T.G.E. Detoxification of Xenobiotics by Plants: Chemical Modification and Vacuolar Compartmentation. *Trends Plant Sci.* **1997**, *2*, 144–151. [CrossRef]
8. Sauvêtre, A.; Eichhorn, P.; Pérez, S. Metabolism of Pharmaceuticals in Plants and Their Associated Microbiota. In *The Handbook of Environmental Chemistry*; Springer: Berlin/Heidelberg, Germany, 2020; Volume 103, pp. 221–264. [CrossRef]
9. Sandermann, H. Higher Plant Metabolism of Xenobiotics: The 'Green Liver' Concept. *Pharmacogenetics* **1994**, *4*, 225–241. [CrossRef]
10. Caragea, G.; Avram, O.; Pauna, A.; Costea, A.; Tudosie, M. Acetaminophen, a Therapeutic or an Extremely Toxic Remedy—A Review. *J. Mind Med. Sci.* **2022**, *9*, 102–110. [CrossRef]

11. Paracetamol Market Research Report [2022–2028] | Industry. Available online: https://www.globenewswire.com/en/news-release/2022/02/15/2384836/0/en/Paracetamol-Market-Research-Report-2022-2028-Industry-Size-Share-Growth-Rate-Business-Strategies-Industry-Revenue-Opportunities-Future-Trends-Leading-Players-Update-Analysis-and-Fo.html (accessed on 14 June 2022).
12. Nieto-Juárez, J.I.; Torres-Palma, R.A.; Botero-Coy, A.M.; Hernández, F. Pharmaceuticals and Environmental Risk Assessment in Municipal Wastewater Treatment Plants and Rivers from Peru. *Environ. Int.* **2021**, *155*, 106674. [CrossRef]
13. Ternes, T.A. Occurrence of Drugs in German Sewage Treatment Plants and Rivers. *Water Res.* **1998**, *32*, 3245–3260. [CrossRef]
14. Eric van Beelen, I.S. *Municipal Waste Water Treatment Plant (WWTP) Effluents a Concise Overview of the Occurrence of Organic Substances Association of River Waterworks*; RIWA: Nieuwegein, The Netherlands, 2007.
15. Moreno-González, R.; Rodriguez-Mozaz, S.; Gros, M.; Barceló, D.; León, V.M. Seasonal Distribution of Pharmaceuticals in Marine Water and Sediment from a Mediterranean Coastal Lagoon (SE Spain). *Environ. Res.* **2015**, *138*, 326–344. [CrossRef]
16. Kummerová, M.; Zezulka, Š.; Babula, P.; Tříska, J. Possible Ecological Risk of Two Pharmaceuticals Diclofenac and Paracetamol Demonstrated on a Model Plant Lemna Minor. *J. Hazard. Mater.* **2016**, *302*, 351–361. [CrossRef]
17. Zheng, Y.; Sun, Z.; Liu, Y.; Cao, T.; Zhang, H.; Hao, M.; Chen, R.; Dzakpasu, M.; Wang, X.C. Phytoremediation Mechanisms and Plant Eco-Physiological Response to Microorganic Contaminants in Integrated Vertical-Flow Constructed Wetlands. *J. Hazard. Mater.* **2022**, *424*, 127611. [CrossRef]
18. Madikizela, L.M.; Ncube, S.; Chimuka, L. Uptake of Pharmaceuticals by Plants Grown under Hydroponic Conditions and Natural Occurring Plant Species: A Review. *Sci. Total Environ.* **2018**, *636*, 477–486. [CrossRef]
19. Nunes, B. Ecotoxicological Effects of the Drug Paracetamol: A Critical Review of Past Ecotoxicity Assessments and Future Perspectives. In *The Handbook of Environmental Chemistry*; Springer: Berlin/Heidelberg, Germany, 2020; Volume 96, pp. 131–145. [CrossRef]
20. Christou, A.; Papadavid, G.; Dalias, P.; Fotopoulos, V.; Michael, C.; Bayona, J.M.; Piña, B.; Fatta-Kassinos, D. Ranking of Crop Plants According to Their Potential to Uptake and Accumulate Contaminants of Emerging Concern. *Environ. Res.* **2019**, *170*, 422–432. [CrossRef] [PubMed]
21. Sungur, Ş. Pharmaceutical and Personal Care Products in the Environment: Occurrence and Impact on the Functioning of the Ecosystem. In *Emerging Contaminants in the Environment*; Elsevier: Amsterdam, The Netherlands, 2022; pp. 137–157. [CrossRef]
22. Wu, X.; Conkle, J.L.; Gan, J. Multi-Residue Determination of Pharmaceutical and Personal Care Products in Vegetables. *J. Chromatogr. A* **2012**, *1254*, 78–86. [CrossRef] [PubMed]
23. Sun, C.; Dudley, S.; McGinnis, M.; Trumble, J.; Gan, J. Acetaminophen Detoxification in Cucumber Plants via Induction of Glutathione S-Transferases. *Sci. Total Environ.* **2019**, *649*, 431–439. [CrossRef] [PubMed]
24. Tăşchină, M.; Copolovici, D.M.; Bungău, S.; Lupitu, A.I.; Copolovici, L.; Iovan, C. The Influence of Residual Acetaminophen on *Phaseolus Vulgaris* L. Secondary Metabolites. *Farmacia* **2017**, *65*, 709–713.
25. Bartha, B.; Huber, C.; Harpaintner, R.; Schröder, P. Effects of Acetaminophen in *Brassica Juncea* L. Czern.: Investigation of Uptake, Translocation, Detoxification, and the Induced Defense Pathways. *Environ. Sci. Pollut. Res.* **2010**, *17*, 1553–1562. [CrossRef]
26. Kotyza, J.; Soudek, P.; Kafka, Z.; Vaněk, T. Phytoremediation of Pharmaceuticals-Preliminary Study. *Int. J. Phytoremed.* **2010**, *12*, 306–316. [CrossRef] [PubMed]
27. Shanableh, A.; Semreen, M.; Semerjian, L.; Abdallah, M.; Mousa, M.; Darwish, N.; Baalbaki, Z. Contaminants of Emerging Concern in Sharjah Wastewater Treatment Plant, Sharjah, UAE. *J. Environ. Eng. Sci.* **2019**, *14*, 225–234. [CrossRef]
28. Semreen, M.H.; Shanableh, A.; Semerjian, L.; Alniss, H.; Mousa, M.; Bai, X.; Acharya, K. Molecules Simultaneous Determination of Pharmaceuticals by Solid-Phase Extraction and Liquid Chromatography-Tandem Mass Spectrometry: A Case Study from Sharjah Sewage Treatment Plant. *Molecules* **2019**, *24*, 633. [CrossRef] [PubMed]
29. Natarajan, R.; Saikia, K.; Ponnusamy, S.K.; Rathankumar, A.K.; Rajendran, D.S.; Venkataraman, S.; Tannani, D.B.; Arvind, V.; Somanna, T.; Banerjee, K.; et al. Understanding the Factors Affecting Adsorption of Pharmaceuticals on Different Adsorbents—A Critical Literature Update. *Chemosphere* **2022**, *287*, 131958. [CrossRef]
30. Gondi, R.; Kavitha, S.; Yukesh Kannah, R.; Parthiba Karthikeyan, O.; Kumar, G.; Kumar Tyagi, V.; Rajesh Banu, J. Algal-Based System for Removal of Emerging Pollutants from Wastewater: A Review. *Bioresour. Technol.* **2022**, *344*, 126245. [CrossRef] [PubMed]
31. Neto, A.D.; Pereira, P.P.; Costa, D.P.; dos Santos, A.C. Chlorophyll fluorescence as a possible tool for salt-stress tolerance screening in the sunflower. *Rev. Ciência Agronômica* **2011**, *42*, 893. [CrossRef]
32. Nunes, F.V.; De Melo, I.S. Isolation and Characterization of Endophytic Bacteria of Coffee Plants and Their Potential in Caffeine Degradation. *WIT Trans. Biomed. Health* **2006**, *10*, 293–297. [CrossRef]
33. Pérez-Cordero, A.; Tuberquia-Sierra, A.; Amell-Jiménez, D. In Vitro Activity of Nitrogen Fixating and Phosphate Solubilizing Bacteria. *Agron. Mesoam.* **2014**, *25*, 214–223.
34. Lopez-Velasco, G.; Welbaum, G.E.; Boyer, R.R.; Mane, S.P.; Ponder, M.A. Changes in Spinach Phylloepiphytic Bacteria Communities Following Minimal Processing and Refrigerated Storage Described Using Pyrosequencing of 16S RRNA Amplicons. *J. Appl. Microbiol.* **2011**, *110*, 1203–1214. [CrossRef]
35. Yang, C.W.; Chen, Y.E.; Chang, B.V. Microbial Communities Associated with Acetaminophen Biodegradation from Mangrove Sediment. *Sustainability* **2020**, *12*, 5410. [CrossRef]

36. Jackson, C.R.; Randolph, K.C.; Osborn, S.L.; Tyler, H.L. Culture Dependent and Independent Analysis of Bacterial Communities Associated with Commercial Salad Leaf Vegetables. *BMC Microbiol.* **2013**, *13*, 274. [CrossRef] [PubMed]
37. Riggs, P.J.; Chelius, M.K.; Iniguez, A.L.; Kaeppler, S.M.; Triplett, E.W. Enhanced Maize Productivity by Inoculation with Diazotrophic Bacteria. *Funct. Plant Biol.* **2001**, *28*, 829. [CrossRef]
38. White, J.F.; Kingsley, K.I.; Kowalski, K.P.; Irizarry, I.; Micci, A.; Soares, M.A.; Bergen, M.S. Disease Protection and Allelopathic Interactions of Seed-Transmitted Endophytic Pseudomonads of Invasive Reed Grass (*Phragmites Australis*). *Plant Soil* **2018**, *422*, 195–208. [CrossRef]
39. Saravanakumar, D.; Samiyappan, R. ACC Deaminase from Pseudomonas Fluorescens Mediated Saline Resistance in Groundnut (*Arachis Hypogea*) Plants. *J. Appl. Microbiol.* **2007**, *102*, 1283–1292. [CrossRef]
40. Sheng, X.F.; Xia, J.J.; Jiang, C.Y.; He, L.Y.; Qian, M. Characterization of Heavy Metal-Resistant Endophytic Bacteria from Rape (*Brassica Napus*) Roots and Their Potential in Promoting the Growth and Lead Accumulation of Rape. *Environ. Pollut.* **2008**, *156*, 1164–1170. [CrossRef]
41. Tian, T.; Tam, N.F.Y.; Zan, Q.; Cheung, S.G.; Shin, P.K.S.; Wong, Y.S.; Zhang, L.; Chen, Z. Performance and Bacterial Community Structure of a 10-Years Old Constructed Mangrove Wetland. *Mar. Pollut. Bull.* **2017**, *124*, 1096–1105. [CrossRef]
42. Ahmed, R.Z.; Ahmed, N.; Gadd, G.M. Isolation of Two Kocuria Species Capable of Growing on Various Polycyclic Aromatic Hydrocarbons. *Afr. J. Biotechnol.* **2010**, *9*, 3611–3617.
43. Habib, I.; Mohamed, M.Y.I. Foodborne Infections in the Middle East. In *Food Safety in the Middle East*; Academic Press: Cambridge, MA, USA, 2022; pp. 71–107. [CrossRef]
44. Scialabba, N.E.-H. Livestock Xenobiotics and Zoonoses. In *Managing Health Livestock Production and Consumption*; Academic Press: Cambridge, MA, USA, 2022; pp. 45–59. [CrossRef]
45. Bugg, T.D.H.; Ahmad, M.; Hardiman, E.M.; Singh, R. The Emerging Role for Bacteria in Lignin Degradation and Bio-Product Formation. *Curr. Opin. Biotechnol.* **2011**, *22*, 394–400. [CrossRef]
46. Villarreal Silva, M. Role of Epiphytic Bacteria in the Colonization of Fruits and Leafy Greens by Foodborne Bacterial Pathogens. Ph.D. Thesis, Texas A & M University, College Station, TX, USA, August 2016.
47. Beyi, A.F.; Hassall, A.; Phillips, G.J.; Plummer, P.J.; Tracking, P.; Kamal, A. Tracking Reservoirs of Antimicrobial Resistance Genes in a Complex Microbial Community Using Metagenomic Hi-C: The Case of Bovine Digital Dermatitis. *Antibiotics* **2021**, *10*, 221. [CrossRef]
48. Pathak, D.; Lone, R.; Nazim, N.; Alaklabi, A.; Khan, S.; Koul, K.K. Plant Growth Promoting Rhizobacterial Diversity in Potato Grown Soil in the Gwalior Region of India. *Biotechnol. Rep.* **2022**, *33*, e00713. [CrossRef]
49. An, J.; Zhou, Q.; Sun, F.; Zhang, L. Ecotoxicological Effects of Paracetamol on Seed Germination and Seedling Development of Wheat (*Triticum Aestivum* L.). *J. Hazard. Mater.* **2009**, *169*, 751–757. [CrossRef] [PubMed]
50. Sun, C.; Dudley, S.; Trumble, J.; Gan, J. Pharmaceutical and Personal Care Products-Induced Stress Symptoms and Detoxification Mechanisms in Cucumber Plants. *Environ. Pollut.* **2018**, *234*, 39–47. [CrossRef] [PubMed]
51. Zhang, H.; Zhu, J.; Gong, Z.; Zhu, J.K. Abiotic Stress Responses in Plants. *Nat. Rev. Genet.* **2022**, *23*, 104–119. [CrossRef] [PubMed]
52. Zezulka, Š.; Kummerová, M.; Babula, P.; Hájková, M.; Oravec, M. Sensitivity of Physiological and Biochemical Endpoints in Early Ontogenetic Stages of Crops under Diclofenac and Paracetamol Treatments. *Environ. Sci. Pollut. Res.* **2018**, *26*, 3965–3979. [CrossRef] [PubMed]
53. Alkimin, G.D.; Daniel, D.; Frankenbach, S.; Serôdio, J.; Soares, A.M.V.M.; Barata, C.; Nunes, B. Evaluation of Pharmaceutical Toxic Effects of Non-Standard Endpoints on the Macrophyte Species *Lemna Minor* and *Lemna Gibba*. *Sci. Total Environ.* **2019**, *657*, 926–937. [CrossRef] [PubMed]
54. Leitão, I.; Martins, L.L.; Carvalho, L.; Oliveira, M.C.; Marques, M.M.; Mourato, M.P.; Spagnuolo, V.; Capozzi, F. Acetaminophen Induces an Antioxidative Response in Lettuce Plants. *Plants* **2021**, *10*, 1152. [CrossRef]
55. Al-Muwayhi, M.A.R. Paracetamol Mediated Changes Modifies the Photosynthetic Efficiency of *Vigna Radiata*. *Legum. Res.* **2018**, *41*, 842–845. [CrossRef]
56. Kudrna, J.; Hnilicka, F.; Kubes, J.; Vachova, P.; Hnilickova, H.; Kuklova, M. Effect of Acetaminophen (Apap) on Physiological Indicators in *Lactuca Sativa*. *Life* **2020**, *10*, 303. [CrossRef]
57. Soares, C.; Pereira, R.; Martins, M.; Tamagnini, P.; Serôdio, J.; Moutinho-Pereira, J.; Cunha, A.; Fidalgo, F. Glyphosate-Dependent Effects on Photosynthesis of *Solanum Lycopersicum* L.—An Ecophysiological, Ultrastructural and Molecular Approach. *J. Hazard. Mater.* **2020**, *398*, 122871. [CrossRef]
58. Saidi, I.; Ayouni, M.; Dhieb, A.; Chtourou, Y.; Chaïbi, W.; Djebali, W. Oxidative Damages Induced by Short-Term Exposure to Cadmium in Bean Plants: Protective Role of Salicylic Acid. *S. Afr. J. Bot.* **2013**, *85*, 32–38. [CrossRef]
59. Guo, J.; Peng, J.; Lei, Y.; Kanerva, M.; Li, Q.; Song, J.; Guo, J.; Sun, H. Comparison of Oxidative Stress Induced by Clarithromycin in Two Freshwater Microalgae *Raphidocelis Subcapitata* and *Chlorella Vulgaris*. *Aquat. Toxicol.* **2020**, *219*, 105376. [CrossRef] [PubMed]
60. Zaheer, I.E.; Ali, S.; Saleem, M.H.; Ali, M.; Riaz, M.; Javed, S.; Sehar, A.; Abbas, Z.; Rizwan, M.; El-Sheikh, M.A.; et al. Interactive Role of Zinc and Iron Lysine on *Spinacia Oleracea* L. Growth, Photosynthesis and Antioxidant Capacity Irrigated with Tannery Wastewater. *Physiol. Mol. Biol. Plants* **2020**, *26*, 2435–2452. [CrossRef] [PubMed]

61. Mezanur Rahman, M.; Golam Mostofa, M.; Kumar Das, A.; Rahman Anik, T.; Sultana Keya, S.; Ahsan, S.M.; Arifur Rahman Khan, M.; Ahmed, M.; Abiar Rahman, M.; Motaher Hossain, M.; et al. Ethanol Positively Modulates Photosynthetic Traits, Antioxidant Defense and Osmoprotectant Levels to Enhance Drought Acclimatization in Soybean. *Antioxidants* **2022**, *2022*, 516. [CrossRef] [PubMed]
62. Kvesitadze, G.; Khatisashvili, G.; Sadunishvili, T.; Agmasheneblis, D. *Mechanisms to Detoxify Selected Organic Contaminants in Higher Plants and Microbes, and Their Potential Use in Landscape Management*; No. 62558; Academy of Sciences Of Georgia (Tbilisi) Durmishidze Institute of Biochemistry and Biotechnology: Tbilisi, Georgia, 2004.
63. Nazaryuk, V.M.; Klenova, M.I.; Kalimullina, F.R. Ecoagrochemical Approaches to the Problem of Nitrate Pollution in Agroecosystems. *Russ. J. Ecol.* **2002**, *33*, 392–397. [CrossRef]
64. Villette, C.; Maurer, L.; Wanko, A.; Heintz, D. Xenobiotics Metabolization in *Salix Alba* Leaves Uncovered by Mass Spectrometry Imaging. *Metabolomics* **2019**, *15*, 122. [CrossRef]
65. Rawat, J.; Pandey, N.; Saxena, J. Nitrate Accumulation, Growth and Leaf Quality of Spinach Beet (*Beta Vulgaris* Linn.) as Affected by NPK Fertilization with Special Reference to Potassium. In *Role of Potassium in Abiotic Stress*; Springer: Singapore, 2022; Volume 2, pp. 1–14. [CrossRef]
66. Graciela Maurino, V.; Narayan Misra, A.; Huang, W.; Prasad, S.M.; Singh, R.; Parihar, P. Sulfur and Calcium Simultaneously Regulate Photosynthetic Performance and Nitrogen Metabolism Status in As-Challenged *Brassica Juncea* L. Seedlings. *Front. Plant Sci.* **2018**, *9*. [CrossRef]
67. Britto, D.T.; Kronzucker, H.J. Futile Cycling at the Plasma Membrane: A Hallmark of Low-Affinity Nutrient Transport. *Trends Plant Sci.* **2006**, *11*, 529–534. [CrossRef]
68. Huber, C.; Bartha, B.; Harpaintner, R.; Schröder, P.; Huber, C.; Bartha, B.; Harpaintner, R.; Schröder, P. Metabolism of Acetaminophen (Paracetamol) in Plants-Two Independent Pathways Result in the Formation of a Glutathione and a Glucose Conjugate. *Environ. Sci Pollut. Res.* **2009**, *16*, 206–213. [CrossRef]
69. Bartha, B. Uptake and Metabolism of Human Pharmaceuticals in Plants: Identification of Metabolites and Specification of the Defense Enzyme Systems under Pharmaceutical Exposure. Ph.D. Thesis, Technische Universität München, Singapore, June 2012.
70. Martins, L.; Teixeira, J. Gene- and Organ-Specific Impact of Paracetamol on *Solanun Nigrum* L.'s γ-Glutamylcysteine Synthetase and Glutathione S-Transferase and Consequent Phytoremediation Fitness. *Acta Physiol. Plant.* **2021**, *43*, 53. [CrossRef]
71. Feng, N.X.; Yu, J.; Zhao, H.M.; Cheng, Y.T.; Mo, C.H.; Cai, Q.Y.; Li, Y.W.; Li, H.; Wong, M.H. Efficient Phytoremediation of Organic Contaminants in Soils Using Plant–Endophyte Partnerships. *Sci. Total Environ.* **2017**, *583*, 352–368. [CrossRef]
72. Chandran, V.; Shaji, H.; Mathew, L. Endophytic Microbial Influence on Plant Stress Responses. In *Microbial Endophytes: Functional Biology and Applications*; Woodhead Publishing: Sawtson, UK, 2020; pp. 161–193. [CrossRef]
73. Ijaz, A.; Shabir, G.; Khan, Q.M.; Afzal, M. Enhanced Remediation of Sewage Effluent by Endophyte-Assisted Floating Treatment Wetlands. *Ecol. Eng.* **2015**, *84*, 58–66. [CrossRef]
74. Ashraf, S.; Afzal, M.; Naveed, M.; Shahid, M.; Zahir, Z.A. Endophytic Bacteria Enhance Remediation of Tannery Effluent in Constructed Wetlands Vegetated with *Leptochloa fusca*. *Int. J. Phytoremed.* **2018**, *20*, 121–128. [CrossRef] [PubMed]
75. Żur, J.; Piński, A.; Marchlewicz, A.; Hupert-Kocurek, K.; Wojcieszyńska, D.; Guzik, U. Organic Micropollutants Paracetamol and Ibuprofen—Toxicity, Biodegradation, and Genetic Background of Their Utilization by Bacteria. *Environ. Sci. Pollut. Res. Int.* **2011**, *25*, 21498–21524. [CrossRef]
76. Kumar, D.; Kumar, A.; Sharma, J. Degradation Study of Lindane by Novel Strains *Kocuria* Sp. DAB-1Y and *Staphylococcus* Sp. DAB-1W. *Bioresour. Bioprocess.* **2016**, *3*, 53. [CrossRef] [PubMed]
77. Zhang, L.; Hu, J.; Zhu, R.; Zhou, Q.; Chen, J. Degradation of Paracetamol by Pure Bacterial Cultures and Their Microbial Consortium. *Appl. Microbiol. Biotechnol.* **2013**, *97*, 3687–3698. [CrossRef]
78. Weyens, N.; Van Der Lelie, D.; Artois, T.; Smeets, K.; Taghavi, S.; Newman, L.; Carleer, R.; Vangronsveld, J. Bioaugmentation with Engineered Endophytic Bacteria Improves Contaminant Fate in Phytoremediation. *Environ. Sci. Technol.* **2009**, *43*, 9413–9418. [CrossRef]
79. Mosa, K.A.; El-Naggar, M.; Ramamoorthy, K.; Alawadhi, H.; Elnaggar, A.; Wartanian, S.; Ibrahim, E.; Hani, H. Copper Nanoparticles Induced Genotoxicty, Oxidative Stress, and Changes in Superoxide Dismutase (SOD) Gene Expression in Cucumber (*Cucumis Sativus*) Plants. *Front. Plant Sci.* **2018**, *9*, 872. [CrossRef]
80. Li, Y.; Xin, J.; Ge, W.; Tian, R. Tolerance Mechanism and Phytoremediation Potential of *Pistia Stratiotes* to Zinc and Cadmium Co-Contamination. *Int. J. Phytoremed.* **2022**, 1–8. [CrossRef]
81. Hussain, M.I.; Reigosa, M.J. Secondary Metabolites, Ferulic Acid and p-Hydroxybenzoic Acid Induced Toxic Effects on Photosynthetic Process in *Rumex Acetosa* L. *Biomolecules* **2021**, *11*, 233. [CrossRef]
82. Lichtenthaler, H.K.; Wellburn, A.R. Determinations of Total Carotenoids and Chlorophylls a and b of Leaf Extracts in Different Solvents. *Biochem. Soc. Trans.* **1983**, *11*, 591–592. [CrossRef]
83. Chuang, Y.H.; Zhang, Y.; Zhang, W.; Boyd, S.A.; Li, H. Comparison of Accelerated Solvent Extraction and Quick, Easy, Cheap, Effective, Rugged and Safe Method for Extraction and Determination of Pharmaceuticals in Vegetables. *J. Chromatogr. A* **2015**, *1404*, 1–9. [CrossRef] [PubMed]
84. Kaynar, S.Ç.A.M.; HİçsÖnmez, Ü.; Özdemİr, A.; Özdemİr, C. Assessment of Element Concentrations in Soil, Root and Leaf of Spinach Plant (*Spinacia Oleracea* L.) Grown in Manisa. *J. Inst. Sci. Tech.* **2018**, *8*, 131–140. [CrossRef]

plants

MDPI

Article

Comparative Plasticity Responses of Stable Isotopes of Carbon ($\delta^{13}C$) and Nitrogen ($\delta^{15}N$), Ion Homeostasis and Yield Attributes in Barley Exposed to Saline Environment

Muhammad Iftikhar Hussain [1,2,*], Zafar Iqbal Khan [3], Taimoor Hassan Farooq [4], Dunia A. Al Farraj [5] and Mohamed Soliman Elshikh [5]

1 Department of Plant Biology & Soil Science, Universidade de Vigo, Campus As Lagoas Marcosende, 36310 Vigo, Spain
2 Research Institute of Science and Engineering, University of Sharjah, Sharjah P.O. Box 27272, United Arab Emirates
3 Department of Botany, University of Sargodha, Sargodha 40100, Pakistan; zafar.khan@uos.edu.pk
4 Bangor College China, A Joint Unit of Bangor University and Central South University of Forestry and Technology, Changsha 410004, China; taimoorhassan2055@gmail.com
5 Department of Botany and Microbiology, College of Science, King Saud University, P.O. Box 2455, Riyadh 11451, Saudi Arabia; dfarraj@ksu.edu.sa (D.A.A.F.); melshikh@ksu.edu.sa (M.S.E.)
* Correspondence: mih786@gmail.com

Abstract: Salinity is a major threat to agricultural productivity worldwide. The selection and evaluation of crop varieties that can tolerate salt stress are the main components for the rehabilitation of salt-degraded marginal soils. A field experiment was conducted to evaluate salinity tolerance potential, growth performance, carbon ($\delta^{13}C$) and nitrogen isotope composition ($\delta^{15}N$), intrinsic water use efficiency (iWUE), harvest index, and yield stability attributes in six barley genotypes (113/1B, 59/3A, N1-10, N1-29, Barjouj, Alanda01) at three salinity levels (0, 7, and 14 dS m^{-1}). The number of spikes m^{-2} was highest in Alanda01 (620.8) while the lowest (556.2) was exhibited by Barjouj. Alanda01 produced the highest grain yield (3.96 t ha^{-1}), while the lowest yield was obtained in 59/3A (2.31 t ha^{-1}). Genotypes 113/1B, Barjouj, and Alanda01 demonstrate the highest negative $\delta^{13}C$ values (−27.10‰, −26.49‰, −26.45‰), while the lowest values were obtained in N1-29 (−21.63‰) under salt stress. The $\delta^{15}N$ was increased (4.93‰ and 4.59‰) after 7 and 14 dS m^{-1} as compared to control (3.12‰). The iWUE was higher in N1-29 (144.5) and N1-10 (131.8), while lowest in Barjouj (81.4). Grain protein contents were higher in 113/1B and Barjouj than other genotypes. We concluded that salt tolerant barley genotypes can be cultivated in saline marginal soils for food and nutrition security and can help in the rehabilitation of marginal lands.

Keywords: *Hordeum vulgare*; stable isotope composition of carbon and nitrogen; saline water stress; isotope ecology; yield stability; ion homeostasis

Citation: Hussain, M.I.; Khan, Z.I.; Farooq, T.H.; Al Farraj, D.A.; Elshikh, M.S. Comparative Plasticity Responses of Stable Isotopes of Carbon ($\delta^{13}C$) and Nitrogen ($\delta^{15}N$), Ion Homeostasis and Yield Attributes in Barley Exposed to Saline Environment. *Plants* **2022**, *11*, 1516. https://doi.org/10.3390/plants11111516

Academic Editor: Anna Aksmann

Received: 17 April 2022
Accepted: 31 May 2022
Published: 5 June 2022

Publisher's Note: MDPI stays neutral with regard to jurisdictional claims in published maps and institutional affiliations.

1. Introduction

Global agriculture is unable to cope with the existing climate change scenario and to feed the worlds growing population that is projected to increase from 6.7 billion (2005) to 9.2 billion by 2050 [1]. Among all these anthropogenic factors, drought, salinity, and climate change are the principal players behind the land degradation and desertification leading to a significant reduction in crop production and yield decline [2–5]. Due to the scarce water resources and drought episodes, the irrigation water requirement in Arabian Gulf countries is mostly fulfilled through salty ground water and treated wastewater that is recruited to irrigate a significant land area (forestry, landscaping, roadside plantation) [5]. To meet the growing need of agriculture, date palm fruit gardens and landscaping, the Gulf countries are using desalinated water (7.2%) and groundwater (91%) to meet their requirements [6].

In this context, appropriate crop accessions that can be well adapted to the marginalized lands and available non-conventional water resources are suitable options for long-term rehabilitation and desertification resistance [3,7–10].

Barley is an important grain crop and ranked fourth among the cereal crops after wheat, rice, and maize [11]. It is mainly used as food, animal fodder, and as a raw material for beer production [12]. Several authors have demonstrated that barley can tolerate a number of environmental stresses, such as drought [13,14], salinity [15], and heavy metals [16]. However, salt tolerance within genotypes of barley under field conditions has not been evaluated intensively. Therefore, the study of genetic diversity and phenotypic plasticity should be integrated in order to evaluate and select the most tolerant genotypes within a wide range of salinity among this plant species. Furthermore, the growth, yield, and productivity of barley are highly variable in the Middle East and North Africa region because the local cultivars do not have sufficient tolerance potential against prevailing environmental constraints, especially drought and salinity. Most researchers have evaluated the variation in salinity tolerance using growth chamber or green house at a single level of salinity and there was no validation of those results under the field setting. Meanwhile, studies conducted in a controlled growth chamber generally involve the determination of salinity stress on seeding growth over a short period of time (often 1–7 days), which does not correspond to salt stress in the field that might indicate a wide variation in the growth, development, physiological, and yield traits [17].

It has been observed that among the population of particular crop genotypes, wide variation exists at various growth and development stages for salinity tolerance. However, it was difficult to predict which salinity range will be appropriate for the screening, selection, and evaluation of genotypes that can best correlate with genetic diversity under field conditions. This kind of study is very important in order to develop efficient breeding programs and tool kits of salt tolerant crop genotypes and to assess the growth, physiological, and yield traits under field conditions [18]. Efforts to enhance crop yields under salinity stress have also had limited success because the underlying mechanisms of salt tolerance have not been turned into useful selection criteria to evaluate a wide range of phenotypic plasticity and genotypes. Several authors have studied the salinity tolerance potential among a wide range of crop plants at the germination and seedling growth stages and showed a large genetic difference among them [19,20]. However, little attention has been paid to show a correlation regarding this early evaluation of salinity tolerance at germination with field condition [21]. However, it is worthwhile to mention that these authors made significant efforts to explain the Na^+ exclusion, K^+ accumulation, and K^+/Na^+ as reliable indicators for selecting suitable genotypes that can tolerate soil and irrigation water salinity [22]. The success of dual-purpose barley in marginal environments is subject to proper agronomic management practices along with the use of improved genotypes.

It is an urgent task of agronomists, plant physiologists, and plant breeders to identify and evaluate the genotypes and plant phenotypic plasticity using non-invasive, rapid, and reliable methods in order to screen the desired traits in a particular environment. The evaluation of the salinity tolerance potential of different genotypes and plant phenotypic attributes is highly necessary in order to understand physiological responses of the target genotypes and concerned traits associated with them [23,24]. The present situation can be changed through the introduction of new salt tolerant and higher yielding barley genotypes that have good yield stability and better salt tolerance potential. This will help to conserve freshwater resources as well as economic and ecological benefits for the sustainable development of salt-degraded marginal lands [6,25,26]. It is important to screen, select, and evaluate the large collection of barley genotypes to check their performance (growth, yield stability, physiological characteristics) and traits are suited to salinity tolerance under field condition. In the present field study, a set of 28 genotypes from a previous trial [27–29] were selected for elucidating the performance of different agronomical attributes (growth, number of tillers, plant biomass), yield traits (number of spikes, number of grains/spike, grain yield, harvest index), and biochemical attributes

(Na^+, Cl^-, K^+), to find more suitable and tolerant genotypes under sandy marginal lands. The current study will provide a basis to promote barley cultivation on a large scale in the salt affected agro-ecosystem environment of the UAE. In addition, genotypes that showed stable yield and salt-tolerance potential will be included in the barley breeding programs for the development and release of salt-tolerant cultivars for seed multiplication and distribution among NARS for multi-location testing and large-scale cultivation.

The phenotypic plasticity, genotype variability, and agronomic adaptation of barley are extremely wide and vary significantly from hot arid to subtropical humid climates. Barley batini land races have not been characterized for salt tolerance on morphological, biochemical, ecophysiological, and isotopic bases. The main aim of the present study was the evaluation of batini barley land races and genotypes through the elucidation of salinity tolerance potential, growth performance, leaf ion homeostasis, leaf carbon and nitrogen isotope discrimination, intrinsic water use efficiency, harvest index, and yield stability attributes on six barley genotypes (113/1B, 59/3A, N1-10, N1-29, Barjouj, Alanda01) at three salinity levels (0, 7 and 14 dS m^{-1}). For this study, it was hypothesized that batini barley land races and genotypes are genetically diverse and vary for salt tolerance potential. The evaluation of the plasticity of physiological attributes, such as number of tillers/m^2, fresh biomass (FW), dry biomass (DW), grain yield, harvest index, and leaf Na^+, K^+, and Cl^- concentration, leaf carbon and nitrogen isotope discrimination, and intrinsic water use efficiency, may help to develop a better understanding of mechanisms of salt tolerance.

2. Materials and Methods

2.1. Experiment Site and Climatic Conditions

The field trials were conducted at an agriculture experiment research station (ICBA, Dubai, UAE) from December 2013 to May 2014. The site is located at N 25°05.847; E 055° 23.464. The experimental field was nutrient-poor, sandy soil (sand 98%, silt 1%, and clay 1%), calcareous (50–60% $CaCO_3$), porous (45% porosity), and moderately alkaline (pH 8.22). The electrical conductivity of saturated extract (Ec) is 1.2 dS m^{-1} and the soil has good drainage capacity and is classified as carbonatic, hyperthermic typic, and torripsamment. To keep the area drained and to control soil salinization at the experimental station, a sub-surface drainage system is installed at 2 m depth from the soil surface. From December to February, the temperature is significantly lower, days are cooler and dry (10 °C, temperature at night), while during the summer season (April to October), the temperature is high, can reach up to 50 °C, and the climate is extremely hot and dry with lots of humidity. During summer, there is almost no chance of rainfall and the sky is mostly cloudless. Average annual temperature, rainfall, and humidity are shown in Figure 1.

2.2. Plant Material and Growth Conditions

Six barley (*Hordeum vulgare* L.) genotypes used in this study (Table 1) include germplasm obtained from ICARDA (27 barley entries from the Barley Observation Nursery (selected from 328 entries), specifically 5 entries from the Heat Nursery Q2-4 (selected from 458 entries) and 11 entries from the Special Heat Nursery (selected from 320 entries), evaluated during the cropping cycle (1999–2003) [30]. A few lines are among the best lines selected from a set of Omani Batini barley landrace from 2308 subpopulations (Batini 1-7 and 1-5) evaluated by Jaradat et al., [27,28] for tolerance to different levels of continuous salinity during germination and seedling growth attributes.

Table 1. The Barley GeneBank accession names and entry number in this study.

S.No.	Accessions Name	Collection Type	Entry Code/Pedigree
1	113/1B	Batini	113/1B
2	59/3A	Batini	59/3A
7	N1-10	nurseries	Manitou//Alanda/Zafraa
8	N1-29	nurseries	Rhn-03//L.527/NK1272
17	Barjouj	varieties	Barjouj

Table 1. *Cont.*

S.No.	Accessions Name	Collection Type	Entry Code/Pedigree
18	Alanda01	varieties	Alanda01

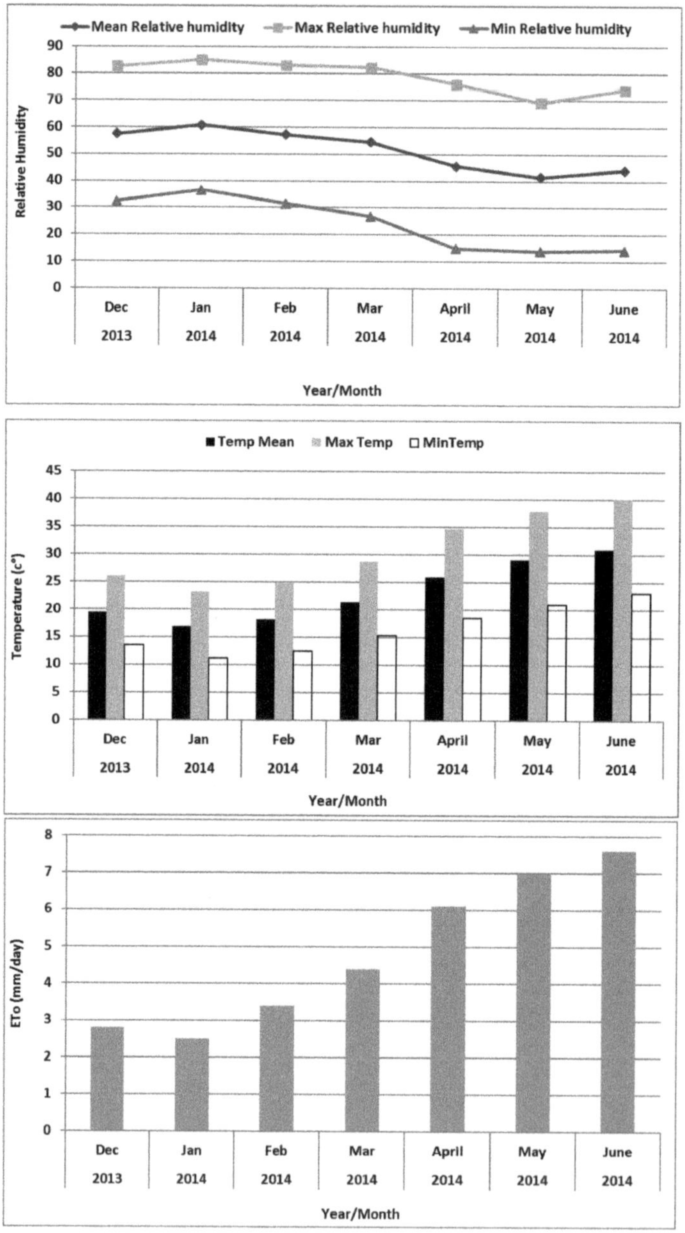

Figure 1. Monthly average values of mean (T mean), maximum (T maxi), and minimum (T min) air temperature and reference evapotranspiration (ETo) in the ICBA weather station, Dubai, UAE from December 2013 to June 2014.

The field plot was prepared by harrowing 1–2 times, followed by planking. Organic fertilizer (N 1.5, K 1.65 and Na 1.22%; pH 7.7, C:N ratio 16.5, organic matter 41% and moisture 1.64%) was applied (30 t ha^{-1}) at the surface before soil was incorporated. The seeds (1600 per each genotype) of each individual barley line were sown (2 November 2013) manually in the rows (0.5 m spacing) in the field with a plot size of 2 m × 4 m (plot area of 8 m^2). The experimental design was a RCBD split plot with three replications. The main-plot factor was the salinity level (0, 7 dS m^{-1}, 14 dS m^{-1}) and the subplot factor was the genotypes that were randomized within each main-plot. The target salinity was maintained throughout the cropping season and a portable EC meter was used to monitor the salinity twice a week. The crop was irrigated using a drip irrigation system, spreading on the soil surface, having a 4 L hr^{-1} flow rate. A distance of 0.5 m was maintained between rows while the drippers were 0.25 m apart (Figure 2). The irrigation period was variable and depended upon the climatic conditions and crop development stage, ranging from full tillering to dough making. The irrigation program was established so that the plant receives total irrigation (net irrigation + effective rainfall) of around 80% crop evapotranspiration (ETc) plus 20% leaching requirement. During the grain filling period, a net (mesh size of c. 15 × 15 mm^2) was used to prevent the entry of small birds and to save the grain losses. The impact of saline water treatments (0, 7, 14 dS m^{-1}) on growth attributes, stable isotope composition of carbon and nitrogen, leaf ion homeostasis, yield components, harvest index, grain protein contents, and yield stability was evaluated on a selected set of 6 barley genotypes (Batini landraces, varieties, and heat nurseries) (Table 2).

Figure 2. (**a**) Barley field plots for sustainable crop production in sandy marginal hyper-arid desert soils at ICBA, Dubai, UAE. (**b**) Irrigation systems, seedling growth, tillering and spike development. (**c**) Barley crop at grain filling stage. (**d**) Barley crop at maturity stage.

Table 2. Physical and chemical characteristics of experimental soil.

	Soil Characteristics										
	Sample Location	**pHs**	**E_{Ce} (dS m^{-1})**	**Total N mg kg^{-1}**	**P mg kg^{-1}**	**K mg kg^{-1}**	**% Organic Matter**	**Sand (%)**	**Silt (%)**	**Clay (%)**	**Textural Class**
Pre-sowing 2013	Control	6.55	1.538	52.62	5.46	79.2	0.83	97.53	2.26	0.2	Sand
Post-harvest (2014)	7 dS/m	7.35	2.04	52	41.51	45.95	1.46	97.6	2.2	0.2	Sand
	14 dS/m	7.89	4.1	51.59	46.74	41.61	1.32	97.6	2.2	0.2	Sand

2.3. *Growth, Agro-Morphological, Leaf Ion Homeostasis and Yield Traits Measurements*

From each subplot, the whole plant was harvested from the middle 1 m of two central rows and data were recorded for different agronomical traits (growth, number of tillers, plant biomass), yield traits (number of spikes, number of grains/spike, grain yield, harvest index), and biochemical attributes (Na^+, Cl^-, K^+). The samples were collected to measure fresh biomass (FW) and dry biomass (DW) after the plant samples were dried at 70 °C for 72 h. Briefly, the dried leaves were ground into a fine powder and then ashed for 6 h at 550 °C. After that, 2 N HCl was added to the cooled ash, and the solution was filtered and tested after 15 min. Inductively coupled plasma optical emission spectrometry (Perkin Elmer Optima 4300DV) was used to determine the concentrations of different elements and expressed as mg/100 g dry weight (DW) [31].

2.4. *Harvest Index (%)*

The harvest index was calculated by using the following formula.

$$\text{Harvest index (\%)} = \text{Grain yield/dry biomass} \times 100 \quad (1)$$

2.5. *Grain Yield*

A sample line of 1 m length was harvested, and seeds were removed from the panicle of plants/plot, threshed, weighed (g m^{-2}), then converted into t ha^{-1}.

2.6. *Stable Carbon and Nitrogen Isotope Analysis*

The leaf samples from each treatment and control were collected, oven dried, and ground into a fine powder. Total N and C contents (% dry matter) were measured by elemental analysis (Flash EA-1112, Swerte, Germany). Dry ground plant material was weighed (1700–2100 µg) using a high precision analytical balance (Metler Toledo GmbH, Greifensee, Switzerland), and filled in tin capsules (5 × 3.5 mm, Elemental Microanalysis Limited, Okehampton, UK). Tin capsules (pressed are in the shape of a microball) were combusted (1600–1800 °C) using an automated elemental analyser coupled to an Isotope Ratio Mass-Spectrometer (Finnegan: Thermo Fisher Scientific, model MAT-253, Swerte, Germany). The Isotopic Ratio Mass Spectrometer has an analytical precision better than 0.3‰ for ^{15}N and 0.05‰ for ^{13}C.

Carbon and nitrogen isotope compositions were calculated as:

$$\delta\ (‰) = [(R_{sample}/R_{standard}) - 1)] \times 1000 \quad (2)$$

where R_{sample} is the ratio of $^{13}C/^{12}C$ or $^{15}N/^{14}N$, and $R_{standard}$ were the standards used. Atmospheric N_2 was the standard for nitrogen while Vienna PeeDee Belemnite (VPDB) was the standard for carbon. The accuracy and reproducibility of the measurements of $\delta^{13}C$ and $\delta^{15}N$ were checked with an internal reference material (NBS 18 and IAEA-C6 for C), and (IAEA-310A and IAEA-N1 for N), and acetanilide for C/N% ratios, respectively.

Carbon isotope discrimination is a measure of the carbon isotopic composition in plant material relative to the value of the same ratio in the air on which plants feed:

$$\Delta\ (‰) = [(\delta a - \delta p)/(1 + \delta p)] \times 1000 \quad (3)$$

where Δ represents carbon isotope discrimination, δa and δp refer to $\delta^{13}C$ of air CO_2 and plant material, respectively.

Farquhar et al. [32] and Farquhar and Richards [33] indicate that carbon isotope discrimination in leaves of plants can be expressed in relationship to CO_2 concentrations inside and outside of leaves in its simplest form as:

$$\Delta = a + (b - a)\ Ci/Ca$$

$$\Delta = 4.4 + (27 - 4.4)\ Ci/Ca \quad (4)$$

where a is discrimination that occurs during the diffusion of CO_2 through the stomata (4.4‰), b is discrimination by RuBisCO (27‰), and Ci/Ca is the ratio of the leaf intercellular CO_2 concentration to that in the atmosphere Ci/Ca- ratio of intercellular to atmospheric CO_2 concentration. Equation (4) establishes a direct and linear relationship between Δ and Ci/Ca. Therefore, the measurement of Δ gives an estimation of the rate-weighed value of Ci/Ca.

Intrinsic Water Use Efficiency (iWUE)

The term "intrinsic water-use efficiency" can be defined as the ratio of the instantaneous rates of CO_2 and transpiration at the stomata. Intrinsic water use efficiency (iWUE) was calculated according to the following equation:

$$iWUE = A/g = Ca\ [1 - (Ci/Ca)] \times (0.625) \quad (5)$$

where A is the rate of CO_2 and "g" is the stomatal conductance.

Carbon isotope discrimination ($\Delta^{13}C$), the ratio of the leaf intercellular CO_2 concentration to that in the atmosphere (Ci/Ca), and intrinsic water use efficiency (iWUE) were determined according to the theory documented by Farquhar et al. [32] and Farquhar and Richards [33]. The close relationship between $\Delta^{13}C$ and Ci/Ca has been explained on the basis that the observed differences reflect the variation of Ci/Ca in the carboxylation step of photosynthesis, in response to environmental constraints that affect stomatal regulation. Both Ci/Ca and iWUE were derived from $\delta^{13}C$ basic data using Equations (4) and (5) as reported previously [34–36].

2.7. Grain Protein Contents Measurements

From each barley genotype, 200 mg FW (three replicates/treatment) were employed for the quantification of grain protein contents using commercial bovine serum albumin (BSA) through Bradford assays [37], as reported previously [38].

2.8. Statistical Analysis

Experiment field data were analyzed through SPSS (version 19.00) using a general linear model. The differences between treatment means, genotypes, and their interaction were determined using Tukey's test ($p \leq 0.05$). The yield stability of different genotypes at different levels of salinity was computed through static yield stability index (S^2i) and dynamic yield stability index (W^2i) [39,40] as reported previously [41].

3. Results

3.1. Impact of Salinity Treatments and Genotypes on Growth Attributes

The present study assessed whether barley could be extended as a crop to more salt-degraded marginal sandy areas in UAE by irrigating with low quality saline water (E_C = 7 and 14 dS m^{-1}). Soil biochemical analysis showed that the soil is sandy loam type. The soil samples showed that soil had low organic matter (OM) content (Table 2) and low contents of nitrogen, phosphorus, and potassium. Mean squares for number of tillers, spike numbers, grain yield, and harvest index were significant (Table 3). The results of the present study demonstrate that both water salinity levels and genotypes in each assessment act independently on the above mentioned attributes. The environmental data (temperature, humidity, and evapotranspiration) during the study period 2013–2014 are shown in Figure 1.

Table 3. Effect of salt stress on biomass and agro-physiological traits, and yield components across 6 barley genotypes.

Salt Stress Level	Plant Dry Biomass (t ha^{-1})	Number of Tillers m^{-2}	Number of Spike m^{-2}	Number of Grain Spike^{-1}	Grain Numbers Per Plant
Control	130.1 a	700.5 a	652.2 a	46.3 a	664.3 a
7 dS m^{-1} Nacl	109.3 b	629.4 b	583.9 b	40 b	482.8 b
14 dS m^{-1} Nacl	89.8 c	572.9 c	519.6 c	34.2 c	357.8 c
Salinity Treatment (T)	**	**	**	**	**
Genotype (G)	**	**	**	**	**
T × G interaction	**	**	**	**	**

Values in a single column sharing the same letter are not significantly different ($p \leq 0.05$) according to Tukey's honestly significant difference (HSD) test. (**) are significant at $p \leq 0.05$ or 0.001, respectively.

3.2. Effect of Salt Stress on Morpho-Physiological Characteristics

Salt water significantly affected the plant dry biomass (PDB) due to irrigation water salinity at all levels. Saline water treatments caused a reduction in PDB from 16% to 31% at 7 and 14 dS m^{-1} respectively (Table 3). Barley genotype 113/1B (of Batini) produced the highest plant dry biomass (116.2 t ha^{-1}), followed by N1-29 and 59/3A (110.3 and 109.1 t/ha). The lowest PDB was produced by Barjouj (105 t ha^{-1}) (Table 4). In addition, the number of tillers m^{-2} significantly reduced following exposure to severe salt stress (572.9) as compared to control (700.5). The percentage reduction in the number of tillers m^{-2} was 18–10% from 14 to 7 dS m^{-1} NaCl stress. Physiological traits, e.g., number of spikes m^{-2}, were also decreased at each salinity stress and the highest reduction (20%) was observed at 14 dS m^{-1} NaCl, respectively, compared to the non-saline treatment (Table 3). Barley genotypes 113/1B, 59/3A, and Alanda01 exhibit the highest tillers m^{-2}, namely 681.1, 635.1, and 616.4, respectively. However, barley genotype N1-29 exhibits the smallest tillers m^{-2} (606) as compared to other genotypes. There was significant variation in the production of the number of spikes^{-2}. The number of spikes m^{-2} was highest in barley variety Alanda01 (620.8), followed by 113/1B (593), while the lowest number of spikes m^{-2} (556.2) was exhibited by Barjouj, respectively (Table 4). Genotype Alanda01 exhibited the highest grain numbers/plant (527.9) followed by 113/1B (508.4) while the lowest grains/plant was produced by Barjouj (480.5) (Table 4). A similar pattern of variation was obtained for number of grains/spike in the corresponding barley genotypes.

Table 4. Barley genotype difference in biomass and agro-physiological traits across all salinity treatments.

Genotypes	Plant Dry Biomass (t ha^{-1})	Number of Tillers m^{-2}	Number of Spike m^{-2}	Number of Grain Spike^{-1}	Grain Number per Plant	Grain Protein Content (mg/g DW)
113/1B	116.2 a	681.1 a	593 b	40.9 b	508.4 b	19.3 a
59/3A	109.1 c	635.1 b	577.5 c	39.5 c	497 c	16.3 b
N1-10	107.6 d	608.6 e	573.1 c	39.1 c	490.3 c	16 b
N1-29	110.3 b	606 f	572 c	39 c	491 c	16.3 b
Barjouj	105 e	610.4 d	556.2 d	37.6 d	480.5 d	19.3 a
Alanda01	107.8 d	616.4 c	620.8 a	43.4 a	527.9 a	16.6 b

Genotype means with different letters within a column for a given trait are significantly different at $p \leq 0.05$ according to Tukey's honestly significant difference (HSD) test.

3.3. Leaf Mineral Analysis

The concentrations of Na^+ and Cl^- ions were significantly higher in the barley leaves grown under saline water irrigation compared to control (Table 5). However, the use of saline water also significantly increased the K^+/Na^+ ratio in the leaf tissues (Table 4). The K^+ content was higher with saline water, while the rest of the elements did not show any changes. A significant difference was observed regarding Na^+ and K^+ concentrations among the barley genotypes (Table 6). Genotypes Barjouj and N1-29 showed the highest grain yield among the salinity treatments as compared to control and at the same time also accumulated higher K^+ levels. It was also noticed that these genotypes have a substantial amount of Na^+ in the leaf tissue that might counterbalance the toxicity effect through the accumulation of K^+ ions. Potassium concentrations varied widely, 2.6-fold, ranging from 599.4 to 639.2 mg/100 g DW. Sodium concentration also varied from 435.9 to 924.3 mg/100 g DW. Genotypes significantly differed for all traits, including Cl^- ions concentration that was significantly higher in Barjouj while the lowest was observed in N1-10 (Table 6). Overall, "N1-10" was the genotype with the highest K^+/Na^+ ratio, followed by N1-29, while Barjouj and Alanda01 exhibit the smallest K^+/Na^+ ratio among all the barley genotypes.

Table 5. Effect of salt stress on biomass and agro-physiological traits, and yield components across 6 barley genotypes.

Salt Stress Level	K^+	Cl^-	Na^+	K^+/Na^+ Ratio	% N	% C	C:N Ratio	Protein
Control	87.7 c	114.5 c	118 c	0.78 b	2.2 a	29.6 a	14.5 a	13.6 b
7 dS m^{-1} Nacl	772.3 b	984.3 b	994.2 b	0.84 b	2.8 a	27.6 b	10.2 b	17.5 a
14 dS m^{-1} Nacl	1024.2 a	1226.1 a	1157.2 a	1.00 a	2.8 a	27.2 b	9.7 c	17.8 a
Salinity Treatment (T)	**	**	**	ns	ns	**	**	**
Genotype (G)	**	**	**	ns	ns	ns	**	ns
T × G interaction	**	**	**	**	ns	ns	**	ns

Values in a single column sharing the same letter are not significantly different ($p \leq 0.05$) according to Tukey's honestly significant difference (HSD) test. ns (non-significant), (**) are significant at $p < 0.05$ or 0.001, respectively.

Table 6. Barley genotype difference in biomass and agro-physiological traits across all salinity treatments.

Genotypes	K^+ (mg 100 g^{-1} DW)	Cl^- (mg 100 g^{-1} DW)	Na^+ (mg 100 g^{-1} DW)	K^+/Na^+ Ratio	Leaf N%	Leaf C%	C:N Ratio
113/1B	615.1 c	760.3 c	526.1 d	1.0 a	3.0 a	28.9 a	9.6 c
59/3A	599.4 d	741.8 d	557.1 c	0.9 b	2.6 b	27.9 b	11.5 a
N1-10	575.2 e	717.7 e	435.9 f	1.12 a	2.6 b	28.2 a	11.3 a
N1-29	638.6 a	779.0 b	507.2 e	1.08 a	2.6 b	27.9 b	11.5 a
Barjouj	639.2 a	784.7 a	924.3 a	0.66 c	3.1 a	28.9 a	9.6 c
Alanda01	620.8 b	760.5 c	899.7 b	0.66 c	2.6 b	27.8 b	10.8 b

Genotype means with different letters within a column for a given trait are significantly different at $p \leq 0.05$) according to Tukey's honestly significant difference (HSD) test.

3.4. *Effect of Salt Stress on Carbon (C%) and Nitrogen (N%) and C:N Ratios*

The level of carbon was reduced at all salinity levels other than control (Table 5). In contrast to C contents, the nitrogen level was elevated at all salty water concentrations. Genotype 113/1B and Barjouj exhibited the highest N% and it was significantly higher than all other genotypes (Table 6). The C% was higher in three barley genotypes, 113/1B, N1-10, and Barjouj, respectively. There was not much difference in the leaf C% among the rest of the barley genotypes (59/3A, N1-29, Alanda01) that exhibit around 27.9%. The C:N value was lowest in 113/1B and Barjouj genotypes while a higher C:N ratio was obtained in 59/3A, N1-10, and N1-29 (Table 6).

3.5. *Effect of Irrigation Water Salinity, and Genotype on Grain Yield, Stable Isotope Composition of Carbon and Nitrogen*

The water salinity generally decreased grain yield among all the genotypes (Table 4). The ANOVA conducted for the carbon isotope data indicated that the Δ values differed among varieties ($p \leq 0.05$). Most of the varieties provided higher dry matter, and grain yield showed, in most cases, higher Δ values. There was a significant reduction in grain yield that decreased from 62.6% and 48.9% following 20 and 10 dS m^{-1} salt water irrigation, respectively, compared to the control (Table 7). In this context, harvest index (HI) values were reduced following increasing salinity level. HI (%) was decreased by 14% and 9.86% at 20 and 10 dS m^{-1} salinity, respectively, as compared to control (Table 7). Genotypes Barjouj and Alanda01 exhibit higher grain yield (3.96 and 3.87 t ha^{-1}), respectively, followed by N1-10 (2.88 t ha^{-1}), than all other genotypes. The lowest yield was produced by 59/3A (2.31 t ha^{-1}), which was 42% less than the salt tolerant genotype Alanda01 (Table 8). Genotypes Barjouj and Alanda01 exhibit higher HI (36.6%, 36.2%), followed by N1-10 (26.8%), while the lowest HI was observed in 59/3A (20.8%) (Table 8).

Table 7. Genotype and treatment effects on seed yield, harvest index, carbon and nitrogen isotope attributes of six barley genotypes grown under different water salinity levels.

Treatments	Grain Yield (t ha^{-1})	Harvest Index (%)	δ ^{13}C	Δ^{13}C	*Ci/Ca*	iWUE	δ N^{15}	Protein
Control	3.8 a	29.4 a	−25.3 a	17.8 a	0.59 a	102.3 a	3.3 c	13.6 b
7 dS m^{-1} Nacl	2.89 b	26.5 b	−24.7 a	17.1 a	0.56 a	109.4 a	4.5 b	17.5 a
14 dS m^{-1} Nacl	2.2 c	25.3 c	−25.5 a	17.9 a	0.60 a	99.9 a	4.8 a	17.8 a
Salinity Treatment (T)	**	**	ns	ns	ns	ns	**	**
Genotype (G)	**	**	ns	ns	ns	ns	**	ns
T × G interaction	**	**	ns	ns	ns	ns	**	ns

SY, Seed yield (t ha^{-1}); HI, harvest index (%); Ci/Ca, ratio of intercellular to ambient CO_2 concentration; iWUE, intrinsic water-use efficiency; δ^{13}C, stable carbon isotope composition (‰); Δ^{13}C carbon isotope discrimination (‰); SY, seed yield (t ha^{-1}); δ^{15}N, stable nitrogen isotope composition. Values in a single column sharing the same letter are not significantly different ($p \leq 0.05$) according to Tukey's honestly significant difference (HSD) test. ns, (**) are non-significant or significant at $p \leq 0.05$ or 0.001, respectively.

Table 8. Genotype and treatment effects on seed yield, harvest index, carbon and nitrogen isotope attributes of six barley genotypes grown under different water salinity levels.

Genotypes	GY (t/ha)	HI	δ^{13}C	Δ^{13}C	*Ci/Ca*	iWUE$_T$	δ^{15}N	Protein
113/1B	2.50 c	21.4 c	−26.49 a	19 a	0.64 b	88.4 d	4.6 a	19.3 a
59/3A	2.31 d	20.8 d	−25.63 b	18.1 b	0.61 b	98.4 c	4.4 a	16.3 b
N1-10	2.88 b	26.8 b	−22.73 c	15.1 c	0.47 c	131.8 b	4.4 a	16.1 b
N1-29	2.49 c	21.9 c	−21.63 d	13.9 d	0.42 d	144.5 a	4.4 a	16.3 b
Barjouj	3.87 a	36.2 a	−27.10 a	19.6 a	0.67 a	81.4 e	4.6 a	19.3 a
Alanda01	3.96 a	36.6 a	−26.45 b	18.9 b	0.64 b	88.9 d	3.13 c	16.6 b

SY, Seed yield (t ha^{-1}); HI, harvest index (%); Ci/Ca, ratio of intercellular to ambient CO_2 concentration; iWUE, intrinsic water-use efficiency; δ^{13}C, stable carbon isotope composition (‰); Δ^{13}C carbon isotope discrimination (‰); SY, seed yield (t ha^{-1}); δ^{15}N, stable nitrogen isotope composition. Values in a single column sharing the same letter are not significantly different ($p \leq 0.05$) according to Tukey's honestly significant difference (HSD) test.

The δ^{13}C was less negative (−25.5‰) and (−24.7‰) after treatment with saline water (14 and 7 dS m^{-1}) as compared to control (−28.88‰), respectively. Genotypes 113/1B and Barjouj demonstrate the highest negative δ^{13}C values (−26.49‰, −27.10‰), followed by 59/3A (−25.63‰) Alanda01 (−26.45‰), while the smallest values were obtained in N1-29 (−21.63‰) under salt stress condition. N1-29 showed the lowest negative value of δ^{13}C (−21.63‰). The carbon isotope discrimination (Δ^{13}C) values were higher in 113/1B and Barjouj (19.6‰ and 19.0‰), while the lowest Δ^{13}C values were observed in N1-29 (13.9‰), respectively. A significant difference (5.7‰) ($p > 0.05$) was observed after salinity treatment in carbon isotope discrimination (Δ^{13}C), that was in the range of 13.9–19.6‰. Genotypic differences for δ^{15}N traits were also examined for salinity treatment, proving higher in treated plants (4.5‰ and 4.8‰) than control treatments (3.3‰). There was not much difference in the barley genotypes for nitrogen isotope composition, which was in the range of 4.4–4.6‰ in most of the genotypes, while Alanda01 exhibit low δ^{15}N (3.13‰) values as compared to other genotypes. The leaf N concentration has significant G × T interaction and the $\delta^{15}N$ of tolerant genotypes was reduced to a greater extent than sensitive ones at all salinity stress, thus causing a significant G × T interaction (Table 8).

The ratio of intercellular to ambient CO_2 concentration (Ci/Ca) was significantly less (0.56 and 0.60) after treatment with 7 and 14 dS m^{-1} as compared to control (0.0.59), indicating the closing of stomata and inhibition of CO_2 (Tables 7 and 8). The maximum value of Ci/Ca was observed in genotype Barjouj (0.67), followed by Alanda01 (0.64), 113/1B (0.64), and 59/3 A (0.61), respectively (Table 8). The intrinsic water use efficiency (iWUE) values significantly increase following salinity treatment. A continuous increase in the values of iWUE was observed with increasing level of salinity. Our results revealed that iWUE was increased to 58.45%, and 37.85% at 14 and 7 dS m^{-1} NaCl treatments, respectively, as compared to non-saline condition (Table 7). The maximum values of iWUE were observed in genotype N1-29 (144.5) followed by N1-10 (131.8). The minimum iWUE value was documented in Barjouj (81.4) (Table 8).

3.6. Impact of Water Salinity on Protein Content in Barley Genotypes

There was a significant impact of saline water stress on the protein contents of barley grains. As compared to control, protein contents in barley grains were enhanced (17.5 and 17.8 mg/g DW) following exposure to both medium and higher salinity. Barley genotypes varied greatly for grain protein contents (Figure 3). GPC was highest in the genotypes 113/1B and Barjouj, ranging from 16.5 to 20.8 mg/g DW. In this regard, the highest GPC was observed in these two genotypes at higher salt stress (14 dS m^{-1}). The lowest GPC was observed in genotype Alanda01 (13.5) (Figure 3) in control treatment. GPC ranged from 14.3 to 16.1 mg/g DW, 14.1 to 17.7 mg/g DW, and 14.3 to 18.6 mg/g DW, respectively, in barley genotypes 59/3A, N1-10, and N1-29.

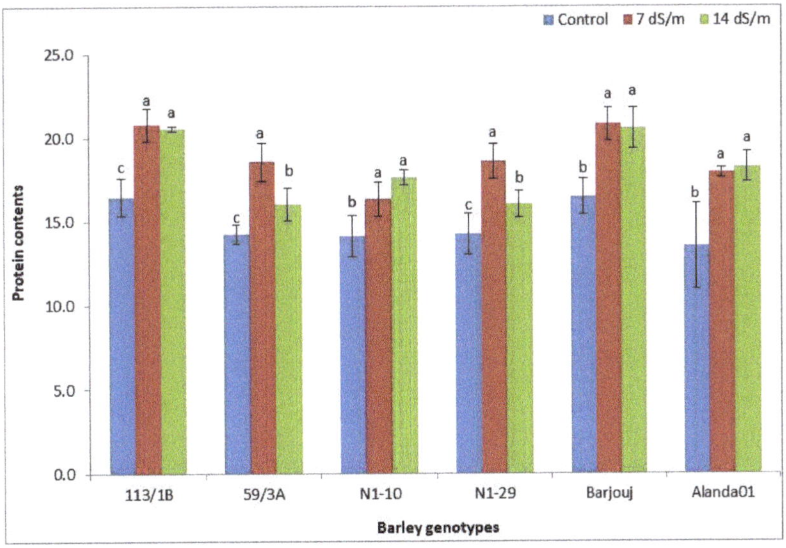

Figure 3. Changes in grain protein contents (mg g^{-1}) in 6 barley genotypes following exposure to three different salinity levels (0, 7, 14 dS m^{-1}). Each bar represents the mean (±S.E.) of three replicates. Bars with different lower case letters indicate significant difference with respect to control at $p \leq 0.05$ according to Tukey's HSD test.

3.7. Grain Yield Stability Evaluation

Barley genotypes, varied greatly for mean grain yield across the treatments (mi) (Table 9). The barley genotypes exhibited very different scores for both static environmental variance (S^2i) and dynamic Wricke's ecovalence (W^2i). The static environment variance for grain yield among the six barley genotypes ranged from 0.122 to 1.031 while Wricke's ecovalence varied from 0.101 to 1.077. In these stability analyses, the lowest values demonstrate the stability in yield over saline environments. The variety 'Barjouj' was static stable and high yielding, ranking first for S^2i grain yield index across all saline environments, and it was followed by Alandra01. The genotype 'Alandra01' showed stable mean yield (W^2i) and ranked first among all the genotypes across all environments. Moreover, variety 'Alandra01' was static stable (S^2i) and high yielding, ranking second for W^2i grain yield index (Table 9).

Table 9. Environmental variance (Si^2) and Wricke's ecovalence (Wi^2) over the saline treatment for the 6 barley genotypes with highest averaged mean yield across treatments (mi).

S.No.	Accessions Name	Collection Type		mi	Si^2	Wi^2
1	113/1B	Batini Landraces	LR	2.533	0.912	0.222
2	59/3A	Batini	LR	2.431	0.538	1.077
7	N1-10	nurseries	NS	2.458	0.542	0.154
8	N1-29	nurseries	NS	2.353	1.031	0.717
17	Barjouj	varieties	VT	3.118	0.122	0.111
18	Alanda01	varieties	VT	3.058	0.349	0.101

VT: varieties; NS: Nurseries; LR: Batini landrace.

3.8. Grain Yield Stability Evaluation

The barley varieties, nurseries, and landraces showed higher mean grain yield across the treatments (mi) (Table 9). The barley genotypes exhibited very different scores for both static environmental variance (S^2) and dynamic Wricke's ecovalence (W^2). The static environment variance for grain yield among the six barley genotypes ranged from 0.122 to 1.031 while Wricke's ecovalence varied from 0.101 to 1.077. In these stability analyses, the lowest values demonstrate the stability in yield over saline environments. The variety 'Barjouj' was static stable and high yielding, ranking 1st for S^2i grain yield index across all saline environments, and it was followed by Alandra01. The genotype 'Alandra01' showed stable mean yield (W^2i) and ranked first among all the genotypes across all environments. Moreover, variety 'Alandra01' was static stable (S^2i) and high yielding, ranking second for W^2i grain yield index (Table 9).

4. Discussion

In hyper arid, salt-degraded, and marginal environments, there are several production constraints that significantly disturb growth, productivity, and crop yield stability. Under the prevailing conditions of the UAE, there is a severe lack of freshwater resources and most of it is only available for domestic purposes and other high value issues. In this situation, the management of available natural water resources (i.e., underground low-quality saline water) and nutrient poor sandy soils, and their conversion to a sustainable production system for food and feed is a most appropriate approach to the rehabilitation of these degraded lands. Soil biochemical analysis indicates that the soil is sandy with almost no organic matter content (Table 3).

Irrigation with saline water decreased the plant dry biomass at all salinity levels, ranging from 16–31%. Meanwhile, genotype 113/1B exhibited the maximum dry biomass (116.2 t/ha) and Barjouj produced the lowest PDB (105 t ha^{-1}) (Table 4). In this context, the number of tillers m^{-2} decreased following exposure to higher salt stress and the reduction was 10–18% at 7–14 dS m^{-1} NaCl stress. According to the reports of Arif et al. [42], sodium stress is a serious global concern for sustainable agriculture that disrupts morphological, cellular, and physiological traits, affecting plant growth and development at all stages of development. Physiological traits, e.g., number of spikes m^{-2}, were also decreased at each salinity stress and the highest reduction (20%) was observed at 14 dS m^{-1} NaCl, respectively, compared to the non-saline treatment (Table 3). Barley genotypes 113/1B, 59/3A, and

Alanda01 exhibit the highest tillers m^{-2} while N1-29 exhibits the smallest tillers m^{-2}. There was significant variation in the production of the number of spikes^{-2}. The highest number of spikes m^{-2} was obtained in barley variety Alanda01 followed by 113/1B while the lowest number of spikes m^{-2} was exhibited by Barjouj, respectively (Table 4). Genotype Alanda01 exhibited the highest grain numbers plant^{-1} followed by 113/1B while the lowest grains/plant was produced by Barjouj. A similar pattern of variation was obtained for number of grains spike^{-1} in the corresponding barley genotypes.

Understanding the biochemical, morphological, and physiological response mechanisms that play a role in improving adaptation to saline water environments is limited and the development of even more salt tolerant barley cultivars is of vital importance [41–45]. This study investigated the salinity tolerance of genetically diverse barley genotypes and landraces based on agro-morphological, biochemical, physiological, and photosynthetic carbon isotope discrimination attributes in order to identify promising genotypes for salt tolerance screening. The current study showed that salt stress reduced PDB from 16% to 31% in field plots that received highly saline water (14 dS m^{-1})(Table 3). Barley genotype 113/1B showed higher dry biomass while Barjouj exhibited the lowest PDB (Table 4). Morpho-physiological traits varied among barley genotypes due to genotypic differences, differences in saline environment, and also genotype by environment interactions. It is critical to understand the scope of such variations in order to develop breeding strategies and improve selection methods.

Salinity stress can cause inhibition of the photosynthetic process and hence agricultural productivity, yield stability, and environmental sustainability. Plants' ability to become photosynthetically active in adverse saline conditions, on the other hand, is largely untapped. Salt stress has been shown to reduce barley yield by interfering with reproductive development and grain filling [46,47]. In barley, both successful seed setting and grain filling processes are critical for determining final grain yield. During the growth, reproductive, and grain filling periods, barley genotypes were exposed to salt stress (14 dS m^{-1}), with an average of number of spikes m^{-2}. However, 113/1B, 59/3A, and Alanda01 showed a greater number of tillers m^{-2} as compared to other genotypes, while genotype N1-29 displayed the lowest tillers m^{-2}. We observed a significant reduction in grains per spike and grain weight across genotypes grown under saline conditions, resulting in a reduction in grain yield of 23% on average when compared to non-saline conditions (Table 4). Genotype Alanda01 revealed highest grain numbers plant^{-1} (527.9) followed by 113/1B (508.4) while Barjouj (480.5) produced the lowest grains plant^{-1}. Meanwhile, severe salinity stress during the grain filling stage may have an impact on other yield components, such as grain filling duration and grain filling process, and hence can cause significant effects in lowering grain weight and yield in barley [46–48].

In response to salt stress, Na^+ and Cl^- levels were significantly higher in the barley leaves while the K^+/Na^+ ratio in the leaf tissues increased consistently. The K^+ content was higher with saline water, while the rest of the elements did not show any changes. The K^+ levels were consistent with K^+ availability, even under saline environment, and they could also be linked to the physiological changes seen in barley. Plant exposure to a saline environment can cause higher Na^+ absorption via roots, which leads to the development of osmotic and water stress [48–50]. In comparison to the control, increased salinity levels resulted in an increase in tissue sodium and chloride content. Under severe salt stress, the increase in tissue sodium affects cell wall integrity and cell expansion, in addition to oxidative damage [51]. In this context, Na^+ stress confines the absorption of other essential nutrient elements (K^+, Ca^{2+}, P, N) [48,52] that trigger the disturbance in the ion homeostasis, physiological, and biochemical cell activities [53].

Genotypes Brjouj and N1-29 showed the highest grain yield among the salinity treatments as compared to control and at the same time also accumulated higher K^+ levels. It was also noticed that these genotypes have a substantial amount of Na^+ in the leaf tissue that might counterbalance the toxicity effect through the accumulation of K^+ ions. Potassium concentrations varied widely, 2.6-fold, ranging from 599.4 to 639.2 mg/100 g DW.

Similar genotypic variation for salinity stress tolerance was demonstrated in barley [54]. Such genotypic variation for salt tolerance might be due to the presence of a discrepancy among physiological traits, such as photosynthetic capacity, ion uptake, and maintenance of plant water status or antioxidant potential [54]. Other researchers also demonstrated that barley exhibits tolerance to medium salinity [55,56]. Our results showed that N concentration increased after salinity treatments. Barley cultivars 113/1B and Barjouj showed highest N% and it was significantly higher than all other genotypes. The C% was higher in three barley genotype, 113/1B, N1-10, and Barjouj, respectively. Several researchers demonstrated that salt stress impedes the plant growth, physiological attributes, and yield contributing factors, such as the number of fertile tillers, grain weight, yield per square meter, and finally grain yield. The carbon metabolism, plant growth, and nutritional deficiency due to excess sodium accumulation in soil and plant tissues will lead to oxidative disorders and lower crop yield [6,7,10,57–59].

Effect of Irrigation Water Salinity, and Genotype on Grain Yield, Stable Isotope Composition of Carbon and Nitrogen

The assessment of stable isotopes of carbon and nitrogen ($\delta^{13}C$ and $\delta^{15}N$) provides a very useful parameter that can help to analyze the impact of the surrounding environment in which the plants are growing. Meanwhile, carbon isotope discrimination can provide an integrated assessment of the stomatal regulation of internal CO_2 content as well as elaborate C_3 plant species' long-term photosynthetic carbon [32,33]. Leaf growth and area development, photosynthesis, and nitrogen use are all closely related to crop yield. Salinity inhibits leaf growth, limiting grain yield and yield characteristics [60]. The current findings show that when salinity increased from 7 to 14 dS m^{-1}, grain yield fell, ranging from 24% to 42.10%. Meanwhile, Ci/Ca was much lower, indicating that the stomata had closed (Table 8). Stomatal closure can reduce CO_2 supply to carboxylation sites, lowering the activity of Ribulose-1,5-bisphosphate carboxylase oxygenase (RuBisCO), carbon synthesis, and translocation [32,33]. Higher $\Delta^{13}C$ is caused by a higher Ci/Ca ratio mainly due to a larger stomatal conductance, which can lead to a higher photosynthetic rate and hence a higher yield, i.e., positive relationship between $\Delta^{13}C$ and yield. Genotypes Barjouj and Alanda01 exhibit higher HI (36.6%, 36.2%), followed by N1-10 (26.8%), while the lowest HI was observed in 59/3A (20.8%) (Table 8). When different barley genotypes were tested for salinity tolerance, they demonstrated better Ci/Ca and yield potential, indicating their adaption to the Dubai climate. There was a substantial difference in seed yield and harvest index between different genotypes which could be attributable to genetic differences. Such variances are much more pronounced in genotypes Barjouj and Alanda01, and 59/3A which had grain yield variation of 1.6%. HI (%) was decreased by 14% and 9.86% at 20 and 10 dS m^{-1} salinity, respectively, as compared to control (Table 7). This is due to some genotypes' superior tolerance to the UAE's agro-climatic conditions. Genotypes Alanda01 and Barjouj had the maximum photosynthetic CO_2 rate (Ci/Ca), yield, and productivity and were the most suited and well-adapted genotypes for the Dubai marginal soil environment. N1-10 and N-29 had the lowest rates (37% and 30% lower Ci/Ca than Barjouj), indicating that they were the least adapted. The Ci/Ca ratio of intercellular to ambient CO_2 concentrations did not differ significantly between the remaining genotypes (113/1B, 59/3A, Barjouj, Alanda01).

Although variation in plant N isotopic composition (^{15}N) does not offer a measure of NUE, it can be used to follow N mobility and infer N sources and/or N cycle dynamics in vegetation at the local, community, and landscape scales. The diffusion gradient for CO_2 into the leaf through the stomata is linked to both the efficiency of water usage (carbon (C) fixed per unit water transpired) and the efficiency of N use (C fixed per unit N absorbed). Plants need the majority of their water to support photosynthesis through transpiration. Photosynthesis accounts for more than half of total leaf N [61], and total leaf N content and photosynthetic capability are frequently associated [62]. If the CO_2 diffusion gradient steepens, reductions in stomatal conductance (gs) or higher investments in foliar N can result in higher water-use efficiency (WUE), while lower intercellular CO_2 concentrations can diminish N-use efficiency (NUE) by reducing rates of C fixation per cell. For salinity treatment, phenotypic differences for ^{15}N characteristics were also investigated, and they were found to be larger in treated plants (4.5 and 4.8) than in control treatments (3.3). In terms of nitrogen isotope composition, most genotypes were in the range of 4.4–4.6, while Alanda01 had low ^{15}N (3.13) values when compared to other genotypes. The leaf N concentration has a substantial G x T interaction, and tolerant genotypes' ^{15}N was lowered to a greater extent than sensitive genotypes under all salinity stress conditions, resulting in a significant GxT interaction (Table 8). Carbon isotope discrimination (Δ^{13}C), the difference in ^{13}C/^{12}C composition between plant C and environmental CO_2, has frequently been used to estimate WUE. Previous studies have demonstrated negative correlations between Δ^{13}C and WUE under a CO_2 in various species, such as barley, cowpea, and wheat [63–66]. Following salinity treatment, the intrinsic water use efficiency (iWUE) values dramatically rise. The genotype N1-29 exhibited highest iWUE values, followed by N1-10, while Barjouj demonstrated the lowest iWUE values (Table 8).

5. Conclusions

In conclusion, we found that barley genotypes exhibited wide genetic variability at various salinity levels tested under UAE desert conditions. We did not find this surprising as the genetic diversity of barley might occur because of large variation among climate and seasonal characteristics, cultivation history, and intensity of selection pressure. These genotypes can be profitable in marginal areas using low quality saline ground water and, through genotypic/phenotypic trials, can be utilized for the growth and production of barley and for the rehabilitation of UAE marginal lands. Most of the barley genotypes that exhibited higher grain yield showed high Δ^{13}C values. Furthermore, stress tolerance indices, static yield stability index, dynamic yield stability index, and physiological characteristics (selective uptake and transport of Na^+ and K^+ and plant vigour) helped us in the assessment of salinity tolerance and comparison of yield from different barley genotypes that will further elucidate adaptation strategies for salt-degraded and marginal lands. Furthermore, the dynamics of this study demonstrated no risk of salt accumulation in these sandy soils of Dubai, UAE, suggesting the sustainability of barley production when irrigated with saline water. Therefore, further investigation is required to certify the genetic variability and adaptive mechanisms of barley for enhancing salt tolerance and crop productivity.

Author Contributions: M.I.H.: conceptualization, methodology, data curation, writing—review & editing. Z.I.K.: methodology, first draft. D.A.A.F.: methodology, first draft, Ms revision. M.S.E.: software, validation, visualization, Ms revision, statistical analysis, T.H.F., investigation, biochemical analysis of soils/plants, Ms revision, reference management. All authors have read and agreed to the published version of the manuscript.

Funding: The authors extend their appreciation to the Researchers Supporting Project number (RSP-2021/190), King Saud University, Riyadh, Saudi Arabia.

Institutional Review Board Statement: Ethics approval and consent to participate Not applicable.

Informed Consent Statement: Not applicable.

Data Availability Statement: All data generated or analyzed during this study are included in this published article.

Acknowledgments: The authors extend their appreciation to the Researchers Supporting Project number (RSP-2021/190), King Saud University, Riyadh, Saudi Arabia.

Conflicts of Interest: The authors declare no competing interests.

References

1. UN. World Population Prospects: The 2015 Revision. 2015. Available online: https://esa.un.org/unpd/wpp (accessed on 21 February 2022).
2. Flowers, T.J.; Colmer, T.D. Salinity tolerance in halophytes. *New Phytol.* **2008**, *179*, 945–963. [CrossRef] [PubMed]
3. Saade, S.; Brien, C.; Pailles, Y.; Berger, B.; Shahid, M.; Russell, J.; Waugh, R.; Negrão, S.; Tester, M. Dissecting new genetic components of salinity tolerance in two-row spring barley at the vegetative and reproductive stages. *PLoS ONE* **2020**, *15*, e0236037. [CrossRef] [PubMed]
4. Shabala, S. Learning from halophytes: Physiological basis and strategies to improve abiotic stress tolerance in crops. *Ann. Bot.* **2013**, *112*, 1209–1221. [CrossRef]
5. Khalil, K.; Khan, Q.; Mohamed, M. Selection criteria of best sites for aquifer storage and recovery in the Eastern District of Abu Dhabi, United Arab Emirates. *Groundw. Sustain. Dev.* **2022**, *18*, 100771. [CrossRef]
6. Hussain, M.I.; Lyra, D.A.; Farooq, M.; Nikoloudakis, N.; Ahmad, N. Salt and drought stresses in safflower: A Review. *Agron. Sust. Develop.* **2016**, *36*, 4. [CrossRef]
7. Hussain, M.I.; Muscolo, A.; Farooq, M.; Ahmad, W. Sustainable use and management of non-conventional water resources for rehabilitation of marginal lands in arid and semiarid environments. *Agric. Water Manag.* **2019**, *221*, 462–476. [CrossRef]
8. Munns, R.; Tester, M. Mechanisms of salinity tolerance. *Ann. Rev. Plant Physiol.* **2008**, *59*, 651–681. [CrossRef]
9. Munns, R.; James, R.A.; Xu, B.; Athman, A.; Conn, S.J.; Jordans, C.; Byrt, C.S.; Hare, R.A.; Tyerman, S.D.; Tester, M.; et al. Wheat grain yield on saline soils is improved by an ancestral Na^+ transporter gene. *Nat. Biotech.* **2012**, *30*, 360–364. [CrossRef]
10. Hussain, M.I.; Al-Dakheel, A.J. Effect of salinity stress on phenotypic plasticity, yield stability and signature of stable isotopes of carbon and nitrogen in Safflower. *Environ. Sci. Pollut. Res.* **2018**, *25*, 23685–23694. [CrossRef]
11. FAO. AQUASTAT Data. Available online: http://www.fao.org/nr/water/aquastat/data/query/index.html?lang=en (accessed on 27 April 2017).
12. Vaezi, B.; Pour-Aboughadareh, A.; Mehraban, A.; Hossein-Pour, T.; Mohammadi, R.; Armion, M.; Dorri, M. The use of parametric and non-parametric measures for selecting stable and adapted barley lines. *Arch. Agron. Soil Sci.* **2018**, *64*, 597–611. [CrossRef]
13. Chloupek, O.; Dostál, V.; Středa, T.; Psota, V.; Dvořáčková, O. Drought tolerance of barley varieties in relation to their root system size. *Plant Breed.* **2010**, *129*, 630–636. [CrossRef]
14. Kebede, A.; Kang, M.S.; Bekele, E. Advances in mechanisms of drought tolerance in crops, with emphasis on barley. *Advan. Agron.* **2019**, *156*, 265–314.
15. Sayed, M.A.; Nassar, S.M.; Moustafa, E.S.; Said, M.T.; Börner, A.; Hamada, A. Genetic mapping reveals novel exotic and elite QTL alleles for salinity tolerance in barley. *Agronomy* **2021**, *11*, 1774. [CrossRef]
16. wa Lwalaba, J.L.; Zvobgo, G.; Gai, Y.; Issaka, J.H.; Mwamba, T.M.; Louis, L.T.; Fu, L.; Nazir, M.M.; Kirika, B.A.; Tshibangu, A.K.; et al. Transcriptome analysis reveals the tolerant mechanisms to cobalt and copper in barley. *Ecotoxicol. Environ. Saf.* **2021**, *209*, 111761. [CrossRef] [PubMed]
17. Rajendran, K.; Tester, M.; Roy, S.J. Quantifying the three main components of salinity tolerance in cereals. *Plant Cell Environ.* **2009**, *32*, 237–249. [CrossRef] [PubMed]
18. El-hendawy, S.E.; Hu, Y.; Schmidhalter, U. Growth, ion content, gas exchange, and water relations of wheat genotypes differing in salt tolerances. *Aust. J. Agric. Res.* **2005**, *56*, 123–134. [CrossRef]
19. Chen, Z.; Pottosin, I.I.; Cuin, T.A.; Fuglsang, A.T.; Tester, M.; Jha, D.; Zepeda-Jazo, I.; Zhou, M.; Palmgren, M.G.; Newman, I.A.; et al. Root plasma membrane transporters controlling K+/Na+ homeostasis in salt-stressed barley. *Plant Physiol.* **2007**, *145*, 1714–1725. [CrossRef]
20. James, R.A.; Munns, R.; Von Caemmerer, S.; Trejo, C.; Miller, C.; Condon, A.G. Photosynthetic capacity is related to the cellular and subcellular partitioning of Na^+, K^+ and Cl^- in salt affected barley and durum wheat. *Plant Cell Environ.* **2006**, *29*, 2185–2197. [CrossRef]
21. Ebrahim, F.; Arzani, A.; Rahimmalek, M.; Sun, D.; Peng, J. Salinity tolerance of wild barley *Hordeum vulgare* ssp. spontaneum. *Plant Breed.* **2020**, *139*, 304–316. [CrossRef]
22. Poustini, K.; Siosemardeh, A. Ion distribution in wheat cultivars in response to salinity stress. *Field Crops Res.* **2004**, *85*, 125–133. [CrossRef]
23. Rozema, J.; Schat, H. Salt tolerance of halophytes, research questions reviewed in the perspective of saline agriculture. *Environ. Exp. Bot.* **2013**, *92*, 83–95. [CrossRef]
24. Wu, D.; Cai, S.; Chen, M.; Ye, L.; Chen, Z.; Zhang, H.; Dai, F.; Wu, F.; Zhang, G. Tissue metabolic responses to salt stress in wild and cultivated barley. *PLoS ONE* **2013**, *8*, e55431. [CrossRef] [PubMed]
25. Rasouli, F.; Pouya, A.K.; Simunek, J. Modeling the effects of saline water use in wheat-cultivated lands using the UNSATCHEM model. *Irrig. Sci.* **2013**, *31*, 1009–1024. [CrossRef]
26. Liang, W.; Ma, X.; Wan, P.; Liu, L. Plant salt-tolerance mechanism: A review. *Biochem. Biophys. Res. Commun.* **2018**, *495*, 286–291. [CrossRef]
27. Jaradat, A.A.; Shahid, M.; Al-Maskri, A.Y. Genetic diversity in the Batini barley landrace from Oman: I. Spike and grain quantitative and qualitative traits. *Crop Sci.* **2004**, *44*, 304–315. [CrossRef]
28. Jaradat, A.A. Saline agriculture in the Arabian Peninsula: Management of marginal lands and saline water resources. *J. Food Agric. Environ.* **2005**, *3*, 302–306.

29. Shahin, S.M.; Salem, M.A. The challenges of water scarcity and the future of food security in the United Arab Emirates (UAE). *Nat. Res. Conserv.* **2015**, *3*, 1–6. [CrossRef]
30. Al-Dakheel, A.J.; Fraj, M.B.; Shabbir, G.M.; Al Gailani, A.Q.M. Evaluation of Batini barley landraces from Oman and breeding lines under various irrigation salinity levels. *Agric. Sci. Res. J.* **2012**, *2*, 42–50.
31. Hussain, M.I.; Reigosa, M.J. Higher peroxidase activity, leaf nutrient contents and carbon isotope composition changes in *Arabidopsis thaliana* are related to rutin stress. *J. Plant Physiol.* **2014**, *171*, 1325–1333. [CrossRef]
32. Farquhar, G.D.; Ehleringer, J.R.; Hubick, K.T. Carbon isotope discrimination and photosynthesis. *Ann. Rev. Plant Physiol.* **1989**, *40*, 503–537. [CrossRef]
33. Farquhar, G.D.; Richards, R.A. Isotope composition of plant carbon correlates with water use efficiency of wheat genotypes. *Aust. J. Plant Physiol.* **1984**, *11*, 539–552.
34. Hussain, M.I.; Reigosa, M.J. Seedling growth, leaf water status and signature of stable carbon isotopes in C_3 perennials exposed to natural phytochemicals. *Aust. J. Bot.* **2012**, *60*, 676–684. [CrossRef]
35. Hussain, M.I.; El-Keblawy, A.; Mitterand Tsombou, F. Leaf age, canopy position, and habitat affect the carbon isotope discrimination and water-use efficiency in three C_3 leguminous prosopis species from a hyper-arid climate. *Plants* **2019**, *8*, 402. [CrossRef] [PubMed]
36. Hussain, M.I.; Reigosa, M.J. Evaluation of photosynthetic performance and carbon isotope discrimination in perennial ryegrass (*Lolium perenne* L.) under allelochemicals stress. *Ecotoxicology* **2017**, *26*, 613–624. [CrossRef]
37. Bradford, M.M. A rapid sensitive method for the quantification of microgram quantities of protein utilizing the principle of protein dye binding. *Ann. Biochem.* **1976**, *72*, 248–254. [CrossRef]
38. Hussain, M.I.; Reigosa, M.J. Allelochemical stress inhibits growth, leaf water relations, PSII photochemistry, non-photochemical fluorescence quenching and heat energy dissipation in three C_3 perennial species. *J. Exp. Bot.* **2011**, *62*, 4533–4545. [CrossRef]
39. Roemer, T. Sind die ertragsreichen Sorten ertragssicherer? Mitteilung Deutsche Landwirtschafts-Gesellschaft. *DGL-Mitt.* **1917**, *32*, 87–89.
40. Wricke, G. Uber eine Methode zur Erfassung der okologischen Streubreite in Feldversuchen. *Z. Pflanzenzuchtg.* **1962**, *47*, 92–96.
41. Hussain, M.I.; Al-Dakheel, A.J.; Reigosa, M.J. Genotypic differences in agro-physiological, biochemical and isotopic responses to salinity stress in quinoa (*Chenopodium quinoa* Willd.) plants: Prospects for salinity tolerance and yield stability. *Plant Physiol. Biochem.* **2018**, *129*, 411–420. [CrossRef]
42. Arif, Y.; Singh, P.; Siddiqui, H.; Bajguz, A.; Hayat, S. Salinity induced physiological and biochemical changes in plants: An omic approach towards salt stress tolerance. *Plant Physiol. Biochem.* **2020**, *156*, 64–77. [CrossRef]
43. Rasel, M.; Tahjib-Ul-Arif, M.; Hossain, M.A.; Hassan, L.; Farzana, S.; Brestic, M. Screening of salt-tolerant rice landraces by seedling stage phenotyping and dissecting biochemical determinants of tolerance mechanism. *J. Plant Growth Regul.* **2021**, *40*, 1853–1868. [CrossRef]
44. Khalil, S.R.; Ashoub, A.; Hussein, B.A.; Brüggemann, W.; Hussein, E.H.; Tawfik, M.S. Physiological and molecular evaluation of ten Egyptian barley cultivars under salt stress conditions. *J. Crop Sci. Biotech.* **2022**, *25*, 91–101. [CrossRef]
45. Hussain, M.I.; Al-Dakheel, A.J. Using alternate water resources for cultivation of salt tolerant perennial grasses under marginal environment. In Proceedings of the TROPENTAG, Management of Land Use Systems for Enhanced Food Security-Conflicts, Controversies and Resolutions, Berlin, Germany, 16–18 September 2015.
46. Djanaguiraman, M.; Prasad, P.V.V.; Stewart, Z.P.; Perumal, R.; Min, D.; Djalovic, I.; Ciampitti, I.A. Agroclimatology of Oats, Barley, and Minor Millets. *Agroclimatol. Link. Agric. Clim.* **2020**, *60*, 243–277.
47. Vasilakoglou, I.; Dhima, K.; Giannakoula, A.; Dordas, C.; Skiada, V.; Papadopoulou, K. Carbon assimilation, isotope discrimination, proline and lipid peroxidation contribution to barley (*Hordeum vulgare*) salinity tolerance. *Plants* **2021**, *10*, 299. [CrossRef]
48. Torun, H.; Novák, O.; Mikulík, J.; Strnad, M.; Ayaz, F.A. The Effects of exogenous salicylic acid on endogenous phytohormone status in *Hordeum vulgare* L. under salt stress. *Plants* **2022**, *11*, 618. [CrossRef]
49. Hussain, M.I.; Farooq, M.; Muscolo, A.; Rehman, A. Crop diversification and saline water irrigation as potential strategies to save freshwater resources and reclamation of marginal soils—A review. *Environ. Sci. Pollut. Res.* **2020**, *27*, 28695–28729. [CrossRef]
50. Yu, H.; Chen, F.; Ma, J.; Khan, Z.I.; Hussain, M.I.; Javaid, I.; Ahmad, K.; Nazar, S.; Akhtar, S.; Ejaz, A.; et al. Comparative evaluation of groundwater, wastewater and canal water for irrigation on toxic metal accumulation in soil and vegetable: Pollution load and health risk assessment. *Agric. Water Manag.* **2022**, *264*, 107515. [CrossRef]
51. Sellami, S.; Le Hir, R.; Thorpe, M.R.; Vilaine, F.; Wolff, N.; Brini, F.; Dinant, S. Salinity effects on sugar homeostasis and vascular anatomy in the stem of the *Arabidopsis thaliana* inflorescence. *Int. J. Mol. Sci.* **2019**, *20*, 3167. [CrossRef]
52. Shahzad, B.; Rehman, A.; Tanveer, M.; Wang, L.; Park, S.K.; Ali, A. Salt stress in brassica: Effects, tolerance mechanisms, and management. *J. Plant Growth Regul.* **2022**, *41*, 781–795. [CrossRef]
53. Altaf, M.A.; Shahid, R.; Ren, M.X.; Altaf, M.M.; Khan, L.U.; Shahid, S.; Jahan, M.S. Melatonin alleviates salt damage in tomato seedling: A root architecture system, photosynthetic capacity, ion homeostasis, and antioxidant enzymes analysis. *Sci. Hortic.* **2021**, *285*, 110145. [CrossRef]
54. Thabet, S.G.; Alomari, D.Z.; Alqudah, A.M. Exploring natural diversity reveals alleles to enhance antioxidant system in barley under salt stress. *Plant Physiol. Biochem.* **2021**, *166*, 789–798. [CrossRef] [PubMed]

55. Akhter, M.S.; Noreen, S.; Mahmood, S.; Ashraf, M.; Alsahli, A.A.; Ahmad, P. Influence of salinity stress on PSII in barley (*Hordeum vulgare* L.) genotypes, probed by chlorophyll-a fluorescence. *J. King Saud Univ. Sci.* **2021**, *33*, 101239. [CrossRef]
56. Sabagh, A.E.; Hossain, A.; Islam, M.S.; Barutcular, C.; Hussain, S.; Hasanuzzaman, M.; Akram, T.; Mubeen, M.; Nasim, W.; Fahad, S.; et al. Drought and salinity stresses in barley: Consequences and mitigation strategies. *Aust. J. Crop Sci.* **2019**, *13*, 810–820. [CrossRef]
57. Javed, S.A.; Arif, M.S.; Shahzad, S.M.; Ashraf, M.; Kausar, R.; Farooq, T.H.; Hussain, M.I.; Shakoor, A. Can different salt formulations revert the depressing effect of salinity on maize by modulating plant biochemical attributes and activating stress regulators through improved N supply? *Sustainability* **2021**, *13*, 8022. [CrossRef]
58. Derakhshani, B.; Jafary, H.; Maleki Zanjani, B.; Hasanpur, K.; Mishina, K.; Tanaka, T.; Kawahara, Y.; Oono, Y. Combined QTL mapping and RNA-Seq profiling reveals candidate genes associated with cadmium tolerance in barley. *PLoS ONE* **2020**, *15*, e0230820. [CrossRef]
59. Derakhshani, Z.; Bhave, M.; Shah, R.M. Metabolic contribution to salinity stress response in grains of two barley cultivars with contrasting salt tolerance. *Environ. Exp. Bot.* **2020**, *179*, 104229. [CrossRef]
60. Taleisnik, E.; Rodriguez, A.; Bustos, D.; Erdei, L.; Ortega, L.; Senn, M.E. Leaf expansion in grasses under salt stress. *J. Plant Physiol.* **2009**, *166*, 1123–1140. [CrossRef]
61. Evans, J.R. Photosynthesis and nitrogen relationships in leaves of C_3 plants. *Oecologia* **1989**, *78*, 9–19. [CrossRef]
62. Liu, J.; Zhang, J.; Estavillo, G.M.; Luo, T.; Hu, L. Leaf N content regulates the speed of photosynthetic induction under fluctuating light among canola genotypes (*Brassica napus* L.). *Physiol. Plant.* **2021**, *172*, 1844–1852. [CrossRef]
63. Hubick, K.; Farquhar, G. Carbon isotope discrimination and the ratio of carbon gained to water lost in barley cultivars. *Plant Cell Environ.* **1989**, *12*, 795–804. [CrossRef]
64. Barbour, M.M.; Warren, C.R.; Farquhar, G.D.; Forrester, G.U.Y.; Brown, H. Variability in mesophyll conductance between barley genotypes, and effects on transpiration efficiency and carbon isotope discrimination. *Plant Cell Environ.* **2010**, *33*, 1176–1185. [CrossRef] [PubMed]
65. Ismail, A.M.; Hall, A. Correlation between water-use efficiency and carbon isotope discrimination in diverse cowpea genotypes and isogenic lines. *Crop Sci.* **1992**, *32*, 7–12. [CrossRef]
66. Bort, J.; Belhaj, M.; Latiri, K.; Kehel, Z.; Araus, J.L. Comparative performance of the stable isotope signatures of carbon, nitrogen and oxygen in assessing early vigour and grain yield in durum wheat. *J. Agric. Sci.* **2014**, *152*, 408–426. [CrossRef]

plants MDPI

Correction

Correction: Alnusairi et al. Exogenous Nitric Oxide Reinforces Photosynthetic Efficiency, Osmolyte, Mineral Uptake, Antioxidant, Expression of Stress-Responsive Genes and Ameliorates the Effects of Salinity Stress in Wheat. *Plants* 2021, *10*, 1693

Ghalia S. H. Alnusairi [1], Yasser S. A. Mazrou [2,3], Sameer H. Qari [4], Amr A. Elkelish [5], Mona H. Soliman [6,*], Mohamed Eweis [6], Khaled Abdelaal [7], Gomaa Abd El-Samad [8], Mohamed F. M. Ibrahim [9] and Nihal ElNahhas [10]

1 Department of Biology, College of Science, Jouf University, Sakaka 72388, Saudi Arabia; gshalnusairi@ju.edu.sa
2 Business Administration Department, Community College, King Khalid University, Guraiger, Abha 62529, Saudi Arabia; ymazrou@kku.edu.sa
3 Faculty of Agriculture, Tanta University, Tanta 31512, Egypt
4 Biology Department, Al-Jumum University College, Umm Al-Qura University, Mecca 21955, Saudi Arabia; shqari@uqu.edu.sa
5 Botany Department, Faculty of Science, Suez Canal University Ismailia, Ismailia 41522, Egypt; amr.elkelish@science.suez.edu.eg
6 Botany and Microbiology Department, Faculty of Science, Cairo University, Giza 12613, Egypt; amradel807080@googelmail.com
7 Plant Pathology and Biotechnology Laboratory, Excellence Center (EPCRS), Faculty of Agriculture, Kafrelsheikh University, Kafr Elsheikh 33516, Egypt; khaled.elhaies@gmail.com
8 Department of Agronomy, Faculty of Agriculture, Ain Shams University, Cairo 11566, Egypt; Gomaa_abdelsamad@agr.asu.edu.eg
9 Department of Agricultural Botany, Faculty of Agriculture, Ain Shams University, Cairo 11566, Egypt; Ibrahim_mfm@agr.asu.edu.eg
10 Department of Botany and Microbiology, Faculty of Science, Alexandria University, Alexandria 21526, Egypt; nihal.elnahhas@alexu.edu.eg
* Correspondence: monahsh1@gmail.com; Tel.: +20-1005-145-454

Citation: Alnusairi, G.S.H.; Mazrou, Y.S.A.; Qari, S.H.; Elkelish, A.A.; Soliman, M.H.; Eweis, M.; Abdelaal, K.; El-Samad, G.A.; Ibrahim, M.F.M.; ElNahhas, N. Correction: Alnusairi et al. Exogenous Nitric Oxide Reinforces Photosynthetic Efficiency, Osmolyte, Mineral Uptake, Antioxidant, Expression of Stress-Responsive Genes and Ameliorates the Effects of Salinity Stress in Wheat. *Plants* 2021, *10*, 1693. *Plants* **2022**, *11*, 576. https://doi.org/10.3390/plants11050576

Received: 20 December 2021
Accepted: 17 January 2022
Published: 22 February 2022

In the original publication [1], the acknowledgements section was not included. The acknowledgements are hereby published as follows:

Acknowledgments: The authors strongly acknowledge the Deanship of Scientific Research at King Khalid University for funding this work through the Program of Research Groups under grant number (RGP 01/258/42). Y.S.A.M. and K.A. would like to extend their appreciation to the Graduate Studies and Research Affairs Sector of Kafrelsheikh University and Tanta University, Egypt. We also thank all staff members of our laboratories for their helpful discussions and comments.

The original publication has also been updated.

Reference

1. Alnusairi, G.S.H.; Mazrou, Y.S.A.; Qari, S.H.; Elkelish, A.A.; Soliman, M.H.; Eweis, M.; Abdelaal, K.; El-Samad, G.A.; Ibrahim, M.F.M.; ElNahhas, N. Exogenous Nitric Oxide Reinforces Photosynthetic Efficiency, Osmolyte, Mineral Uptake, Antioxidant, Expression of Stress-Responsive Genes and Ameliorates the Effects of Salinity Stress in Wheat. *Plants* **2021**, *10*, 1693. [CrossRef] [PubMed]

plants MDPI

Article

Variations in Morphological Characters and Antioxidant Potential of Different Plant Parts of Four *Ziziphus* Mill. Species from the Cholistan

Muhammad Umair Riaz [1], Muhammad Ali Raza [2], Amjad Saeed [1], Mukhtar Ahmed [3,*] and Tanveer Hussain [1,*]

[1] Department of Forestry, Range and Wildlife Management, Faculty of Agriculture and Environment, The Islamia University of Bahawalpur, Bahawalpur 63100, Pakistan; cumair84@yahoo.com (M.U.R.); amjadsaeed101@gmail.com (A.S.)
[2] National Research Center of Intercropping, The Islamia University of Bahawalpur, Bahawalpur 63100, Pakistan; razaali0784@yahoo.com
[3] Department of Agronomy, PMAS-Arid Agriculture University, Rawalpindi 46300, Pakistan
* Correspondence: ahmadmukhtar@uaar.edu.pk (M.A.); dr.tanveer@iub.edu.pk (T.H.)

Abstract: Genus *Ziziphus* (Z.) contains various important species in tropical and subtropical regions that are globally famous for their food and medicinal uses. However, no comprehensive study was available on the morphology and phytochemistry of *Ziziphus* species, mainly under different growth conditions, i.e., irrigated and desert (Cholistan). Therefore, this study was carried out to evaluate the morphological and phytochemical characteristics of *Ziziphus* species, i.e., *Z. jujuba*, *Z. mauritiana*, *Z. spina-christi*, and *Z. nummularia*, found in the irrigated and desert conditions. Our results revealed significant variations for most of the measured parameters, showing a large-scale diversity among *Ziziphus* species under irrigated and desert conditions. Specifically, *Ziziphus* species showed better morphology of all measured parameters of leaves and fruits under irrigated conditions compared to desert conditions, indicating that the optimum water availability in irrigated conditions improved the morphological parameters of Z. species. Meanwhile, among all *Ziziphus* species, the maximum leaf length (7.4 cm), leaf width (4.1 cm), leaf area (30.6 cm^2), and leaf petiole length (1.3 cm) were observed for *Z. jujuba*, and the highest leaf dry weight (55.4%) was recorded for *Z. mauritiana*. Similarly, the highest fruit length (3.9 cm), fruit stalk length (1.5 cm), fruit diameter (3.6 cm), fruit width (3.8 cm), fruit area (66.1 cm^2), seed length (2 cm), and seed diameter (1.1 cm) were measured for species *Z. jujuba*, while the maximum fruit dry weight (49.9%) and seed width (1.4 cm) were recorded for species *Z. nummularia*. Interestingly, compared to irrigated conditions, higher values of bioactive contents, i.e., phenol, flavonoid, and antioxidant activity, in fruits and leaves of *Ziziphus* species under desert conditions indicated the positive impact of desert climate on the phytochemistry of the Z. plants. Among *Ziziphus* species, *Z. nummularia* accumulated the maximum fruit phenols (304.4 mg GAE/100 g), leaf phenols (314.2 mg GAE/100 g), fruit flavonoids (123.7 mg QE/100 g), and leaf flavonoids (113.4 mg QE/100 g). Overall, this study demonstrated the significant morphological and phytochemical variations of the *Ziziphus* species under irrigated and desert conditions, which could be utilized for future studies to improve the production and medicinal potential of the *Ziziphus*, especially in desert areas.

Keywords: bioactive; desert; irrigated; flavonoid; phenol; phytochemistry

Citation: Riaz, M.U.; Raza, M.A.; Saeed, A.; Ahmed, M.; Hussain, T. Variations in Morphological Characters and Antioxidant Potential of Different Plant Parts of Four *Ziziphus* Mill. Species from the Cholistan. *Plants* **2021**, *10*, 2734. https://doi.org/10.3390/plants10122734

Academic Editor: Biancaelena Maserti

Received: 11 November 2021
Accepted: 9 December 2021
Published: 12 December 2021

Publisher's Note: MDPI stays neutral with regard to jurisdictional claims in published maps and institutional affiliations.

1. Introduction

Genus *Ziziphus* Mill., commonly called Ber, comprises deciduous or evergreen trees and shrubs, widely distributed in the tropical and sub-tropical areas of the world [1]. *Ziziphus* belongs to the family *Rhamnaceae*, which contains 100 species in the Old-world (i.e., African, European, and Asian) and New-world (i.e., America). *Ziziphus* species are pantropical and paraphyletic and generally split into three different geographical lineages [2].

Recent studies revealed that these lineages are categorized into genus *Condaliopsis* and *Sarcomphalus*, regarded as new world *Ziziphus* and genus *Ziziphus*, *considered* old-world *Ziziphus* [3,4]. *Ziziphus* species are reported as a good food source, especially for those who live in desert areas, because it fulfills nutritional needs and maintains health and food security and economic welfare [5]. In addition, these species contain a high content of several minerals (i.e., vitamin C, phosphorus, calcium, and iron), amino acids (i.e., asparagine, glutamic acid, arginine, aspartic acid, serine, glycine, and threonine), and sugars (i.e., sucrose, fructose, glucose, and starch) [6]. Furthermore, local communities eat *Ziziphus* fruit (Ber) directly in fresh or dry form. Numerous companies use its fruit to make butter, pastes, and flour [7]. Its leaves are the source of vegetables and medicine for treating various diseases (i.e., diabetes, asthma, and depression). It is also a great source of fodder for livestock due to its high nutritional value [1]. Notably, *Ziziphus* species are drought-tolerant plants usually found as natural vegetation in the deserts [8] and exist in pastures, coastal, and wet mountain regions [9]. These species comprise diverse plants from shrubs to small or medium-sized trees, having an extensive range of canopy morphology such as spreading, semi-erect, and erect, and its height ranges from 4 to 16 m [10]. Z. leaves, fruits, and flowers have great diversity in shape, color, and taste [11–13].

Environmental factors, i.e., temperature, light, and rainfall, affect plant growth and development in several ways [14,15]. However, trees can cope with changing environmental conditions by alternating their organs and tissues [16]. In this perspective, due to *Ziziphus* species' drought-tolerant and heat resistant characteristics [17], they can grow well under stress conditions, i.e., high temperature and low rainfall, thus, are responsible for diverse morphological patterns to mitigate these effects [18]. Such interspecific patterns between plant traits and climatic factors were previously reported by many scientists [19–21], which strongly correlate mean annual rainfall, nutrient availability, and light intensity with leaf shape, leaf area, and leaf size. However, there is no study available in the current literature on determining the morphological characteristics of *Ziziphus* plants under different growing conditions, especially in Cholistan (high-temperature region, where total rainfall is less than total evapotranspiration). Morphological characteristics are the most vital determining indices for better characterization, taxonomic classification, and agronomic value of the plants while examining the genetic diversity. Notably, the information about the phenotypic diversity of *Ziziphus*, particularly the native desert species (e.g., *Ziziphus jujuba*, *Ziziphus mauritiana*, *Ziziphus spina-christi*, and *Ziziphus nummularia*) is minimal. Therefore, a comprehensive study is needed to determine the morphological plasticity of each component of the *Ziziphus* species present in the desert. Such information is also crucial for the commercial production of Beri with better yield, lower inputs, and improved nutrition [2,22]. In addition, this type of study could also enhance our understanding of adaptations and responses of morphological traits in different desert environments, which could help us better utilize the desert lands under changing climatic conditions.

Ziziphus species are mainly used as traditional medicine and possess important phytochemicals, i.e., flavonoids, phenols, triterpenoids, and alkaloids [23–26], which play a promising role in the treatments of many diseases, i.e., cancer, ulcer, and inflammation [27,28], due to their therapeutic prospective, i.e., antioxidant properties [29,30]. Previously, scientists reported that the antioxidant compound in *Ziziphus* species can inhibit or delay the oxidative damage caused by reactive oxygen species in humans and animals, thus reducing several ailments, i.e., inflammation and aging [31–33]. Similarly, flavonoids and phenols are natural antioxidants, which possess a broad range of multiple biological activities and unique structures, playing important roles in minimizing the risks of oxidative stress-based complaints and infectious ailments [34–36]. These findings conclude that *Ziziphus* species are a rich source of phytochemicals for humans and animals. However, how desert conditions affect the quantity of phytochemicals in *Ziziphus* species is rarely investigated. Therefore, we hypothesized that desert conditions could affect the phytochemistry of *Ziziphus* species, ultimately influencing the therapeutic prospective of these species. A morphological characterization is an important tool in identifying species and determining phenotypic diversity among various plant species. The present research was designed to investigate the variations of (i) morphological characters (i.e., qualitative and quantitative), (ii) bioactive contents (i.e., phenols, flavonoids), and antioxidant potential of different parts of *Ziziphus* species (*Z. jujuba*, *Z. mauritiana*, *Z. spina-christi*, and *Z. nummularia*) under different growth conditions, i.e., irrigated and desert (Cholistan). In addition, these types of phytochemical investigation of *Ziziphus* species can help to quantify the different bioactive compounds and their potential use as medicine and will reduce the pressure on other medicinal plant species, which are being used in traditional medicines, thus, providing an alternative against different ailments.

2. Materials and Methods

2.1. Characteristics of the Study Area

This study was conducted for one year in the Cholistan desert with a hot and arid climate, extended over a vast area of 2600 km^2, adjacent to the district Bahawalpur, Southern Punjab, Pakistan (Figure 1). Cholistan desert lies between 75°24′ E and 27°42′ N [37]. The average rainfall of this region ranges from 100 to 250 mm, falling mainly during the monsoon season. Mean summer and winter temperature ranges from 34 to 38 °C and 14 to 16 °C, respectively. During summer (May–July), the maximum temperature reaches 52 °C while the minimum temperature during the winter falls below zero (Metrology Department, Government of Punjab, Pakistan). The soil of the Cholistan desert is a mixture of sandy silt and sandy clay [38].

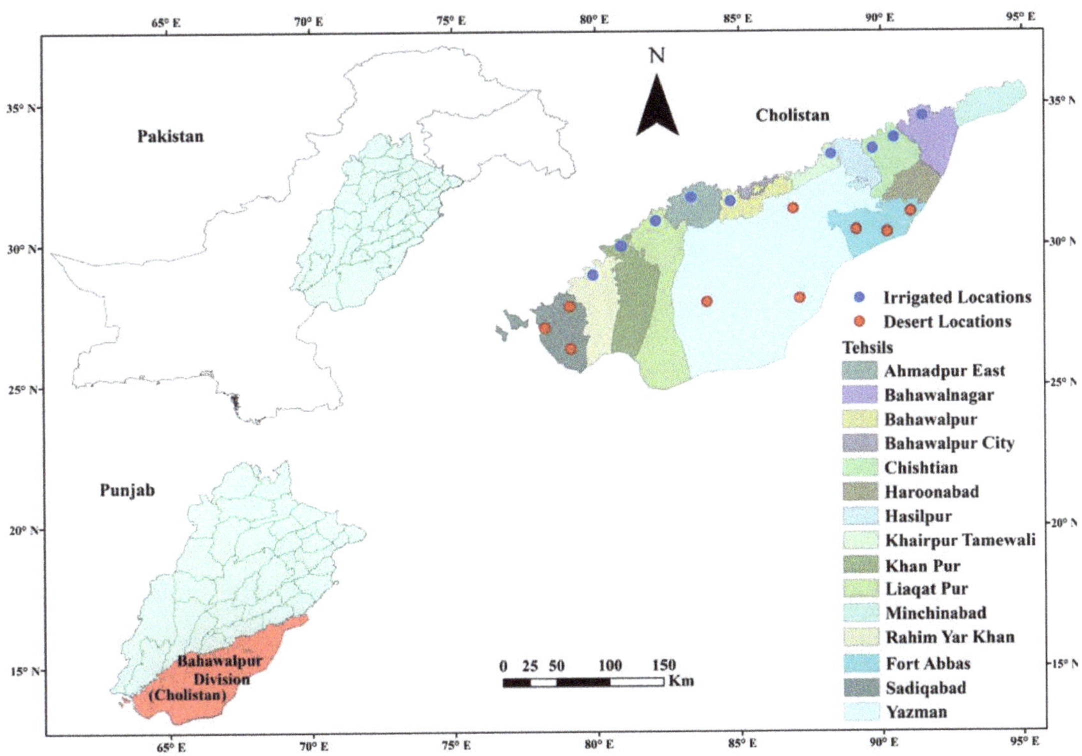

Figure 1. Study area map.

2.2. Morphological Characterization

Seventy-two samples (24 plants; 12 plants from the desert and 12 plants from the irrigated area) of fruits, leaves, and seeds of four *Ziziphus* species were collected randomly from three districts, i.e., Bahawalnagar, Bahawalpur, and Rahim Yar Khan, of the Cholistan desert and Irrigated plains, providing the opportunity to study the relationship between environmental factors and morphological traits of different *Ziziphus* species under the desert and irrigated conditions. Plants with similar morphological characters were selected for the measurement of morphological parameters. Fruit and seed collection was done at the final stage of fruit ripening (March) from the naturally grown old plants, while leaves were collected before the fruit set (October–November). At each location in the irrigated and desert conditions, three individual plants of each *Ziziphus* species (*Z. jujuba, Z. mauritiana, Z. spina-christi*, and *Z. nummularia*) were selected to study the 27 morphological characters [2]. Fruit, leaves, and seed were collected to measure the quantitative and qualitative characteristics of *Ziziphus* species. For qualitative measurements, selected plant parts of each species were subjected to visual analysis to identify traits according to the available key for each parameter. Fruit shape, fruit apex, fruit base, fruit color, stalk color, leaf color, leaf shape, leaf veins, leaf base, leaf apex, leaf margin, leaf petiole color, and leaf petiole surface were recorded to determine the qualitative morphological characters of *Ziziphus* species. Furthermore, for the measurement of quantitative traits, three random fresh samples of fruits and three random fresh samples of leaves of each species from each growing condition were sampled [18]. To calculate fruit and leaf dry weight, 100 g of fresh fruit samples and leaves were used. Fruit weight (g), fruit length (cm), fruit width (cm), fruit diameter (cm), fruit area (cm^2), stalk length (cm), leaf length (cm), leaf width (cm), leaf area (cm^2), leaf weight (g), and petiole length were recorded to evaluate the quantitative characters of *Ziziphus* species [8]. Fruit, leaves, and seed sizes were measured using a vernier caliper, while the electrical weigh-balance was used to measure the weights of fruits and leaves [39]. Furthermore, leaf area was measured by multiplying leaf length with leaf width. The following equation determined fruit area:

$$\text{Fruit Area} = 2\pi r^2 + 2\pi rl$$

where, $\pi = 3.14$, r = fruit radius, and l = fruit length.

2.3. Preparation of Fruits and Leaves for Chemical Analysis

Fresh fruit and leaf samples of four *Ziziphus* (Z) species (*Z. jujuba, Z. mauritiana, Z. spina-christi*, and *Z. nummularia*) were collected from the different growing conditions, i.e., irrigated and desert (Cholistan). Collected samples were analyzed for the determination of phytochemicals and antioxidant potential. Firstly, the collected samples were thoroughly washed with clean water to remove any contamination and then dried under the shade for about 15 to 17 days. After that, the leaf and fruit samples were ground and stored in zipped plastic bags for analysis in the laboratory. Subsequently, 200 g of plant material powder was mixed with 1000 mL of *N*-hexane (solvent) and kept soaked for two weeks at room temperature. During this duration, the soaked solution was shaken regularly every three days. A muslin cloth was used to filter out any coarse material from soaked material, and the obtained filtrate was boiled to evaporate the menstruum. Finally, a Whatman filter paper was used to filter the mixture, and the final extract was collected in a petri dish. A similar extraction method was repeated with methanol in place of *N*-hexane as a solvent [40]. *N*-hexane extract was used to measure the total phenolic and flavonoid contents, while methanolic extract was used to evaluate free radical scavenger DPPH (1, 1-diphenyl-2-picrylhydrazyl) activities.

2.4. Total Phenolic Content

The Folin-Ciocalteu reagent method was used for the determination of phenol quantity in plant extract (sample). *N*-hexane (0.2 mL) extract was mixed with the Folin reagent

(750 μL) and 6% sodium carbonate Na_2CO_3 (750 μL), followed by a dark incubation of the mixture for 90 min. After dark treatment, the absorbance of the sample was calculated at 765 nm through a spectrophotometer. Gallic acid was used for the preparation of the standard curve (0.250 mg/L). Total phenolic content was measured from the standard curve and stated as milligrams (mg) of gallic acid equivalents (GAE) per 100 g of extract (mgGAE/100 g) [41].

2.5. Total Flavonoid Content

For the determination of flavonoids, the aluminum chloride colorimetric method was used; 1 mL of 2% aluminum chloride solution ($AlCl_3$, $6H_2O$) was mixed with 1 mL *N*-hexane plant extract (sample). This mixture was incubated for 10 min at room temperature, and the absorbance of the sample was calculated at 420 nm through a spectrophotometer. Quercetin was used as standard (0–100 mg/L). Quantity of flavonoid was measured from the standard curve and stated as milligram (mg) of quercetin equivalents (QE) per 100 g of extract (mg QE/100 g) [42].

2.6. Free Radical Scavenger DPPH (1,1-Diphenyl-2-picrylhydrazyl) Activity

DPPH radical scavenging assay was implemented to measure the antioxidant potential of the plant extract [41]. A solution containing 0.004% 1,1-diphenyl-2-picrylhydrazyl (DPPH), 1 mL methanol, and 3 mL of plant extract was prepared and kept in the dark for 30 min. Then, the absorbance of the mixture was calculated at 517 nm with a spectrophotometer, and an extract-free solution was used as a control. A low absorbance of the reaction mixture highlighted a vast radical scavenging activity. For measuring the percentage inhibition of DPPH radical of the samples, the calculation was done through the following equation:

$$\text{DPPH Inhibition (\%)} = \{(\text{Ab}-\text{As})/\text{Ab}\} \times 100$$

where Ab is the absorbance of blank, and As is the absorbance of the sample.

2.7. Statistical Analysis

Data were analyzed using computer software Statistix 8.1. Growing conditions were allocated to main plots while *Ziziphus* species were maintained in the sub-plots. Significant differences were determined using ANOVA in combination with the LSD (least significance difference) test to determine the differences among the plant species. The significance of differences was evaluated at $p < 0.05$ levels. Tables report the means of calculated means based on the three replicates per treatment.

3. Results

3.1. Qualitative Morphological Characters

3.1.1. Fruit Morphological Parameters of *Ziziphus* Species under Irrigated and Desert Conditions

The morphological characterization of *Ziziphus* species in the present study revealed significant changes for various qualitative morphological traits among *Ziziphus* species under different growing conditions, i.e., irrigated and desert (Cholistan). Overall, the fruit shape was noticed as oval, orbicular, rounded, and ovate, and the fruit stalk color was observed as light green and dark green in all the studied species under both conditions. The fruit apex was found as retuse and truncate in *Z. jujuba*, oblate and rounded in *Z. mauritiana*, obtuse in *Z. spina-christi*, and oblate in *Z. nummularia* in both growing conditions. The fruit base was observed as rounded in *Z. jujuba* and *Z. spina-christi*, while in *Z. mauritiana* and *Z. nummularia*, it showed a flat pattern in desert conditions. In irrigated lands, the fruit base was shown as rounded and oval in *Z. jujuba*, orbicular in *Z. mauritiana*, rounded in *Z. spina-christi*, and flattened in *Z. nummularia*. Moreover, the color of the ripened fruit was yellowish-green to red in *Z. jujuba*, dark green in *Z. mauritiana*, light-green to yellow in

Z. spina-christi, and yellow-green to red in Z. nummularia under both growing conditions (Table 1).

Table 1. Qualitative morphological characters of fruits of *Ziziphus* species.

	Desert				Irrigated			
Parameters	***Z. jujuba***	***Z. mauritiana***	***Z. spina-christi***	***Z. nummularia***	***Z. jujuba***	***Z. mauritiana***	***Z. spina-christi***	***Z. nummularia***
Shape	Oval	Orbicular	Oval	Orbicular	Oval, rounded	Rounded, ovate	Oval, rounded	Orbicular
Apex	Retuse	Oblate	Obtuse	Oblate	Retuse, truncate	Rounded, oblate	Obtuse	Oblate
Base	Rounded	Flattened	Rounded	Flattened	Rounded, oval	Orbicular	Rounded	Flattened, rounded
Color	Yellowish green to red	Dark green	Light green to yellow	Red, yellow, green	Yellowish green to red	Dark green	Light green to yellow	Red, yellow, green
Stalk color	Light green	Green	Light green	Light green	Light green	Light green, green	Light green	Light green, green

3.1.2. Leaf Morphological Parameters of *Ziziphus* Species under Irrigated and Desert Conditions

Variations in the leaf morphological characters were recorded for the studied *Ziziphus* species in different growth conditions. Leaf shape was observed as oval and oblong in *Z. jujuba*, oval and lanceolate in *Z. mauritiana*, oval, ovate, and cordate in *Z. spina-christi* and, oval and rounded in *Z. nummularia* in desert conditions. In contrast, leaf shape was found as oval and oblong in *Z. jujuba*, oval, ovate, and cordate in *Z. mauritiana*, oval, ovate, and cordate in *Z. spina-christi* and, rounded and ovate in *Z. nummularia* in irrigated conditions. Leaf apex was an obtuse shape for all the studied species, except for the acute leaf apex in the *Z. spina-christi* in both growing conditions. In the desert conditions, the shape of the leaf base was measured as round and acute in *Z. jujuba* and *Z. mauritiana*, rounded and cordate in *Z. spina-christi*, and round only in *Z. nummularia*. Moreover, the shape of the leaf base was determined as cuneate and rounded in *Z. jujuba*, acute and rounded in *Z. mauritiana*, rounded and cordate in *Z. spina-christi*, and round in *Z. nummularia* in irrigated plains. Leaf margins showed no variation as all the species exhibited an entire shape in both growing conditions. However, leaf color varied among the studied species as *Z. jujuba* showed shining dark green leaves, *Z. mauritiana* showed light green leaves, and both *Z. spina-christi* and *Z. nummularia* showed dull green leaves in both growth conditions. Moreover, variations were shown in the surface of the leaf petiole, even sometimes in the same species of *Ziziphus*. Leaf petiole surface was detected as glabrous in *Z. nummularia* and pubescent in *Z. jujuba*. At the same time, some species (*Z. mauritiana* and *Z. spina-christi*) were covered with hairs in both growing conditions (Table 2).

3.2. *Quantitative Morphological Characters*

3.2.1. Fruit Morphological Parameters of *Ziziphus* Species under Irrigated and Desert Conditions

In the current study, different growing conditions significantly ($p < 0.05$) changed the fruit length, fruit width, fruit diameter, fruit area, dry fruit weight, seed length, seed diameter, and seed width, while both growing conditions showed non-significant differences for fruit stalk length of *Ziziphus* species; data are shown in Table 3. The maximum fruit length (2.7 cm), fruit width (2.8 cm), fruit diameter (2.8 cm), fruit area (38.1 cm^2), seed length (1.6 cm), seed diameter (1.0 cm), and seed width (1.4 cm) were noticed under irrigated conditions. In contrast, minimum fruit length (2.5 cm), fruit area (36.1 cm^2), fruit width (2.6 cm), fruit diameter (2.6 cm), seed length (1.4 cm), seed diameter (0.9 cm), and seed width (1.2 cm) were observed in desert conditions. In addition, the highest (32%) and lowest (27%) values of dry fruit weight were measured in the irrigated and desert conditions, respectively.

Table 2. Qualitative morphological characters of leaves of Ziziphus species.

	Desert				Irrigated			
Parameters	***Z. jujuba***	***Z. mauritiana***	***Z. spina-christi***	***Z. nummularia***	***Z. jujuba***	***Z. mauritiana***	***Z. spina-christi***	***Z. nummularia***
Shape	Oval, oblong	Oval, lanceolate	Oval, ovate, cordate	Oval, rounded	Oval, oblong	Oval, ovate, cordate	Oval, ovate	Rounded, ovate
Apex	Obtuse	Obtuse	Obtuse, acute	Obtuse	Obtuse	Obtuse	Obtuse, acute	Obtuse
Base	Rounded, acute	Rounded, acute	Cordate, rounded	Rounded	Cuneate, rounded	Acute, rounded	Cordate, rounded	Rounded
Margin	Entire	Entire	Entire	Entire	Entire	Entire	Entire	Entire
Veins	Prominent	Branched	Branched	Less prominent	Prominent	Much branched	Less branched	Less prominent
Color	Shining dark green	Light green	Dull green	Less dull green	light and dark shining green	Light and dark green	Dull green	Less dull shining green
Petiole surface	Pubescent	Hairs	Hairs	Glabrous	Pubescent	Hairs	Hairs	Glabrous
Petiole color	Green	Red, green	Green, red and yellow	Red	Green	Red, green	Green, red and yellow	Red

Table 3. Mean value of some quantitative characters for fruits of *Ziziphus* species.

		Fruit Length (cm)	**Fruit Width (cm)**	**Fruit Diameter (cm)**	**Fruit Area (cm^3)**	**Fruit Stalk Length (cm)**	**Fruit dry Weight (%)**	**Seed Length (cm)**	**Seed Width (cm)**	**Seed Diameter (cm)**
Conditions (C)	**Irrigated**	2.7 a	2.8 a	2.8 a	38.1 a	1.2 a	32.2 a	1.6 a	1.4 a	1.0 a
	Desert	2.5 b	2.6 b	2.6 b	36.1 b	1.2 a	26.5 b	1.4 b	1.2 b	0.9 b
LSD (0.05)		7.2	6.1	0.02	1.6	0.02	1.3	0.03	0.03	0.02
Species (S)	**S1**	3.9 a	3.8 a	3.6 a	66.1 a	1.5 a	14.9 d	2.0 a	1.3 b	1.1 a
	S2	2.6 b	2.5 b	2.5 c	31.0 c	1.0 c	29. 1 b	1.5 b	1.3 b	1.1 a
	S3	2.5 c	2.3 c	2.7 b	33.1 b	1.2 b	24.4 c	1.3 c	1.3 b	0.8 b
	S4	1.5 d	2.0 d	2.0 d	17.8 d	1.0 c	48.9 a	1.0 d	1.4 a	0.8 b
LSD (0.05)		0.03	0.02	0.03	1.9	0.01	0.9	0.02	0.01	0.2
Interaction	**(C × S)**	*	*	NS	NS	*	NS	NS	*	*

S1, S2, S3, and S4 represent four *Ziziphus* species, i.e., *Z. jujuba*, *Z. mauritiana*, *Z. spina-christi*, and *Z. nummularia*, respectively. Means are averaged over three replicates. Means that do not share the same letters in the column differ significantly at $p \leq 0.05$. While the symbol * represents significant interaction.

Among *Ziziphus* species, the highest fruit length (3.9 cm), fruit stalk length (1.5 cm), fruit diameter (3.7 cm), fruit width (3.8 cm), fruit area (66.2 cm^2), seed length (2.1 cm), and seed diameter (1.2 cm) were measured for species *Z. jujuba*. In comparison, the maximum dry fruit weight (50%) and seed width (1.3 cm) were recorded for species *Z. nummularia*, respectively. Whereas the minimum fruit length (1.6 cm), fruit width (2.1 cm), fruit area (17.9 cm^2), fruit diameter (2.1 cm), seed length (1.1 cm), and seed diameter (0.8 cm) were noted for species *Z. nummularia*; while the lowest fruit stalk length (1.0 cm), and dry fruit weight (15%) and seed width (1.3 cm) were determined for *Z. mauritiana* and *Z. jujuba*, respectively, indicating that the different growing conditions caused a significant change in the morphology of *Ziziphus* fruit. Furthermore, the interactive results for *Ziziphus* species and growing conditions were significant for fruit length and seed diameter, while a non-significant interaction was found for fruit diameter, fruit width, fruit area, dry fruit weight, fruit stalk length, seed length, and seed width (Table 3).

3.2.2. Leaf Morphological Parameters of *Ziziphus* Species under Irrigated and Desert Conditions

Growing conditions and Ziziphus species exhibited significant ($p < 0.05$) variations for leaf area, leaf length, leaf petiole length, leaf width, and leaf weight, as presented in Table 4. The highest leaf area (30.6 cm^2), leaf length (7.4 cm), and leaf width (4.1 cm) were determined for *Z. jujuba;* while the maximum leaf petiole length (1.3 cm) and leaf dry weight (55%) were obtained for *Z. mauritiana* under irrigated conditions, suggesting that optimum water conditions can improve the tree physiology (net photosynthetic rate, nutrient, and water uptake from soil) of *Ziziphus* species by improving the leaf morphology, as observed in this study. Whereas the lowest leaf area (5.5 cm^2), leaf length (2.6 cm), leaf petiole length (0.5 cm) and leaf width (2.1 cm), and leaf dry weight (38%) were noticed for *Z. nummularia* and *Z. spina-christi*, respectively, in desert conditions. Moreover, the interactive results for *Ziziphus* species and growing conditions were found significant for leaf dry weight, while it was found non-significant for leaf area, leaf length, leaf petiole length, and leaf width (Table 4).

Table 4. Mean value of some quantitative characters for leaves of *Ziziphus* species.

		Leaf Length (cm)	Leaf Width (cm)	Leaf Area (cm^3)	Leaf Petiole Length (cm)	Leaf Dry Weight (%)
Conditions (C)	Irrigated	4.9 a	3.2 a	17.9 a	1.0 a	50. 5 a
	Desert	4.8 a	3.1 b	16.3 b	0.9 b	43.9 b
LSD (0.05)		0.4	0.07	0.8	0.03	0.9
Species (S)	S1	7.4 a	4.2 a	30.6 a	1.3 a	45.6 c
	S2	5.1 b	3.0 c	16.5 b	1.3 a	55. 4 a
	S3	4.5 c	3.4 b	15.8 b	0.8 b	37.7 d
	S4	2.6 d	2.1 d	5.5 c	0.5 c	50.1 b
LSD (0.05)		0.2	0.1	0.9	0.04	1.2
Interaction	(C × S)	NS	NS	NS	NS	NS

S1, S2, S3, and S4 represent four *Ziziphus* species, i.e., *Z. jujuba, Z. mauritiana, Z. spina-christi,* and *Z. nummularia*, respectively. Means are averaged over three replicates. Means that do not share the same letters in the column differ significantly at $p \leq 0.05$.

3.3. *Total Phenol and Flavonoid Contents in Fruits and Leaves of Ziziphus Species under Irrigated and Desert Conditions*

In the current study, different growing conditions significantly ($p < 0.05$) altered the phytochemical (i.e., total phenol and flavonoid) contents in fruits and leaves of *Ziziphus* species; data are shown in Table 5. Compared with irrigated conditions, the relative increase in the total phenol and flavonoid contents in leaves and fruits of *Ziziphus* species were 3.6% and 3.9%, respectively, under desert conditions, suggesting that drought conditions favor the production and accumulation of phenol and flavonoid *Ziziphus* species.

In addition, among *Ziziphus* species, the highest fruit phenols (304.4 mg GAE/100 g), leaf phenols (314.2 mg GAE/100 g), fruit flavonoids (123.7 mg QE/100 g), and leaf flavonoids (113.4 mg QE/100 g) were measured for species *Z. nummularia*. Meanwhile, the lowest leaf phenols (234.5 mg GAE/100 g) and fruit phenols (207.6 mg GAE/100 g) were determined for species *Z. mauritiana*, while the minimum leaf flavonoids (87.9 mg QE/100 g) and fruit flavonoids (95.6 mg QE/100 g) were noticed for species *Z. spina-christi*. Furthermore, the interactive results for Ziziphus species and growing conditions were found significant for leaf flavonoids and fruit flavonoids, whereas it was found non-significant for leaf phenols and fruit phenols.

Table 5. Total phenol contents, total flavonoid contents, and % inhibition of DPPH scavenger activities in the fruits and leaves of *Ziziphus* species.

		Total Phenol Fruit (mg GAE/100 g)	Total Phenol Leaves (mg GAE/100 g)	Total Flavonoids Fruit (mg QE/100 g)	Total Flavonoids Leaves (mg QE/100 g)	Fruits (% Inhibition of DPPH)	Leaves (% Inhibition of DPPH)
Conditions (C)	Irrigated	261.8 b	271.1 b	102.4 b	98.0 b	53.2 b	58.5 b
	Desert	272.0 a	280.9 a	108.0 a	104.1 a	65.4 a	69.0 a
LSD (0.05)		0.3	0.9	1.2	1.4	11.0	3.6
Species (S)	S1	284.6 b	296.9 b	98.7 c	94.3 c	59.6 b	64.8 b
	S2	207.6 d	234.4 d	102.9 b	108.9 b	57.7 c	62.5 c
	S3	271.0 c	258.4 c	95.6 d	87.9 d	66.7 a	70.7 a
	S4	304.4 a	314.2 a	123.7 a	113.4 a	53.1 d	57.1 d
LSD (0.05)		1.4	1.2	1.9	2.2	1.4	1.7
Interaction	(C × S)	*	*	NS	NS	*	NS

S1, S2, S3, and S4 represent four *Ziziphus* species, i.e., *Z. jujuba*, *Z. mauritiana*, *Z. spina-christi*, and *Z. nummularia*, respectively. Means are averaged over three replicates. Means that do not share the same letters in the column differ significantly at $p \leq 0.05$. While the symbol * represents significant interaction.

3.4. DPPH Scavenger Activities for the Fruits and Leaves of Ziziphus Species under Irrigated and Desert Conditions

In the current study, different climatic conditions significantly ($p < 0.05$) changed the antioxidant potential of leaves and fruits of *Ziziphus* species; data are presented in Table 5. The maximum % inhibition of DPPH in leaves (69%) and fruits (65%) were calculated under desert conditions. In contrast, minimum % inhibition of DPPH in leaves (58%) and fruits (53%) were measured in irrigated conditions.

Among Ziziphus species, the maximum % inhibition of DPPH in fruits (67%) and leaves (71%) were measured for species *Z. spina-christi*. Whereas the minimum % inhibition of DPPH in fruits (53%) and leaves (57%) were determined for species *Z. nummularia*, indicating that the different levels of water availability directly altered the antioxidant activity potential of *Ziziphus* plants. Furthermore, the interactive results for *Ziziphus* species and climatic conditions were significant and non-significant for % inhibition of DPPH in leaves and fruits, respectively.

4. Discussions

This study reveals that improved water conditions under irrigated areas, influenced by irrigation, have significant impacts on various morphological parameters of *Ziziphus* species. Compared to desert *Ziziphus* plants, the better morphology of all studied *Ziziphus* species in irrigated conditions resulted in large leaves and fruits of *Ziziphus* species. Similarly, variations in the fruit size of different *Ziziphus* species (*Z. jujuba* and *Z. spina-christi*) were also reported by Zeinelabdin [43] under different types of growing conditions in Sudan. However, the observed differences among the different *Ziziphus* species with respect to petiole surface were not matched with the previous findings reported by Almalki and Alzahrani [44]; these ontogenic changes in petiole surface were associated with genetic characters of *Ziziphus* species [13]. In line with this, Li et al. [18] also studied the significant effects of different environmental conditions on the leaf shape (leaf morphology) for the different varieties of *Z. jujuba*, which also suggests that *Ziziphus* species respond to changing environmental factors, i.e., nutrient provision and light intensity, in terms of their leaf morphology. Our results showed the morphological variation in the studied qualitative traits that could be utilized to optimize *Ziziphus* species in the Cholistan desert. Previously, Akter and Rehman [1] also characterized the morphological traits such as fruit shape, fruit apex, fruit base, and fruit color, leaf shape, leaf apex, leaf base, leaf margin, and leaf color for the determination of morphological differences in the different genotypes of *Z. mauritiana*. Moreover, Ivanišová et al. [45] performed an analysis regarding the morphological characteristics of the fruits and leaves for different genotypes of *Z. jujuba* grow in agroecological conditions in Ukraine.

In our results, the mean values of fruit length and fruit width of *Z. mauritiana* were noted as 2.53 cm and 2.55 cm, respectively. In similar research, the average fruit length and fruit width of *Z. mauritiana* were recorded as 1.60 cm and 1.16 cm by Yahia et al. [42]. They studied the features of *Z. mauritiana* under arid conditions. Variations in *Ziziphus* morphology could be attributed to different sampling sites and ecological effects. Similarly, the fruit size, leaf size, and petiole length of *Z. spina-christi* and *Z. nummularia* were larger than the previous findings under semi-arid conditions. This higher phenotypic diversity among the *Ziziphus* species can result from their cross and self-incompatibility characteristics [2]. Furthermore, Baghazadeh-Daryaiia et al. [46] also evaluated the petiole length, leaf size, and fruit pulp weight for *Z. spina-christi* in high rainfall conditions to understand the differences between the plant morphologies. In addition, Sabaghzadeh and Morid [47] also measured some values for the fruit diameter (23.9 mm) and fruit length (39.4 mm) for *Ziziphus mauritiana* under a cold semi-arid climate. These morphological differences were allied with changing climatic conditions, immigration, and proximity. Kumar et al. [48] found that fruit content depends on environmental and genetic factors, as the good quality seed is responsible for higher pulp weight and fruit size, which otherwise relies on the size of the seed. In *Ziziphus* plants, fruit with high weight is a supreme fruit character, while ecological conditions and cultivars can greatly influence the fruit weight. Our results were in accordance with the previous findings, which reported 78–83% moisture content in fruits of *Z. jujuba* cultivars [49]. Similarly, the difference in the fruit weight in *Ziziphus* plants was also reported by [50,51], who calculated the different weights of fruit and seed, which estimates high variability in *Ziziphus* plants. Variations in the fruit weight of *Z. jujuba* genotypes in different agroecological zones were reported previously by many scientists [52,53]. The difference in the weight of fruits from similar geographical areas may be a consequence of genotypic effects [54]. Prasad and Bankar [55] reported that genotypes with bigger and smaller sizes were responsible for varying the fruit weight in *Ziziphus*. In addition, the leaves of terrestrial plants are highly diverse and very sensitive to climate (high temperature and low rainfall). For instance, the variations in leaf traits, i.e., leaf size and leaf shape, are strongly correlated with the availability of light quantity and quality, and mean annual precipitation [18]. Therefore, we can conclude that (1) the extreme weather conditions, i.e., high temperature and low rainfall, significantly affect the morphological, physiological, and biochemical traits of *Ziziphus* species present in deserts; and (2) these types of extreme weather conditions favor the accumulation of some important biochemicals (flavonoids and phenols) in fruits and leaves of *Ziziphus* species, which are the primary source of medicine for local people living in these regions.

However, interestingly, the stressful conditions (desert) significantly improved the phytochemistry of *Ziziphus* species and increased the accumulation of phenols and flavonoids in leaves and fruits; it also enhanced the antioxidant activity (measured as DPPH) in all *Ziziphus* species, suggesting that *Ziziphus* species cope with water stress under desert conditions through increased antioxidant activity, which ultimately increases their medicinal and nutritional value. Results revealed that desert conditions significantly affect the total phenol and flavonoid content in *Ziziphus* plants. Our findings with phenolic content (fruit 214.20–309.23 mg GAE/100 g) and flavonoid content (fruits 98.06–127.41 mg QE/100 g and leaves 91.13–177.57 mg QE/100 g) were much higher than the previous results reported by Yahia et al. [42]; they determined the range of phenolic content in fruit as 148.75 to 293.46 mg GAE/100 g. In comparison, flavonoid content ranged from 21.21 to 46.51 mg QE/100 g and 90.26 to 92.22 mg QE/100 g in the fruits and leaves of two *Ziziphus* species, respectively. In *Ziziphus* leaves, flavonoid contents are located in the cuticula and epidermis, resulting in higher flavonoid content in leaves than fruits. Similarly, our results with flavonoid content in fruits ranged between 98.06 and 127.41 mg QE/100 g are greater than the previous findings of Cosmulescu et al. [41], where the total flavonoid content in full mature fruit was given as 26.27 and 19.9 mg QE/100 g in two cultivars of *Ziziphus jujuba*. Variations in the bioactive contents can be affected by many factors, i.e., fruit maturity, geographical locations, cultivar type, and other horticulture factors. *Ziziphus* species can grow well under dry conditions and high temperatures. Their fruit quality is best under sunny, dry, and hot conditions. In contrast, tropical plants with a higher content of flavonoids rely on the availability of light, while the amount of phenol depends mainly on different geographical conditions [42]. Furthermore, the amount of flavonoid content is related to the size of the *Ziziphus* fruits [46].

Similarly, Memon et al. [56] also reported the different values of the phenolic amount in ber fruit (*Z. mauritiana*), i.e., 12.8 mg GAE/1 g. Moreover, Ashraf et al. [57] also recorded the variation of flavonoid and phenol content in different leaf extracts of *Ziziphus* plants. They found the highest value for phenol in chloroform extract, while the maximum amount of flavonoid was noted in methanol extract. Differences in the quantity of flavonoids and phenols in various parts of *Ziziphus* species may be linked to nature, availability, and solubility of the chemical compound being extracted. Water and carbon fluctuations in the leaves can influence the morphological, physiological, developmental, and biochemical mechanisms of the plants, which are important for the preparation of primary and secondary metabolites, and also provide mechanical stability to the plants. Photosynthesis in the leaves of *Ziziphus* plants is very sensitive to water scarcity, which significantly affects the productivity and development of the plant [18]. Therefore, combined with the previous findings, our results also showed that the different *Ziziphus* species bear different phytochemical profiles as the phenols and flavonoid levels varied between all the studied species.

Antioxidant activity differences were also reported by Ashraf et al. [57], where the leaf extracts of *Ziziphus mauritiana* were analyzed to evaluate antioxidants. Different solvents (methanol, chloroform, and hexane) with different concentrations (mg/mL) were used to prepare leaf extracts. Methanol extract with 0.4 and 0.8 (mg/mL) concentrations revealed the highest percentage inhibition of DPPH. The leaf extract of *Ziziphus mauritiana* was observed to be endowed with the highest DPPH activities with an increase in concentration (mg/mL). At the same time, the percentage inhibition of DPPH can also be strongly influenced by the procedure of solvent extraction. Besides, variations in the antioxidants potential were also attributed to the different ripening stages of *Ziziphus* fruits measured by Cosmulescu et al. [41], where the antioxidant values ranged from 1661.4 to 1154.6 (mg acid ascorbic/100 g) in different cultivars of *Ziziphus jujuba*. Current findings agree well with the previous results reported by Yahia et al. [42], who indicated higher values of antioxidant potential in leaves than fruits of *Ziziphus lotus* and *Ziziphus mauritiana* due to the high number of bioactive compounds, i.e., polyphenols including flavonoids and tannins in the genus *Ziziphus*. Moreover, our findings follow the earlier studies, which endorse the high degree of variability between the *Ziziphus* species [58–60]. Such variation of antioxidants contents in different parts of *Ziziphus* plants mainly rely on the diverse weather conditions, i.e., temperature and rainfall [58].

5. Conclusions

Present inquiries outspread the different arrangements of morphological characters to recognize the distinctive morpho-types among the *Ziziphus* species. The results indicated the significant impact of different water conditions on the morphology of four *Ziziphus* species. Under irrigated conditions, *Ziziphus* species showed better morphology as compared to desert conditions. Observed results explored the high interspecific morphological differences among the species studied. *Ziziphus* includes highly marked and easily recognized plant individuals. Intraspecific diversity within species was also highly variant, which could be attributed to ecological effects and different growing conditions. We find that this study provides a vast range of variations in the phenols, flavonoids, and antioxidant activities in the fruits and leaves of the understudied *Ziziphus* species. Further, a higher amount of bioactive compounds in fruits and leaves of *Ziziphus* species under desert conditions than irrigated conditions greatly highlights the impact of the desert on the phytochemistry of the *Ziziphus* plants. However, further research is required to understand the resource capture mechanism of *Ziziphus* species in desert conditions, especially under the changing climate (high-temperature and low rainfall regions). A series of comprehensive studies are required to identify the physiological and biochemical mechanisms responsible for the accumulation of important phytochemicals in *Ziziphus* species.

Author Contributions: Conceptualization, M.U.R. and T.H.; data curation, M.U.R.; formal analysis, M.U.R.; funding acquisition, T.H.; investigation, M.U.R. and T.H.; methodology, M.U.R.; project administration, T.H.; resources, T.H. and A.S.; supervision, T.H. and A.S.; validation, M.U.R.; visualization, M.A.R.; writing—original draft, M.U.R., T.H., and M.A.R.; writing—review and editing, T.H., M.A.R., A.S., and M.A. All authors have read and agreed to the published version of the manuscript.

Funding: Not applicable.

Institutional Review Board Statement: Not applicable.

Informed Consent Statement: Not applicable.

Data Availability Statement: Not applicable.

Conflicts of Interest: The authors have declared that no competing interests exist.

References

1. Akter, A.; Rahman, H. Characterization of Ber (*Ziziphus mauritiana*) Genotypes. *J. Agric. Sci. Technol.* **2019**, *8*, 11–23.
2. Norouzi, E.; Erfani-Moghadam, J.; Fazeli, A.; Khadivi, A. Morphological variability within and among three species of *Ziziphus* genus using multivariate analysis. *Sci. Hortic.* **2017**, *222*, 180–186. [CrossRef]
3. Islam, M.B.; Guralnick, R.P. Generic placement of the former *Condaliopsis* (Rhamnaceae) species. *Phytotaxa* **2015**, *236*, 25–39. [CrossRef]
4. Hauenschild, F.; Matuszak, S.; Muellner-Riehl, A.N.; Favre, A. Phylogenetic relationships within the cosmopolitan buckthorn family (Rhamnaceae) support the resurrection of *Sarcomphalus* and the description of *Pseudoziziphus* gen. nov. *Taxon* **2016**, *65*, 47–64. [CrossRef]
5. Suliman, M.B.; Mohammed, A.A. Preliminary phytochemical screening and antibacterial activity of ethanolic and aqueous extracts of Sudanese medicinal plant *Ziziphus spina-christi* L leaves. *Arab. J. Med. Aromat. Plants* **2018**, *4*, 36–44.
6. Nyanga, L.K.; Gadaga, T.H.; Nout, M.J.; Smid, E.J.; Boekhout, T.; Zwietering, M.H. Nutritive value of masau (*Ziziphus mauritiana*) fruits from Zambezi Valley in Zimbabwe. *Food Chem.* **2013**, *138*, 168–172. [CrossRef]
7. Orwa, C. Agroforestree Database: A Tree Reference and Selection Guide, Version 4.0. 2009. Available online: http://www.worldagroforestry.org/sites/treedbs/treedatabases (accessed on 15 August 2021).
8. Razi, M.; Anwar, R.; Basra, S.; Khan, M.M.; Khan, I.A. Morphological characterization of leaves and fruit of jujube (*Ziziphus mauritiana* Lamk.) germplasm in Faisalabad, Pakistan. *Pak. J. Agric. Sci.* **2013**, *50*, 211–216.
9. Alhassan, K.A.; Indabawa, A.S.; Shah, M.M. Phytochemical analysis, proximate composition and antibacterial activities of *Ziziphus* species (*Z. jujube* and *Z. spina-christi*). *J. Appl. Adv. Res.* **2019**, *4*, 42–46. [CrossRef]
10. Beg, M.A.; Teotia, U.; Farooq, S. In vitro antibacterial and anticancer activity of Ziziphus. *J. Med. Plants Stud.* **2016**, *4*, 230–233.
11. Arbonnier, M. *Trees, Shrubs and Lianas of West African Dry Zones*; Magraf Publishers CIRAD: Paris, France, 2004; p. 574.
12. El Amin, H.M. *Trees and Shrubs of the Sudan*; Ithaca Press: Exeter, UK, 1990; ISBN 0863721168.
13. Saied, A.S.; Gebauer, J.; Hammer, K.; Buerkert, A. Ziziphus spina-christi (L.) Willd.: A multipurpose fruit tree. *Genet. Resour. Crop Evol.* **2008**, *55*, 929–937.
14. Diaz, S.; Cabido, M.; Casanoves, F. Plant functional traits and environmental filters at a regional scale. *J. Veg. Sci.* **1998**, *9*, 113–122. [CrossRef]
15. Garnier, J.; Billen, G. The Riverstrahler modelling approach applied to a tropical case study (The Red-Hong-River, Vietnam): Nutrient transfer and impact on the Coastal Zone. *Coll. Mar. Res. Works* **2002**, *12*, 51–65.
16. Walter, A.; Schurr, U. Dynamics of leaf and root growth: Endogenous control versus environmental impact. *Ann. Bot.* **2005**, *95*, 891–900. [CrossRef] [PubMed]
17. Paroda, R.; Mal, B. New plant sources for food and industry in India. In *New Crops for Food and Industry*; Wickens, G.E., Haq, N., Day, P., Eds.; Chapman and Hall: London, UK, 1989; pp. 135–149.
18. Li, X.; Li, Y.; Zhang, Z.; Li, X. Influences of environmental factors on leaf morphology of Chinese jujubes. *PLoS ONE* **2015**, *10*, e0127825. [CrossRef]
19. Wolfe, A.D.; Liston, A. Contributions of PCR-based methods to plant systematics and evolutionary biology. In *Molecular Systematics of Plants II*; Springer: Boston, MA, USA, 1998; pp. 43–86.

20. Royer, D.L.; McElwain, J.C.; Adams, J.M.; Wilf, P. Sensitivity of leaf size and shape to climate within *Acer rubrum* and *Quercus kelloggii*. *New Phytol.* **2008**, *179*, 808–817. [CrossRef]
21. Jones, C.S. Does shade prolong juvenile development? A morphological analysis of leaf shape changes in *Cucurbita argyrosperma* Subsp. *Sororia* (Cucurbitaceae). *Am. J. Bot.* **1995**, *82*, 346–359. [CrossRef]
22. Du, N.; Wang, R.; Liu, J.; Zhang, X.; Tan, X.; Wang, W.; Chen, H.; Guo, W. Morphological response of *Vitex negundo* var. *heterophylla* and *Ziziphus jujuba* var. *spinosa* to the combined impact of drought and shade. *Agrofor. Syst.* **2013**, *87*, 403–416.
23. Goswami, P.; Banerjee, R.; Mukherjee, A. Potential antigenotoxicity assessment of *Ziziphus jujuba* fruit. *Heliyon* **2019**, *5*, e01768. [CrossRef]
24. Masullo, M.; Montoro, P.; Autore, G.; Marzocco, S.; Pizza, C.; Piacente, S. Quali-quantitative determination of triterpenic acids of *Ziziphus jujuba* fruits and evaluation of their capability to interfere in macrophages activation inhibiting NO release and iNOS expression. *Food Res. Int.* **2015**, *77*, 109–117. [CrossRef]
25. Song, L.; Liu, P.; Yan, Y.; Huang, Y.; Bai, B.; Hou, X.; Zhang, L. Supercritical CO_2 fluid extraction of flavonoid compounds from Xinjiang jujube (*Ziziphus jujuba* Mill.) leaves and associated biological activities and flavonoid compositions. *Ind. Crops Prod.* **2019**, *139*, 111508. [CrossRef]
26. Bai, L.; Zhang, H.; Liu, Q.; Zhao, Y.; Cui, X.; Guo, S.; Zhang, L.; Ho, C.-T.; Bai, N. Chemical characterization of the main bioactive constituents from fruits of *Ziziphus jujuba*. *Food Funct.* **2016**, *7*, 2870–2877. [CrossRef] [PubMed]
27. Ahmad, M.; Zafar, M.; Sultana, S. *Salvadora persica*, *Tamarix aphylla* and *Zizyphus mauritiana*-Three woody plant species mentioned in Holy Quran and Ahadith and their ethnobotanical uses in north western part (DI Khan) of Pakistan. *Pak. J. Nutr.* **2009**, *8*, 542–547.
28. Hussain, S.M.; Khan, A.; Khan, A.-u.; Qazi, N.G.; Ali, F. Pharmacological basis for medicinal use of *Ziziphyus nummularia* (Rhamnaceae) leaves in gastrointestinal disorders. *Trop. J. Pharm. Res.* **2017**, *16*, 2379–2385. [CrossRef]
29. Zhang, R.; Chen, J.; Shi, Q.; Li, Z.; Peng, Z.; Zheng, L.; Wang, X. Phytochemical analysis of Chinese commercial *Ziziphus jujube* leaf tea using high performance liquid chromatography–electrospray ionization-time of flight mass spectrometry. *Food Res. Int.* **2014**, *56*, 47–54. [CrossRef]
30. Li, D.; Yue, D.; Liu, D.; Zhang, L.; Song, S. Phytochemical and chemotaxonomic study on *Ziziphus jujuba* Mill.(Rhamnaceae). *Biochem. Syst. Ecol.* **2020**, *91*, 104058. [CrossRef]
31. Dalleau, S.; Baradat, M.; Guéraud, F.; Huc, L. Cell death and diseases related to oxidative stress: 4-hydroxynonenal (HNE) in the balance. *Cell Death Differ.* **2013**, *20*, 1615–1630. [CrossRef]
32. Lobo, V.; Patil, A.; Phatak, A.; Chandra, N. Free radicals, antioxidants and functional foods: Impact on human health. *Pharmacogn. Rev.* **2010**, *4*, 118. [CrossRef]
33. Khadri, A.; Neffati, M.; Smiti, S.; Falé, P.; Lino, A.R.L.; Serralheiro, M.L.M.; Araújo, M.E.M. Antioxidant, antiacetylcholinesterase and antimicrobial activities of *Cymbopogon schoenanthus* L. Spreng (lemon grass) from Tunisia. *LWT-Food Sci. Technol.* **2010**, *43*, 331–336. [CrossRef]
34. Duthie, G.; Morrice, P. Antioxidant capacity of flavonoids in hepatic microsomes is not reflected by antioxidant effects in vivo. *Oxidative Med. Cell. Longev.* **2012**, *2012*, 165127. [CrossRef]
35. Huang, W.-Y.; Cai, Y.-Z.; Zhang, Y. Natural phenolic compounds from medicinal herbs and dietary plants: Potential use for cancer prevention. *Nutr. Cancer* **2009**, *62*, 1–20. [CrossRef]
36. Pandey, K.B.; Rizvi, S.I. Plant polyphenols as dietary antioxidants in human health and disease. *Oxidative Med. Cell. Longev.* **2009**, *2*, 270–278. [CrossRef]
37. Farooq, Z.; Iqbal, Z.; Mushtaq, S.; Muhammad, G.; Iqbal, M.Z.; Arshad, M. Ethnoveterinary practices for the treatment of parasitic diseases in livestock in Cholistan desert (Pakistan). *J. Ethnopharmacol.* **2008**, *118*, 213–219. [CrossRef] [PubMed]
38. Ahmed, M.J.; Murtaza, G. A study of medicinal plants used as ethnoveterinary: Harnessing potential phytotherapy in Bheri, district Muzaffarabad (Pakistan). *J. Ethnopharmacol.* **2015**, *159*, 209–214. [CrossRef] [PubMed]
39. Hernández, F.; Noguera-Artiaga, L.; Burló, F.; Wojdyło, A.; Carbonell-Barrachina, Á.A.; Legua, P. Physico-chemical, nutritional, and volatile composition and sensory profile of Spanish jujube (*Ziziphus jujuba* Mill.) fruits. *J. Sci. Food Agric.* **2016**, *96*, 2682–2691. [CrossRef] [PubMed]
40. Hussain, T.; Fatima, I.; Rafay, M.; Shabir, S.; Akram, M.; Bano, S. Evaluation of antibacterial and antioxidant activity of leaves, fruit and bark of Kigelia africana. *Pak. J. Bot.* **2016**, *48*, 277–283.
41. Cosmulescu, S.; Trandafir, I.; Violeta, N.; Achim, G.; Mihai, B.; Iordanescu, O. Variation of bioactive compounds and antioxidant activity of jujube (*Ziziphus jujuba*) fruits at different stages of ripening. *Not. Bot. Horti Agrobot. Cluj-Napoca* **2018**, *46*, 134–137. [CrossRef]
42. Yahia, Y.; Benabderrahim, M.A.; Tlili, N.; Bagues, M.; Nagaz, K. Bioactive compounds, antioxidant and antimicrobial activities of extracts from different plant parts of two *Ziziphus* Mill. species. *PLoS ONE* **2020**, *15*, e0232599. [CrossRef]
43. Zeinelabdin, M. Variation of some morphological characteristics of *Ziziphus spina-chriti* (A.) Ritch. fruits among four provenances. *Int. J. Sci. Res. Publ.* **2016**, *6*, 433–439.
44. Almalki, R.A.; Alzahrani, D.A. Morphological Investigation of Genus *Ziziphus* Mill.(Rhamnaceae) in Saudi Arabia. *Am. J. Plant Sci.* **2018**, *9*, 2644. [CrossRef]
45. Ivanišová, E.; Grygorieva, O.; Abrahamova, V.; Schubertova, Z.; Terentjeva, M.; Brindza, J. Characterization of morphological parameters and biological activity of jujube fruit (*Ziziphus jujuba* Mill.). *J. Berry Res.* **2017**, *7*, 249–260. [CrossRef]

46. Baghazadeh-Daryaii, L.; Sharifi-Sirchi, G.-R.; Samsampoor, D. Morphological, phytochemical and genetic diversity of *Ziziphus spina–christi* (L) Des. in South and Southeastern of Iran. *J. Appl. Res. Med. Aromat. Plants* **2017**, *7*, 99–107. [CrossRef]
47. Sabaghzadeh, F.; Morid, A. Comparison above genotypes of *Ziziphus* species (*Ziziphus spina-christi* & *Ziziphus mauritania*) extant in Dezfoul fadak botanical garden used fruit and seed traits. *Agric. Sci. J.* **2014**, *38*, 113–123.
48. Kumar, V.; Ajeesh, R.; Vidyasagaran, K.; Babu, S. Variation in pulp content and physical characters of *Elaeocarpus serratus* L. Fruits in different landuse patterens of Western Ghats, India. *Ecol. Environ. Conserv.* **2015**, *21*, 1–4.
49. Gao, Q.-H.; Wu, P.-T.; Liu, J.-R.; Wu, C.-S.; Parry, J.W.; Wang, M. Physico-chemical properties and antioxidant capacity of different jujube (*Ziziphus jujuba* Mill.) cultivars grown in loess plateau of China. *Sci. Hortic.* **2011**, *130*, 67–72. [CrossRef]
50. Markovski, A.; Velkoska-Markovska, L. Investigation of the morphometric characteristics of jujube types (*Zizyphus jujuba* Mill.) fruits in Republic of Macedonia. *Genetika* **2015**, *47*, 33–43. [CrossRef]
51. Azam-Ali, S.; Bonkoungou, E.; Bowe, C.; DeKock, C.; Godara, A.; Williams, J. *Fruits for the Future-2 (Revised Edition), Ber and Other Jujubes*; International Centre for Underutilised Crops, University of Southampton: Southampton, UK, 2001.
52. Karnatovska, M.; Brindza, J.; Grygorieva, O.; Derevjanko, V.; Kochanova, Z.; Birova, D. Jujube Fruit (*Zizyphus jujuba* Mill.) Variability Determination. In Proceedings of the 1st International Scientific Conference on Medicinal, Aromatic and Spice Plants Book of Scientific Papers and Abstracts, Nitra, Slovakia; 2007; p. 219.
53. Brindza, J.; Karnatovska, M.; Grygorieva, O.; Vietoris, V.; Kucelová, L.; Erdélyová, G. Morphological and organoleptic nature of *Ziziphus jujuba* Mill. *Potravin. Slovak J. Food Sci.* **2011**, *5*, 1–11. [CrossRef]
54. Karadeniz, T. Selection of native 'Cornelian' cherries grown in Turkey. *J. Am. Pomol. Soc.* **2002**, *56*, 164. [CrossRef]
55. Prasad, R.; Bankar, G. Evaluation of pomegranate cultivars under arid conditions. *Indian J. Hortic.* **2000**, *57*, 305–308.
56. Memon, A.A.; Memon, N.; Luthria, D.L.; Pitafi, A.A.; Bhanger, M.I. Phenolic compounds and seed oil composition of *Ziziphus mauritiana* L. fruit. *Pol. J. Food Nutr. Sci.* **2012**, *62*, 15–21. [CrossRef]
57. Ashraf, A.; Sarfraz, R.A.; Anwar, F.; Shahid, S.A.; Alkharfy, K.M. Chemical composition and biological activities of leaves of *Ziziphus mauritiana* L. native to Pakistan. *Pak. J. Bot.* **2015**, *47*, 367–376.
58. Hossain, M.A. A phytopharmacological review on the Omani medicinal plant: *Ziziphus jujube*. *J. King Saud Univ. Sci.* **2019**, *31*, 1352–1357. [CrossRef]
59. Esteki, T.; Urooj, A. Antioxidant components and activity in the peel of *Ziziphus Jujuba* mill. *J. Pharm. Res.* **2012**, *5*, 2705–2709.
60. Al-Saeedi, A.H.; Al-Ghafri, M.T.H.; Hossain, M.A. Comparative evaluation of total phenols, flavonoids content and antioxidant potential of leaf and fruit extracts of Omani *Ziziphus jujuba* L. *Pac. Sci. Rev. Nat. Sci. Eng.* **2016**, *18*, 78–83. [CrossRef]

plants

MDPI

Article

Responses of Lowland Rice Genotypes under Terminal Water Stress and Identification of Drought Tolerance to Stabilize Rice Productivity in Southern Thailand

Tajamul Hussain [1], Nurda Hussain [1,2], Mukhtar Ahmed [3], Charassri Nualsri [4] and Saowapa Duangpan [1,*]

1 Laboratory of Plant Breeding and Climate Resilient Agriculture, Agricultural Innovation and Management Division, Faculty of Natural Resources, Prince of Songkla University, Songkhla 90110, Thailand; 6110630006@psu.ac.th (T.H.); 6310120017@psu.ac.th (N.H.)
2 Energy Technology Program, Faculty of Engineering, Prince of Songkla University, Songkhla 90110, Thailand
3 Department of Agronomy, Faculty of Crop and Food Sciences, PMAS Arid Agriculture University, Rawalpindi 46300, Punjab, Pakistan; ahmadmukhtar@uaar.edu.pk
4 Agricultural Innovation and Management Division, Faculty of Natural Resources, Prince of Songkla University, Songkhla 90110, Thailand; ncharass@yahoo.com
* Correspondence: saowapa.d@psu.ac.th; Tel.: +66-74-286-138

Abstract: Lowland rice is an important cereal crop that plays a key role in the food security and the economy of Thailand. Terminal water stress (TWS) in rainfed lowland areas poses threats to rice productivity due to stress occurrence at terminal crop stages and extreme sensitivity of rice to TWS. A two-year study was conducted to characterize the performance of yield and yield attributes of twelve Thai lowland rice genotypes under TWS, to identify stress-tolerant genotypes using stress response indices and to identify promising stress indices which are correlated with grain yield (GY) under well-watered (WW) and TWS conditions for their use as rapid identifiers in a rice crop breeding program for enhancing drought stress tolerance. Measurements were recorded under WW and TWS conditions. Highly significant variations were observed amongst assessed genotypes for their yield productivity responses. According to stress response indices, genotypes were categorized into stress-tolerant and stress susceptible genotypes. Genotypes Hom Pathum, Sang Yod, Dum Ja and Pathum Thani-1 were found highly stress tolerant and relatively high yielding; genotypes Look Pla and Lep Nok were stress tolerant, whereas genotypes Chor Lung, Hom Nang Kaew and Hom Chan were moderately tolerant genotypes. Hence, stress-tolerant genotypes could be potentially used for cultivation under rainfed and water-limited conditions, where TWS is predicted particularly in southern Thailand to stabilize rice productivity. Stress tolerance indices, including stress tolerance index (STI), geometric mean productivity (GMP), mean productivity index (M_{PRO}) and harmonic mean index (M_{HAR}), indicated strong and positive associations with GY under WW and TWS; thus, these indices could be used to indicate stress tolerance in rice crop breeding program aimed at a rapid screening of lowland rice genotypes for stress tolerance.

Keywords: lowland rice; terminal water stress; grain yield; stress indices; stress tolerance

Citation: Hussain, T.; Hussain, N.; Ahmed, M.; Nualsri, C.; Duangpan, S. Responses of Lowland Rice Genotypes under Terminal Water Stress and Identification of Drought Tolerance to Stabilize Rice Productivity in Southern Thailand. *Plants* **2021**, *10*, 2565. https://doi.org/10.3390/plants10122565

Academic Editor: Shigeto Morita

Received: 21 September 2021
Accepted: 20 November 2021
Published: 24 November 2021

1. Introduction

Rice is an important cereal after wheat that contributes to food security worldwide [1]. However, water stress has limited the production of both cereal crops [2]. Lowland rice systems contribute a major portion of rice production [3], and rainfed lowland rice is cultivated on approximately 6.2 million hectares worldwide [4]. In Thailand, rice is a major crop contributing to the food security and economy of the country. Even though rice production in southern Thailand contributes only 6% of the total rice production [5], it is of great importance to the regional food security. Rainfed lowland rice is a major production system in southern Thailand. However, rainfed lowland rice production systems are extremely vulnerable and variable in nature as water stress can occur at any crop growth stages. Climate change has also caused an increase in temperature and fluctuations in rainfall occurrence leading to regular heat and drought stress intervals [6,7]. Water stress is considered an important abiotic stress deleteriously affecting field crop productivity [6,8]. Rainfed lowland rice is cultivated in the rainy season in Thailand [7,9]. Due to seasonal variations in rainfall and occurrence of WS at different crop developmental stages, lowland rice production is drastically affected.

Water stress occurrence is critical under rainfed conditions as it affects plant growth and development [10]. Occurrence of water stress at various crop growth stages negatively influences the performance of specific attributes [11], leading to declined yield [12]. Timing of stress occurrence during early growth, mid-season and at terminal stages impact on severity of yield losses [13]. A stress event at early rice growth stages has an influence on leaf numbers and size, tillering capacity and stem height and affects panicle development, ultimately resulting in a reduced yield [14,15]. Water availability after the stress interval at the early growth stage helps plants recover, leading to lesser loss in yield. However, terminal water stress (TWS) intervals highly influence plant performance and lessens the chances of recovery to occur, leading to increased yield losses as rice is extremely sensitive to TWS [16]. TWS delays various plant development stages including panicle initiation and flowering [17], leading to spikelet sterility and reduction in number of panicles [18]. In addition, TWS causes abortion of ovules, deteriorates the grain filling process and alters source to sink distribution of assimilates, leading to reduced grain yield (GY) [19,20].

Stress-tolerant genotypes are genotypes that have the potential to maintain higher productivity under water stress [21]. Due to the extreme sensitivity of rice to TWS, different rice genotypes exhibit differential responses [10,18,22]. In the perspective of farmers, a stress-tolerant genotype is that which is highly capable of maintaining yield under limited water availability [23]. Therefore, high yielding genotypes under a diverse range of environments are desired and the cultivation of such genotypes could help to maintain rice productivity [2]. The GY of stress-tolerant genotypes is less affected under water stress as compared to stress susceptible genotypes. Cha-um et al. [24] reported that panicle size and filled grains of two stress tolerant rice genotypes were not significantly reduced as compared to two stress susceptible genotypes. According to Ichsan et al. [2], there are various local genotypes used by farmers around the world that have tolerance against water stress, in addition to stress-tolerant genotypes developed by research institutions and organizations. To enhance the resistance of rice against water stress, these genotypes are potential sources of germplasm, which are available in each growing season. In addition, it was observed that wild genotypes exhibited less decline and maintained GY under water stress as compared to cultivated genotypes [25]. Therefore, the identification and cultivation of stress tolerant genotypes from local germplasm could help to stabilize productivity under terminal water stressed environments.

Several techniques and procedures are used to study water stress tolerance in rice genotypes at different crop growth stages [14,18,26,27]. A drought stress scoring method was used as the main criteria for the assessment and selection of rice cultivars for stress tolerance at reproductive crop growth stages in field trials [28] and genotypes producing high yields under water stress were selected as stress-tolerant genotypes. Numerous stress tolerance indices have been used [6,29–38] based on mathematical association among yield production under well-watered (WW) and water stressed conditions. According to Clarke et al. [38] and Fernandez [32], stress indices are generally based on the stress sensitivity or stress tolerance of tested genotypes. In the selection of stress tolerant genotypes, these indices provide the effect of water stress based on yield losses occurring under stress as compared to optimal or WW conditions [39]. The relative yield performance of a specific genotype in comparison to other tested genotypes under the same water stress indicates stress tolerance [40], and measure of reduction in yield under stress refers to the stress susceptibility of a genotype [41]. The stress susceptibility index (SSI) for a genotype was suggested by Fischer and Maurer [37], whereas geometric mean productivity (GMP) and stress tolerance index (STI) were proposed by Fernandez [32]. The mean productivity

(M_{PRO}) index is an average yield under WW and water stressed conditions [33]. Harmonic mean index (M_{HAR}) was suggested by Schneider et al. [34]. The tolerance index (TI) is the difference in productivity between WW and water stressed conditions [35]. The yield stability index (YSI) was defined by Bouslama and Schapaugh [36]. All these indices have been used widely and are proposed in drought stress tolerance studies. However, the positive or negative associations of these indices with GY may vary. The significant differences among various indices were reported by Golabadi et al. [42] and Saba et al. [43] except SSI. Significant positive associations for GY under WW and stress indices (GMP, MP, STI, YSI, TOL and YI) and GY under water stressed conditions and stress indices (STI, GMP, MP, YSI and YI) have been observed by Golabadi et al. [42] and Arif et al. [44]. Hence, evaluating the associations of stress indices with GY under different environments is necessary. Therefore, the objectives of the current study were to (i) evaluate the performance of yield and yield attributes of Thai lowland rice genotypes under TWS and identify stress tolerant genotypes using stress indices; (ii) to identify promising stress indices which are correlated with GY under WW and TWS conditions for their use as rapid identifiers in rice crop breeding program for enhancing drought stress tolerance.

2. Results

2.1. Effect of Water Stress on Yield Performance and Productivity

In this study, different lowland rice genotypes were assessed based on the performance of yield and yield attributes in response to terminal water stress (TWS) applied at the terminal crop growth stage. In both years, treatment and genotype effect resulted as highly significant different ($p < 0.001$) for most of the yield attributes except a non-significant difference for days to maturity (DM) under treatment effect in 2018–2019 (Table 1). Interactions of genotype and treatment effects indicated non-significant differences in both years, except for a significant difference for days to flowering (DF) ($p < 0.05$) and a highly significant difference for DM ($p < 0.001$) in 2018–2019 (Table 1). DF, number of tillers (NT), number of panicles (NP), grain yield (GY) and biomass were highly significant different; DM was moderately significantly different, whereas no significant difference was observed for plant height (PH) under the effect of years. Mean comparisons indicated that all tested genotypes differed and a significant variability in performance prevailed under well-watered (WW) and TWS conditions. TWS resulted in a delay in flowering duration (Figure 1a,b) of all genotypes except genotype 9 in the first year (Figure 1a). Flowering occurred 4 days earlier in genotype 9 (Table 2). Delay in flowering duration ranged 2–19 days in the first year while 1–4 days in the second year (Table 2). The maximum delay in flowering was observed for the top three genotypes 7, 12 and 6 by 19, 8 and 6 days in the first year and for 11, 8, 3, 4 and 5 by 7 and 4 days in the second year, respectively. TWS caused delays in the maturity duration (Figure 1c,d) of most of the genotypes except for genotypes 7, 9 and 10 in the first year (Figure 1a). Genotypes 7, 9 and 10 matured earlier in the first year by 19, 5 and 11 days (Table 2). In the second year, maturity duration was increased for all genotypes under TWS (Figure 1d). The delay in maturity duration ranged 4–14 days in the first year while 3–8 days in the second year (Table 2). PH was reduced under TWS for all genotypes in both years (Figure 1e,f). PH was reduced 4–13% in the first year and 2–14% in the second year (Table 2). Reduction in PH was higher than 10% for genotypes 1, 2, 4, 7, 8 and 11 (Table 2). NT (Figure 2a,b) and NP (Figure 2c,d) were reduced under TWS (Figure 2). However, reduction in NT and NP ranged one–two tillers and panicles per plant (Table 2). No change was observed in NT of genotypes 1, 5 and 6 in the first year and genotypes 2, 5, 6, 8, 9 and 10 in the second year (Table 2). Genotypes 1 and 3 maintained their NP under TWS in the first year, whereas the NP of all genotypes were affected in the second year (Table 2). TWS caused decline in GY (Figure 3a,b) and biomass (Figure 3c,d) of all genotypes in both years (Figure 3). GY was decreased 17–45% in the first year, whereas 21–52% in the second year (Table 2). The GY of genotypes 1, 7, 9, 11 and 12 in the first year and GY of genotypes 2, 9, 11 and 12 in the second year decreased more than 30%, indicating a major decline in GY under TWS (Table 2). Similarly, biomass was reduced

20–41% in the first year and 15–38% in the second year (Table 2). Biomass reduction of genotypes 4 and 12 in the first year and genotypes 1, 3 and 10 in the second year was more than 30%, indicating a major decline in biomass under TWS (Table 2).

Table 1. The analysis of variance for days to flowering (DF), days to maturity (DM), plant height (PH), number of tillers (NT), number of panicles (NP), grain yield (GY) and biomass (BM) of twelve lowland rice genotypes.

Year	Traits	Treatment (T) Effect	Genotype (G) Effect	Interaction (T × G)	Year Effect
2018–2019	DF	***	***	*	***
	DM	ns	***	***	**
	PH	***	***	ns	ns
	NT	***	***	ns	***
	NP	***	***	ns	***
	GY	***	***	ns	***
	BM	***	***	ns	***
2019–2020	DF	***	***	ns	
	DM	***	***	ns	
	PH	***	***	ns	
	NT	**	***	ns	
	NP	***	***	ns	
	GY	***	***	ns	
	BM	***	***	ns	

***: highly significant ($p < 0.001$), **: moderately significant ($p < 0.01$), *: significant ($p < 0.05$), ns: non-significant.

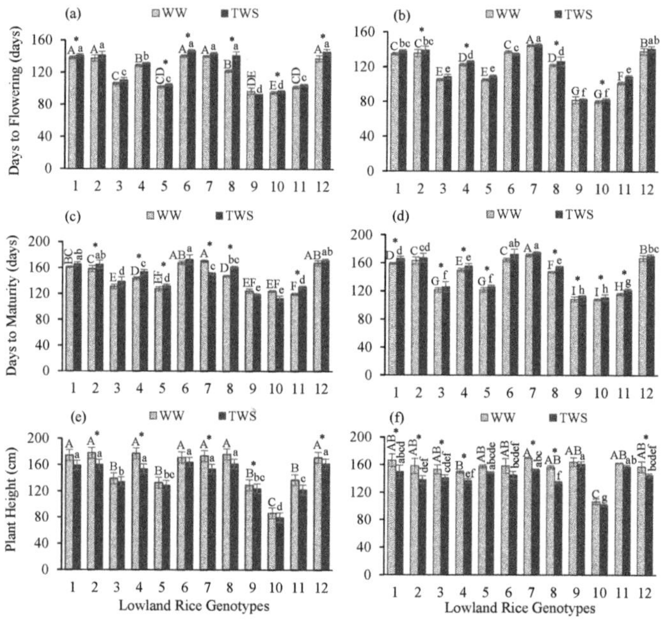

Figure 1. Days to flowering (**a**,**b**), days to maturity (**c**,**d**) and plant height (**e**,**f**) of twelve lowland rice genotypes under well-watered (WW) and terminal water stressed (TWS) conditions during 2018–2019 (**a**,**c**,**e**) and 2019–2020 (**b**,**d**,**f**). Vertical bars show ± standard errors for means of three repetitions. Capital letters represent the significant ($p < 0.05$) differences among genotypes in WW condition. Small letters represent the significant ($p < 0.05$) differences among genotypes in TWS condition. Centered stars above each pair of the bars represent the significance of parameters for each genotype under WW and TWS conditions.

Table 2. Changes in performance of yield and yield attributes of twelve lowland rice genotypes under terminal water stressed conditions. Changes in days to flowering (DF) and days to maturity (DM) are presented by difference in days. Changes in number of tillers (NT) and number of panicles (NP) are presented by difference in numbers (no.), whereas changes in plant height (PH), grain yield (GY) and biomass (BM) are presented by % difference.

Genotypes	2018–2019							2019–2020						
	DF Days	DM Days	PH %	NT no.	NP no.	GY %	BM %	DF Days	DM Days	PH %	NT no.	NP no.	GY %	BM %
1	3	5	−9	0	0	−39	−20	3	7	−10	−1	−1	−25	−38
2	4	7	−10	−1	−1	−26	−24	3	4	−12	0	−1	−43	−20
3	5	8	−4	1	−0	−28	−21	4	4	−8	−2	−2	−26	−30
4	3	10	−13	−1	−1	−18	−41	4	6	−8	−1	−1	−26	−24
5	2	5	−3	0	−1	−21	−21	4	5	−5	0	−1	−24	−19
6	6	5	−4	0	−1	−23	−28	−2	8	−8	0	−1	−22	−24
7	4	−19	−11	−1	−1	−31	−20	1	4	−10	−1	−1	−21	−23
8	19	14	−8	−1	−1	−17	−25	4	7	−14	0	−1	−25	−22
9	−4	−5	−4	−1	−2	−30	−28	1	4	−2	0	−1	−52	−17
10	2	−11	−8	−1	−1	−26	−26	2	3	−6	0	−1	−36	−38
11	3	11	−11	−1	−1	−45	−29	7	6	−3	−1	−1	−34	−19
12	8	4	−5	−1	−1	−36	−38	3	3	−7	−1	−1	−33	−15

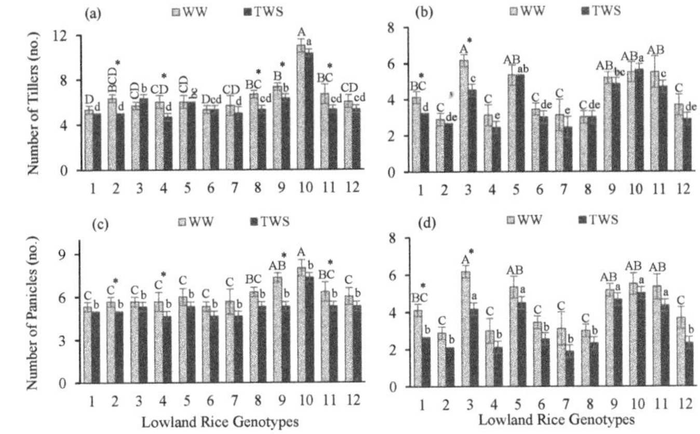

Figure 2. Number of tillers (**a**,**b**), and number of panicles (**c**,**d**) of twelve lowland rice genotypes under well-watered (WW) and terminal water stressed (TWS) conditions during 2018–2019 (**a**,**c**) and 2019–2020 (**b**,**d**). Vertical bars show ± standard errors for means of three repetitions. Capital letters represent the significant ($p < 0.05$) differences among genotypes in WW condition. Small letters represent the significant ($p < 0.05$) differences among genotypes in TWS condition. Centered stars above each pair of the bars represent the significance of parameters for each genotype under WW and TWS conditions.

2.2. Association among Yield and Yield Attributes under Terminal Water Stress

Figure 4 indicates combined correlations among yield and yield attributes, including the DF, DM, PH, NT, NP, GY and biomass of twelve lowland rice genotypes. Under WW condition, highly positive associations among DF and biomass (0.89), DF and DM (0.98), DM and biomass (0.86), NT and NP (0.95), moderately positive associations among DF and PH (0.82), DM and PH (0.76), PH and biomass (0.82) and positive associations among PH and GY (0.56) and GY and biomass (0.64) were observed. Whereas highly negative associations among DF and NP (−0.94), DM and NP (−0.90), DM and NT (−0.84), PH and NP (−0.87), PH and NT (−0.97), NT and biomass (−0.88) and NP and biomass (−0.87) were detected. Under the TWS condition, highly positive associations among DF and biomass (0.89), DF and DM (0.99), DM and biomass (0.91), PH and biomass (0.86), NT and

NP (0.97) and moderately positive associations among DF and PH (0.73), DM and PH (0.74) and GY and biomass (0.73) were observed. Whereas highly negative associations among DF and NP (−0.92), DF and NT (−0.85), DM and NP (−0.91), DM and NT (−0.84), PH and NP (−0.85), PH and NT (−0.92), NT and biomass (−0.83) and NP and biomass (−0.86) were detected (Figure 4).

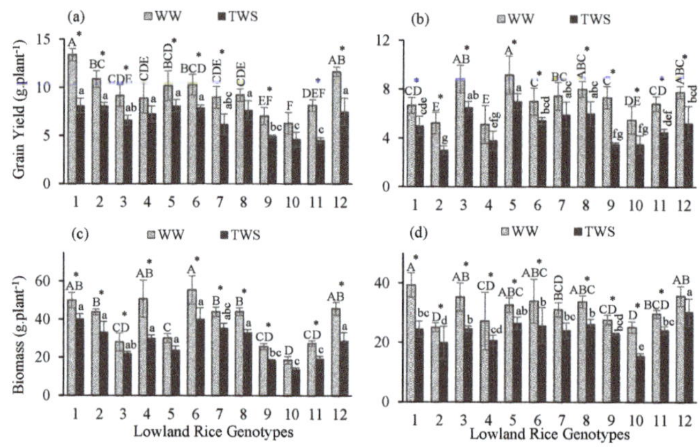

Figure 3. Grain yield (**a**,**b**) and biomass (**c**,**d**) of twelve lowland rice genotypes under well-watered (WW) and terminal water stressed (TWS) conditions during 2018–2019 (**a**,**c**) and 2019–2020 (**b**,**d**). Vertical bars show ± standard errors for means of three repetitions. Capital letters represent the significant ($p < 0.05$) differences among genotypes in WW condition. Small letters represent the significant ($p < 0.05$) differences among genotypes in TWS condition. Centered stars above each pair of the bars represent the significance of parameters for each genotype under WW and TWS conditions.

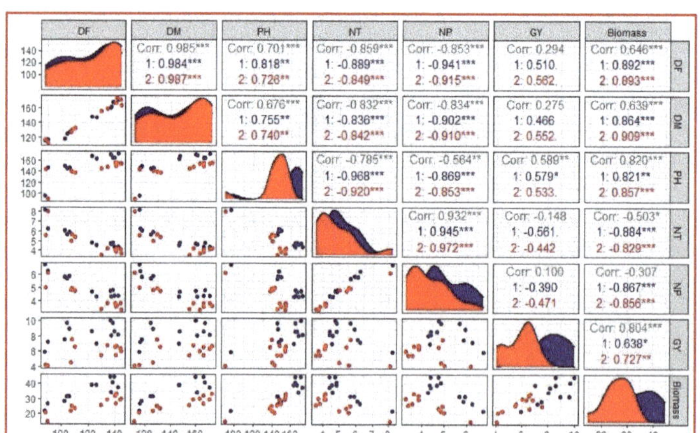

Figure 4. Combined correlation matrix, scatter plot and data distribution for yield and yield attributes of twelve lowland rice genotypes under well-watered (WW) and terminal water stressed (TWS) conditions. Diagonals indicate the distribution of each parameter. Scatter plots are shown in the bottom of diagonals. Values of correlations and significance are indicated with stars and are shown on the top of the diagonal. Values and stars in the blue color (1) indicate correlation among parameters in WW whereas, values and stars in the red color (2) indicate correlation among parameters in TWS conditions. DF: days to flowering, DM: days to maturity, PH: plant height, NT: number of tillers, PN: number of panicles, GY: grain yield, ***: highly significant ($p < 0.001$), **: moderately significant ($p < 0.01$), *: significant ($p < 0.05$).

2.3. Genotypic Classification Corresponding to Stress Indices

Seven stress tolerance indices, including SSI, GMP, STI, M_{PRO}, M_{HAR}, TI and YSI, were computed to distinguish stress-tolerant genotypes from stress-sensitive ones based on GY and RY and the promising values of stress indices under TWS conditions (Table 3). In addition, stress tolerance indices were also studied for hierarchical clustering using a heatmap (Figure 5) and the assessed genotypes were categorized into two main groups: (1) stress tolerant and (2) stress susceptible group and four subgroups (A–D). Subgroup A consisted of four genotypes with the highest GY, RY and stress indices values under TWS; hence, these genotypes could be considered as highly tolerant genotypes. Subgroup B consisted of two genotypes with higher GY, RY and higher stress indices values under TWS; hence, they could be considered as stress-tolerant genotypes. Subgroup C was moderate stress tolerant (three genotypes), as they exhibited intermediate values for GY, RY and stress indices. Subgroup D also consisted of three genotypes that exhibited lower values for GY, RY and stress indices; hence, these genotypes were considered stress susceptible genotypes correspondingly.

Table 3. Values of seven stress tolerance indices for lowland rice genotypes based on grain yield observed under well-watered and terminal water stressed conditions. (Values taken as average from two growing years 2018–2019 and 2019–2020).

	Lowland Rice Genotypes	Y_{WW}	Y_{TWS}	RY_{TWS}	SSI	GMP	STI	M_{PRO}	M_{HAR}	TI	YSI
1	Look Pla	10.02	6.55	0.87	1.19	8.10	6.75	8.29	7.92	3.47	0.65
2	Hom Nang Kaew	8.04	5.54	0.73	1.07	6.67	4.58	6.79	6.56	2.50	0.69
3	Pathum Thani-1	9.00	6.56	0.87	0.93	7.68	6.07	7.78	7.59	2.43	0.73
4	Hom Chan	7.00	5.52	0.73	0.72	6.21	3.97	6.26	6.17	1.48	0.79
5	Hom Pathum	9.68	7.55	1.00	0.75	8.54	7.51	8.61	8.48	2.13	0.78
6	Dum Ja	8.64	6.68	0.89	0.78	7.60	5.94	7.66	7.54	1.96	0.77
7	Chor Lung	8.22	6.03	0.80	0.91	7.04	5.10	7.12	6.96	2.18	0.73
8	Sang Yod	8.61	6.83	0.90	0.71	7.66	6.04	7.72	7.61	1.78	0.79
9	Khao Dawk Mali-105	7.19	4.22	0.56	1.41	5.51	3.12	5.71	5.32	2.97	0.59
10	RD-15	5.91	4.08	0.54	1.06	4.91	2.48	5.00	4.83	1.82	0.69
11	Tia Malay Dang	7.51	4.52	0.60	1.36	5.82	3.49	6.01	5.64	2.99	0.60
12	Lep Nok	9.72	6.37	0.84	1.18	7.87	6.37	8.04	7.69	3.35	0.66

Y_{WW} is mean yield under well-watered conditions, Y_{TWS} is mean yield under terminal water stressed conditions, RY_{TWS} is relative yield under water stressed conditions, SSI is stress susceptibility index, GMP is geometric mean productivity, STI is stress tolerance index, M_{PRO} is mean productivity index, M_{HAR} is harmonic mean index, TI is tolerance index and YSI is yield stability index.

2.4. Association among Stress Tolerance Indices and Grain Yield

Highly positive associations were observed among Y_{WW} and Y_{TWS} (0.85), Y_{WW} and GMP (0.95), Y_{WW} and STI (0.95), Y_{WW} and M_{PRO} (0.97), Y_{WW} and M_{HAR} (0.94), Y_{TWS} and GMP (0.97), Y_{TWS} and STI (0.97), Y_{TWS} and M_{PRO} (0.96) and Y_{TWS} and M_{HAR} (0.98). Whereas Y_{TWS} and YSI (0.64) were positively and Y_{TWS} and SSI (−0.64) were negatively correlated (Figure 6). Correlation assessment among stress indices revealed that there were highly positive associations among GMP, STI, M_{PRO} and M_{HAR} (1.00), whereas there was a moderate positive association among SSI and TI (0.81). In contrast, a highly negative association among SSI and YSI (−1.00) and moderate negative association among TI and YSI (−0.81) were observed (Figure 6).

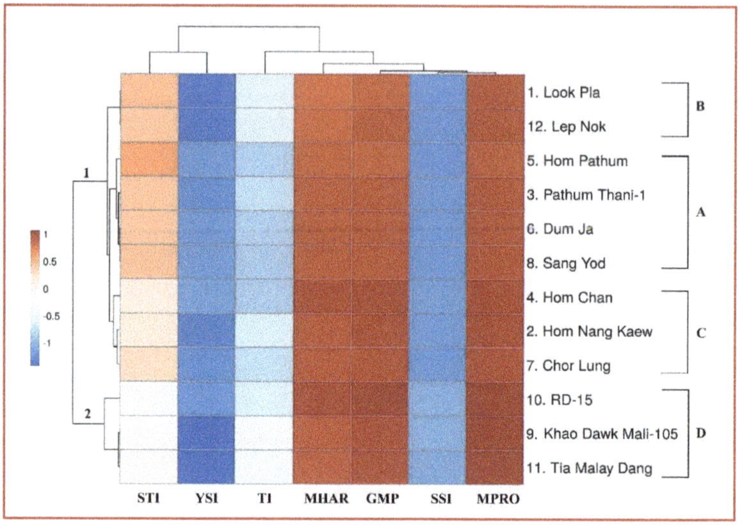

Figure 5. Heatmap of stress indices among twelve lowland rice genotypes under well-watered and terminal water stressed conditions. Group 1 refers to stress-tolerant genotypes, whereas group 2 refers to stress susceptible genotypes. Subgroup A is highly stress tolerant; subgroup B is stress tolerant; subgroup C is moderately stress tolerant, whereas subgroup group D is stress susceptible. Dark red and dark blue colors indicate higher correlation followed by light red and light blue with minimum or no correlation among genotypes and indices.

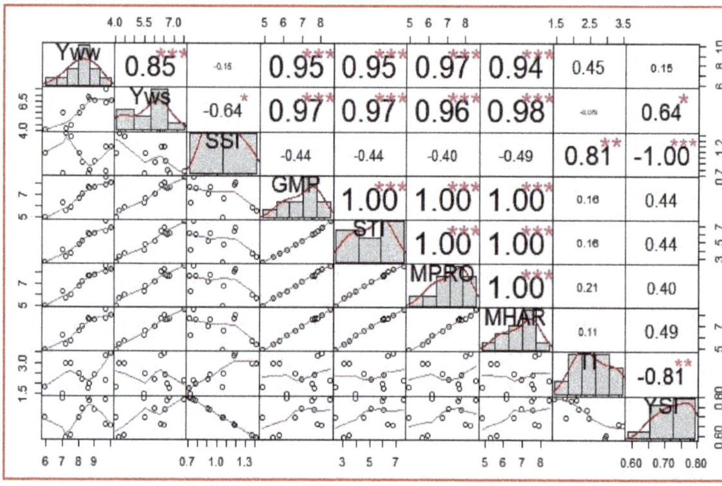

Figure 6. Correlation matrix (Pearson's) of grain yield under well-watered (Y_{WW}), grain yield under terminal water stress (Y_{WS}), stress susceptibility index (SSI), geometric mean productivity (GMP), stress tolerance index (STI), mean productivity index (M_{PRO}), harmonic mean index (M_{HAR}), tolerance index (TI) and yield stability index (YSI) for lowland rice genotypes. Values were taken as average from two growing years 2018–2019 and 2019–2020. Diagonals indicate the distribution of each parameter. Scatter plots with lines are shown in the bottom of diagonals. Values of correlations and significance levels indicated with stars are shown on the top of diagonals. Correlation coefficients are proportional to intensity of color and size of correlation values. ***: highly significant ($p < 0.001$), **: moderately significant ($p < 0.01$), *: significant ($p < 0.05$).

3. Discussion

Water stress is critical to rice crop productivity, especially in rainfed lowland environments. Rainfed lowland rice is vulnerable as it is dependent upon natural precipitation. Variability in seasonal rainfalls and the occurrence of hot, dry spells have increased in rainfed areas. According to Campozano et al. [45] and Spinoni et al. [46], water stress occurrence is expected to be more common, severe and extended as a result of variations in rainfalls due to climate change. Water stress due to climate change would impact on rainfed rice crop productivity. Rice is extremely sensitive to water stress [2,14,15] and rice productivity is significantly affected under terminal water stress (TWS). Different rice genotypes exhibit differential response to TWS, producing a range of grain yield (GY). Hence, it becomes critical to evaluate the performance of yield attributes and yield productivity of rice genotypes under TWS and to identify stress-tolerant genotypes. This strategy will help to stabilize the rice productivity under TWS occurrence and provide sufficient information for genotypic stress tolerance. Furthermore, identification of promising stress tolerance indices under well-watered (WW) and TWS could be useful for their use in rapid selection process for water stress tolerance in the rice crop breeding program.

Twelve lowland rice genotypes were evaluated under WW and TWS conditions in the current experimental study to examine their responses and identify stress-tolerant genotypes. It was observed that all genotypes indicated significant variations in their performance for yield and yield attributes under WW and TWS conditions. Generally, in our study, day to flowering (DF) and day to maturity (DM) were increased and DF and DM were significantly positive and strongly correlated. TWS caused delay in panicle emergence; hence, delaying the flowering time of most of genotypes. Delayed flowering in rice was also observed under water stress by Davatgar et al. [47], Saikumar et al. [48] and Hussain et al. [49]. Late flowering in rice under TWS is considered as a common impact of TWS [50,51]. Delayed panicle emergence and longer grain filling duration increased the time to maturity, thus increasing the total irrigation water input under TWS. All genotypes consumed more water input in delayed maturity under TWS after resuming irrigation. Plant height (PH) was decreased for all genotypes possibly due to limited water availability resulting in reduced cell elongation. Reduction in the PH of rice genotypes under water stress has been reported in numerous studies [47–49,52,53]. Significant positive correlation was observed among PH and GY and biomass while significant negative associations were indicated among PH and number of panicles (NP) and number of tillers (NT). NT and NP were reduced for all genotypes under TWS in both years. Increase in tiller mortality with increased duration of water stress has been reported by Zain et al. [54]. According to Davatgar et al. [47], water stress at terminal crop stages alters the source to sink association, which results in a reduced number of panicles. NT and NP were highly correlated, which indicated that more tillers produced more panicles. Stress induced at the terminal stage significantly reduced GY and biomass of all genotypes. TWS increases spikelet sterility and reduced grain weight resulting in declined final GY. Reduction in final GY under various water stress levels have been reported in several studies [19,48,55,56]. Biomass of all genotypes was reduced under TWS. However, genotypes with higher biomass produced higher GY. Strong positive association among GY and biomass was observed, and our results were in line with the findings of Torres and Henry [53], Torres et al. [56] and Kumar et al. [55]. High variability among genotypes for their performance of yield and yield attributes indicated that the genotypes could be used in the rice crop breeding program to exploit specific plant attributes such as early maturity, shorter plant height, higher tillering capacity and better GY under TWS for improvement in drought tolerance.

Explored genotypes exhibited highly significant variability in their GY productivity under WW and TWS conditions, which demonstrated that studied genotypes possessed significant genetic variability. Genotypes were differentiated based on GY productivity, relative yield (RY) and performance of computed stress indices which were further categorized into stress tolerant, and stress susceptible groups based on hierarchical clustering. Subgroup A was highly stress tolerant; subgroup B was stress tolerant; subgroup C was moderately stress tolerant, whereas subgroup group D was found stress susceptible. Highly stress-tolerant genotypes indicated the highest GY, RY and improved indices under TWS, whereas tolerant genotypes indicated higher GY, RY and better indices. However, stress-susceptible genotypes indicated lowered GY, RY and inadequate performance for stress indices. According to GY and performance of stress indices, hierarchical clustering helped to identify similarly acting genotypes under evaluation. Highly significant and positive correlation observed among GY under WW and GY under TWS exhibited that genotypes that performed better in WW conditions also produced well under TWS. Similar findings were also reported by Raman et al. [57]. Strongly significant and positive associations of stress indices, GMP, STI, M_{PRO}, M_{HAR} with GY under WW and TWS were observed, which indicated that GMP, STI, M_{PRO} and M_{HAR} were better performer and promising indices to evaluate rice genotypes under WW and TWS conditions. Raman et al. [57] found that GMP and STI were suitable indices in identifying entries under non−stressed and extreme water stressed conditions. GMP has also been reported [31] as a better predictor for GY under water stress when stress was applied at the flowering stage. SSI, TI and YSI were not correlated with GY under WW. SSI was negatively correlated, YSI was significant and positively correlated, whereas TI was not correlated with GY under TWS. Weak associations of SSI, TI and YSI indicated that these indices were not adequate for evaluating lowland rice genotypes under TWS. Anwar et al. [29] also found that SSI, TI and YSI were not appropriate predictors of GY under WW and stressed conditions for evaluating wheat genotypes for drought stress tolerance. GMP, STI, M_{PRO} and M_{HAR} have been found to be suitable stress indices to evaluate genotypes under WW and stressed conditions for various crops including rice, wheat, maize and soyabean. Therefore, it was concluded that GMP, STI, M_{PRO} and M_{HAR} were appropriate indices for their use as rapid selection criteria for screening stress tolerant lowland rice genotypes grown under water stressed conditions, especially when stress is applied at reproductive or terminal crop stages.

4. Materials and Methods

4.1. Plant Material

Twelve commonly cultivated Thai lowland rice genotypes including Look Pla (1), Hom Nang Kaew (2), Pathum Thani-1 (3), Hom Chan (4), Hom Pathum (5), Dum Ja (6), Chor Lung (7), Sang Yod (8), Khao Dawk Mali-105 (9), RD-15 (10), Tia Malay Dang (11) and Lep Nok (12) were used for assessment in this study. Germplasm for genotypes 2, 4, 6, 7, 8 and 11 were collected from Phatthalung Rice Research Center, Phatthalung, Thailand (7°33′59.0″ N, 100°07′32.7″ E) (https://ptl-rrc.ricethailand.go.th/address.php (accessed on 21 September 2021)). Germplasm for genotypes 3, 9 and 10 was collected from commercial seed market. Whereas seeds for genotypes 1, 5 and 12 were collected from farmers in Songkhla province, Thailand.

4.2. Site Description and Crop Management

This research study was conducted in the sheds located at field research area (7°00′14.5″ N, 100°30′14.7″ E) of Faculty of Natural Resources, Prince of Songkla University, Hat Yai, Songkhla Province, in Southern Thailand (Figure 7) for two consecutive years during 2018–2019 and 2019–2020. Topsoil was prepared and a uniform soil sample was collected prior to soil filling in planting containers for soil properties analysis. Soil physicochemical properties observed for both years are indicated in Table S1. Planting was performed on 12 September 2018 and 2 September 2019 for 2018–2019 and 2019–2020, respectively. Completely randomized design (CRD) with three repeats was used to design the experiments for both years. Seeds were sown at 5 cm soil depth by direct seeding in containers having the capacity of 12 kg soil. Three plants were maintained in each container after thinning at seedling stage. Experiments were subjected to two treatments, including control under well-watered (WW) and drought under terminal water stressed (TWS) conditions. Each genotype in treatments was placed in separate group of containers. Automatic drip irrigation system, having the dripper head water flow capacity of 8 litters of water per hour, was installed to apply irrigation for specified time for each day. Plants in both treatments were irrigated equally till 75 days after planting (DAP). To induce TWS, irrigation was stopped at 75th DAP in TWS treatment only for 13 days until temporary wilting was observed, following which irrigation was resumed till maturity. Irrigation water amount as total water consumption for each genotype in each treatment for both growing years was calculated by dripper water flow capacity, irrigation time duration for each day and size of container used in experiments. Total water consumption for genotypes in WW and TWS conditions for each year is shown in Figure S1. Thinning, weeding, fertilization and insect pest management was completed through standard crop management practices.

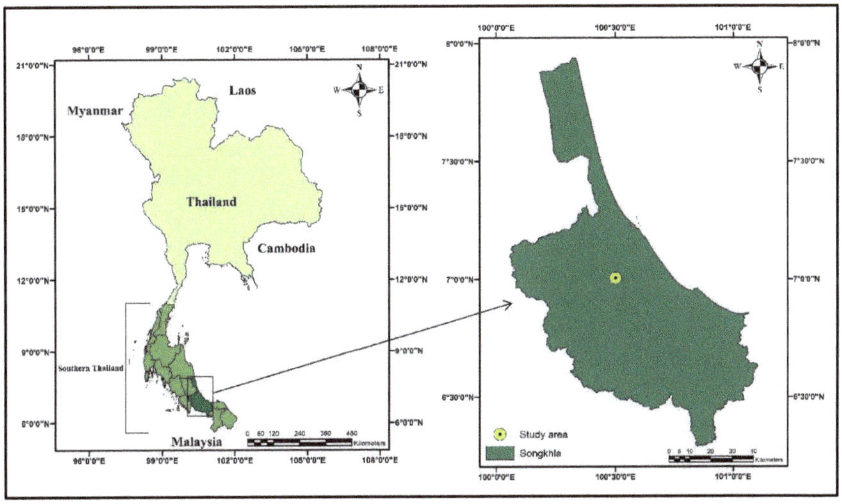

Figure 7. Experimental location at Faculty of Natural Resources, Prince of Songkla University, Songkhla, Thailand (Source: adapted from ArcGIS: $v-10.5$).

4.3. Crop Data Collection

Days to flowering (DF) and days to maturity (DM) were recorded at 50% of panicle emergence and 50% plants at physiological maturity, respectively, from planting date. Plant height (PH) was measured from base of the stems to the flag leaf tip. GY and biomass were recorded by randomly selected three plants for each genotype from each replication as well as each treatment. Plants were hand−harvested, and number of tillers (NT) and number of panicles (NP) were counted per plant as an average from three plants. Grain and plant biomass samples were dried to obtain dry weight in an oven at 70 °C for different time durations till constant weight was observed.

4.4. Computation of Stress Tolerance Indices

Stress tolerance indices were computed to differentiate and identify stress tolerant genotypes from stress susceptible genotypes. GY under WW and TWS conditions was taken as average over 2 years of data to compute stress indices according to methodology adopted by Mansour et al. [6]. Seven different stress tolerance indices comprising stress susceptibility index (SSI) (1) [37], geometric mean productivity (GMP) (2) [32], stress tolerance index (STI) (3) [32], mean productivity index (M_{PRO}) (4) [33], harmonic mean index (M_{HAR}) (5) [34], tolerance index (TI) (6) [35] and yield stability index (YSI) (7) [36] were computed. Mean relative yield (RY) indicates the performance of specific genotype in relation to other examined genotypes under similar level of water stress. Hence, RY under TWS was calculated as GY of each genotype under TWS divided by highest GY achieved in all genotypes. Genotypes with higher GY under WW and TWS, higher RY and exhibiting promising values for stress tolerance indices were classified as stress tolerant genotypes.

$$\text{Stress Suceptibility Index (SSI)} = \left(1 - \frac{Y_{TWS}}{Y_{WW}}\right) / \text{D} \quad (1)$$

$$\text{Geometric Mean Productivity (GMP)} = \sqrt{Y_{WW} \times Y_{TWS}} \quad (2)$$

$$\text{Stress Tolerance Index (STI)} = (Y_{TWS} \times Y_{WW}) \ / \ \text{aww} \quad (3)$$

$$\text{Mean Productivity Index } (M_{PRO}) = (Y_{TWS} + Y_{WW}) \ / \ 2 \quad (4)$$

$$\text{Harmonic Mean Index } (M_{HAR}) = 2(Y_{WW} \times Y_{TWS}) \ / \ (Y_{WW} + Y_{TWS}) \quad (5)$$

$$\text{Tolerance Index (TI)} = (Y_{WW} - Y_{TWS}) \quad (6)$$

$$\text{Yield Stability Index (YSI)} = Y_{TWS} \ / \ Y_{WW} \quad (7)$$

where, Y_{TWS} = mean yield under terminal water stressed (TWS) condition, Y_{WW} = mean yield under well-watered (WW) condition, D = environmental stress intensity, which is 1 (mean yield of all genotypes under TWS/mean yield of all genotypes under WW condition) and aww is an average value for all examined genotypes for grain yield under WW conditions.

4.5. Analysis of Data

Data collected from 2 years of experiments was used to test the significance of results and mean comparisons in R software. Two-way analysis of variance (ANOVA) was performed for yield and yield attributes of all genotypes from three replicates with effect to applied treatments. The effect of years among 2018–2019 and 2019–2020 was also examined. Mean comparisons were made by using the least significant difference (LSD) and p-value < 0.05 was considered as significantly different [58], which was represented using capital and small letters and stars. Pearson's correlation analysis was used to correlate yield and yield attributes as well as computed stress tolerance indices. "Corr" and "GGally" packages of R program were used to compute correlation matrices and visuals. ClustVis [59] software was used to create heatmap and hierarchical clustering [58] for various stress indices taken as an average over two years.

5. Conclusions

Terminal water stress (TWS) significantly reduced the performance of yield and yield attributes. Studied genotypes were found unique in their yield potential as they reflected different responses under well-watered (WW) and TWS conditions. Genotypes Look Pla (1), Pathum Thani-1 (3) Hom Pathum (5), Dum Ja (6) Sang Yod (8), and Lep Nok (12) were found water stress tolerant as they produced relatively higher grain yield (GY), promising values for stress indices and improved performance under TWS. The performance of stress tolerant genotypes was less affected under TWS as compared to stress susceptible genotypes. Hence, these genotypes are potentially recommended for sustaining yield productivity in such environments where TWS occurrence is predicted, especially in southern Thailand. Stress-tolerant genotypes could be used in obtaining better GY under TWS and for improvement in drought tolerance. Strong associations of GMP, STI, M_{PRO} and M_{HAR} with GY under WW and, especially under TWS conditions, indicated that these indices could be used to indicate stress tolerance in rice crop breeding programs for a rapid selection process.

Supplementary Materials: The following are available online at https://www.mdpi.com/article/10.3390/plants10122565/s1, Table S1: Details of soil properties analyzed for experimental soil for 2018–2019 and 2019–2020. Figure S1: Total amount of irrigation water consumed by lowland rice genotypes under well-watered (WW) and terminal water stressed (TWS) conditions during 2018–2019 (a) and 2019–2020 (b).

Author Contributions: T.H. and S.D. conceived and conceptualized the idea. T.H. and N.H. performed the literature review. S.D. and M.A. provided technical expertise to strengthen the basic idea. T.H. conducted experiments and N.H. helped in the collection of data and its analysis. S.D., acquired funds, proofread, and provided intellectual guidance. C.N. proofread the manuscript. All authors read the first draft, helped in revision and approved the article. All authors have read and agreed to the published version of the manuscript.

Funding: This research was partially funded by Graduate School, Prince of Songkla University and Faculty of Natural Resources, Prince of Songkla University. Research was part of Ph.D. study, supported by Higher Education Research Promotion and the Thailand's Education Hub for Southern Region of ASEAN Countries, Project Office of the Higher Education Commission.

Data Availability Statement: The data presented in this study are available in this article and supplementary materials.

Acknowledgments: Authors are thankful to anonymous reviewers for providing valuable comments.

Conflicts of Interest: The authors declare no conflict of interest.

References

1. FAOSTAT. FAO (Online). 2020. Available online: http://www.fao.org/faostat/en/#home (accessed on 18 August 2021).
2. Ichsan, C.N.; Basyah, B.; Zakaria, S.; Efendi, E. Differences of water status and relationship with roots growth and yield of rice under water stress. *Syst. Rev. Pharm.* **2020**, *11*, 611–618.
3. Varinruk, B. Thailand rice production and rice research and development on climate change. In Proceedings of the Workshop on Strengthening APEC Cooperation on Food Security and Climate Change, Ha Noi, Vietnam, 19–21 April 2017.
4. USDA Thailand: Irrigation Shortage Reduces 2015/16 Rice Production. 2015. Available online: https://ipad.fas.usda.gov/highlights/2015/10/th/index.htm (accessed on 25 June 2020).
5. GRiSP. *Rice Almanac*, 4th ed.; International Rice Research Institute: Los Baños, Philippines, 2013.
6. Mansour, E.; Desoky, E.S.M.; Ali, M.M.A.; Abdul-Hamid, M.I.; Ullah, H.; Attia, A.; Datta, A. Identifying drought-tolerant genotypes of faba bean and their agro-physiological responses to different water regimes in an arid Mediterranean environment. *Agric. Water Manag.* **2021**, *247*, 106754. [CrossRef]
7. Ullah, H.; Santiago−Arenas, R.; Ferdous, Z.; Attia, A.; Datta, A. *Improving Water Use Efficiency, Nitrogen Use Efficiency, and Radiation Use Efficiency in Field Crops under Drought Stress: A Review*, 1st ed.; Elsevier Inc.: Amsterdam, The Netherlands, 2019; Volume 156, ISBN 9780128175989.
8. Ray, D.K.; West, P.C.; Clark, M.; Gerber, J.S.; Prishchepov, A.V.; Chatterjee, S. Climate change has likely already affected global food production. *PLoS ONE* **2019**, *14*, e0217148. [CrossRef] [PubMed]
9. Watcharin, S.; Kanrob, P.; Charassri, N.; Nattapon, J. Improving photoperiod insensitivity of a Thai upland rice variety by marker−assisted foreground selection. *Int. J. Agric. Technol.* **2020**, *16*, 199–206.
10. Chutia, J.; Borah, S.P. Water stress effects on leaf growth and chlorophyll content but not the grain yield in traditional rice (*Oryza sativa* Linn.) genotypes of Assam, India II. Protein and proline status in seedlings under PEG induced water stress. *Am. J. Plant Sci.* **2012**, *3*, 971–980. [CrossRef]
11. Ahmadikhah, A.; Marufinia, A. Effect of reduced plant height on drought tolerance in rice. *Biotech* **2016**, *6*, 221. [CrossRef]
12. Zhou, J.; Wang, X.; Jiao, Y.; Qin, Y.; Liu, X.; Zhang, Q.; Fan, L.; Wang, X. Global genome expression analysis of rice in response to drought and high-salinity stresses in shoot, flag leaf, and panicle. *Plant Mol. Biol.* **2007**, *63*, 591–608. [CrossRef]
13. Fischer, K.; Lafitte, R.; Fukai, S.; Atlin, G.; Hardy, B. *Breeding Rice for Drought-Prone Environments*; International Rice Research Institute: Los Baños, Philippines, 2003; ISBN 9712201899.
14. Singh, B.; Reddy, K.R.; Redoña, E.D.; Walker, T. Screening of rice cultivars for morpho-physiological responses to early-season soil moisture stress. *Rice Sci.* **2017**, *24*, 322–335. [CrossRef]
15. Swain, P.; Raman, A.; Singh, S.P.; Kumar, A. Breeding drought tolerant rice for shallow rainfed ecosystem of eastern India. *Field Crops Res.* **2017**, *209*, 168–178. [CrossRef]
16. Agarwal, P.; Parida, S.K.; Raghuvanshi, S.; Kapoor, S.; Khurana, P.; Khurana, J.P.; Tyagi, A.K. Rice improvement through genome-based functional analysis and molecular breeding in India. *Rice* **2016**, *9*, 5. [CrossRef]
17. Rahman, M.T.; Islam, M.T.; Islam, M.O. Effect of water stress at different growth stages on yield and yield contributing characters of transplanted Aman rice. *Pak. J. Biol. Sci.* **2002**, *5*, 169–172. [CrossRef]
18. Yue, B.; Xue, W.; Xiong, L.; Yu, X.; Luo, L.; Cui, K.; Jin, D.; Xing, Y.; Zhang, Q. Genetic basis of drought resistance at reproductive stage in rice: Separation of drought tolerance from drought avoidance. *Genetics* **2006**, *172*, 1213–1228. [CrossRef] [PubMed]
19. Pantuwan, G.; Fukai, S.; Cooper, M.; Rajatasereekul, S. Yield response of rice (*Oryza sativa* L.) genotypes to drought under rainfed lowlands 4. Vegetative stage screening in the dry season. *Field Crops Res.* **2004**, *89*, 281–297. [CrossRef]
20. Ovenden, B.; Milgate, A.; Lisle, C.; Wade, L.J.; Rebetzke, G.J.; Holland, J.B. Selection for water—Soluble carbohydrate accumulation and investigation of genetic × environment interactions in an elite wheat breeding population. *Theor. Appl. Genet.* **2017**, *130*, 2445–2461. [CrossRef] [PubMed]

21. Farooq, M.; Kobayashi, N.; Ito, O.; Wahid, A.; Serraj, R. Broader leaves result in better performance of indica rice under drought stress. *J. Plant Physiol.* **2010**, *167*, 1066–1075. [CrossRef]
22. Wadhwa, R.; Kumari, N.; Sharma, V. Varying light regimes in naturally growing Jatropha curcus: Pigment, proline and photosynthetic performance. *J. Stress Physiol. Biochem.* **2010**, *6*, 67–80.
23. Blum, A. Drought adaptation in cereal crops. In *Drought Adaptation in Cereals*; Haworth Food Products Press: New York, NY, USA, 2006; pp. 301–333.
24. Cha-um, S.; Yooyongwech, S.; Supaibulwatana, K. Water deficit stress in the reproductive stage of four indica rice (*Oryza sativa* L.) genotypes. *Pak. J. Bot.* **2010**, *42*, 3387–3398.
25. Majidi, M.M.; Tavakoli, V.; Mirlohi, A.; Sabzalian, M.R. Wild safflower species (*Carthamus oxyacanthus* Bieb.): A possible source of drought tolerance for arid environments. *Aust. J. Crop Sci.* **2011**, *5*, 1055–1063.
26. Todaka, D.; Shinozaki, K.; Yamaguchi-shinozaki, K. Recent advances in the dissection of drought-stress regulatory networks and strategies for development of drought-tolerant transgenic rice plants. *Front. Plant Sci.* **2015**, *6*, 84. [CrossRef]
27. Hussain, N.; Ahmed, M.; Duangpan, S.; Hussain, T.; Taweekun, J. Potential impacts of water stress on rice biomass composition and feedstock availability for bioenergy production. *Sustainability* **2021**, *13*, 10449. [CrossRef]
28. IRRI. *Standard Evaluation System for Rice*, 5th ed.; The Philippines International Rice Research Institute: Los Banos, CA, USA, 2014.
29. Anwar, J.; Subhani, G.M.; Hussain, M.; Ahmad, J.; Hussain, M.; Munir, M. Drought tolerance indices and their correlation with yield in exotic wheat genotypes. *Pak. J. Bot.* **2011**, *43*, 1527–1530.
30. Rashid, A.; Saleem, Q.; Nazir, A.; Kazım, H.S. Yield potential and stability of nine wheat varieties under water stress conditions. *Int. J. Agric. Biol.* **2003**, *5*, 7–9.
31. Wasae, A. Evaluation of drought stress tolerance based on selection indices in Haricot Bean varieties exposed to stress at different growth stages. *Int. J. Agron.* **2021**, *2021*, 6617874. [CrossRef]
32. Fernandez, G.C. Effective selection criteria for assessing plant stress tolerance. In Proceedings of the International Symposium on Adaptation of Vegetables and other Food Crops in Temperature and Water Stress, Shanhua, Taiwan, 13–16 August 1992; pp. 257–270.
33. Hossain, A.B.S.; Sears, R.G.; Cox, T.S.; Paulsen, G.M. Desiccation tolerance and its relationship to assimilate partitioning in winter wheat. *Crop Sci.* **1990**, *30*, 622–627. [CrossRef]
34. Schneider, K.A.; Rosales-serna, R.; Ibarra-perez, F.; Cazares-enriquez, B.; Acosta-gallegos, J.A.; Ramirez-vallejo, P.; Wassimi, N.; Kelly, J.D. Improving common bean performance under drought stress. *Crop Sci.* **1997**, *37*, 43–50. [CrossRef]
35. Rosielle, A.A.; Hamblin, J. Theoretical aspects of selection for yield in stress and non-stress environments. *Crop Sci.* **1981**, *21*, 943–946. [CrossRef]
36. Bouslama, M.; Schapaugh, W.T. Stress tolerance in soybeans. I. Evaluation of three screening techniques for heat and drought tolerance. *Crop Sci.* **1984**, *24*, 933–937. [CrossRef]
37. Fischer, R.A.; Maurer, R. Drought resistance in spring wheat cultivars. I. Grain yield responses. *Aust. J. Agric. Res.* **1978**, *29*, 897–912. [CrossRef]
38. Clarke, J.M.; Townley-Smith, F.; McCaig, T.N.; Green, D.G. Growth analysis of spring wheat cultivars of varying drought resistance. *Crop Sci.* **1984**, *24*, 537–541. [CrossRef]
39. Mitra, J. Genetics and genetic improvement of drought resistance in crop plants. *Curr. Sci.* **2001**, *80*, 758–762.
40. Hall, A.E. Is dehydration tolerance relevant to genotypic differences in leaf senescence and crop adaptation to dry environments. In *Plant Responses to Cellular Dehydration during Environmental Stress*; Close, T.J., Bray, E.A., Eds.; Current Topics in Plant Physiology: Washington, DC, USA, 1993; pp. 1–10.
41. Blum, A. *Plant Breeding for Stress Environments*; CRC Press: Boca Raton, FL, USA, 2018; p. 212.
42. Golabadi, M.; Arzani, A.; Maibody, M. Assessment of drought tolerance in segregating populations in Durum Wheat. *African J. Agric. Res.* **2006**, *1*, 162–171.
43. Saba, J.; Moghadam, M.; Ghasemi, K.; Nishabouri, M. Genetic properties of drought resistance indices. *J. Agric. Sci. Technol.* **2001**, *3*, 43–49.
44. Arif, A.; Parveen, N.; Waheed, M.Q.; Atif, R.M.; Waqar, I.; Shah, T.M. A comparative study for assessing the drought-tolerance of Chickpea under varying natural growth environments. *Front. Plant Sci.* **2021**, *11*, 2228. [CrossRef]
45. Campozano, L.; Ballari, D.; Montenegro, M.; Avilés, A. Future meteorological droughts in ecuador: Decreasing trends and associated spatio-temporal features derived from CMIP5 models. *Front. Earth Sci.* **2020**, *8*, 17. [CrossRef]
46. Spinoni, J.; Vogt, J.V.; Naumann, G.; Barbosa, P.; Dosio, A. Will drought events become more frequent and severe in Europe? *Int. J. Climatol.* **2018**, *1736*, 1718–1736. [CrossRef]
47. Davatgar, N.; Neishabouri, M.R.; Sepaskhah, A.R.; Soltani, A. Physiological and morphological responses of rice (*Oryza sativa* L.) to varying water stress management strategies. *Int. J. Plant Prod.* **2009**, *3*, 19–32. [CrossRef]
48. Saikumar, S.; Varma, C.M.K.; Saiharini, A.; Kalmeshwer, G.P.; Nagendra, K.; Lavanya, K.; Ayyappa, D. Grain yield responses to varied level of moisture stress at reproductive stage in an interspecific population derived from Swarna/*O. glaberrima* introgression line. *NJAS Wagening. J. Life Sci.* **2016**, *78*, 111–122. [CrossRef]
49. Hussain, T.; Anothai, J.; Nualsri, C.; Soonsuwon, W. Application of CSM-CERES-Rice in scheduling irrigation and simulating effect of drought stress on upland rice yield. *Indian J. Agric. Res.* **2018**, *52*, 140–145. [CrossRef]

50. Pantuwan, G.; Fukai, S.; Cooper, M.; Rajatasereekul, S.; O'Toole, J.C. Yield response of rice (*Oryza sativa* L.) genotypes to drought under rainfed lowland 3. Plant factors contributing to drought resistance. *Field Crops Res.* **2002**, *73*, 181–200. [CrossRef]
51. Zhao, D.L.; Atlin, G.N.; Amante, M.; Cruz, M.T.S.; Kumar, A. Developing aerobic rice cultivars for water-short irrigated and drought-prone rainfed areas in the tropics. *Crop Sci.* **2010**, *50*, 2268–2276. [CrossRef]
52. Anantha, M.S.; Patel, D.; Quintana, M.; Swain, P.; Dwivedi, J.L.; Torres, R.O.; Verulkar, S.B.; Variar, M.; Mandal, N.P.; Kumar, A.; et al. Trait combinations that improve rice yield under drought: Sahbhagi Dhan and new drought-tolerant varieties in South Asia. *Crop Sci.* **2016**, *56*, 408–421. [CrossRef]
53. Torres, R.O.; Henry, A. Yield stability of selected rice breeding lines and donors across conditions of mild to moderately severe drought stress. *Field Crops Res.* **2018**, *220*, 37–45. [CrossRef] [PubMed]
54. Zain, N.A.M.; Ismail, M.R.; Puteh, A.; Mahmood, M.; Islam, M.R. Impact of cyclic water stress on growth, physiological responses and yield of rice (*Oryza sativa* L.) grown in tropical environment. *Ciência Rural* **2014**, *44*, 2136–2141. [CrossRef]
55. Kumar, A.; Verulkar, S.; Dixit, S.; Chauhan, B.; Bernier, J.; Venuprasad, R.; Zhao, D.; Shrivastava, M.N. Yield and yield-attributing traits of rice (*Oryza sativa* L.) under lowland drought and suitability of early vigor as a selection criterion. *Field Crops Res.* **2009**, *114*, 99–107. [CrossRef]
56. Torres, R.O.; McNally, K.L.; Cruz, C.V.; Serraj, R.; Henry, A. Screening of rice genebank germplasm for yield and selection of new drought tolerance donors. *Field Crops Res.* **2013**, *147*, 12–22. [CrossRef]
57. Raman, A.; Verulkar, S.B.; Mandal, N.P.; Variar, M.; Shukla, V.D.; Dwivedi, J.L.; Singh, B.N.; Singh, O.N.; Swain, P.; Mall, A.K.; et al. Drought yield index to select high yielding rice lines under different drought stress severities. *Rice* **2012**, *5*, 2. [CrossRef] [PubMed]
58. Duangpan, S.; Buapet, P.; Sujitto, S.; Eksomtramage, T. Early assessment of drought tolerance in oil palm D $\times$ P progenies using growth and physiological characters in seedling stage. *Plant Genet. Resour.* **2018**, *16*, 544–554. [CrossRef]
59. Metsalu, T.; Vilo, J. ClustVis: A web tool for visualizing clustering of multivariate data using principal component analysis and heatmap. *Nucleic Acid Res.* **2015**, *43*, W566–W570. [CrossRef]

plants

MDPI

Article

Co-Application of Biochar and *Arbuscular mycorrhizal* Fungi Improves Salinity Tolerance, Growth and Lipid Metabolism of Maize (*Zea mays* L.) in an Alkaline Soil

Ndiaye Ibra Ndiate [1,2], Qudsia Saeed [3], Fasih Ullah Haider [1,2], Cai Liqun [1,2,*], Jackson Nkoh Nkoh [4,5] and Adnan Mustafa [6]

1. College of Resources and Environmental Sciences, Gansu Agricultural University, Lanzhou 730070, China; ibra@st.gsau.edu.cn (N.I.N.); fasihullahhaider281@gmail.com (F.U.H.)
2. Gansu Provincial Key Laboratory of Arid Land Crop Science, Gansu Agricultural University, Lanzhou 730070, China
3. College of Natural Resources and Environment, Northwest Agriculture and Forestry University, Xianyang 712100, China; syedaqudsia.saeed@yahoo.com
4. Organization of African Academic Doctors, Off Kamiti Road, Nairobi 25305-00100, Kenya; nkohjackson@issas.ac.cn
5. State Key Laboratory of Soil and Sustainable Agriculture, Institute of Soil Science, Chinese Academy of Sciences, P.O. Box 821, Nanjing 210008, China
6. Biology Center CAS, SoWa RI, Na Sadkach 7, 370-05 České Budějovice, Czech Republic; adnanmustafa780@gmail.com

* Correspondence: cailq@gsau.edu.cn; Tel.: +86-138-9327-3886

Citation: Ndiate, N.I.; Saeed, Q.; Haider, F.U.; Liqun, C.; Nkoh, J.N.; Mustafa, A. Co-Application of Biochar and *Arbuscular mycorrhizal* Fungi Improves Salinity Tolerance, Growth and Lipid Metabolism of Maize (*Zea mays* L.) in an Alkaline Soil. *Plants* **2021**, *10*, 2490. https://doi.org/10.3390/plants10112490

Academic Editors: M. Iftikhar Hussain, Adele Muscolo and Mukhtar Ahmed

Received: 29 September 2021
Accepted: 15 November 2021
Published: 17 November 2021

Abstract: This study reports the mitigating strategy against salinity by exploring the potential effects of biochar (5%), *Arbuscular mycorrhizal* fungi (20 g/pot, AMF), and biochar + AMF on maize (*Zea mays* L.) plants grown under saline stress in a greenhouse. The maize was grown on alkaline soil and subjected to four different saline levels; 0, 50, 100, and 150 mM NaCl. After 90 d for 100 mM NaCl treatment, the plant's height and fresh weight were reduced by 17.84% and 39.28%, respectively, compared to the control. When the saline-treated soil (100 mM NaCl) was amended with AMF, biochar, and biochar + AMF, the growth parameters were increased by 22.04%, 26.97%, 30.92% (height) and 24.79%, 62.36%, and 107.7% (fresh weight), respectively. Compared to the control and single AMF/biochar treatments, the combined application of biochar and AMF showed the most significant effect in improving maize growth under saline stress. The superior mitigating effect of biochar + AMF was attributed to its effective ability in (i) improving soil nutrient content, (ii) enhancing plant nutrient uptake, (iii) increasing the activities of antioxidant enzymes, and (iv improving the contents of palmitoleic acid (C16:1), oleic acid (C18:1), linoleic acid (C18:2), and linolenic acid (C18:3). Thus, our study shows that amending alkaline and saline soils with a combination of biochar-AMF can effectively mitigate abiotic stress and improve plant growth. Therefore, it can serve as a reference for managing salinity stress in agricultural soils.

Keywords: biochar; alkaline soils; abiotic stress; *Arbuscular mycorrhizal* fungi; fatty acids; *Zea mays* L.

1. Introduction

Biochar is an alkaline byproduct of the fast or slow pyrolysis of different biomass in a limited oxygen environment. Due to its high pH, biochar application to soils is mostly popular for the management of acidic soils [1,2]. Studies have shown that when applied to alkaline soils, biochar decreased the soil bulk density, cation exchange capacity, organic carbon content, nitrate retention, and bioavailable potassium [3,4]. The negative effect of biochar on the availability of phosphorus in alkaline soils was reported by Baigorri et al. [5]. The authors observed that the adsorption of phosphorous to Al/Fe-modified biochar decreased its bioavailability and partially explained the observed negative effects of pristine biochar applied to alkaline soils. Understanding the role of pristine (unmodified) biochar in alkaline soils has been extended to studying their interactions with traditional NPK fertilizers. Mete et al. [6] showed that when biochar was applied together with NPK fertilizer, then it significantly improved seed yield and total biomass in three genotypes of soybean in alkaline soils. These different studies reveal that under the right conditions, biochar can induce beneficial effects on crops grown on alkaline soils.

In soils having a large concentration of soluble salts (e.g., containing Na^+ ions), plants suppress growth due to the negative effects of high saline concentration on osmotic balance, ion homeostasis, and oxidative stress due to reactive oxygen species (ROS) [7,8]. According to Foyer and Noctor [9], ROS can disrupt cellular functions and negatively affect nucleic acids, activities of oxidant proteins, and induce lipid peroxidation. Thus, researchers have suggested that the accumulation of ROS under salinity stress in plants is one of the major causes of reduced global agricultural productivity [10]. The study of Farhangi-Abriz and Torabian [10] showed that under increased saline stress, the activities of catalase (CAT), ascorbate peroxidase, peroxidase (POD), and superoxide dismutase (SOD) in the leaves and roots of common bean (*Phaseolus vulgaris* L. cv. Derakhshan) were significantly increased. According to the authors, amending the soils with biochar suppressed saline-induced oxidative stress and improved the growth of the bean plant.

The management of saline soils with appropriate rates of biochar has a mitigating effect on N leaching, enhances N retention, and reduces volatilization of ammonia [11]. Also, the application of biochar to alkaline soils has been shown to favor the colonization of ammonia-oxidizing microorganisms and inhibit N leaching [12]. Cui et al. [13] reported that co-application of biochar and effective microorganisms significantly inhibited salinization, improved soil fertility, increased soil nutrient content, enhanced enzymes activities, thereby improving the growth of *S. cannabin*. The inoculation of saline contaminated soils with *Arbuscular mycorrhizal* fungi (AMF) strain *Glomus mosseae* improved the growth of tomato (*Lycopersicon esculentum* L. cv. Zhongzha105) plants by improving root colonization, contents of chlorophyll, fruit fresh weight, fruit yield, and total plant growth [14]. Mycorrhizal colonization can be enhanced when AMF is co-applied with biochar to mitigate the adverse effects of drought-related stress on plant growth [15]. The few studies that have reported on the combined application of biochar and AMF suggest that biochar modifies the physicochemical properties of soils which improves AMF colonization [16]. Due to the limited number of studies on the co-application of biochar and AMF in alkaline-saline soils, the specific mechanisms for improving plant growth in these soils is not clear. Thus, this study was designed to study the growth of maize (*Zea mays* L.) in alkaline soils under saline induced stress and to evaluate the effects of the combined application of AMF and biochar on maize growth parameters. It was hypothesized that AMF and biochar would have additive effects on plant growth, lipid metabolism, and nutrient availability, meaning that combining the two treatments would result in greater plant growth than either treatment alone under salinity stress.

2. Results

2.1. *Effect of AMF and Biochar on Fatty Acids Composition of Maize Leaves*

Table 1 shows the contents of fatty acids in the leaves of 90 days old plant under saline-induced stress. The contents of myristic acid (C14:0), palmitic acid(C16:0), palmitoleic acid (C16:1), stearic acid (C18:0), oleic acid (C18:1), linoleic acid (C18:2), linolenic acid (C18:3), arachidic acid (C20:0), and behenic acid (C22:0) were affected differently for the different treatments. The contents of C14:0, C16:0, and C18:0 were not significantly altered for different treatments compared to S0. With increasing salinity, the content of unsaturated fatty acids (C16:1, C18:1, C18:2, and C18:3) was observed to decrease for control (S1, S2, S3), biochar (BS1, BS2, BS3), and AMF (AS1, AS2, AS3) treatments. Compared to S0, the contents of C16:1, C18:1, C18:2, C18:3 was significantly ($p < 0.05$) decreased by 44.02%, 33.49%, 25.99%, and 0.53% in S3 treatment, respectively. Nevertheless, the contents of C20:0 (52.82%) and C22:0 (87.77%) were increased significantly ($p < 0.05$) improved for the same treatment. Also, relative to S3 treatment, BS3 treatment positively impacted the contents of C16:1, C18:2, C20:0, and C22:0while negatively affecting C18:1 and C18:3 (Table 1). For AS3 treatment, the contents of C16:1, C18:1, C20:0, and C22:0 were significantly decreased while C18:2 and C18:3 were only slightly increased. The combined application of biochar and AMF (ABS3) demonstrated the most significant positive effect on the contents of all fatty compared to S0 while C22:0 was negatively affected relative to S3. Specifically, the contents of C16:1, C18:1, C18:2, C18:3 and C20:0 were increased by 161.36%, 101.3%, 65.07%, 12.17%, 18.23% relative to S3 or 46.31%, 33.89%, 22.17%, 11.58%, 88.31%, and 56.46% relative to S0.

2.2. *Influence of AMF and Biochar on Soil Nutrient Content*

The following acronyms will be used as defined in the materials and method section (Table 2): S0, S1, S2 and S3 (0, 50, 100, and 150 Mm NaCl treatment, respectively); BS0, BS1, BS2 and BS3 (biochar + 0, 50, 100, and 150 Mm NaCl treatment, respectively); AS0, AS1, AS2 and AS3 (AMF + 0, 50, 100, and 150 Mm NaCl treatment, respectively); ABS0, ABS1, ABS2 and ABS3 (biochar + AMF + 0, 50, 100, and 150 Mm NaCl treatment, respectively). Figure 1 gives a summary of the selected physicochemical properties of the soil after the growth of maize. After maize growth, the soil pH was decreased in the S0, S1, BS0, AS0, ABS0, ABS1 treatments by 0.42, 0.03, 0.48, 0.38, 0.44, 0.38 units while S2, S3, and A3 had an increase of 0.06, 0.20, and 0.05 units compared to the original soil (pH 8.25), respectively (Figure 1A). Compared to the S0 treatment, only the BS0 and ABS0 treatments experienced a slight decrease in pH. For different treatments, the content of potassium was either increased or decreased with an increment in the concentration of NaCl (Figure 1B). For instance, under no amendment, the content of K was decreased by 3.83%, 23.76%, and 23.8% when the concentration of NaCl was 50 (S1), 100 (S2) and 150 mM (S3), respectively. This decrement was also observed for AS0, AS1, AS2, AS3, and ABS3 with values lowered by 24.68%, 20.72%, 8.82%, 11.64%, and 21.37%, respectively. This observation shows that increase in the concentration of Na^+ ions has a negative impact on available K^+ ions needed for plant growth, which has been sighted as a major concern for saline soils [17]. Nevertheless, amending the alkaline soil with biochar (BS0, BS1, BS2, BS3) and biochar + AMF (ABS0, ABS1, and ABS2) significantly ($p < 0.05$) improved the content of K; with the largest increment recorded for BS3 (32.49%), ABS0 (24.62%), ABS1 and (23.31%) relative to S0. When compared with the saline-treated control (S1, S2, S3), biochar (BS1, BS2, BS3), AMF (A1, A3, except A1), and biochar + AMF (ABS1, ABS2, ABS3) treatments significantly ($p < 0.05$) improved K in soil. Thus, biochar amendment with/without AMF can effectively mitigate K^+ ion loss in high pH saline soils. Phosphorus and nitrogen are important nutrient requirements for plant growth which can become deficient when the soil health and fertility are threatened. For the control (S0, S1, S2, and S3) and AMF (AS0, AS1, AS2, and AS3) treatments, the content of P was decreased as the concentration of NaCl was increased (Figure 1C). Relative to the S0 treatment, the content of P was increased by 22.58% for B_0 but was decreased by 38.45% and 30.41% for A_0 and AB_0 treatments, respectively. Also, relative to the saline (S1, S2, S3) treatments, biochar (BS1, BS2, BS3), AMF (AS1, AS2, AS3), and biochar + AMF (ABS1, ABS2, ABS3) amendments showed contrasting effects on available P. For instance, the available P was increased by 21.43%, 12.78%, 11.97%, 4.23%, 7.43%, 46.43%, 18.94% for BS1, BS2, BS3, ABS1, ABS2, ABS3, and AS1, respectively, but decreased for AS2 and AS3 treatments. From this result, it is evident that salinity negatively impacts available P and the individual or co-application of biochar with AMF can mitigate this negative effect.

Table 1. Fatty acid composition in lipids of the maize plant.

Treatment	Fatty Acids (mol %)								
	C14:0	C16:0	C16:1	C18:0	C18:1	C18:2	C18:3	C20:0	C22:0
S0	1.84 ± 0.138a	12.7 ± 0.013a	0.786 ± 0.035bcd	11.5 ± 0.611a	10.8 ± 0.017b	27.4 ± 0.141b	28.4 ± 0.107b	2.48 ± 0.103i	2.29 ± 0.049g
S1	1.75 ± 0.151a	12.1 ± 0.216a	0.701 ± 0.011cdef	11.2 ± 0.064a	11.0 ± 1.07b	26.9 ± 0.06b	28.1 ± 0.017b	3.09 ± 0.005g	3.02 ± 0.002e
S2	1.81 ± 0.053a	11.9 ± 0.63a	0.540 ± 0.059ghi	11.5 ± 0.236a	8.92 ± 0.016c	21.8 ± 0.948cd	28.5 ± 1.18b	3.79 ± 0.077de	3.85 ± 0.15c
S3	1.90 ± 0.073a	12.4 ± 0.199a	0.440 ± 0.033ij	11.0 ± 1.13a	7.18 ± 0.556d	20.3 ± 0.425e	28.3 ± 0.587b	3.95 ± 0.055c	4.3 ± 0.167b
BS0	1.93 ± 0.056a	11.5 ± 1.51a	0.840 ± 0.081bc	11.7 ± 1.03a	10.7 ± 0.022b	27.5 ± 0.578b	28.8 ± 1.71b	2.46 ± 0.005i	2.32 ± 0.006g
BS1	1.84 ± 0.188a	11.8 ± 2.18a	0.640 ± 0.076defgh	11.0 ± 1.0a	11.1 ± 1.03b	27.2 ± 0.647b	28.0 ± 1.07b	3.70 ± 0.154e	2.79 ± 0.056f
BS2	1.84 ± 0.286a	12.0 ± 2.81a	0.560 ± 0.044fghi	12.2 ± 1.1a	8.06 ± 1.001cd	22.2 ± 1.53c	28.9 ± 1.04b	4.63 ± 0.017b	3.54 ± 0.006d
BS3	1.78 ± 0.161a	11.7 ± 0.793a	0.510 ± 0.055hi	11.7 ± 0.519a	6.92 ± 1.70d	20.8 ±0.054de	28.0 ± 2.31b	5.66 ± 0.111a	5.61 ± 0.053a
AS0	1.83 ± 0.031a	12.6 ± 0.574a	0.710 ± 0.071cdef	11.0 ± 0.049a	10.8 ± 0.577b	28.1 ± 0.021b	28.8 ±0.941b	2.49 ± 0.053i	2.29 ± 0.037g
AS1	1.79 ± 0.172a	12.2 ± 1.61a	0.680 ± 0.108defg	11.1 ± 1.29a	10.8 ± 0.585b	27.4 ± 0.584b	28.7 ± 0.58b	2.84 ± 0.063h	2.62 ± 0.05f
AS2	1.74 ± 0.057a	12.0 ± 1.47a	0.630 ± 0.044efgh	11.5 ± 1.1a	8.88 ± 0.028c	22.0 ± 1.217cd	28.1 ± 0.988b	2.71 ± 0.0005h	2.78 ± 0.052f
AS3	1.69 ± 0.112a	12.3 ± 1.14a	0.330 ± 0.249j	11.7 ± 1.77a	6.72 ± 0.527d	20.7 ± 0.051de	28.3 ± 0.498b	2.79 ± 0.059h	2.67 ± 0.302f
ABS0	1.84 ± 0.063a	11.9 ± 1.41a	0.760 ± 0.039cde	11.9 ± 1.24a	10.6 ± 0.06b	27.9 ± 0.554b	28.2 ± 1.00b	2.42 ± 0.025i	2.26 ± 0.051g
ABS1	1.90 ± 0.074a	12.4 ± 0.557a	0.920 ± 0.03b	12.0 ± 1.77a	10.3 ± 0.052b	27.9 ± 0.101b	28.6 ± 1.01b	3.53 ± 0.025f	2.72 ± 0.296f
ABS2	1.94 ± 0.042a	12.2 ± 0.641a	1.16 ± 0.001a	11.9 ± 1.24a	14.4 ± 1.16a	33.9 ± 1.52a	28.9 ± 0.973b	3.91 ± 0.051cd	3.043 ± 0.058e
ABS3	1.91 ± 0.184a	12.1 ± 1.15a	1.15 ± 0.01a	11.2 ± 1.02a	14.5 ± 0.579a	33.5 ± 0.57a	31.7 ± 1.53a	4.67 ± 0.194b	3.56 ± 0.034d

Data are presented as the means± standard deviation of three replicates and mean values followed by different letters are significantly different at $p < 0.05$ according to Duncan multiple range test.

Table 2. Treatment arrangement.

Code	Description
S0	0 mM NaCl without soil amendment
S1	50 mM NaCl without soil amendment
S2	100 mM NaCl without soil amendment
S3	150 mM NaCl without soil amendment
BS0	biochar without NaCl and AMF
BS1	biochar and 50 mM NaCl without AMF
BS2	biochar and 100 mM NaCl without AMF
BS3	biochar and 150 mM NaCl without AMF
AS0	AMF inoculation without NaCl and biochar
AS1	AMF inoculation with 50 mM NaCl but without biochar
AS2	AMF inoculation with 100 mM NaCl but without biochar
AS3	AMF inoculation with 150 mM NaCl but without biochar
ABS0	combined AMF inoculation and biochar but without NaCl
ABS1	combined AMF inoculation, biochar and 50 mM NaCl
ABS2	combined AMF inoculation, biochar and 100 mM NaCl
ABS3	combined AMF inoculation, biochar and 150 mM NaCl

From Figure 1, it can be observed that the content of N significantly decreases in the control treatment as the saline content was increased from 0–150 mM; except for S1 treatment (Figure 1D). For instance, compared to the S0 treatment, N was increased by 26.4 mg kg^{-1} and decreased by 47.3 and 74.9 mg kg^{-1} when the soil was treated with 50 (S1), 100 (S2), and 150 Mm (S3) of NaCl, respectively. The application of biochar alone (BS0, BS1, BS2, BS3) and in combination with AMF (ABS0, ABS1, ABS2, ABS3) significantly improved the content of N under increasing salt stress. For no salt treatments (BS0, AS0, and ABS0), N content was improved by up to 74.25%, 16.61%, and 90.89% for BS0, AS0, and ABS0 amendments relative to S0 treatment, respectively. Under increasing salt stress, N content was improved progressively when biochar and/or AMF were applied. Specifically, for BS1, BS2, and BS3 treatments, the amount of N (relative to S1, S2, and S3) increased by 37.3%, 88.91%, and 105.6% (or 50.25, 57, and 50.57% relative to S0), respectively. Similarly, the recorded increment in the content of N for ABS1, ABS2, and ABS3 was 50.88%, 95.79%, and 102.44% (or 65.11, 62.71, and 48.29% relative to S0), respectively. Additionally, applying AMF alone did not mitigate N loss (−18.05%) when 50 mM NaCl was added but showed a positive effect at saline concentrations of 100 (8.25%) and 150 mM (26.08%). Thus, amending alkaline soils with biochar, biochar + AMF, and AMF alone can mitigate the negative effect of Na^+ ions on N content, with biochar and biochar + AMF exhibiting the most significant effects at 100 and 150 mM of Na^+ ions.

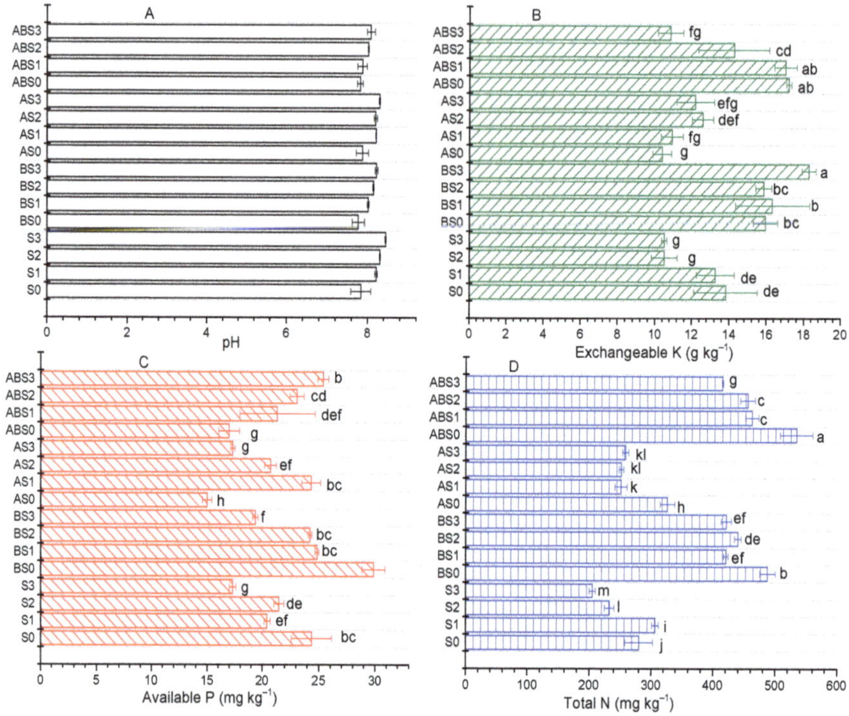

Figure 1. Influence of *Arbuscular mycorrhizal* fungi (AMF) and biochar on (**A**) soil pH, (**B**) exchangeable K, (**C**) available P, and (**D**) total N after harvesting. Data present the mean ± standard deviation of three replicates and mean values followed by different letters are significantly different at $p < 0.05$ according to Duncan multiple range test. S0, S1, S2, and S3 (0, 50, 100, and 150 Mm NaCl treatment, respectively); BS0, BS1, BS2 and BS3 (biochar + 0, 50, 100, and 150 Mm NaCl treatment, respectively); AS0, AS1, AS2 and AS3 (AMF + 0, 50, 100, and 150 Mm NaCl treatment, respectively); ABS0, ABS1, ABS2 and ABS3 (biochar + AMF + 0, 50, 100, and 150 Mm NaCl treatment, respectively).

2.3. The Effects of Salinity, Biochar, and AMF on Plant Growth Parameters

Plant growth parameters were evaluated based on the maize height after 45 and 90 days of growth, total fresh weight, and the total number of leaves after 90 days (Figures 2 and 3). The plant height after 45 days of growth was 2.73%, 16.02%, and 4.92% shorter compared to S0 when the saline concentration was 50 (S1), 100 (S2), and 150 mM (S3), respectively (Figure 2A). After 90 days of growth for the same treatments (Figure 2B), the plant height was 6.22%, 17.84%, and 17.3% shorter, respectively. When the soil was amended with biochar and/or AMF without NaCl, the plant height was increased by 23.67%, 28.67%, 2.11% after 45 days and by 21.08%, 18.65%, 7.3% after 90 days for BS0, ABS0, and AS0 compared to S0, respectively. Under increasing saline stress, biochar enhanced plant growth by 19.68% (BS1), 24.19% (BS2), 14.38% (BS3) after 45 days and by 24.21% (BS1), 26.97% (BS2), 13.4% (BS3) after 90 days when compared to S1, S2, and S3, respectively. This significant ($p < 0.05$) increment in the plant height was also recorded when the soil was amended with biochar + AMF (ABS1, ABS2, and ABS3). Comparatively, the biochar treatments demonstrated a significant effect in improving the plant's height compared to single AMF treatment; with biochar + AMF being a better option.

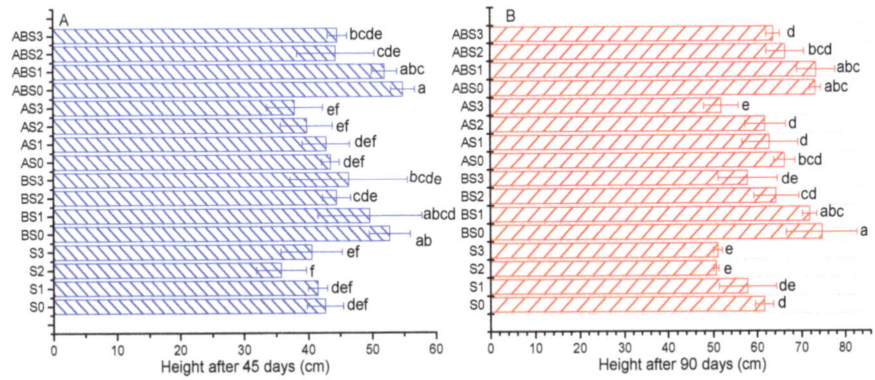

Figure 2. Impact of AMF and biochar on the height of corn plants under different levels of salinity after 45 (**A**) and 90 (**B**) days of growth. Data present the mean ± standard deviation of three replicates and mean values followed by different letters are significantly different at $p < 0.05$ according to Duncan multiple range test. The significance of acronyms is the same as in Figure 1.

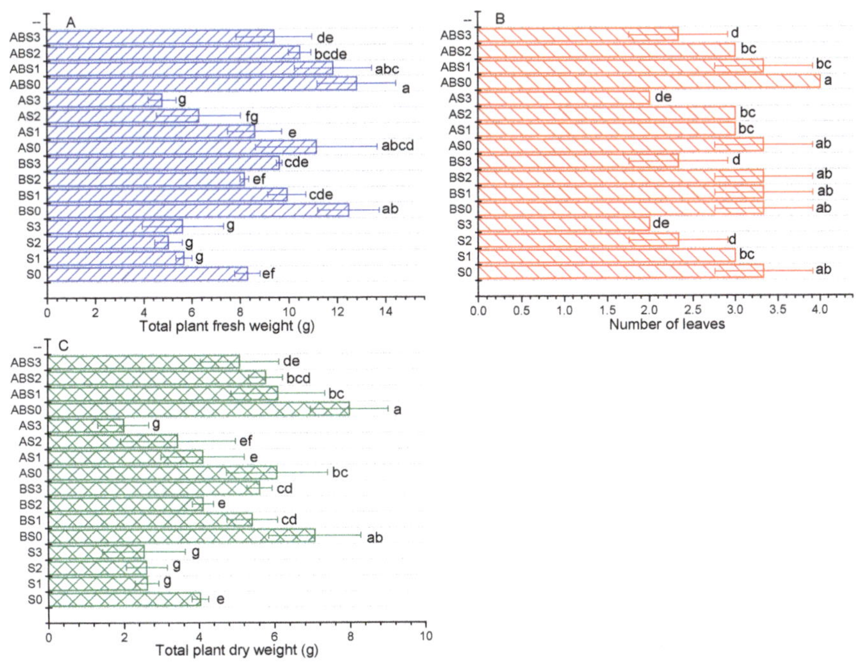

Figure 3. Impact of AMF and biochar on plant fresh weight (**A**), number of healthy leaves (**B**), and dry weight (**C**) under different concentrations of salinity. Data present the mean ± standard deviation of three replicates and mean values followed by different letters are significantly different at $p < 0.05$ according to Duncan multiple range test. The significance of acronyms is the same as in Figure 1.

Figure 3 shows the difference in the total plant fresh weight (Figure 3A), the number of leaves (Figure 3B), and the total plant dry weight (Figure 3C). As shown, increasing the concentration of NaCl negatively affects the total plant fresh weight (Figure 3A). Increasing the NaCl concentration by 50 (S1), 100 (S2), and 150 mM (S3) significantly decreased the plant's fresh weight by 31.62%, 39.28%, and 32.28% relative to S0, respectively. When no salt was applied to the soil, biochar and AMF amendments improved the plant's fresh weight by 50.21%, 54.28%, and 34.4% for BS0, ABS0, and AS0, respectively. Compared to the saline treatments (S1, S2, and S3), biochar amendments improved the plant's fresh weight by 74.84%, 62.36%, 71.14% for BS1, BS2, BS3 and 108.5%, 107.7%, 67.48% for ABS1, ABS2, and ABS3, respectively. For the treatments containing only AMF, the plant fresh weight was increased for AS1 (51.49%) and AS2 (24.76%) but was decreased as the salt content was increased to 150 mM for AS3 (−14.91%). Compared to the S0 treatment, all the treatments with biochar induced a significant improvement in plant fresh weight and demonstrates the individual ability of biochar or when combined with AMF to alleviate saline-related stress on plant growth.

The mean number of healthy leaves decreased with increasing salt stress for the control treatment (S1, S2, S3) (Figure 3B). Amending the soil with biochar (BS0, BS1, BS3) or combined with AMF (ABS0, ABS1, ABS2) or with AMF alone (AS0, AS1, AS2) mitigated the adverse effect of increasing saline concentration on the number of healthy leaves. Nevertheless, at a higher saline concentration (150 mM), the different amendments had fewer leaves when compared to S0 but higher when compared to the S3 treatment.

The total plant dry weight (Figure 3C) was negatively affected by increasing NaCl concentration. This observation corroborates the negative impact of saline stress on plant water content observed in Figure 3A. The plant dry weight was 4.03 g for S0 treatment but was decreased by 35.06%, 35.21%, and 36.95% in the S1, S2, and S3 treatments, respectively. The different amendments demonstrated positive impacts on the plant dry weight, both in the absence and presence of saline stress. For instance, biochar (BS0), AMF (AS0), and biochar + AMF (ABS0) increased the plant dry weight by up to 75.38%, 50.74%, and 90.31% when compared to S0, respectively. Under increased NaCl concentration, biochar, AMF (except AS3), and biochar + AMF treatments significantly improved the plant dry weight. For example, at the highest saline concentration and when compared to S0, the plant dry weight was 76.4% (BS3) and 63.6% (ABS3) larger than the corresponding untreated S3 sample.

2.4. The Effects of Salinity, Biochar, and AMF on the Nutrient Uptake Ability of the Maize Plant during Growth

Figure 4 shows the nutrient uptake ability of maize plants under different saline concentrations before and after amending the soil with biochar and/or AMF. It can be observed that increasing the saline concentration from 0 to 150 mM (S0 to S3) induced a negative effect on the nutrient uptake ability of the plant. Specifically, %N was decreased from 3.481 to 1.694, %P from 0.265 to 0.189, and %K from 1.562 to 0.937. This corroborates the observed negative effect of increasing salt stress on maize growth observed above (Figures 2 and 3). Under no salt stress, the biochar and AMF treatments (AS0, ABS0, and BS0) enhanced the nutrient uptake ability of the plant compared to S0 treatment. Under increasing saline stress, all treatments containing biochar exhibited a superior effect on promoting nutrient uptake compared to single AMF treatments, with a combined application of biochar + AMF showing the best effect relative to S0 and saline controls (S1, S2, and S3). From this result, it can be inferred that the application of biochar and/or AMF to alkaline and saline soils can enhance plant growth by suppressing salt-related stress thereby improving soil fertility, improving nutrient availability, and promoting the uptake of essential nutrients required for growth.

2.5. Influence of AMF and Biochar on Photosynthetic Pigments in Plant

The effect of increasing saline concentration and the mitigating effects of different amendments on photosynthetic pigments were evaluated and the results are shown in Figure 5. Compared to S0, the content of chlorophyll a was decreased by 27.11%, 51.56%, and 65.78% (S_3) as the salt content was increased in S1, S2, and S3, respectively (Figure 5A). Similarly, the contents of chlorophyll b were decreased by 26.8%, 30.93%, and 51.55% (Figure 5B) while carotenoid was decreased by 16.08%, 44.48%, and 44.41% (Figure 5C) in S1, S2, and S3, respectively. From this result, it is evident that under increasing salt stress, plant growth becomes inhibited as the required nutrients are made less available which results in a reduced photosynthetic rate.

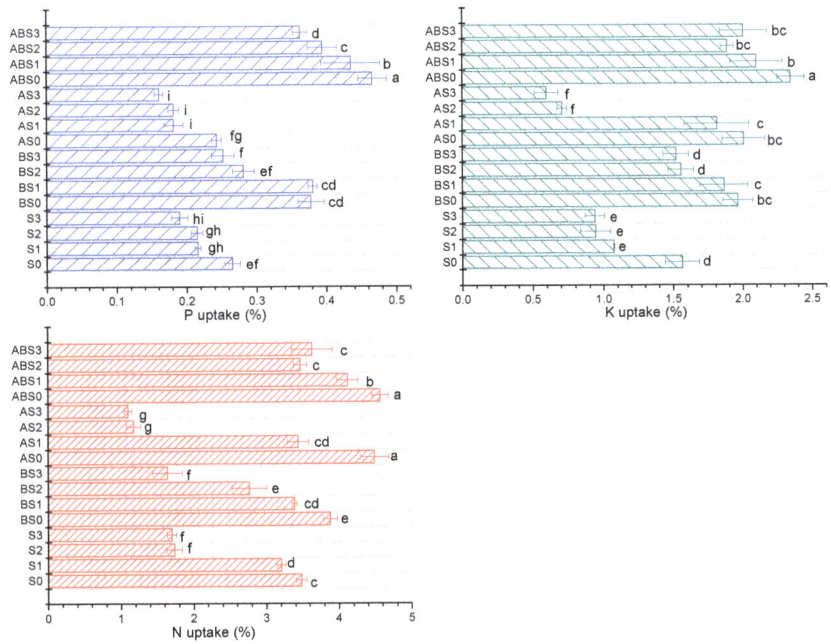

Figure 4. Effect of biochar and/or AMF on percentage of N, P, and K uptake by the maize plant under salinity stress. Data present the mean ± standard deviation of three replicates and mean values followed by different letters are significantly different at $p < 0.05$ according to Duncan multiple range test. The significance of acronyms is the same as in Figure 1.

Amending the alkaline soil with biochar alone or combined with AMF significantly improved the plant's photosynthetic ability as chlorophyll a, chlorophyll b, and carotenoids were significantly ($p < 0.05$) improved. The contents of chlorophyll a, chlorophyll b, and carotenoids were increased by 52.89%, 15.46%, 25.78% for B_0 and 68%, 52.58%, 89.46% for AB_0 compared to S0, respectively. Similarly, the A_0 treatment improved the contents of chlorophyll a and carotenoids by 34.22% and 50.34%, respectively, while decreasing that of chlorophyll b by 6.19%. Under increasing saline stress, all amendments containing biochar showed positive effects on improving the photosynthetic ability of the maize plant while the single AMF treatment demonstrated negative effects on chlorophyll b at 100 (AS2) and 150 mM NaCl (AS3). Of all treatments, the combined application of biochar + AMF induced the most significant effect on improving the contents of photosynthetic pigments under salt stress. Relative to the S0 treatment, the content of chlorophyll a was 61.33%, 78.67%, 86.22%; chlorophyll b was 38.14%, 19.59%, 19.59%; and carotenoid was 130.6%, 95.89%, 71.18% higher in ABS1, ABS2, ABS3 treatments than in the S1, S2, S3 treatments, respectively. This observation is in agreement with the superior ability of biochar + AMF treatments in significantly enhancing nutrient uptake by the plant (Figure 4) under increased saline stress relative to other treatments. Therefore, amending alkaline and saline soils with AMF, biochar, or biochar + AMF can play an important role in mitigating stress-related adverse effects in these soils that inhibit plant growth.

2.6. Impact of AMF and Biochar on Antioxidant Enzyme Activities

The activities of SOD and POD in the leaves of the maize plant were significantly ($p < 0.05$) affected after 90 days of growth under different saline conditions (Figure 6). Compared to S0, increasing the saline concentration reduced the activity of SOD by 25.41%,

32.26%, and 34.04% in S1, S2, and S3 treatments, respectively (Figure 6A). For similar treatments, the content of POD was increased by 15.41% for S1butdecreased by 38.44% for S2 and 44.09% for S3. After amending the soil with biochar (BS0), AMF (AS0), and biochar + AMF (ABS0), the activity of SOD was increased by 1.55%, 10.01%, and 15.18%, respectively. Also, while the BS0 (7.14%) and ABS0 (21.04%) treatments enhanced the activity of POD, the AS0 treatment induced a 4.41% decrease in its activity compared to S0.

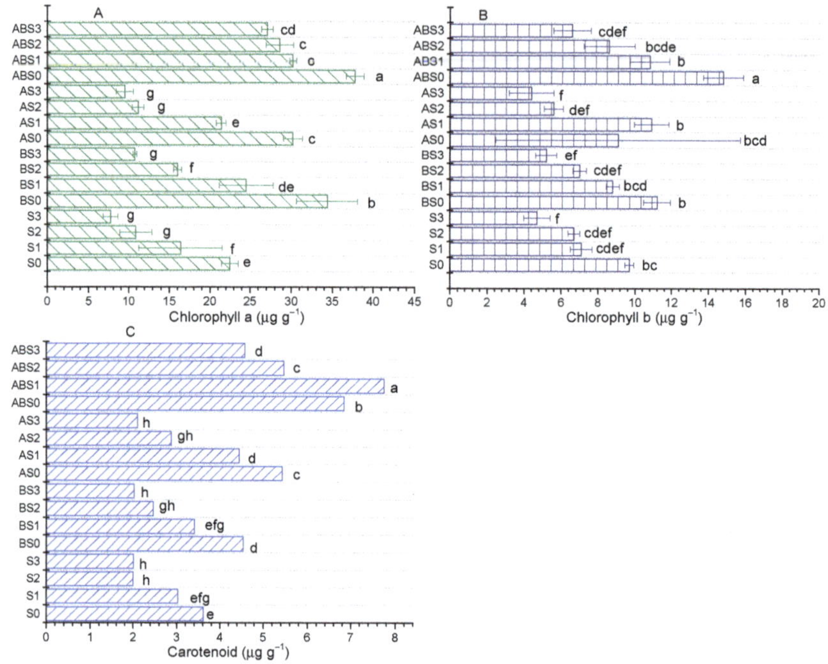

Figure 5. Effect of biochar and AMF on photosynthetic pigments: (**A**) Chlorophyll a; (**B**) Chlorophyll b; (**C**) Carotenoid. Data present the mean ± standard deviation of three replicates and mean values followed by different letters are significantly different at $p < 0.05$ according to Duncan multiple range test. The significance of acronyms is the same as in Figure 1.

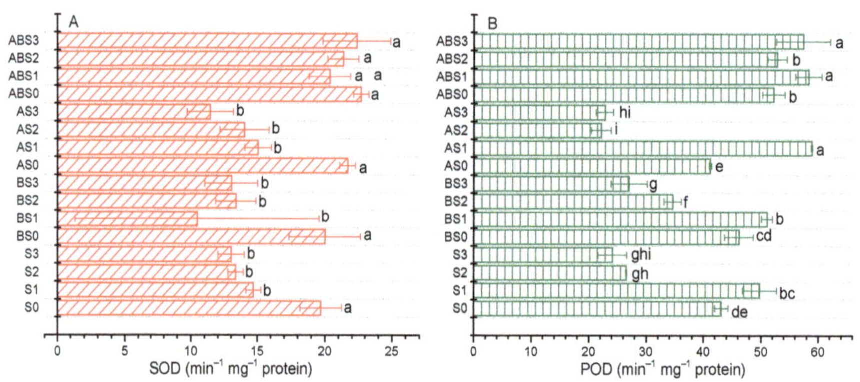

Figure 6. Impact of AMF and biochar on the activity of SOD (**A**) and POD (**B**) in maize under different levels of salinity. Data present the mean ± standard deviation of three replicates and mean values followed by different letters are significantly different at $p < 0.05$ according to Duncan multiple range test. The significance of acronyms is the same as in Figure 1.

Under different saline conditions, the amendments showed diverse effects on the activities of SOD and POD. Relative to S0 treatment, the content of SOD in biochar amendments was lower by 21.52% (BS1) or insignificantly higher by 0.08% (BS2) and 0.17% (BS3) when compared to S1, S2, and S3, respectively. Similarly, an insignificant increase was observed for AS1 (1.69%) and AS2 (3.42%) and a decrease for AS3 (7.91%). The application of biochar mitigated the adverse effect of saline stress on the activity of POD (BS1, BS2, BS3) while AMF only had a positive effect at a lower saline concentration (AS1). Of all treatments, the combined application of biochar and AMF showed the most significant ($p < 0.05$) effect on the activities of SOD and POD under saline stress. Compared to S0, the activity of SOD was 28.74%, 40.63%, and 47.5% higher for ABS1, ABS2, ABS3 than for S1, S2, S3, respectively. Similarly, the activity of POD was19.86%, 61.03%, and 77.21% higher in ABS1, ABS2, ABS3 treatments than for S1, S2, S3, respectively. This result shows that biochar + AMF had the most significant effect on the activities of SOD and POD under increasing saline stress, making this combination the best for improving antioxidant activities in alkaline and saline soils.

3. Discussion

The application of biochar to alkaline soils have demonstrated both positive and negative effects on the availability of nutrients such as K, P, and N. Chen et al. [18] showed that biochar nanoparticles can reduce P retention in alkaline soils by up to 23% and 18% and the authors associated this effect to increased leaching of P associated to Fe/Al oxides in soils. Contrarily, Cui et al. [13] observed that the co-application of biochar and effective microorganisms enhanced the growth of *Sesbania cannabina* in coastal saline-alkali soil. Our study shows that the application of biochar without/without *Arbuscular mycorrhizal* fungi (AMF) (*Glomus mosseae*) had a positive effect on the available nutrients (N, K, and P) in alkaline soil (Figure 1). Although the overall effect of biochar and/or AMF on soil pH was not significant, a significant increase was observed in the nutrient uptake ability of maize plants (Figure 4) and the measured growth parameters (Figures 2 and 3).

Salt-affected soils are prone to nutrient deficiency and excess Na^+ ions in soils affect the availability of plant nutrients either by direct competition or indirectly by increasing the osmotic pressure of the soil solution and retarding mass uptake of nutrients by the roots [19,20]. For saline soils, nutrients such as N, P, and K are in low quantities mostly due to low levels and rapid loss of organic matter [21]. This observation agrees with our results (Figure 1) given that the contents of N, P, and K were respectively reduced by 26.75%, 28.94%, and 23.8% when the NaCl concentration was increased to 150 mM for the control (S3). Under similar conditions, the individual application of biochar and combined application with AMF demonstrated the most significant effect in alleviating the adverse effects of saline stress on nutrient availability. Specifically, when compared to S0, the contents of N, P, and K were higher by 77.32%, 8.51%, and 56.3% in BS3 compared to S3 treatment, while in biochar + AMF (ABS3) treatment, it was 75.04%, 32.99%, 2.44% higher, respectively (Figure 1).

Nutrient deficiency and low use efficiency have negative impacts on the photosynthetic potentials of plants and their subsequent resistance to stress [22]. When plants are under stress, they may suffer chlorophyll degradation which results in a reduction of the plant's photosynthetic ability [23]. From Figure 5, it can be observed that increasing salinity reduced the plant's photosynthetic ability. The saline-related stress significantly reduced the contents of photosynthetic pigments, with the reduction increasing with NaCl concentrations and most severe for chlorophylls a and b. Amending the soils with biochar and/or AMF alleviated the negative effects of increasing salinity on maize photosynthetic ability, with the biochar treatments being more effective than single AMF treatments. The colonization of plant roots by AMF has been reported to be beneficial for plant growth in low pH soils, and this effect was significantly improved when AMF was co-applied with biochar [15]. Thus, biochar application to soils played a modifying role in improving the physicochemical characteristics for effective colonization by AMF and it can be inferred that by improving soil fertility and nutrient uptake, biochar and/or AMF provided a conducive environment for plant growth.

The secretion of antioxidant enzymes is one key mechanism by which plants mitigate the adverse effects of biotic and abiotic stress [24,25]. By scavenging ROS, antioxidants can mitigate the adverse effects of oxidative stress induced by high saline concentrations thereby promoting plant growth [14]. As observed by Farhangi-Abriz and Torabian [10], an increase in the concentration of NaCl causes an increase in antioxidant enzymes such as CAT, SOD, and POD in the leaves and roots of bean seedlings. In our study, increasing the saline concentration decreased the contents of both SOD and POD (Figure 6) while co-application with biochar and/or AMF showed diverse effects. Nevertheless, applying biochar alone or combined with AMF mostly demonstrated an enhancing effect on the activities of SOD and POD; with the combined application of biochar + AMF showing the most significant ($p < 0.05$) increase. Our result may differ from that of Farhangi-Abriz and Torabian [10] primarily due to the difference in the plant types and growth stages considered.

The ability of plants to alter the contents of unsaturated fatty acids is an important mechanism through which they adapt to stress. Through the activities of fatty acid desaturases, plants modify their membranes to provide a suitable environment for the functioning of photosynthetic proteins [26]. Oleic acid (18:1) has been reported to be critical for resistance against pathogens [27] while linolenic acid (C18:3) is an important stress signal [28]. The importance of C18:1 and C18:2 levels have been documented in several studies and it was reported that they play important roles in the regulation of fungal development, seed colonization, and mycotoxin production by *Aspergillus* spp. [29,30]. When plants are subjected to saline stress, the content of C18:3 may decrease to indicate damages caused by the stress [26]. Under drought stress, *Brassica napus* leaves showed decreased levels of C18:3 and C18:2 [31]. Pál et al. [32] studied the effect of cadmium contamination on the content of membrane lipids of maize plants. They found that while the content of C16:0 in the leaves decreased that of C18:2 and C18:3 increased with increasing Cd levels. Also, they observed that for the roots, the levels of C18:0 and C18:1 decreased while those of C18:2 and C18:3 increased. We observed that saturated fatty acids (C14:0, C16:0, and C18:0) that play an important role in controlling the hydrophobicity of membrane proteins [33] were not significantly altered under saline stress and in the presence of biochar. Nevertheless, the levels of saturated acids with longer chains (C20:0 and C22:0) were significantly improved under increasing saline stress and in the presence of biochar. Conversely, increasing salinity negatively affected unsaturated fatty acid concentrations (C16:1, C18:1, C18:2) while the application of biochar demonstrated a mitigating effect and increased the levels of unsaturated fatty acids. The combined application of biochar and AMF showed the most significant effect in increasing the concentrations of unsaturated fatty acids. Thus, by improving the levels of membrane unsaturation, the co-application of biochar and AMF improved the growth of maize in alkaline soil under increasing saline stress. Therefore, we infer that under conditions of stress, plants become better adapted to handle different levels of stress by adjusting the fatty acid unsaturation levels. Also, soil amendments that can improve the content of fatty acid unsaturation may enhance plants' adaptability to stress.

4. Materials and Methods

4.1. Experimental Materials

The agricultural soil used in this study was a loess soil collected from Lanzhou, China, with a bulk density of 1.25 g cm^{-3} and sampled from the surface 0–20 cm. The soil was air-dried and sieved with a 2 mm mesh sieve. On average, the annual rainfall of the study site was 415 mm and the annual temperature was about 6.2 °C. For soil characterization, five soil samples were collected and properly mixed to make a representative composite soil sample that had TN content of 1.74 ± 0.01 g kg^{-1}, available phosphorus 25.1 ± 0.19 mg kg^{-1}, soil organic carbon of 9.08 ± 0.09 g kg^{-1}, and soil pH (in solution)

8.25 ± 0.11. Soil pH was determined using a METTLER TOLEDO Desktop pH meter after the soil sample was equilibrated in distilled water (1:5). Soil organic carbon was estimated by the Wet Oxidation method [34]. TN and phosphorus were determined by the Kjeldahl method [35] and Olsen method [36], respectively. The exchangeable K was evaluated by the Ammonium Acetate extraction method [37]. The soil test was performed in the laboratory of the College of Resources and Environmental Sciences at Gansu Agricultural University, Lanzhou, China. For this study, a pot experiment was conducted in greenhouse conditions (air temperature 30 °C; relative humidity 50%) from July to October 2020 at the Dingxi experimental station of Gansu Academy of Agricultural Sciences (Tangjiao Town, Dingxi city, 35°350 N, 104°360 E, 1970 m a.s.l), Gansu, China.

The biochar was prepared from corn straws collected from a cornfield in Tangjiao Town, Dingxi city, and transported to the College of Resources and Environmental Sciences of Gansu Agricultural University, Lanzhou, China. The straws were cut into 10–15 cm pieces and pyrolyzed at a temperature ranging from 350 to 500 °C for 1 h under an oxygen-deficient condition in an Oven-Electric Furnace (Heraeus MR 170, Meinerzhagen, Germany). After pyrolysis and cooling, the biochar produced was ground to pass a 250 µm mesh sieve and characterized. Total Organic Carbon (TOC) of biochar samples was measured after oxidization with potassium dichromate following Nelson and Sommers [38]. The total K contents were determined with a flame photometer (Jenway Flame Photometer, Bibby Scientific Ltd-Stone-Staffs-St15 0SA–UK.), P was estimated by a spectrophotometer as described by Sparks [39], and TN was determined by the micro-Kjeldahl method [40]. The %TN, %P, %K, and %TOC of biochar were 0.43 ± 0.01, 0.17 ± 0.004, 0.46 ± 0.003, and 41.2 ± 0.41, respectively. The seeds of corn (*Zea mays* L.) were collected from Gansu Provincial Key Laboratory of Arid Land Crop Science and kept at 4 °C for 24 h and later washed with running distilled water for 30 min before planting. Round plastic pots (21 cm in diameter * 16 cm in height) were used in this experiment. At the harvest period, the plant nutrient uptake (N, P, and K) were estimated following Hashem et al. [15]. To determine the contents of N and K in maize shoot, dried and ground samples were digested with H_2SO_4-H_2O_2 at 260–270 °C. N contents were determined by an Auto-analyzer 3 digital colorimeter (AA3, Bran + Luebbe, Hamburg, Germany) and K contents were estimated using Flame Photometry (FP6400, Shanghai Precision Scientific Instrument, Shanghai, China). The phosphorus was extracted by nitric-perchloric acid digestion and measured using the Vanado-molybdophosphoric colorimetric method. Standard curve of each mineral (10–100 μg^{-1} mL) used as reference.

4.2. Arbuscular Mycorrhizal Fungi (AMF) Inoculum

The Arbuscular mycorrhizal fungi (AMF) strain used was a single genus of *Glomus mosseae* provided by the Gansu Provincial Key Laboratory of Arid Land Crop Science, Lanzhou, China. It was multiplied with *Zea mays* (L.) as the host plant for 4 months in sterilized soil in the greenhouse of the College of Resources and Environmental Sciences of Gansu Agricultural University, Lanzhou, China. The methods previously described by Gerdemann and Nicolson [41] and Giovannetti and Mosse [42] were used to determine AMF inoculum characteristics. Root mycorrhizal colonization, soil spores content, arbuscules, and extraradical hyphae in the roots samples were observed accordingly. The AMF inoculum consisted of mycorrhizal roots (80% root mycorrhizal colonization), soil containing spores (50–80 per 10 g inoculum), and extraradical hyphae (2.5 m per 1 g soil) mixed with soil.

4.3. Experimental Treatments and Design

The greenhouse pot experiment was conducted in a completely randomized design with five replications per treatment and increasing salt stress. In total, there were 16 treatments as mentioned in Table 2 The NaCl concentration in the pots was gradually increased from 50 to 150 mM at a rate of 50 mM per 24 h. Frequent irrigations (thrice a week) with saline solution permitted the various salt concentrations in the pots to be maintained at a constant level. There was a total of 80 pots in the present greenhouse experiment. The biochar (5%) and/or AMF (20 g) were applied and thoroughly mixed with soil in each treated pot. This was followed by the sowing of eight imbibed seeds in each pot. After germination, the seedlings were thinned to four per pot.

4.4. Measurement of Plant Growth Parameters

The maize plant height was measured just before the final harvest. The height was measured by holding a measuring tape close to the stem of the plant. Plant height was recorded from the ground level to the base of the highest fully expanded leaf. The plant fresh weight (FW) was also determined by weighing the different parts of the harvested plant. The pots were cut away on two sides to permit careful root separation from the soil. For each treatment, plants were removed from the soil and washed with distilled water. After adsorbing residual water using tissue paper, an electronic balance was used to measure plant shoot and roots FW.

4.5. Lipid Extraction and Analysis

3 g 12-week-old maize leaves were harvested from plants grown on saline soils and transferred into 1.5 mL polypropylene reaction tubes. Fresh leaves were flash-frozen in liquid nitrogen and stored at −80 °C. To each sample, 300 µL extraction solvent composed of methanol, chloroform and formic acid (20:10:1, *v/v/v*) was used. The mixture was then vigorously shaken (using a paint shaker) for 5 min. Briefly, 150 µL of 0.2 M phosphoric acid, 1 M potassium chloride was added, and samples were centrifuged at 13,000× *g* at room temperature for 1 min. Lipids were extracted and followed by separation using two-dimensional thin-layer chromatography (TLC) on silica gel plates (pre-coated silica gel plates, Merck 5626) according to Xu and Siegenthaler [43]. The first developing solvent was acetone/toluene/H_2O (91:30:8, by volume) and the second was chloroform/methanol/25% NH_3/H_2O (65:35:3:2, by volume). The plates were shortly air-dried before being delicately sprayed with 0.01% primuline and viewed under ultraviolet radiation. The trans-esterification of individual lipids previously separated by TLC was performed with 5% H_2SO_4 in methanol at 85 °C for 1 h. A Hewlett-Packard 5890 gas chromatography system equipped with a hydrogen flame ionization detector and an FFAP capillary column (30 m; i.d. 0.53 mm) was used to separate the fatty acid methyl esters. The column was run isothermally at 190 °C, while the detector was kept at 230 °C. The internal standard was heptadecanoic acid provided by Sigma Aldrich Chemical Co. (St. Louis, MO, USA). All other used chemicals were purchased from Merck (Darmstadt, Germany) and were of analytical purity.

4.6. Antioxidant Enzyme Activities

For protein extraction, young fresh leaves were collected from 6- and 12-week-old maize plants. The leaves (1 g) were immediately frozen in liquid nitrogen, lyophilized, and homogenized in 2-mL of 0.1 mM potassium phosphate (pH 7.8). The samples were centrifuged for 15 min at 12,000× *g* (4 °C). The supernatants were collected into tubes and stored at −20 °C until needed for enzymatic activity assays. The activity of superoxide dismutase (SOD) was measured spectrophotometrically by determining the inhibition of blue diformazan formation in the presence of riboflavin/nitroblue tetrazolium (NBT) and light [44]. The modified assay solution was prepared with 1-mL of 50 mM sodium phosphate (pH 7.8), 0.1 mM EDTA, 0.3 mM riboflavin, and 30 L of leaf extract. After 5 min at room temperature, the solution was mixed with NBT to obtain a final concentration of 0.03 mM NBT. The reaction mixture was then illuminated for 3 min with a fluorescent light (75 W, 20 cm above the mixture) and absorbance was determined at 560 nm. The reaction mixture without extract was used to calculate the control rate. NBT absorption was insignificant. The activity of SOD is presented in min^{-1} mg^{-1} protein with one unit described as 50% inhibition of blue diformazan formation. Also, the activity of Peroxidase (POD) was determined by measuring the increase rate in absorbance at 470 nm with o-dianisidine as the substrate [45]. The assay solution was 1 mL of 0.01 M sodium phosphate (pH 6.0) containing 1.3 mM H_2O_2, 1 mM o-dianisidine and 5 μL of extract. The activity was expressed as $\Delta OD_{470\ nm}$ min^{-1} mg^{-1} protein.

4.7. Photosynthetic Pigments

Fresh leaves from 12-week-old maize plants were sampled for photosynthetic pigments assessment [46,47]. The leaves were finely cut into small sections (~0.1 g) and ground to a powder in 80% acetone (10 mL) and then centrifuged for 5 min at 10,000 rpm. After collecting the supernatant, the procedure was repeated until the residue was colorless. The absorbance of the solution was recorded at 480, 645 nm, and 663 nm. 80% acetone was included as the blank solution. The photosynthetic pigments of leaves were determined by estimating the contents of chlorophylls (Equations (1)–(3)) and carotenoids content by (Equation (3)).

$$\text{Chlorophyll a (mg g}^{-1}\text{ FW)} = (0.0127 * A_{663}) - (0.00269 * A_{645}) * V/W \quad (1)$$

$$\text{Chlorophyll b (mg g}^{-1}\text{ FW)} = (0.0229 * A_{645}) - (0.00468 * A_{663}) * V/W \quad (2)$$

$$\text{Carotenoids (mg g}^{-1}\text{ FW)} = [A_{480} + (0.114 * A_{663}) - (0.638 - A_{645})] * V/W \quad (3)$$

where A_{663}, A_{645}, A_{480} are absorbance at 663, 645, and 480 nm, respectively, and V is the total volume of sample solution and W is the sample weight.

4.8. Statistical Analysis

Data were analyzed by one-way analysis of variance (ANOVA) using Genstat statistical software (ver.12). Significant differences among treatments were computed by Duncan's multiple range tests ($p < 0.05$).

5. Conclusions

Maize cultivated on alkaline soils suffer from nutrient deficiency and this situation may become worse if such soils are affected by salinization. The use of biochar or AMF alone in amending such soils can mitigate the negative impacts of nutrient deficiency and low levels of salinization but become inefficient under high saline concentrations. Our study demonstrates that the combined application of biochar and AMF can effectively alleviate the adverse effects of saline stress on plant growth by (a) improving soil fertility, (b) increasing antioxidant enzymes activities and (c) increasing the levels of unsaturated fatty acid. Nevertheless, biochar application in alkaline and saline soils has shown contrasting results in different studies and this study adds to the limited literature on this. It is suggested that (1) broader studies be conducted in the greenhouse with different plants and different alkaline soils to have a larger picture of the effect of this amendment on alkaline soils under saline stress, (2) researchers should perform small scale field experiments with actual alkaline-saline soils to compare the mechanisms of different amendments in improving soil fertility, and (3) researchers should evaluate the difference in plant-growth improvement mechanisms of these amendments between actual alkaline-saline soils and alkaline soils which are salinized in laboratory studies.

Author Contributions: Conceptualization, N.I.N. and C.L.; methodology, N.I.N.; software, N.I.N.; validation, N.I.N., Q.S. and J.N.N.; formal analysis, N.I.N.; writing—original draft preparation, N.I.N.; writing—review and editing, Q.S., F.U.H., A.M., N.I.N. and J.N.N.; supervision, C.L.; funding acquisition, C.L. All authors have read and agreed to the published version of the manuscript.

Funding: This research was financially supported by the National Natural Science Foundation of China (41661049).

Institutional Review Board Statement: Not applicable.

Informed Consent Statement: Not applicable.

Data Availability Statement: Not applicable.

Acknowledgments: We thank Richard Dormatey, Cheikh Ndiaye, Babakar Thioye, and S. Ndour for their assistance during the preparation of this manuscript.

Conflicts of Interest: The authors declare no conflict of interest.

References

1. Shi, R.Y.; Hong, Z.N.; Li, J.Y.; Jiang, J.; Baquy, M.A.A.; Xu, R.K.; Qian, W. Mechanisms for Increasing the pH Buffering Capacity of an Acidic Ultisol by Crop Residue-Derived Biochars. *J. Agric. Food Chem.* **2017**, *65*, 8111–8119. [CrossRef]
2. Shi, R.Y.; Hong, Z.N.; Li, J.Y.; Jiang, J.; Kamran, M.A.; Xu, R.K.; Qian, W. Peanut straw biochar increases the resistance of two Ultisols derived from different parent materials to acidification: A mechanism study. *J. Environ. Manag.* **2018**, *210*, 171–179. [CrossRef]
3. Abrishamkesh, S.; Gorji, M.; Asadi, H.; Bagheri-Marandi, G.H.; Pourbabaee, A.A. Effects of rice husk biochar application on the properties of alkaline soil and lentil growth. *Plant Soil Environ.* **2015**, *62*, 475–482. [CrossRef]
4. Zhang, Q.-Z.; Wang, X.-H.; Du, Z.-L.; Liu, X.-R.; Wang, Y.-D.; Zhang, Q.-Z.; Wang, X.-H.; Du, Z.-L.; Liu, X.-R.; Wang, Y.-D. Impact of biochar on nitrate accumulation in an alkaline soil. *Soil Res.* **2013**, *51*, 521–528. [CrossRef]
5. Baigorri, R.; Francisco, S.S.; Urrutia, Ó.; García-Mina, J.M. Biochar-Ca and Biochar-Al/-Fe-Mediated Phosphate Exchange Capacity are Main Drivers of the Different Biochar Effects on Plants in Acidic and Alkaline Soils. *Agronomy* **2020**, *10*, 968. [CrossRef]
6. Mete, F.Z.; Mia, S.; Dijkstra, F.A.; Abuyusuf, M.; Hossain, A.S.M.I. Synergistic Effects of Biochar and NPK Fertilizer on Soybean Yield in an Alkaline Soil. *Pedosphere* **2015**, *25*, 713–719. [CrossRef]
7. Naveed, M.; Sajid, H.; Mustafa, A.; Niamat, B.; Ahmad, Z.; Yaseen, M.; Kamran, M.; Rafique, M.; Ahmar, S.; Chen, J.T. Alleviation of salinity-induced oxidative stress, improvement in growth, physiology and mineral nutrition of canola (*Brassica napus* L.) through calcium-fortified composted animal manure. *Sustainability* **2020**, *12*, 846. [CrossRef]
8. Niamat, B.; Naveed, M.; Ahmad, Z.; Yaseen, M.; Ditta, A.; Mustafa, A.; Rafique, M.; Bibi, R.; Sun, N.; Xu, M. Calcium-Enriched Animal Manure Alleviates the Adverse Effects of Salt Stress on Growth, Physiology and Nutrients Homeostasis of *Zea mays* L. *Plants* **2019**, *8*, 480. [CrossRef]
9. Foyer, C.H.; Noctor, G. Redox homeostasis and antioxidant signaling: A metabolic interface between stress perception and physiological responses. *Plant Cell* **2005**, *17*, 1866–1875. [CrossRef]
10. Farhangi-Abriz, S.; Torabian, S. Antioxidant enzyme and osmotic adjustment changes in bean seedlings as affected by biochar under salt stress. *Ecotoxicol. Environ. Saf.* **2017**, *137*, 64–70. [CrossRef] [PubMed]
11. Sun, H.; Lu, H.; Chu, L.; Shao, H.; Shi, W. Biochar applied with appropriate rates can reduce N leaching, keep N retention and not increase NH3 volatilization in a coastal saline soil. *Sci. Total Environ.* **2017**, *575*, 820–825. [CrossRef]
12. Song, Y.; Zhang, X.; Ma, B.; Chang, S.X.; Gong, J. Biochar addition affected the dynamics of ammonia oxidizers and nitrification in microcosms of a coastal alkaline soil. *Biol. Fertil. Soils* **2013**, *50*, 321–332. [CrossRef]
13. Cui, Q.; Xia, J.; Yang, H.; Liu, J.; Shao, P. Biochar and effective microorganisms promote Sesbania cannabina growth and soil quality in the coastal saline-alkali soil of the Yellow River Delta, China. *Sci. Total Environ.* **2021**, *756*, 143801. [CrossRef]
14. Latef, A.A.H.A.; Chaoxing, H. Effect of arbuscular mycorrhizal fungi on growth, mineral nutrition, antioxidant enzymes activity and fruit yield of tomato grown under salinity stress. *Sci. Hortic.* **2011**, *127*. [CrossRef]
15. Hashem, A.; Kumar, A.; Al-Dbass, A.M.; Alqarawi, A.A.; Al-Arjani, A.B.F.; Singh, G.; Farooq, M.; Abd_Allah, E.F. Arbuscular mycorrhizal fungi and biochar improves drought tolerance in chickpea. *Saudi J. Biol. Sci.* **2019**, *26*, 614–624. [CrossRef]
16. Yusif, S.A.; Dare, M.O. Effect of Biochar Application and Arbuscular Mycorrhizal Inoculation on Root Colonization and Soil Chemical Properties. *Int. Ann. Sci.* **2016**, *1*, 33–38. [CrossRef]
17. Qadir, M.; Ghafoor, A.; Murtaza, G. Amelioration strategies for saline soils: A review. *Land Degrad. Dev.* **2000**, *11*, 501–521. [CrossRef]
18. Chen, M.; Alim, N.; Zhang, Y.; Xu, N.; Cao, X. Contrasting effects of biochar nanoparticles on the retention and transport of phosphorus in acidic and alkaline soils. *Environ. Pollut.* **2018**, *239*, 562–570. [CrossRef] [PubMed]
19. Abrar, M.M.; Saqib, M.; Abbas, G.; Atiq-ur-Rahman, M.; Mustafa, A.; Shah, S.A.A.; Mehmood, K.; Maitlo, A.A.; Sun, N.; Xu, M. Evaluating the contribution of growth, physiological, and ionic components towards salinity and drought stress tolerance in Jatropha curcas. *Plants* **2020**, *9*, 1574. [CrossRef]
20. Saifullah; Dahlawi, S.; Naeem, A.; Rengel, Z.; Naidu, R. Biochar application for the remediation of salt-affected soils: Challenges and opportunities. *Sci. Total Environ.* **2018**, *625*, 320–335. [CrossRef]
21. Lakhdar, A.; Rabhi, M.; Ghnaya, T.; Montemurro, F.; Jedidi, N.; Abdelly, C. Effectiveness of compost use in salt-affected soil. *J. Hazard. Mater.* **2009**, *171*, 29–37. [CrossRef]
22. Umar, W.; Ayub, M.A.; ur Rehman, M.Z.; Ahmad, H.R.; Farooqi, Z.U.R.; Shahzad, A.; Rehman, U.; Mustafa, A.; Nadeem, M. Nitrogen and Phosphorus Use Efficiency in Agroecosystems. In *Resources Use Efficiency in Agriculture*; GATEWAY EAST Springer: Singapore, 2020; pp. 213–257.
23. Gajewska, E.; Skłodowska, M.; Słaba, M.; Mazur, J. Effect of nickel on antioxidative enzyme activities, proline and chlorophyll contents in wheat shoots. *Biol. Plant.* **2006**, *50*. [CrossRef]

24. Sabir, A.; Naveed, M.; Bashir, M.A.; Hussain, A.; Mustafa, A.; Zahir, Z.A.; Kamran, M.; Ditta, A.; Núñez-Delgado, A.; Saeed, Q.; et al. Cadmium mediated phytotoxic impacts in Brassica napus: Managing growth, physiological and oxidative disturbances through combined use of biochar and *Enterobacter* sp. MN17. *J. Environ. Manag.* **2020**, *265*, 110522. [CrossRef] [PubMed]
25. Kamran, M.; Malik, Z.; Parveen, A.; Zong, Y.; Abbasi, G.H.; Rafiq, M.T.; Shaaban, M.; Mustafa, A.; Bashir, S.; Rafay, M.; et al. Biochar alleviates Cd phytotoxicity by minimizing bioavailability and oxidative stress in pak choi (*Brassica chinensis* L.) cultivated in Cd-polluted soil. *J. Environ. Manag.* **2019**, *250*, 109500. [CrossRef] [PubMed]
26. Upchurch, R.G. Fatty acid unsaturation, mobilization, and regulation in the response of plants to stress. *Biotechnol. Lett.* **2008**, *30*, 967–977. [CrossRef]
27. Kachroo, P.; Shanklin, J.; Shah, J.; Whittle, E.J.; Klessig, D.F. A fatty acid desaturase modulates the activation of defense signaling pathways in plants. *Proc. Natl. Acad. Sci. USA* **2001**, *98*, 9448–9453. [CrossRef] [PubMed]
28. Blée, E. Impact of phyto-oxylipins in plant defense. *Trends Plant Sci.* **2002**, *7*, 315–322. [CrossRef]
29. Wilson, R.A.; Calvo, A.M.; Chang, P.K.; Keller, N.P. Characterization of the Aspergillus parasiticus Δ 12-desaturase gene: A role for lipid metabolism in the Aspergillus-seed interaction. *Microbiology* **2004**, *150*. [CrossRef]
30. Xue, H.Q.; Upchurch, R.G.; Kwanyuen, P. Ergosterol as a quantifiable biomass marker for Diaporthe phaseolorum and Cercospora kikuchii. *Plant Dis.* **2006**, *90*. [CrossRef]
31. Dakhma, W.S.; Zarrouk, M.; Cherif, A. Effects of drought-stress on lipids in rape leaves. *Phytochemistry* **1995**, *40*, 1383–1386. [CrossRef]
32. Pál, M.; Leskó, K.; Janda, T.; Páldi, E.; Szalai, G. Cadmium-induced changes in the membrane lipid composition of maize plants. *Cereal Res. Commun.* **2007**, *35*, 1631–1642. [CrossRef]
33. Trinidad, J.L.; Pabuayon, I.C.M.; Kohli, A. Harnessing protein posttranslational modifications for plant improvement. *Adv. Crop. Improv. Tech.* **2020**, 385–401. [CrossRef]
34. Walkley, A.; Black, D.R. An examination of the digestion method for determination soil organic matter and proposed modification of the chronic acid titration method. *Soil Sci.* **1935**, *37*, 29–38. [CrossRef]
35. Jackson, M.L. Soil Chemical Analysis Prentice Hall of India. *Pvt. Ltd. New Delhi* **1967**, *151–154*, 183–192.
36. Olsen, S.R.; Sommers, L.E. *1982 Phosphorus*; Page, A.L., Miller, R.H., Keeney, D.R., Eds.; American Society of Agronomy: Madison, WI, USA, 1973; pp. 403–430.
37. Pansu, M.; Gautheyrou, J. *Handbook of Soil Analysis. Mineralogical, Organic and Inorganic Methods*; Springer: Berlin/Heidelberg, Germany; New York, NY, USA, 2006; p. 993.
38. Nelson, D.W.; Sommers, L.E. Total Carbon, Organic Carbon, and Organic Matter. In *Methods of Soil Analysis Part 3—Chemical Methods*; SSSA Book Series: Madison, WI, USA, 1996; pp. 961–1010.
39. Sparks, D.L.; Page, A.L.; Helmke, P.A.; Loeppert, R.H.; Soltanpour, P.N.; Tabatabai, M.A.; Johnston, C.T.; Summer, M.E. *Methods of Soil Analysis, Parts 2 and 3 Chemical Analysis*; Soil Science Society of America Inc.: Madison, WI, USA, 1996.
40. A.O.A.C. (Association of Official Agricultural Chemists). *Official Methods of Analysis*, 10th ed.; A.O.A.C.: Washington, WA, USA, 1965; pp. 308–764.
41. Gerdemann, J.W.; Nicolson, T.H. Spores of mycorrhizal Endogone species extracted from soil by wet sieving and decanting. *Trans. Br. Mycol. Soc.* **1963**, *46*, 235–244. [CrossRef]
42. Giovannetti, M.; Mosse, B. An Evaluation of Techniques for Measuring Vesicular Arbuscular Mycorrhizal Infection in Roots. *New Phytol.* **1980**, *84*, 489–500. [CrossRef]
43. Xu, Y.; Siegenthaler, P.A. Low Temperature Treatments Induce an Increase in the Relative Content of Both Linolenic and Δ3-trans-Hexadecenoic Acids in Thylakoid Membrane Phosphatidylglycerol of Squash Cotyledons. *Plant Cell Physiol.* **1997**, *38*, 611–618. [CrossRef]
44. Beauchamp, C.O.; Fridovich, I. Isozymes of superoxide dismutase from wheat germ. *BBA Protein Struct.* **1973**, *317*, 50–64. [CrossRef]
45. Maehly, A.C.; Chance, B. The assay of catalases and peroxidases. *Methods Biochem. Anal.* **1954**, *1*, 357–424. [CrossRef]
46. Arnon, D.I. Copper Enzymes in Isolated Chloroplasts. Polyphenoloxidase in Beta vulgaris. *Plant Physiol.* **1949**, *24*, 1–15. [CrossRef]
47. Khan, M.Y.; Haque, M.M.; Molla, A.H.; Rahman, M.M.; Alam, M.Z. Antioxidant compounds and minerals in tomatoes by Trichoderma-enriched biofertilizer and their relationship with the soil environments. *J. Integr. Agric.* **2017**, *16*, 691–703. [CrossRef]

plants

Article

Roles of Exogenous α-Lipoic Acid and Cysteine in Mitigation of Drought Stress and Restoration of Grain Quality in Wheat

Amr Elkelish [1], Mohamed M. El-Mogy [2,*], Gniewko Niedbała [3], Magdalena Piekutowska [4], Mohamed A. M. Atia [5], Maha M. A. Hamada [6], Mostafa Shahin [6], Soumya Mukherjee [7], Ahmed Abou El-Yazied [8], Mohamed Shebl [9], Mohammad Shah Jahan [10,11], Ali Osman [12], Hany G. Abd El-Gawad [8], Hatem Ashour [13], Reham Farag [13], Samy Selim [14] and Mohamed F. M. Ibrahim [13,*]

1 Botany Department, Faculty of Science, Suez Canal University Ismailia, Ismailia 41522, Egypt; amr.elkelish@science.suez.edu.eg
2 Vegetable Crops Department, Faculty of Agriculture, Cairo University, Giza 12613, Egypt
3 Department of Biosystems Engineering, Faculty of Environmental and Mechanical Engineering, Poznań University of Life Sciences, Wojska Polskiego 50, 60-627 Poznań, Poland; gniewko.niedbala@up.poznan.pl
4 Department of Geoecology and Geoinformation, Institute of Biology and Earth Sciences, Pomeranian University in Słupsk, Partyzantów 27, 76-200 Słupsk, Poland; magdalena.piekutowska@apsl.edu.pl
5 Molecular Genetics and Genome Mapping Laboratory, Genome Mapping Department, Agricultural Genetic Engineering Research Institute (AGERI), Agricultural Research Center (ARC), Giza 12619, Egypt; matia@ageri.sci.eg
6 Department of Agronomy, Faculty of Agriculture, Ain Shams University, Cairo 11566, Egypt; Maha_hamada@agr.asu.edu.eg (M.M.A.H.); mostafa_shahin@agr.asu.edu.eg (M.S.)
7 Department of Botany, Jangipur College, University of Kalyani, Kalyani 742213, India; soumobios@gmail.com
8 Department of Horticulture, Faculty of Agriculture, Ain Shams University, Cairo 11566, Egypt; ahmed_abdelhafez2@agr.asu.edu.eg (A.A.E.-Y.); hany_gamal2005@agr.asu.edu.eg (H.G.A.E.-G.)
9 Food Science Department, Faculty of Agriculture, Ain Shams University, Cairo 11566, Egypt; Mohamed6080@agr.asu.edu.eg
10 College of Horticulture, Nanjing Agricultural University, Nanjing 210095, China; shahjahansau@gmail.com
11 Department of Horticulture, Sher-e-Bangla Agricultural University, Dhaka 1207, Bangladesh
12 Biochemistry Department, Faculty of Agriculture, Zagazig University, Zagazig 44511, Egypt; aokhalil@zu.edu.eg
13 Department of Agricultural Botany, Faculty of Agriculture, Ain Shams University, Cairo 11566, Egypt; Hatem_ashour@agr.asu.edu.eg (H.A.); Reham_hassan@agr.asu.edu.eg (R.F.)
14 Department of Clinical Laboratory Sciences, College of Applied Medical Sciences, Jouf University, Sakaka 72388, Saudi Arabia; sabdulsalam@ju.edu.sa
* Correspondence: elmogy@agr.cu.edu.eg (M.M.E.-M); Ibrahim_mfm@agr.asu.edu.eg (M.F.M.I.); Tel.: +20-1068027607 (M.M.E.-M); +20-1123403173 (M.F.M.I.)

Citation: Elkelish, A.; El-Mogy, M.M.; Niedbała, G.; Piekutowska, M.; Atia, M.A.M.; Hamada, M.M.A.; Shahin, M.; Mukherjee, S.; El-Yazied, A.A.; Shebl, M.; et al. Roles of Exogenous α-Lipoic Acid and Cysteine in Mitigation of Drought Stress and Restoration of Grain Quality in Wheat. *Plants* **2021**, *10*, 2318. https://doi.org/10.3390/plants10112318

Academic Editors: M. Iftikhar Hussain, Adele Muscolo and Mukhtar Ahmed

Received: 11 September 2021
Accepted: 15 October 2021
Published: 28 October 2021

Abstract: Cysteine (Cys) and α-lipoic acid (ALA) are naturally occurring antioxidants (sulfur-containing compounds) that can protect plants against a wide spectrum of environmental stresses. However, up to now, there are no conclusive data on their integrative roles in mitigation of drought stress in wheat plants. Here, we studied the influence of ALA at 0.02 mM (grain dipping pre-cultivation treatment) and Cys (25 and 50 ppm as a foliar application) under well watered and deficit irrigation (100% and 70% of recommended dose). The results showed that deficit irrigation markedly caused obvious cellular oxidative damage as indicated by elevating the malondialdehyde (MDA) and hydrogen peroxide content (H_2O_2). Moreover, water stressed plants exhibited multiple changes in physiological metabolism, which affected the quantitative and qualitative variables of grain yield. The enzymatic antioxidants, including superoxide dismutase (SOD), ascorbate peroxidase (APX), catalase (CAT) and peroxidase (POX) were improved by Cys application. SOD and APX had the same response when treated with ALA, but CAT and POX did not. Moreover, both studied molecules stimulated chlorophyll (Chl) and osmolytes' biosynthesis. In contrast, the Chl a/b ratio was decreased, while flavonoids were not affected by either of the examined molecules. Interestingly, all above-mentioned changes were associated with an improvement in the scavenging capacity of reactive oxygen species (ROS), leaf relative water content (RWC), grain number, total grain yield, weight of 1000 kernels, gluten index, falling number, and alveographic parameters (P, W, and P/L values). Furthermore, heatmap plot analysis revealed several significant correlations between different studied parameters, which may explore the importance of applied Cys and ALA as effective compounds in wheat cultivation under water deficit conditions.

Keywords: wheat; water stress; antioxidant capacity; grain quality; alveographic parameters; alpha-lipoic acid; cysteine

1. Introduction

Wheat (*Triticum aestivum* L.) is an extremely important cereal crop for human nutrition and animal feed worldwide. Though wheat cultivation is pervasive in a wide spectrum of ecosystems, many parts of the world (particularly in the arid and semiarid regions) suffer from drought stress as a major constraint hindering its expansion from meeting the growing demands day by day for the population [1,2].

Frequent climate change scenarios and rareness of fresh water are two major limiting constraints to sustainable agriculture and preserving food security worldwide. These two major problems may be exacerbated in the semiarid regions where the annual precipitation (200 to 750 mm/year) is not sufficient to meet the needs for farming throughout the year [3,4]. In plants, drought stress or water deficit can trigger a wide spectrum of successive events at morphological, biochemical, and molecular levels that may contribute to the adaptation or tolerance processes to such conditions [5–8]. Among these responses is the excessive release of reactive oxygen species (ROS) that induce oxidative damage to plant cell components, i.e., protein, lipids and, nucleic acids [9–11], and the development of several efficient non-enzymatic and enzymatic antioxidant systems [1,12,13]. Plant resilience to drought stress is mostly attained by accumulating compatible solutes, ion homeostasis, and redox management [5,14–16]. Additionally, water shortage can trigger a cascade of interconnected responses that coordinate the physiological metabolisms of plants, such as altering the signaling pathways, gene expression, hormonal homeostasis, and photosynthetic machinery, leading eventually to affecting crop productivity [8,17–21].

Sulfur-containing molecules and a number of non-protein and protein thiols have been evidenced to play a fundamental role in plant tolerance to various abiotic stresses [22–24]. These molecules can work together, representing a crucial network of responses that enable plants to cope with environmental stresses [24]. Therefore, special attention should be given to reveal a complete picture in this respect.

Alpha-lipoic acid (1,2-dithiolane-3-pentanoic acid; ALA) is a dithiol short-chain fatty acid that is ubiquitous both in prokaryotic or eukaryotic organisms [25]. It has received great attention as a dietary supplement for humans because of its antioxidant and therapeutic properties [26,27]. This antioxidant capacity depends on its two sulfhydryl moieties [25] which enable it to scavenge free radicals and chelate metals [28]. In cereal crops, a few studies during the seedling stage have reported that exogenous ALA can enhance the photosynthetic performance in maize under drought stress [29], ameliorate lipid peroxidation, and induce the antioxidant systems of maize under osmotic stress [25]. Furthermore, it can regulate the ion homeostasis and osmotic potential in wheat under saline conditions [30]. In addition, it has been found that there is a close connection between the endogenous ALA and the cellular redox status of wheat seedlings grown under excess copper [31].

Cysteine (Cys) is an essential thiol-containing amino acid that comprises amino group (NH_2), carboxylic acid (COOH) and sulfhydryl group (SH) as reactive centers. This distinct structure enables Cys to be a potent antioxidant and efficient scavenger for reactive oxygen species (ROS). This is due to the presence of the thiol side chain, which smoothly oxidizes, protecting against the oxidative damage induced by biotic and abiotic stresses [32,33]. Moreover, Cys is involved in synthesizing a wide array of vital and defensive molecules such as protein, glutathione, phytoalexins, thionins, glucosinolates, metallothioneins, and phytochelatins [23,32,34–36]. Additionally, Cys is implicated in different sulfur metabolism pathways and the synthesis of methionine [33]. These responses can regulate plant growth and development via the importance of methionine and/or S-adenosylmethionine (SAM) as precursors for some vital phytohormones, i.e., ethylene and polyamines [37,38]. Several lines of evidence have suggested that exogenous Cys can alleviate the deleterious effects of heavy metals in plants through its roles in metal sequestration and detoxification [35,36]. Similarly, applied Cys has been shown to delay the senescence of postharvest green leafy vegetables by decreasing the rate of Chl degradation, cellular respiration, and ethylene biosynthesis [39]. Up to now, there are no conclusive data to elucidate the role of ALA and/or Cys in wheat plants subjected to water deficit. The current research explores novel approaches to learn more about the combined effect of biomolecules on the diverse parameters of wheat plants under water deficit.

2. Materials and Methods

2.1. Plant Material, Treatments, and Experimental Design

A field experiment was conducted during the period from November to April in 2018/2019 and 2019/2020 at a private farm at Beheira Governorate, Egypt (latitude: 30.8481 N; longitude: 30.34 E; mean altitude: 60 m above sea level). Healthy and uniform wheat grains (*Triticum aestivum* L, Cv. Giza 168) were divided into two groups. The first group of grains was soaked in distilled water for 4 h and then sown in the experimental soil, whereas the second group was soaked for 4 h in a solution containing 0.02 mM α-lipoic acid (ALA, Cayman Chemical Company, Michigan, USA). The experimental design was Split-Split-Plot arranged into a randomized complete block design (RCBD) with three replicates. Irrigation with two application levels of 70% and 100% of the recommended dose for wheat plants was set up in the main plots, while ALA-treated and non-treated grains were distributed in the sub plots. Cys foliar applications (Cys, Alpha Chemika, Mumbai, India) at 0, 25 and 50 ppm plus 0.05% (*v*/*v*) Tween 20 as a non-ionic surfactant were randomly distributed in the experimental units. The experimental unit was 15 m^2 (3 × 5 m) with sowing space 12 cm. Applied Cys was repeated three times during the growing season, starting from the tillering to stem extension stage at 25, 45, and 65 days after sowing. At the anthesis stage (85 days after sowing), leaves were collected for different biochemical estimations, whereas the total yield ($ton \cdot ha^{-1}$), various physical/chemical properties and alveographic parameters of grains were determined at the end of each season (158 days after sowing).

2.2. Irrigation Requirements

A local weather station registered meteorological data. The mean daily temperature data, wind speed at the height of 2 m, precipitation, daily solar radiation, and relative humidity were used in developing the Reference Evapotranspiration (ET_o) in mm/day. The ET_o was created according to the calculation procedure given in FAO paper n. 56 [40]. Estimation of irrigation requirement (IR) in m^3/ha is derived from crop evapotranspiration (ETc), which can be calculated by multiplying the ET_o with the crop coefficient (Kc). It has been taken into account that the sprinkler irrigation efficiency is 75% when calculating the IR. Based on the quantities of IR calculated in Table 1, the irrigation quantities in the main experimental plots were determined with two application levels of 70% and 100% from IR. The period considered in this study was calculated ET_o for two consecutive seasons (2018/2019 and 2019/2020). The ET_o equation can be expressed as:

$$ET_o = \frac{0.408\,\Delta\,(R_n - G) + \gamma\left(\frac{900}{(T+273)}\right)U_2(e_s - e_a)}{\Delta + (1 + 0.34\,U_2)} \quad (1)$$

where Δ is the slope of the vapor pressure curve ($kPa \cdot {}^{\circ}C^{-1}$), R_n the surface net radiation ($MJ \cdot m^{-2} \cdot day^{-1}$), G soil heat flux density ($MJ \cdot m^{-2} \cdot day^{-1}$), γ psychometric constant (kPa·°C^{-1}), T mean daily air temperature (°C), U_2 wind speed ($m \cdot s^{-1}$), e_s saturated vapor pressure, e_a actual vapor pressure.

Kc = Crop Coefficient, ET_o = Reference Evapotranspiration (mm), IR = Irrigation Requirements (m^3/ha). (2)

Table 1. Details of irrigation requirements of two consecutive seasons (2018/2019 and 2019/2020) for wheat based on the reference evapotranspiration calculated by FAO 56 Penman–Monteith equation. The irrigation system efficiency of 75% is considered in the calculations.

Growth Stage	Kc	Season 2018/2019		Season 2019/2020	
		Mean ET_o, mm	Total IR, m^3/ha	Mean ET_o, mm	Total IR, m^3/ha
Initial	0.7	3.1	578.7	3.9	728.0
Crop development	1.15	2.8	2037.4	2.8	2025.6
Mid season	1.15	3.8	4081.7	3.4	3687.7
Late season	0.25	5.9	1105.0	5.3	984.9
Total IR, m^3/ha			7802.8		7426.2

2.3. *Agricultural Management*

Wheat grains were sown using 178.57 kg ha^{-1} at the 10th and 8th of November in the first and second season, respectively. The recommended doses of mineral fertilizers (NPK) were applied as follows: P (36 kg ha^{-1}) was supplemented during the soil preparation as calcium super phosphate 15.5% P_2O_5, while N (187.5 kg ha^{-1}) was added as ammonium nitrate (33.5% N) into seven equal portions, at 10, 20, 30, 40, 50, 60 and 70 days after sowing. Additionally, K (79.2 kg ha^{-1}) was supplemented as potassium sulfate (48% K_2O) in two equal portions at sowing and heading stages.

2.4. *Determination of Chlorophyll (Chl)*

The leaf content of Chl a and Chl b was determined as described by Costache et al. [41] with some modification, small pieces of fresh leaves (0.5 g) were submerged into 10 mL pure acetone for 24 h/4 °C in small dark bottles to avoid pigment degradation. The homogenate was centrifuged at 4000 rpm for 15 min then the absorbance was measured at 645 and 663 nm, respectively. The concentration was calculated using the following equations:

$$\text{Chl a (mg/g FW)} = 11.75\ A_{662} - 2.350\ A_{645} \times (V/1000 \times W) \quad (3)$$

$$\text{Chl b (mg/g FW)} = 18.61\ A_{645} - 3.960\ A_{662} \times (V/1000 \times W) \quad (4)$$

where, A is the absorbance at 645 and 663 nm, V is the final volume of Chl extract in pure acetone and W is the fresh weight of tissue extract. Additionally, Chl a+b and Chl a/b ratio were calculated.

2.5. *Determination of Osmolytes and Leaf Water Status*

Proline concentration was determined with ninhydrin reagent as described by Bates et al. [42]. Total soluble sugars were estimated by phenol-sulfuric acid method as described by Chow and Landhäusser [43]. Leaf relative water content was determined according to Ünyayar et al. [44]. Leaf discs from 10 leaves were weighed (FW) and placed immediately in distilled water for 2 h at 25 °C and then their turgid weights (TW) were recorded. The samples were then dried in an oven at 110 °C for 24 h (DW). Relative water content (RWC) was calculated by using the following formula: RWC = (FW − DW)/(TW − DW) × 100.

2.6. *Determination of Oxidative Damage and Scavenging Capacity*

The level of lipid peroxidation was measured by the determination of malondialdehyde (MDA) as described by Heath and Packer [45]. One gram of fresh leaves was homogenized in 10 mL diluted trichloroacetic acid (TCA; 0.1% w/v). The homogenate was centrifuged at 4500 rpm for 15 min. The reaction mixture contained 1 mL from the supernatant and 4 mL 0.5% (w/v) thiobarbituric acid (TBA) dissolved in 20% (w/v) TCA. The mixture was heated in boiling water for 30 min then the mixture was cooled at room temperature and centrifuged at 4500 rpm for 15 min. The absorbance of the supernatant was measured at 535 nm and corrected for non-specific turbidity at 600 nm using a spectrophotometer (UV-1601PC; Shimadzu, Tokyo, Japan). The MDA concentration (nmol·g^{-1} FW) was calculated using Δ OD (A532-A600) and the extinction coefficient (ε = 155 mm^{-1} cm^{-1}).

Hydrogen peroxide (H_2O_2) concentration was determined according to Velikova et al. [46] with some modifications. Leaf samples of 0.5 g were homogenized in 3 mL of 1% (w/v) tri-chloroacetic acid (TCA). The homogenate was centrifuged at 10,000 rpm and 4 °C for 10 min. Subsequently, 0.75 mL of the supernatant was added to 0.75 mL of 10 mM K-phosphate buffer (pH 7.0) and 1.5 mL of 1M KI. H_2O_2 concentration was evaluated by comparing its absorbance at 390 nm to a standard calibration curve. The concentration of H_2O_2 was calculated from a standard curve plotted in the range from 0 to 15 nmol mL^{-1}.

The scavenging capacity of free radicals was estimated by the reduction of the reaction color method of 1,1-Diphenyl-2-picrylhydrazyl (DPPH) with sample extract as described by Huang et al. [47]. One gram of fresh leaves was homogenized in 20 mL of 70% methanol (v/v) for 15 h and kept in dark until assay. A final concentration (0.15 mM) of DPPH solution (3.9 mL) was mixed with sample solution (0.1 mL). The mixture was kept in the dark at ambient temperature. The absorbance of the mixtures was recorded at 515 nm for exactly 30 min. Blank was made from 3.9 mL of DPPH and 0.1 mL methanol and measured absorbance at T_0. The scavenging of DPPH was calculated according to the following equation:

$$\%\ \text{DPPH scavenging} = [(\text{Abs } T_0 - \text{Abs } T_{30})/\text{Abs } T_0] \times 100 \quad (5)$$

where Abs (T_0) = absorbance of DPPH at 0 time, T_{30} absorbance at 30 min.

2.7. Determination of Non-Enzymatic Antioxidants

Carotenoids were quantified using the acetone and petroleum ether method as described by de Carvalho et al. [48] using the following formula:

$$\text{Carotenoids (mg/g FW)} = A_{450} \times \text{V (mL)} \times 10/[A^{1\%}{}_{1\text{cm}} \times \text{W (g)}] \quad (6)$$

where A_{450} = absorbance at 450 nm, V = total extract volume; W = sample weight; $A^{1\%}{}_{1\text{cm}}$ = 2592 (β-carotene coefficient in petroleum ether). The specific wavelengths for all estimated leaf pigments were determined using UV visible spectrophotometer (UV-1601PC; Shimadzu, Tokyo, Japan).

Aluminum chloride colorimetric method was used for flavonoids determination [49]. Each plant was extracted (0.5 mL of 1:10 g mL^{-1}) in methanol and was separately mixed with 1.5 mL of methanol, 0.1 mL of 10% aluminum chloride, 0.1 mL of 1 M potassium acetate, and 2.8 mL of distilled water. It was kept at room temperature for 30 min; the reaction mixture's absorbance was measured at 415 nm. The calibration curve was obtained by preparing quercetin solutions at concentrations of 12.5 to 100 µg/mL.

The extraction of fresh leaves in cold MeOH 85% was used to determine total soluble phenols according to the method of Folin–Denis as described by Shahidi et al. [50]. One milliliter of crude extract was mixed with 0.5 mL of Folin–Denis reagent and were well mixed in dry test tube, the tube was thoroughly shaken for 3 min, 1.0 mL of saturated Na_2CO_3 solution was added and well mixed then 3 mL of distilled water was added. After one hour, phenolic compounds were determined by reading the developed blue color at 725 nm by using a spectrophotometer. Ascorbic acid (AsA) was determined using 2, 6-Dichloroindophenol titrimetric method [51].

2.8. Determination of Antioxidant Enzymes Activity

To prepare the extraction of enzyme and soluble proteins, fresh leaves (0.5 g) were homogenized in 4 mL 0.1 M sodium phosphate buffer (pH 7.0) containing 1% (*w*/*v*) polyvinylpyrrolidone (PVP) and 0.1 mM EDTA, centrifuged at 10,000× *g* for 20 min at 4 °C and then the supernatant was used for assays. Soluble proteins were evaluated by the method of Bradford [52]. All studied enzyme activities and protein concentration in the crude enzyme extract were measured using a spectrophotometer (UV-1601PC; Shimadzu, Tokyo, Japan) as follows.

Superoxide dismutase (SOD) assay was based on the method described by Beyer et al. [53]. The reaction mixture with a total volume of 3 mL contained 100 µL crude enzyme, 50 mM phosphate buffer (pH 7.8), 75 µM NBT, 13 mM L-methionine, 0.1 mM EDTA and 0.5 mM riboflavin. The addition of riboflavin initiated the reaction then the reaction mixture was illuminated for 20 min with a 20 W fluorescent lamp. One enzyme activity unit was defined as the amount of enzyme required to result in a 50% inhibition in the rate of nitro blue tetrazolium (NBT) reduction at 560 nm.

Catalase (CAT) activity was measured by monitoring the decrease in absorbance at 240 nm as described by Cakmak et al. [54]. The reaction mixture with a total volume of 3 mL contained 15 mM H_2O_2 in 50 mM phosphate buffer (pH = 7). The reaction was initiated by adding 50 µL crude enzyme. The activity was calculated from the extinction coefficient (ε = 40 mm^{-1} cm^{-1}) for H_2O_2. One unit of enzyme activity was defined as the decomposition of 1 µmol of H_2O_2 per minute.

The activity of ascorbate peroxidase (APX) was determined according to Nakano and Asada [55]. The decrease of absorbance at 290 nm was monitored for 3 min. The reaction mixture with a total volume of 3 mL included 100 µL crude enzyme, 50 mM phosphate buffer (pH 7), 0.1 mM EDTA, 0.5 mM ascorbic acid, and 0.1 mM H_2O_2. The addition of H_2O_2 initiated the reaction. One enzyme activity unit was defined as the amount of enzyme required for oxidation of 1 µmol of ascorbate per minute. The rate of ascorbate oxidation was calculated using the extinction coefficient (ε = 2.8 mm^{-1} cm^{-1}).

Peroxidase (POX) activity was quantified by the method of Dias and Costa [56] with some minor modifications. The assay mixture (100 mL) contained 10 mL of 1% (*v*/*v*) guaiacol, 10 mL of 0.3% H_2O_2 and 80 mL of 50 mM phosphate buffer (pH = 6.6). The volume of 100 µL of the crude enzyme was added to 2.9 mL of the assay mixture to start the reaction. The absorbance was recorded every 30 s for 3 min at 470 nm.

2.9. Properties of Wheat Grains

2.9.1. Physical/Chemical Characteristics

Moisture was determined according to AACC standard method No 44-15.02 [57]. Weight per 1000 kernels and hectoliter weight were determined according to AACC standard method No 55-10.01 [57]. Wet gluten and gluten index were determined for wheat meal. Ground whole grain was performed by laboratory mill (3100 with 0.8-mm screen). Wet gluten was washed from whole-grain wheat meal by an automatic gluten washing apparatus (Glutomatic system 2200, Perten Instruments AB, Huddinge, Sweden). Then, it was centrifuged on a specially constructed sieve under standardized conditions. The weight of wet gluten was forced through the sieve then total wet gluten was measured according to AACC standard method No 38-12.01 [57]. The total wet gluten was expressed as percent of sample, and the gluten index was expressed as percentage of wet gluten remaining on the sieve after centrifuging. Hagberg falling number was determined according to AACC standard method No 56-81.03 [57] using FN 1500 (Perten Instruments AB, Hagersten, Sweden). This method is based on the ability of α-amylase to liquefy a starch gel. The activity of the enzyme is measured by falling number (FN), defined as time in sec required to stir and allow stirrer to fall a measured distance through a hot aqueous flour or meal gel undergoing liquefaction. α-Amylase activity is associated with kernel sprouting, and both of these are inversely correlated with FN. Protein (N×5.7) content was analyzed according to AOAC standard method No 920.87 [58].

2.9.2. Alveograph Test

Alveographic properties were evaluated using Chopin Alveograph (Chopin, Villeneuve La Garenne cedex—France) according to AACC standard method No 54-30.02 [57]. Alveograph test was carried out to determine resistance of dough to extension (maximum overpressure, *P*; this is average of maximum ordinates, measured in mm), extensibility (average abscissa at rupture, *L*; abscissa at rupture of each curve is measured in mm on zero line, from origin of curve to point corresponding vertically with clear drop due to rupture of bubble; average of abscissae at rupture points of curves, expressed to nearest unit, represents length), curve configuration ratio, *P*/*L*, and Deformation energy of dough, *W*. Average curve is drawn based on average of ordinates up to *L*, expressed in 10^{-4} J.

2.10. Statistical Analysis and Data Visualization

A combined analysis of variance over the two seasons was carried out (Table 2). The significance of difference among the studied varieties was tested by analyzing variance (ANOVA) test as outlined by Snedecor and Cochran [59]. Mean comparisons for variables were made among genotypes using least significant differences (LSD at 5%) tests. Moreover, silhouette analysis was performed to evaluate the quality of the irrigation, seed soaking, and foliar application treatments measurements clustering by testing the cluster distances within and between each cluster [60]. Additionally, we performed a multidimensional preference analysis to disclose the interrelationships amongst parameters in addition to the similarity classification of in terms of dependent and independent variables [61]. Finally, hierarchical clustering based on the correlation analysis was conducted and two-dimensional heatmap plotting was constructed. Boxplot of the parameter's classes and "seed soaking and foliar application" levels are represented graphically, showing distribution of the data, and the X inside the box represents the mean. Colored Mosaic plot was displayed to summarizing the high dimensional data levels. The heatmap, Mosaic plot and Boxplot were drawn with R software.

Table 2. Analysis of variance (mean square) of leaf chlorophyll concentration.

Source	Df	Chl a	Chl b	Chl a+b	Chl a/b
		(mg·g^{-1} FW)	(mg·g^{-1} FW)	(mg·g^{-1} FW)	Ratio
Irrigation (I)	1	3.561 ***	1.746 ***	10.304 ***	0.399 ***
Seed soaking (S)	1	0.669 ***	0.783 ***	2.901 ***	0.708 ***
Foliar appl. (F)	2	0.067 ***	0.033 ***	0.196 ***	0.010 **
I × S	2	0.031 ***	0.101ns	0.019 **	0.664 ***
I × F	2	0.004ns	0.001ns	0.003ns	0.014 **
S × F	2	0.001ns	0.001ns	0.000ns	0.011 **
I × S × F	2	0.001ns	0.002ns	0.000ns	0.000ns

ns, **, *** not significant or significant at $p \leq 0.01$, $p \leq 0.001$, analysis of variance.

3. Results

3.1. Changes in Chlorophyll Composition

Analysis of variance (Table 2) shows that all leaf chlorophyll compositions including Chl a, Chl b, Chl a+b, and Chl a/b ratio were dramatically and significantly affected by irrigation level (I), ALA as seed soaking (S) and Cys as foliar application (F) treatments. However, the interaction between the irrigation level and seed soaking (I × S) was significant in respect to the Chl a, Chl a+b and Chl a/ b ratios. Furthermore, the Chl a/b ratio was significantly affected by the interaction treatments of I × F and S × F. Generally, deficit irrigation caused an obvious and significant ($p \leq 0.05$) decrease in Chl a, Chl b and Chl a+b, whereas the Chl a/b ratio was significantly increased particularly in the plants not treated with ALA (Figure 1). This trend indicates the positive effect of ALA as a seed soaking treatment on maintaining the content of Chl b under water shortage. The highest significant increase in Chl a, Chl b and Chl a+b was obtained by the treatment of 0.02 mM ALA + 50 ppm Cys under well watered conditions. In contrast, under limited water supply, the treatments of 0.02 mM ALA and/or 50 ppm Cys significantly reduced the Chl a/b ratio compared to the untreated plants.

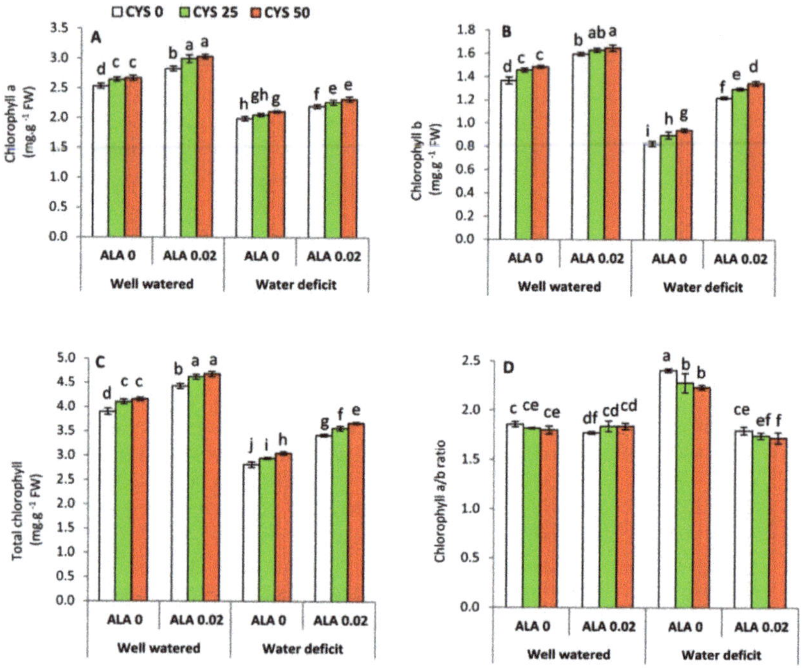

Figure 1. Effect of irrigation level, seed soaking by α-lipoic acid (ALA; 0 and 0.02 mM) and foliar application by cysteine (Cys; 0, 25 and 50 ppm) on the chlorophyll concentration of wheat plants. (**A**) chlorophyll a, (**B**) chlorophyll b, (**C**) total chlorophyll, (**D**) chlorophyll a/b ratio. For each parameter, the mean values ± SD followed by a different letter are significantly different according to Tukey's range test ($p \leq 0.05$).

3.2. Oxidative Stress and Scavenging Capacity

To assess the efficiency of plants to scavenge the cytotoxic molecules which lead to the oxidative damage, antioxidant activity by DPPH, and lipid peroxidation as indicated by malondialdehyde (MDA) and H_2O_2 were determined (Table 3, Figure 2). The results showed that all above-mentioned variables were significantly affected by irrigation level (I), ALA as seed soaking (S) and Cys as foliar application (F) treatments. The interaction treatment of I × S was significant with H_2O_2, while the interaction treatment of I × F was significant with MDA and H_2O_2. The general tendency was that deficit irrigation resulted in a significant increase ($p \leq 0.05$) in the antioxidant activity, MDA and H_2O_2 compared to the well watered plants. Despite being under water deficit, the treatment of 0.02 mM ALA individually reduced the concentration of MDA and H_2O_2, and the increase of antioxidant activity by DPPH did not reach the level of significance compared to ALA-untreated plants. The foliar applications of Cys (25/50 ppm) either individually or combined with ALA led to improve significantly ($p \leq 0.05$) the antioxidant activity by DPPH in parallel with reducing the concentration of MDA and H_2O_2 under water deficit conditions.

Table 3. Analysis of variance (mean square) of ROS and scavenging capacity.

Source	Df	Antioxidant Activity by DPPH (%)	MDA (μmol g^{-1} FW)	H_2O_2 (μmol g^{-1} FW)
Irrigation (I)	1	1229.671 ***	15.484 ***	1015.590 ***
Seed soaking (S)	1	27.737 ***	0.697 *	15.906 ***
Foliar appl. (F)	2	239.534 ***	2.350 ***	7.408 ***
I × S	2	15.210ns	0.148ns	5.313 ***
I × F	2	10.787ns	1.528 ***	6.077 ***
S × F	2	15.547ns	0.045ns	0.314ns
I × S × F	2	10.570ns	0.176ns	0.036ns

ns, *, *** not significant or significant at $p \leq 0.05$, $p \leq 0.001$, analysis of variance.

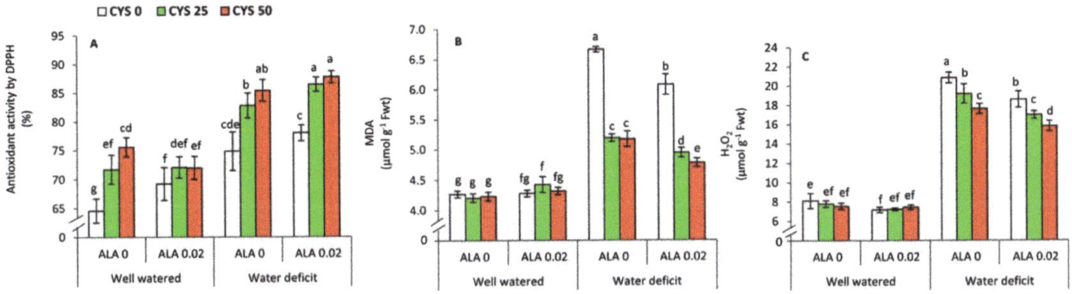

Figure 2. Effect of irrigation level, seed soaking by α-lipoic acid (ALA; 0 and 0.02 mM) and foliar application by cysteine (Cys; 0, 25 and 50 ppm) on the oxidative stress and scavenging capacity of wheat plants. (**A**) total antioxidant activity, (**B**) Lipid peroxidation (MDA), (**C**) Hydrogen peroxide content. For each parameter, the mean values ± SD followed by a different letter are significantly different according to Tukey's range test ($p \leq 0.05$).

3.3. Non-Enzymatic Antioxidants

Non-enzymatic antioxidants including carotenoids, total soluble phenols and ascorbic acid were significantly influenced by the irrigation level (I), ALA as seed soaking (S) and Cys as foliar application (F) treatments, whereas flavonoids were significantly affected by reducing the irrigation level (Table 4). The interactive treatments revealed that flavonoids and total soluble phenols were affected by I × S. Ascorbic acid was also significantly changed by I × F interaction treatment. Generally, carotenoids were significantly ($p \leq 0.05$) decreased under water deficit. Conversely, flavonoids, total soluble phenols and ascorbic acid were increased with reducing the irrigation level (Figure 3). The highest significant ($p \leq 0.05$) increase in carotenoids was obtained by the treatment of 0.02 mM ALA + 50 ppm Cys under well watered conditions. On the other hand, the treatment of 0.02 mM ALA + 50 ppm Cys achieved the maximum significant ($p \leq 0.05$) increase in flavonoids, total soluble phenols and ascorbic acid compared to the untreated plants under water shortage.

3.4. Changes in the Activities of Antioxidant Enzymes

The activities of antioxidant enzymes including SOD, CAT, POX, and APX were significantly affected by the irrigation level (I), ALA as seed soaking (S) and Cys as foliar application (F) treatments (Table 5). Moreover, POX was affected by all the interactions except the irrigation × seed soaking × Cys interaction. All antioxidant enzymes were increased under water deficit condition compared to well watered condition (Figure 4). Under water deficit conditions, SOD and APX were significantly ($p \leq 0.05$) increased at higher ALA concentration (0.02 mM) compared to the untreated plants with ALA (0 mM). Conversely, the activity of CAT was significantly decreased at higher ALA concentration, whereas POX did not show any significant changes. All antioxidant enzymes were increased by increasing the concentrations of Cys except POX at higher ALA under well water condition. Under water shortage, the treatment of 50 ppm Cys without ALA and all Cys treatments with ALA recorded the maximum significant ($p \leq 0.05$) activities in SOD. Additionally, the highest activity of CAT was achieved by the treatment of 50 ppm Cys without ALA under water deficit conditions, whereas the highest POX value was recorded at zero ALA concentration and 25 ppm Cys. Moreover, the treatment of 0.02 ALA + 50 ppm Cys achieved the highest significant ($p \leq 0.05$) activity regarding to APX.

Table 4. Analysis of variance (mean square) of non-enzymatic antioxidants.

Source	Df	Carotenoids	Flavonoids	Total Soluble Phenols	Ascorbic Acid
		(mg·g $^{-1}$ FW)	(mg·g^{-1} FW)	(mg·g^{-1} FW)	(µg·g^{-1} FW)
Irrigation (I)	1	0.055 ***	0.225 ***	1.046 ***	1836.1225 ***
Seed soaking (S)	1	0.026 ***	0.000ns	0.101 ***	592.922 ***
Foliar appl. (F)	2	0.021 ***	0.009ns	0.034 ***	771.960 ***
I × S	2	0.001ns	0.026 *	0.043 ***	0.233ns
I × F	2	0.001ns	0.006ns	0.001ns	77.160 **
S × F	2	0.001ns	0.008ns	0.006ns	37.307ns
I × S × F	2	0.001ns	0.005ns	0.004ns	4.423ns

ns, *, **, *** not significant or significant at $p \leq 0.05$, $p \leq 0.01$, $p \leq 0.001$, analysis of variance.

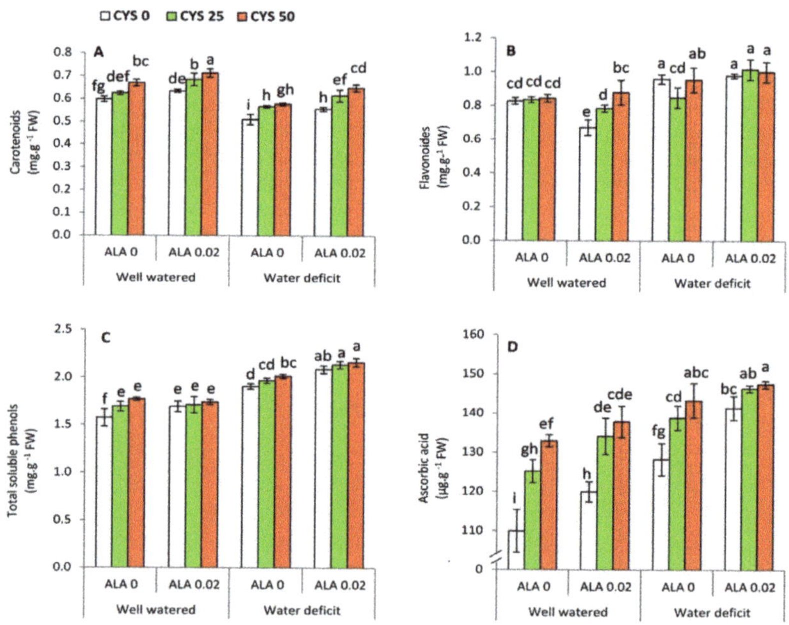

Figure 3. Effect of irrigation level, seed soaking by α-lipoic acid (ALA; 0 and 0.02 mM) and foliar application by cysteine (Cys; 0, 25 and 50 ppm) on the concentration of non-enzymatic antioxidants in wheat plants. (**A**) Carotenoids, (**B**) Flavonoids, (**C**) Total soluble phenols, (**D**) Ascorbic acid. For each parameter, the mean values ± SD followed by a different letter are significantly different according to Tukey's range test ($p \leq 0.05$).

Table 5. Analysis of variance (mean square) of antioxidant enzymes.

Source	Df	SOD (unit. mg^{-1} protein)	CAT (unit. mg^{-1} protein)	POX (unit. mg^{-1} protein)	APX (unit. mg^{-1} protein)
Irrigation (I)	1	1207.313 ***	560.742 ***	32852.022 ***	13.788 ***
Seed soaking (S)	1	129.925 ***	191.730 ***	2352.249 *	1.904 ***
Foliar appl. (F)	2	51.214 ***	140.043 ***	4573.021 **	1.713 ***
I × S	2	0.0270ns	1.343ns	19834.022 ***	0.024ns
I × F	2	1.470ns	2.729ns	7253.027 ***	0.051ns
S × F	2	5.391ns	8.817ns	5868.582 **	0.005ns
I × S × F	2	1.894ns	2.335ns	2776.361ns	0.009ns

ns, *, **, *** not significant or significant at $p \leq 0.05$, $p \leq 0.01$, $p \leq 0.001$, analysis of variance.

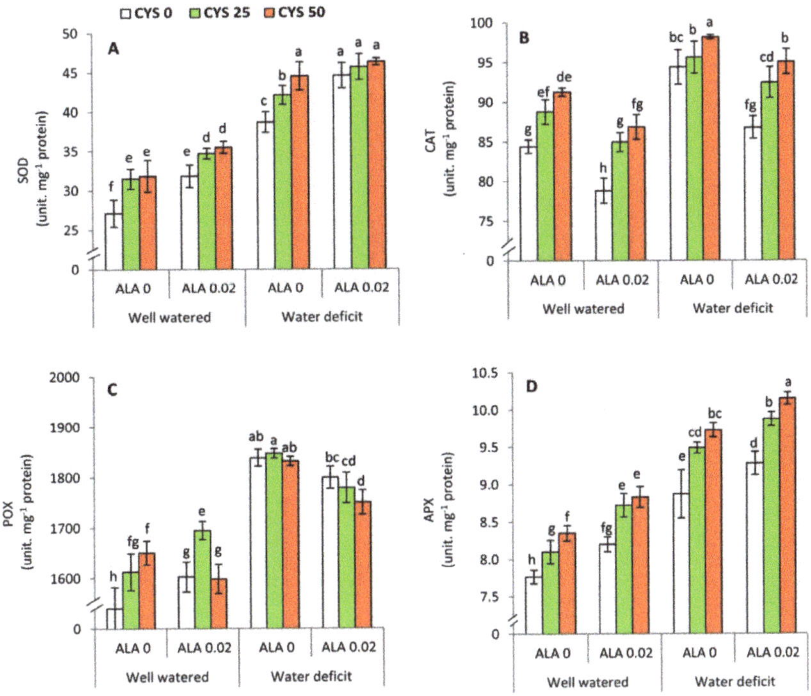

Figure 4. Effect of irrigation level, seed soaking by α-lipoic acid (ALA; 0 and 0.02 mM) and foliar application by cysteine (Cys; 0, 25 and 50 ppm) on the activities of antioxidant enzymes in wheat plants. (**A**) SOD, (**B**) CAT, (**C**) POX, (**D**) APX. For each parameter, the mean values ± SD followed by a different letter are significantly different according to Tukey's range test ($p \leq 0.05$).

3.5. Changes in Osmolytes and Leaf Relative Water Content

Proline, total soluble sugars, and leaf relative water content (RWC) were significantly affected by the irrigation level (I), ALA as seed soaking (S) and Cys as foliar application (F) treatments (Table 6). The interaction treatments revealed that there was a significant relationship between the irrigation level and the seed soaking in ALA in respect to all traits (proline, total soluble sugars and RWC), as well as the interaction treatment of irrigation x foliar applications by Cys demonstrated a significant effect in proline. Generally, plants subjected to water deficit exhibited an obvious and significant ($p \leq 0.05$) increase in both investigated osmolytes (proline and sugars). However, an opposite trend was observed in respect to RWC (Figure 5). The foliar applications of Cys at both examined concentrations (25/50 ppm) were showed to display a significant ($p \leq 0.05$) increase in proline in ALA-treated plants under well watered conditions. In this respect, the treatment of 50 ppm Cys + 0.02 mM ALA recorded the highest significant values under water deficit conditions. In addition, total soluble sugars were significantly ($p \leq 0.05$) improved by increasing the concentration of Cys either in ALA-treated or non-treated plants under both investigated levels of irrigation. Leaf relative water content (RWC) demonstrated remarkable and significant ($p \leq 0.05$) improvement in the plants that exposed to both investigated concentrations of Cys under water deficit conditions.

Table 6. Analysis of variance (mean square) of osmolytes and leaf water status.

Source	Df	Proline	Sugars	Leaf RELATIVE Water Content (RWC)
		($\mu g \cdot g^{-1}$ FW)	($mg \cdot g^{-1}$ FW)	(%)
Irrigation (I)	1	202155.146 ***	28.178 ***	973.856 ***
Seed soaking (S)	1	8405.833 ***	2.002 ***	37.576 ***
Foliar appl. (F)	2	1919.463 ***	0.295 ***	18.286 ***
I × S	2	10332.723 ***	0.258 ***	21.098 **
I × F	2	414.193 ***	0.005ns	1.639ns
S × F	2	21.095ns	0.003ns	0.157ns
I × S × F	2	2.801ns	0.006ns	0.735ns

ns, **, *** not significant or significant at $p \leq 0.01$, $p \leq 0.001$, analysis of variance.

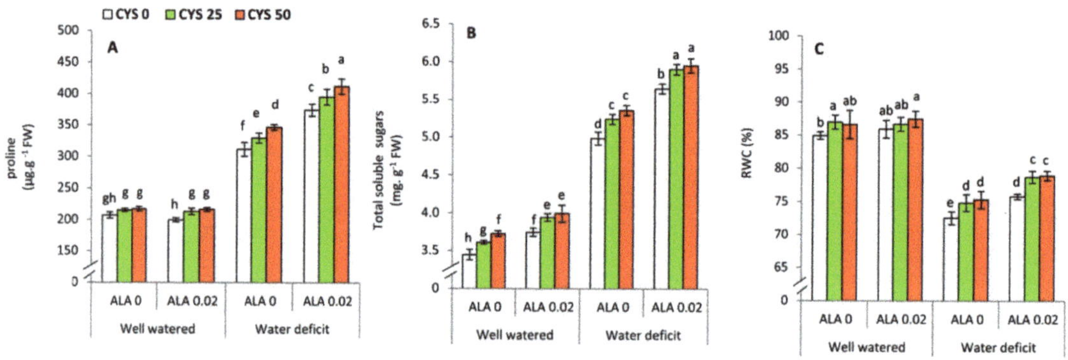

Figure 5. Effect of irrigation level, seed soaking by α-lipoic acid (ALA; 0 and 0.02 mM) and foliar application by cysteine (Cys; 0, 25 and 50 ppm) on the concentration of osmolytes and leaf relative water content (RWC) of wheat plants. (**A**) Proline, (**B**) Total soluble sugars, (**C**) RWC. For each parameter, the mean values ± SD followed by a different letter are significantly different according to Tukey's range test ($p \leq 0.05$).

3.6. Grain Yield and Its Components

Analysis of variance (Table 7) shows that number of spikes per m^2, number of grains per plant and total grain yield (ton/ha) were significantly affected by the irrigation level (I), ALA as seed soaking (S) and Cys as foliar application (F) treatments. The interaction treatments demonstrated that number of spikes and total grain yield were affected by the interaction between the irrigation level and seed soaking treatment by ALA. Additionally, total grain yield was affected by the interaction between the irrigation level and Cys treatments. The highest significant ($p \leq 0.05$) number of spikes/m^2 was obtained by the treatment of ALA 0.02 + Cys 50 under well watered conditions (Figure 6A). Under limited water supply, plants that were treated by Cys (25 or 50 ppm) displayed a significant ($p \leq 0.05$) increase in the number of spikes compared to the untreated plants regardless of the ALA treatments. Additionally, the highest number of grains per plant was achieved by the treatment of ALA 0.02 + Cys 50 under well watered conditions (Figure 6B). A similar trend was observed under water deficit conditions. Enhancement of the number of spikes and number of grains by ALA and Cys treatments was eventually reflected on the total grain yield (ton/ha) (Figure 6C).

Table 7. Analysis of variance (mean square) of grain yield and its components.

Source	Df	Spike Number per m^2	Grain Number per plant	Grain Yield ($ton \cdot ha^{-1}$)
Irrigation (I)	1	15604.17 ***	6445.413 ***	18.261 ***
Seed soaking (S)	1	447.32 ***	238.702 ***	0.224 ***
Foliar appl. (F)	2	733.24 ***	190.810 ***	0.230 ***
I × S	2	433.33 ***	20.400ns	0.054 **
I × F	2	60.02ns	32.418ns	0.021 *
S × F	2	24.92ns	1.740ns	0.001ns
I × S × F	2	30.95ns	7.463ns	0.002ns

ns, *, **, *** not significant or significant at $p \leq 0.05$, $p \leq 0.01$, $p \leq 0.001$, analysis of variance.

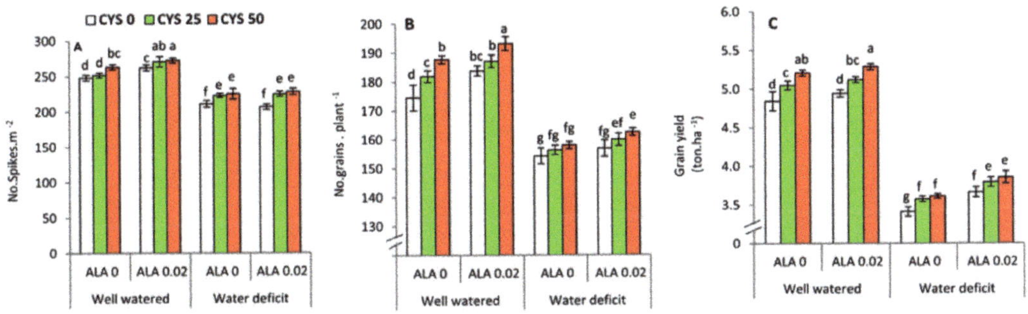

Figure 6. Effect of irrigation level, seed soaking by α-lipoic acid (ALA; 0 and 0.02 mM) and foliar application by cysteine (Cys; 0, 25 and 50 ppm) on grain yield and its components of wheat plants. (**A**) number of spikes, (**B**) number of grains, (**C**) Grain yield. For each parameter, the mean values ± SD followed by a different letter are significantly different according to Tukey's range test ($p \leq 0.05$).

3.7. The Physical/Chemical Properties of Grains

Quality assessment of wheat grains was performed by evaluating some simple physical/chemical parameters (Table 8). All studied parameters including moisture content, wet gluten, gluten index, falling number, weight of 1000 kernels and hectoliter weight were significantly affected by reducing the irrigation level. Moreover, ALA as a seed soaking treatment significantly affected gluten index, whereas applied Cys significantly affected moisture content, gluten index, falling number and weight of 1000 kernels. No significant effects were detected between all possible interaction treatments except irrigation x Cys treatments in respect to the moisture content and falling number.

The highest significant ($p \leq 0.05$) moisture content (Figure 7A) was obtained by the treatment of 50 ppm Cys under well watered conditions compared to the control. The wet gluten (Figure 7B) revealed an obvious and significant ($p \leq 0.05$) increase under water deficit conditions especially with increasing the concentration of Cys up to 50 ppm. The highest significant ($p \leq 0.05$) value in the gluten index (Figure 7C) was achieved by the treatment of 0.02 mM ALA + 25 ppm Cys under deficit irrigation conditions. Falling number displayed a significant ($p \leq 0.05$) decrease in all treatments of Cys at both examined concentrations compared to the Cys-untreated plants (Figure 7D). This increase was more pronounced under deficit irrigation compared to the well watered conditions. The weight of 1000 kernels (Figure 7E) was significantly ($p \leq 0.05$) increased in parallel with rising the concentration of Cys up to 50 ppm under both investigated levels of irrigation, but the treatment of 0.02 mM ALA did not show any significant results. Furthermore, no significant changes were observed in the hectoliter weight by all investigated treatments under the same level of irrigation (Figure 7F). However, the treatment of 0.0 mM ALA + 50 ppm Cys under well watered condition revealed a significant increase compared to the treatments of zero ALA + zero Cys, zero ALA + 25 ppm Cys and 0.02 mM ALA + 50 ppm Cys under water limited supply.

Table 8. Analysis of variance (mean square) of physical and chemical properties of grains.

Source	Df	Moisture	Wet Gluten	Gluten Index	Falling Number	Weight per 1000	Hectoliter Weight
		(%)	(%)	(%)	Sec.	kernels (g)	kg/L
Irrigation (I)	1	0.122 *	9.506 ***	58.778 ***	413.444 **	1534.027 ***	6.250 *
Seed soaking (S)	1	0.003ns	0.034ns	32.111 **	0.444ns	0.251ns	1.361ns
Foliar appl. (F)	2	0.112 *	0.280ns	10.333 *	3270.361 ***	11.861 *	0.027ns
I × S	2	0.063ns	0.466ns	1.000ns	4.000ns	0.694ns	0.027ns
I × F	2	0.081 *	1.041ns	4.777ns	130.527 *	0.027ns	0.583ns
S × F	2	0.017ns	0.181ns	5.444ns	3.694ns	1.750ns	0.861ns
I × S × F	2	0.017ns	0.031ns	4.333ns	20.083ns	0.361ns	0.861ns

ns, *, **, *** not significant or significant at $p \leq 0.05$, $p \leq 0.01$, $p \leq 0.001$, analysis of variance.

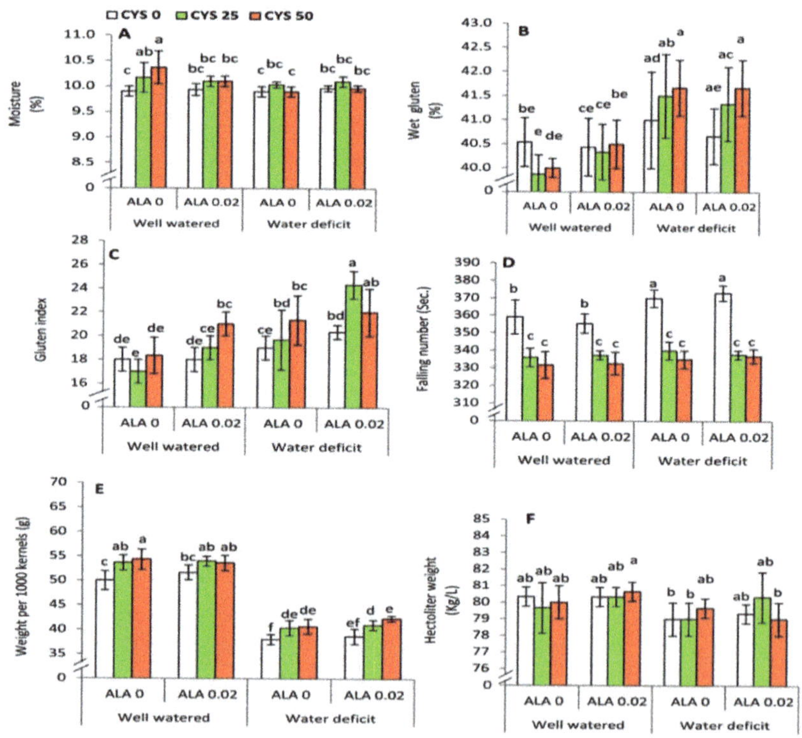

Figure 7. Effect of irrigation level, seed soaking by α-lipoic acid (ALA; 0 and 0.02 mM) and foliar application by cysteine (Cys; 0, 25 and 50 ppm) on the physical/chemical properties of wheat grains. (**A**) Moisture content, (**B**) Wet gluten, (**C**) Gluten index, (**D**) Falling number, (**E**) Weight per 1000 kernels, (**F**) Hectoliter weight. For each parameter, the mean values ± SD followed by a different letter are significantly different according to Tukey's range test ($p \leq 0.05$).

3.8. Alveographic Parameters of Wheat Grains

All alveographic parameters of wheat grains were significantly affected by irrigation, seed soaking, and foliar application treatments interaction except seed soaking of *L* (Table 9, Figure 8). All alveographic parameters of wheat grains were significantly increased by water deficit, high concentration of ALA (except *L*), and the highest Cys concentration (except *L*). *L* values were decreased with increasing Cys concentration.

Table 9. Analysis of variance (mean square) of alveographic parameters of wheat grains.

Source	Df	P	L	W	P/L
		(mm H_2O)	(mm)	(10^{-4} J)	
Irrigation (I)	1	210.254 ***	110.250 ***	306.250 ***	0.005 **
Seed soaking (S)	1	132.255 ***	12.250ns	140.027 ***	0.010 ***
Foliar appl. (F)	2	2186.111 ***	582.750 ***	7596.694 ***	0.214 ***
I × S	2	38.022ns	1.361ns	0.694ns	0.002ns
I × F	2	2.333ns	5.250ns	3.083ns	0.001ns
S × F	2	7.001ns	0.750ns	13.527ns	0.001ns
I × S × F	2	0.777ns	9.027ns	3.694ns	0.001ns

ns, **, *** not significant or significant at $p \leq 0.01$, $p \leq 0.001$, analysis of variance.

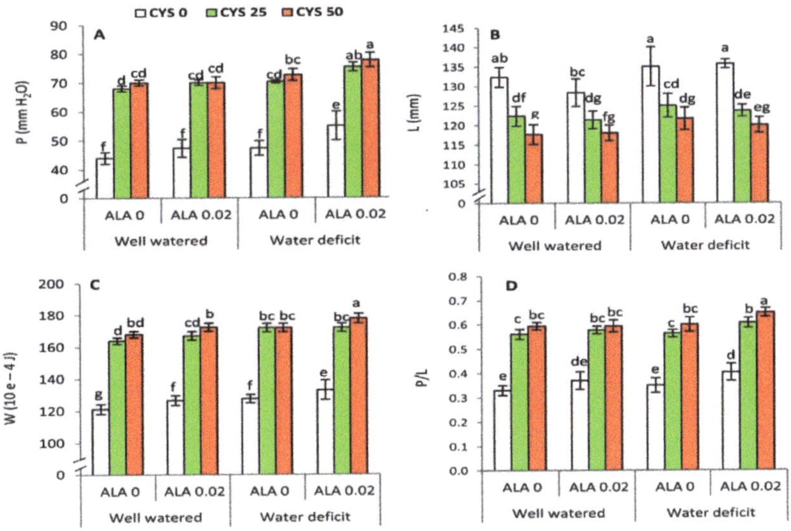

Figure 8. Effect of irrigation level, seed soaking by α-lipoic acid (ALA; 0 and 0.02 mM) and foliar application by cysteine (Cys; 0, 25 and 50 ppm) on the alveographic parameters of wheat grains. (**A**) P, (**B**) L, (**C**) W, (**D**) P/L. For each parameter, the mean values ± SD followed by a different letter are significantly different according to Tukey's range test ($p \leq 0.05$).

3.9. Correlation between Irrigation, Seed Soaking and Foliar Application Parameters

To assess the cluster quality of treatments based on the irrigation, seed soaking, and foliar application treatments measurements by testing the cluster distances within and between each cluster, the silhouette analysis plot was calculated and generated based on the Euclidean distance metric (Figure 9). The results revealed that all the parameters except P exhibited positive values indicating that clustering of treatments based on the irrigation and seed soaking and foliar application showed mostly similar behavior with different levels of effect, and the clustering configuration may have few clusters. Meanwhile, two-dimensional heatmap plotting based on all parameters clustered the APX, P, and RWC (%) in a separate cluster, indicating that these parameters almost have the similar power to elucidate the water deficit effects when compared to the well watered condition. While the second cluster comprised two distinctive sub-clusters, the first comprised MDA, sugars, grain yield, weight per 1000 kernels, SOD, Chl a+b, wet gluten (%), and falling number parameters/traits. At the same time, the second sub-cluster comprised Chl a, Chl a/b, proline, ascorbic acid, Chl b, protein (%), POX, H_2O_2, CAT, carotenoids, moisture (%), P/L, and flavonoids parameters/traits (Figure 10). Boxplot analysis revealed that the average of the parameter-classes leaf pigments, grain yield, and physical/chemical properties of wheat grain classes decreased significantly under the water deficit compared to well watered condition. The classes of non-enzymatic antioxidants, ROS and scavenging capacity, antioxidant enzymes, osmolytes, leaf water status, and alveographic parameters of wheat grains increased the water deficit significantly compared to well watered condition (Supplementary Table S1).

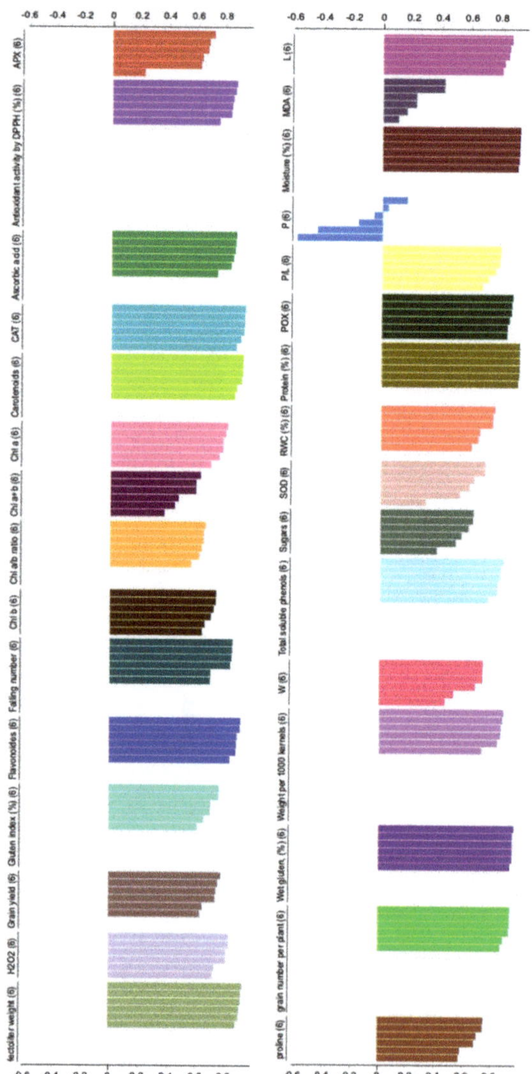

Figure 9. Plot of silhouette analysis values for clustering of all parameters based on "seed soaking and foliar application" treatments variables. On the y-axis each cluster is ordered by decreasing silhouette value. The silhouette value can range between −1 and 1.

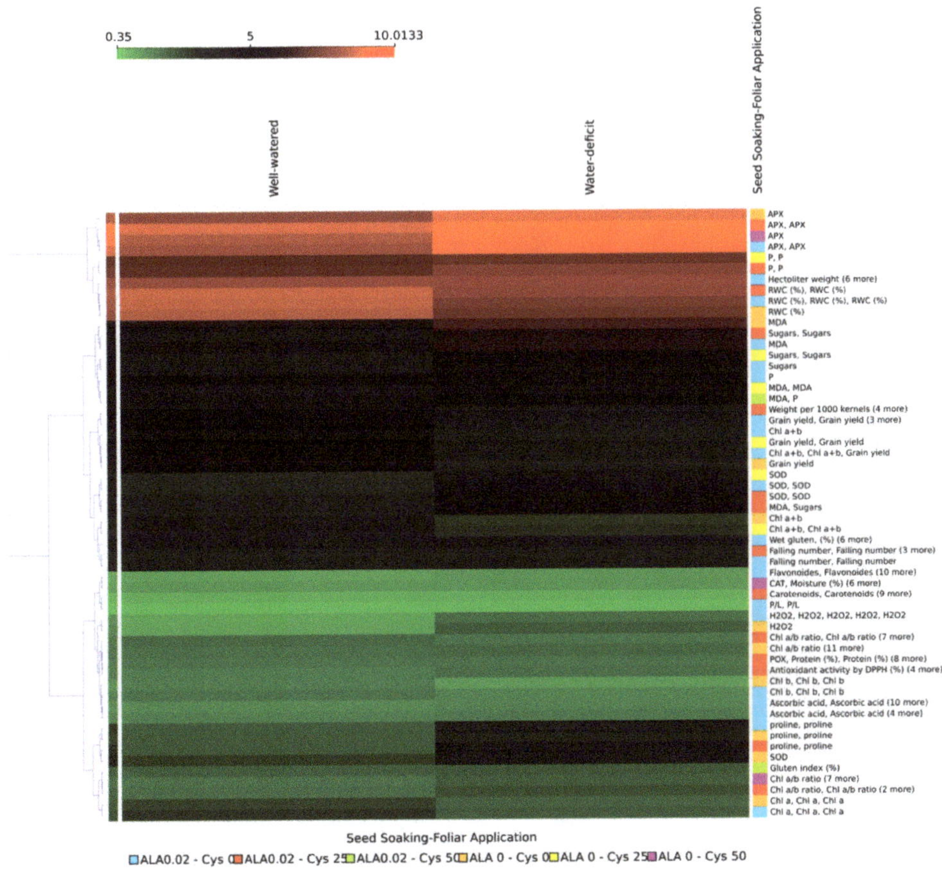

Figure 10. Two-dimensional heatmap visualization shows the interaction between the irrigation treatments and both the 31 measured parameters included in the study and the six "seed soaking and foliar application" types.

Meanwhile, Boxplot representation of the "seed soaking and foliar application" treatments on both water deficit and well watered conditions revealed that the average of all treatments except treatments ALA 0.02 - Cys 25 and ALA 0.02 - Cys 50 showed minor variations under the water deficit compared well watered condition (Supplementary Table S1). Moreover, multidimensional preference analysis was performed to summarize the interrelationships amongst treatments, parameters, and classes. The plot shows a high level of consistency and interrelationships between each parameter and treatment type. Ultimately, to summarize and visualize such high-dimensional data levels, a Mosaic plot representing a contingency matrix of the water deficit and well watered treatments versus the 31 parameters and their classes included in the study was developed (Figure 11). The plot further confirmed that the six "seed soaking and foliar application" treatments applied on both water deficit and well watered plant groups exhibited different levels of effectiveness (moderate to minor) to alleviate the negative effect of water deficit when compared to well watered condition.

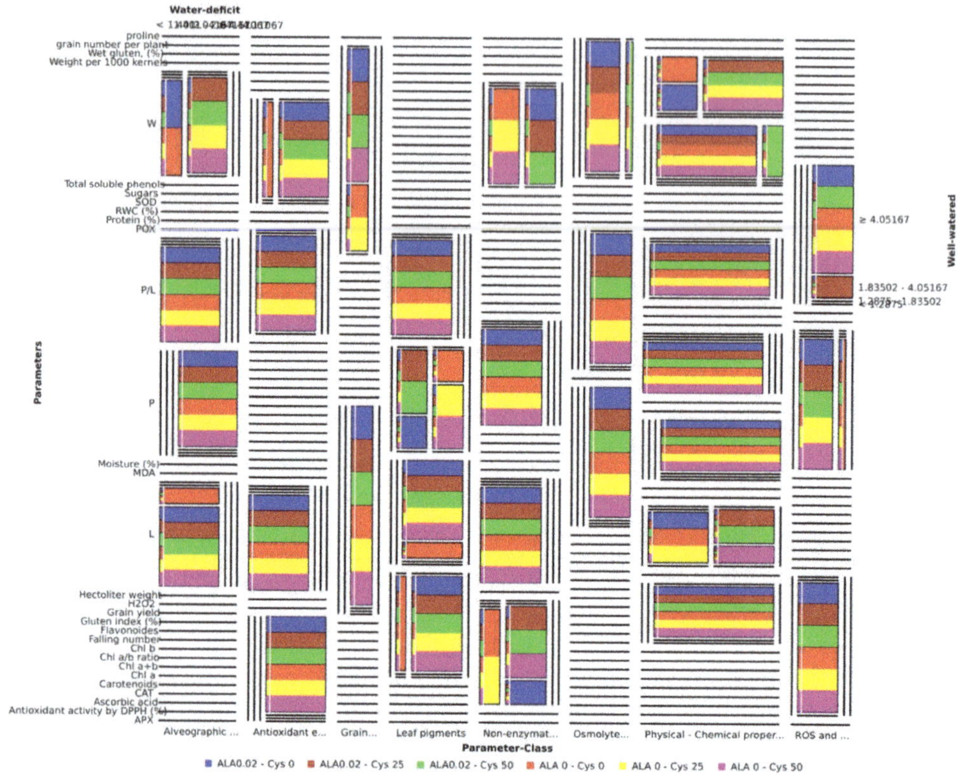

Figure 11. Mosaic plot representing a contingency table of the water deficit and well watered treatments variants versus the 31 parameters included in the study. The vertical size of the cells is proportional to the number of variants found in the respective parameter; the horizontal size of the cells is proportional to the effect level of the six "seed soaking and foliar application" types for each parameter under each parameter-class. The colors of the variants (seed soaking and foliar application) are indicated at the bottom of the Mosaic plot in colored squares similar to the respective bars in the Mosaic plot. Variants were not found at all possible locations of each parameter-class, which causes the reduction of several bars to dashed lines drawn as place holders and indicating that at the particular location no variant has been found in the parameter-class.

4. Discussion

In the present study, exposing wheat plants to deficit irrigation manifestly altered the leaf chlorophyll composition. In this respect, it can be observed that Chl a, Chl b, Chl a+b contents were significantly decreased in the water stressed plants compared to the well watered ones (Figure 1A–C). This reduction could be attributed to increase the activity of chlorophyllase [62], as well as water deficiency induced oxidative damage (Figure 2B,C) which can collapse the membranes and chloroplast structure where the leaf pigments are localized [63,64]. In contrast, the Chl a/b ratio (Figure 1D) in ALA/Cys-untreated plants was significantly increased under water deficit conditions compared to the well watered plants. It is well known that Chl a is the major cofactor for the photochemical reactions in the plastid because it is required for the assembly of pigment–protein complexes, while Chl b can also act as one of the accessory pigments in light-harvesting chlorophyll complexes (LHCs) [65,66]. In this study, increasing the Chl a/b ratio may act as a protective and adaptive behavior to maintain the function of photosynthetic apparatus under water stress conditions.

On the other hand, applied ALA and/or Cys were shown to enhance the content of photosynthetic pigments (Chl a, Chl b and Chl a+b) under well watered and water stressed conditions. However, an opposite trend was observed in Chl a/b ratio (Figure 1). Sulfur-containing biomolecules (ALA and Cys) are considered very potent antioxidants that can protect plants against abiotic stresses [24,30]. In this context, exogenous ALA was shown to stimulate photosystem II and gene expression of carbon fixation enzymes (Rubisco and PEP carboxylase) in maize seedlings exposed to drought stress [29]. These responses were associated with a simultaneous down-regulation in the chlorophyllase gene (Chlase) [29]. Moreover, Cys has been reported to improve Chl a and Chl b content in oat plants subjected to drought stress [67]. This effect could be attributed to enhancing the uptake of N in the plants, which is essential for Chl biosynthesis [68]. In contrast, the Chl a/b ratio (Figure 1D) differentially responded to ALA treatments under water deficit conditions. It was significantly decreased in the ALA-treated plants compared to the ALA-untreated plants. This effect implies that ALA as an efficient antioxidant has a more positive impact on Chl b than Chl a.

In the present study, water shortage led to an increase of H_2O_2 and the rate of lipid peroxidation as indicated by malondialdehyde (MDA) (Figure 2B,C). The excessive generation of reactive oxygen species (ROS) under drought stress conditions has been evidenced in several plant species [69–72]. This increase of toxic molecules should be restricted by enhancing the overall antioxidant activities and scavenging capacity of plants (Figure 2A). In this context, the treatments of ALA/Cys revealed beneficial effects as protectant antioxidants by reducing MDA and H_2O_2 in parallel with enhancing the antioxidant capacity by DPPH (Figure 2A–C). Previous investigations on the antioxidative role of ALA during Pb toxicity revealed ALA-induced accumulation of various antioxidant molecules in wheat plants [73]. Furthermore, ALA can enhance the photosynthetic performance in maize under drought stress [29], ameliorate lipid peroxidation, and induce the antioxidant systems of maize under osmotic stress [25]. Cysteine (Cys) has also been suggested as an important precursor for the synthesis of glutathione in plants [24].

Carotenoids were significantly decreased in the water stressed plants compared to the well watered plants (Figure 3A). This response could be explained by that carotenoids are involved in the biosynthesis of ABA, the major signaling hormone in response to drought stress [74]. In contrast, the treatments of ALA/Cys improved the content of carotenoids under deficit irrigation. This synergistic effect may have occurred to keep the carotenoids that mediate the xanthophyll cycle, leading to dissipating the exceeding light energy and enhancing the photosynthetic capacity under drought stress. In this regard, earlier evidence has confirmed that Cys can hinder energy transfer to prevent photooxidation [75].

On the other hand, flavonoids and total soluble phenols revealed a significant increase in the water stressed plants compared to the well watered ones (Figure 3B,C). This over-accumulation of secondary metabolites can enhance plant tolerance to drought stress and alleviate the induced oxidative damage due to their higher antioxidant capacity [76,77]. In addition to their antioxidant properties, phenolic compounds can be involved in plant tolerance to drought stress as a sink for carbon under stress conditions [78]. Furthermore, flavonoids are widely distributed secondary metabolites that are synthesized through the phenylpropanoid pathway, transforming phenylalanine into 4-coumaroyl-CoA, which finally enters the flavonoid biosynthesis pathway [79]. The carbon skeleton of all secondary metabolites including the compound-mediated phenylpropanoid pathway basically relies on photosynthesis, since several changes in the stressed plants happen in carbon metabolism to achieve the balance between the biomass production and formation of defensive secondary compounds [76]. In this study, flavonoids and total soluble phenols were relatively enhanced by the combined ALA/Cys treatments. This enhancement could be attributed to the increase of the rate of photosynthesis by both compounds [29,80].

Alpha lipoic acid (ALA) is soluble in water and lipid phases, connects the activity of antioxidants in the cell membrane (α-tocopherol) with other antioxidants in the cytoplasm, i.e., ascorbic acid (AsA) and glutathione (GSH), leading to reinforcing the antioxidant power of plants [31]. Moreover, Cys is an important precursor of GSH biosynthesis in plants [24]. In the current work, the treatments of ALA/Cys increased the concentration of AsA under both investigated levels of irrigation (Figure 3D). These findings indicate that the treatments of ALA/Cys can play a modulatory role in the operation of the GSH-AsA cycle under the circumstances of this study [81].

The activities of antioxidant enzymes including superoxide dismutase (SOD), catalase (CAT), peroxidase (POX) and ascorbate peroxidase (APX) were observed to be significantly affected by deficit irrigation and ALA/Cys treatments (Table 5; Figure 4). Interestingly, ALA and Cys exhibited differential effects on the activity of antioxidative enzymes, whereas SOD and APX activity were positively upregulated by ALA treatment while CAT and POX activity were decreased. Thus, ALA functions through the activation of the major ROS scavenging system that includes APX and SOD. Reduction of H_2O_2 content was accompanied by ALA and Cys-mediated elevation in APX, POX, and CAT activity. Further investigations are necessary to reveal the mechanism of ALA and Cys signaling in the modulation of antioxidative enzymes in specific organelles in response to water stress in wheat leaves. Interestingly, Cys treatment during water deficit elevated the activity of all antioxidative enzymes, which was all the higher for POX. Thus, the present work highlights the integrative and synergistic role of ALA and Cys application in the upregulation of the enzymatic antioxidative defense systems. It is further advocated that both ALA and Cys are beneficial to instigate redox management and ROS scavenging activity [25,30,67,73]. Exogenously applied ALA and Cys have been suggested to mitigate the oxidative stress and confer osmotic tolerance to wheat and soybean plants subjected to NaCl stress [30,68].

In the present work, ALA and Cys treatments also increased the proline, total soluble sugars and leaf relative water content (RWC) of drought stressed wheat plants (Table 6, Figure 5). Several lines of evidence demonstrated that to regulate the osmotic potential and enhance water uptake under osmotic stress (drought or salinity), plants usually accumulate considerable concentrations of some organic molecules such as proline and sugars that function as osmotic regulators and enhance plant water relation [64,82,83].

It is noteworthy that the improved osmotic regulation in leaves was also associated with the amelioration of yield attributes (Table 7, Figure 6) including number of spikes/m^2, number of grains/plant and grain yield (ton/ha). Several previous studies reported that water deficit has a destructive effect on the yield of wheat plants [84,85]. Generally, water stress can affect many aspects related to the field performance of grain crops such as the photosynthesis and translocation of carbohydrates, which are responsible for the filling of grains [86,87]. In this study, besides the amelioration of drought induced oxidative damage, the treatments of ALA/Cys were evidenced to enhance photosynthetic pigments, accumulation of osmolytes and improving the non-enzymatic and enzymatic antioxidants. These positive effects could explain the increase of grain yield attributes.

ALA and Cys treatments improved moisture percentage and gluten index in grains obtained from water stressed plants. Cys treatment further improved kernel weight (per 1000 kernels) under water deficit conditions. Analysis of alveographic characteristics indicated that ALA and Cys applications were effective in improving flour and dough quality of wheat grains. Total amino acid content, water relation, and growth characteristics are known to be associated with physiochemical properties, as well as grain quality in drought stressed wheat plants [88].

Additionally, Cys and ALA treatments (sulfur-containing molecules) significantly improved grain attributes. In recent years, investigations have revealed the emerging role of ALA (dithiol) as a potential antioxidant in plants [89]. Reduced ALA possesses 2 sulfhydryl moieties which function as potential ROS scavenging sites. ALA is an important sulfur-bearing compound that exerts a plethora of effects in plants subjected to abiotic stress [73]. ALA further activates a number of mitochondria-localized metabolic enzymes [90]. Sulfur compounds are important regulators of signaling pathways associated with abiotic stress tolerance in plants [91,92]. In the present work, application of the two sulfur-containing priming molecules (ALA and Cys) led to several synergistic effects that enhanced the tolerance of wheat plants to water shortage.

Various physiochemical attributes have been improved in the presence of ALA and Cys treatment during water stress in wheat plants. Future investigations are necessary to establish the possible role of ALA and Cys in the modulation of the glutathione-ascorbate cycle in water stressed wheat plants. It is noteworthy that application of the two biomolecules has led to an improvement in the gluten index and quality of grains (Table 8; Figure 7). Thus, using ALA and Cys are effective in the grain filling stage of wheat plants under water stress. This effect may indicate the presence of long distance signaling of ALA and Cys from leaves to grains. Correlation analysis revealed significant interaction among irrigation, seed soaking, and ALA/Cys treatments in wheat plants raised in the absence and presence of drought stress. Thus, ALA and Cys are effective priming molecules and bear agronomic importance in improving wheat yield in the arid cultivable regions. Priming for drought or salinity allows stress tolerance and offers several benefits in many crops [93]. This improvement was extended to the grain processing technology by affecting the alveographic parameters (Tables 8 and 9). Analysis of alveographic characteristics indicated that ALA and Cys application were effective in improving flour and dough quality of wheat grains. In our present work, Cys and ALA treatment (sulfur-containing molecules) significantly improved grain attributes. Various physiochemical attributes of grains have improved in the presence of ALA and Cys treatment during water stress in wheat plants, and alveographic parameters of wheat flour dough also was improved. This improvement extended to the grain processing technology by affecting the alveographic parameters (technological quality parameters). Such parameters are strongly related to the flour yield and flour properties by enhancing the viscoelastic properties of dough [94]. In this context, it has been found that altering water relation and total amino acid content is concomitant with substantial changes in the physiochemical properties and grain quality of drought stressed wheat plants [88].

5. Conclusions

The present work provides novel insights into the synergistic action of ALA and Cys as effective stress priming molecules in the mitigation of drought stress in wheat plants. Amelioration of drought stress in wheat plants is primarily attained by the enhancement in the function of the antioxidative system and regeneration of osmotic tolerance in leaves. Apart from their antioxidative role, these sulfur-containing compounds also appear beneficial in the improvement of alveographic characteristics of wheat grain which determine various attributes of dough quality. Therefore, applications of these two biomolecules provide both physiological tolerance and restore yield attributes in wheat. Future investigations are necessary to decipher the signaling mechanisms of these two biomolecules in relation to various plant growth regulators in wheat plants. The molecular mechanism of ALA-Cys crosstalk shall also provide new insights to wheat management in arid regions. From our obtained results it can be concluded that applied ALA at 0.02 mM as seed soaking treatment combined by Cys at 50 ppm as a foliar application could be recommended as potent compounds in wheat cultivation under water deficit conditions.

Supplementary Materials: The following are available online at https://www.mdpi.com/article/10.3390/plants10112318/s1, Table S1: Germination assay of Wheat seed after soaking with different concentration of α-lipoic acid.

Author Contributions: Conceptualization, A.E., M.M.E.-M., H.A., R.F., S.S. and M.F.M.I.; methodology, A.E., M.M.E.-M., G.N., M.P., M.A.M.A., M.M.A.H., M.S. (Mostafa Shahin), S.M., A.A.E.-Y., M.S. (Mohamed Shebl), M.S.J., A.O., H.G.A.E.-G., H.A., R.F., S.S. and M.F.M.I.; software, A.E., M.M.E.-M., G.N., M.P., M.A.M.A., M.M.A.H., M.S. (Mostafa Shahin), S.M., A.A.E.-Y., M.S. (Mohamed Shebl), M.S.J., A.O., H.G.A.E.-G., H.A., R.F., S.S. and M.F.M.I.; validation, A.E., M.M.E.-M., G.N., M.P., M.A.M.A., M.M.A.H., M.S. (Mostafa Shahin), S.M., A.A.E.-Y., M.S. (Mohamed Shebl), M.S.J., A.O., H.G.A.E.-G., H.A., R.F., S.S. and M.F.M.I.; formal analysis, A.E., M.M.E.-M., G.N., M.P. and M.F.M.I.; investigation, S.M., A.A.E.-Y., M.S. (Mohamed Shebl), M.S.J., A.O., H.G.A.E.-G., H.A., R.F., S.S.

and M.F.M.I.; resources, A.E., M.M.E.-M., G.N., M.P., M.A.M.A., M.M.A.H., M.S. (Mostafa Shahin), S.M., A.A.E.-Y., M.S. (Mohamed Shebl), M.S.J., A.O., H.G.A.E.-G., H.A., R.F., S.S. and M.F.M.I.; data curation, G.N., M.P., M.A.M.A., M.M.A.H., M.S. (Mostafa Shahin), S.M., A.A.E.-Y., M.S. (Mohamed Shebl), M.S.J., A.O., H.G.A.E.-G., H.A., R.F., S.S. and M.F.M.I.; writing—original draft preparation, A.E., M.M.E.-M., S.M., A.A.E.-Y., M.S. (Mohamed Shebl), M.S.J., A.O., H.G.A.E.-G., R.F., S.S. and M.F.M.I.; writing—review and editing, A.E., M.M.E.-M., G.N., M.P., M.A.M.A., M.M.A.H., M.S. (Mostafa Shahin), S.M., A.A.E.-Y., M.S. (Mohamed Shebl), M.S.J., A.O., H.G.A.E.-G., H.A., R.F., S.S. and M.F.M.I.; visualization, M.A.M.A., M.M.A.H., M.S. (Mostafa Shahin), S.M., A.A.E.-Y., H.A., R.F., S.S. and M.F.M.I.; funding acquisition, A.E., M.M.E.-M., H.G.A.E.-G., H.A., R.F., S.S. and M.F.M.I., All authors have read and agreed to the published version of the manuscript.

Funding: This research received no external funding.

Institutional Review Board Statement: Not applicable.

Informed Consent Statement: Not applicable.

Data Availability Statement: Not applicable.

Acknowledgments: The authors are grateful to Amr Mossad, Agricultural Engineering Department, Faculty of Agriculture, Ain Shams University, Cairo, Egypt for his great helpful efforts and revising the irrigation requirements of wheat plants at the site of the experiments.

Conflicts of Interest: The authors declare no conflict of interest in this investigation.

References

1. Elkeilsh, A.; Awad, Y.M.; Soliman, M.H.; Abu-Elsaoud, A.; Abdelhamid, M.T.; El-Metwally, I.M. Exogenous application of β-sitosterol mediated growth and yield improvement in water-stressed wheat (*Triticum aestivum*) involves up-regulated antioxidant system. *J. Plant Res.* **2019**, *132*, 881–901. [CrossRef] [PubMed]
2. Sallam, A.; Alqudah, A.M.; Dawood, M.F.; Baenziger, P.S.; Börner, A. Drought stress tolerance in wheat and barley: Advances in physiology, breeding and genetics research. *Int. J. Mol. Sci.* **2019**, *20*, 3137. [CrossRef]
3. Sharar, M.; Saied, E.M.; Rodriguez, M.C.; Arenz, C.; Montes-Bayón, M.; Linscheid, M.W. Elemental Labelling and Mass Spectrometry for the Specific Detection of Sulfenic Acid Groups in Model Peptides: A Proof of Concept. *Anal Bioanal Chem* **2017**, *409*, 2015–2027. [CrossRef] [PubMed]
4. Lal, R. Soil carbon sequestration impacts on global climate change and food security. *Science* **2004**, *304*, 1623–1627. [CrossRef]
5. Ahmad, P.; Jamsheed, S.; Hameed, A.; Rasool, S.; Sharma, I.; Azooz, M.; Hasanuzzaman, M. Drought stress induced oxidative damage and antioxidants in plants. In *Oxidative Damage to Plants*; Elsevier: Amsterdam, The Netherlands, 2014; pp. 345–367. [CrossRef]
6. Alhaithloul, H.A.; Soliman, M.H.; Ameta, K.L.; El-Esawi, M.A.; Elkelish, A. Changes in Ecophysiology, Osmolytes, and Secondary Metabolites of the Medicinal Plants of Mentha piperita and Catharanthus roseus Subjected to Drought and Heat Stress. *Biomolecules* **2020**, *10*, 43. [CrossRef] [PubMed]
7. Habib, N.; Ali, Q.; Ali, S.; Javed, M.T.; Zulqurnain Haider, M.; Perveen, R.; Shahid, M.R.; Rizwan, M.; Abdel-Daim, M.M.; Elkelish, A. Use of Nitric Oxide and Hydrogen Peroxide for Better Yield of Wheat (*Triticum aestivum* L.) under Water Deficit Conditions: Growth, Osmoregulation, and Antioxidative Defense Mechanism. *Plants* **2020**, *9*, 285. [CrossRef]
8. Harb, A.; Krishnan, A.; Ambavaram, M.M.; Pereira, A. Molecular and physiological analysis of drought stress in Arabidopsis reveals early responses leading to acclimation in plant growth. *Plant Physiol.* **2010**, *154*, 1254–1271. [CrossRef]
9. Gaber, A.; Refat, M.S.; Belal, A.A.M.; El-Deen, I.M.; Hassan, N.; Zakaria, R.; Alhomrani, M.; Alamri, A.S.; Alsanie, W.F.; Saied, E.M. New Mononuclear and Binuclear Cu(II), Co(II), Ni(II), and Zn(II) Thiosemicarbazone Complexes with Potential Biological Activity: Antimicrobial and Molecular Docking Study. *Molecules* **2021**, *26*, 2288. [CrossRef]
10. Miller, G.; Suzuki, N.; Ciftci-Yilmaz, S.; Mittler, R. Reactive oxygen species homeostasis and signalling during drought and salinity stresses. *Plant Cell Environ.* **2010**, *33*, 453–467. [CrossRef] [PubMed]
11. Noctor, G.; Mhamdi, A.; Foyer, C.H. The roles of reactive oxygen metabolism in drought: Not so cut and dried. *Plant Physiol.* **2014**, *164*, 1636–1648. [CrossRef]
12. Elkelish, A.; Ibrahim, M.F.; Ashour, H.; Bondok, A.; Mukherjee, S.; Aftab, T.; Hikal, M.; El-Yazied, A.A.; Azab, E.; Gobouri, A.A. Exogenous Application of Nitric Oxide Mitigates Water Stress and Reduces Natural Viral Disease Incidence of Tomato Plants Subjected to Deficit Irrigation. *Agronomy* **2021**, *11*, 87. [CrossRef]
13. Ibrahim, M.F.; Elbar, O.H.A.; Farag, R.; Hikal, M.; El-Kelish, A.; El-Yazied, A.A.; Alkahtani, J.; Abd El-Gawad, H.G. Melatonin counteracts drought induced oxidative damage and stimulates growth, productivity and fruit quality properties of tomato plants. *Plants* **2020**, *9*, 1276. [CrossRef] [PubMed]
14. Bin-Jumah, M.; Abdel-Fattah, A.-F.M.; Saied, E.M.; El-Seedi, H.R.; Abdel-Daim, M.M. Acrylamide-Induced Peripheral Neuropathy: Manifestations, Mechanisms, and Potential Treatment Modalities. *Environ. Sci. Pollut. Res.* **2021**, *28*, 13031–13046. [CrossRef] [PubMed]

15. Nawaz, F.; Shehzad, M.A.; Majeed, S.; Ahmad, K.S.; Aqib, M.; Usmani, M.M.; Shabbir, R.N. Role of Mineral Nutrition in Improving Drought and Salinity Tolerance in Field Crops. In *Agronomic Crops*; Springer: Berlin/Heidelberg, Germany, 2020; pp. 129–147. [CrossRef]
16. Gaber, A.; Alsanie, W.F.; Kumar, D.N.; Refat, M.S.; Saied, E.M. Novel Papaverine Metal Complexes with Potential Anticancer Activities. *Molecules* **2020**, *25*, 5447. [CrossRef]
17. Huang, G.-T.; Ma, S.-L.; Bai, L.-P.; Zhang, L.; Ma, H.; Jia, P.; Liu, J.; Zhong, M.; Guo, Z.-F. Signal transduction during cold, salt, and drought stresses in plants. *Mol. Biol. Rep.* **2012**, *39*, 969–987. [CrossRef] [PubMed]
18. Ibrahim, M.; Ibrahim, H.A. Assessment of Selenium Role in Promoting or Inhibiting Potato Plants under Water Stress. *J. Hortic. Sci. Ornam. Plants* **2016**, *8*, 125–139.
19. Ibrahim, M.F.; El-Samad, A.; Ashour, H.; El-Sawy, A.M.; Hikal, M.; Elkelish, A.; El-Gawad, H.A.; El-Yazied, A.A.; Hozzein, W.N.; Farag, R. Regulation of agronomic traits, nutrient uptake, osmolytes and antioxidants of maize as influenced by exogenous potassium silicate under deficit irrigation and semiarid conditions. *Agronomy* **2020**, *10*, 1212. [CrossRef]
20. Reddy, A.R.; Chaitanya, K.V.; Vivekanandan, M. Drought-induced responses of photosynthesis and antioxidant metabolism in higher plants. *J. Plant Physiol.* **2004**, *161*, 1189–1202. [CrossRef]
21. Zou, J.-J.; Wei, F.-J.; Wang, C.; Wu, J.-J.; Ratnasekera, D.; Liu, W.-X.; Wu, W.-H. Arabidopsis calcium-dependent protein kinase CPK10 functions in abscisic acid-and Ca^{2+}-mediated stomatal regulation in response to drought stress. *Plant Physiol.* **2010**, *154*, 1232–1243. [CrossRef] [PubMed]
22. Colville, L.; Kranner, I. Desiccation tolerant plants as model systems to study redox regulation of protein thiols. *Plant Growth Regul.* **2010**, *62*, 241–255. [CrossRef]
23. Meyer, A.J.; Hell, R. Glutathione homeostasis and redox-regulation by sulfhydryl groups. *Photosynth. Res.* **2005**, *86*, 435–457. [CrossRef]
24. Zagorchev, L.; Seal, C.E.; Kranner, I.; Odjakova, M. A central role for thiols in plant tolerance to abiotic stress. *Int. J. Mol. Sci.* **2013**, *14*, 7405–7432. [CrossRef]
25. Terzi, R.; Saruhan, G.N.; Güven, F.G.; Kadioglu, A. Alpha lipoic acid treatment induces the antioxidant system and ameliorates lipid peroxidation in maize seedlings under osmotic stress. *Arch. Biol. Sci.* **2018**, *70*, 503–511. [CrossRef]
26. Fogacci, F.; Rizzo, M.; Krogager, C.; Kennedy, C.; Georges, C.M.; Knežević, T.; Liberopoulos, E.; Vallée, A.; Pérez-Martínez, P.; Wenstedt, E.F. Safety evaluation of α-lipoic acid supplementation: A systematic review and meta-analysis of randomized placebo-controlled clinical studies. *Antioxidants* **2020**, *9*, 1011. [CrossRef]
27. Tibullo, D.; Volti, G.L.; Giallongo, C.; Grasso, S.; Tomassoni, D.; Anfuso, C.D.; Lupo, G.; Amenta, F.; Avola, R.; Bramanti, V. Biochemical and clinical relevance of alpha lipoic acid: Antioxidant and anti-inflammatory activity, molecular pathways and therapeutic potential. *Inflamm. Res.* **2017**, *66*, 947–959. [CrossRef]
28. Packer, L.; Witt, E.H.; Tritschler, H.J. Alpha-lipoic acid as a biological antioxidant. *Free Radic. Biol. Med.* **1995**, *19*, 227–250. [CrossRef]
29. Sezgin, A.; Altuntaş, C.; Demiralay, M.; Cinemre, S.; Terzi, R. Exogenous alpha lipoic acid can stimulate photosystem II activity and the gene expressions of carbon fixation and chlorophyll metabolism enzymes in maize seedlings under drought. *J. Plant Physiol.* **2019**, *232*, 65–73. [CrossRef] [PubMed]
30. Gorcek, Z.; Erdal, S. Lipoic acid mitigates oxidative stress and recovers metabolic distortions in salt-stressed wheat seedlings by modulating ion homeostasis, the osmo-regulator level and antioxidant system. *J. Sci. Food Agric.* **2015**, *95*, 2811–2817. [CrossRef]
31. Sgherri, C.; Quartacci, M.F.; Izzo, R.; Navari-Izzo, F. Relation between lipoic acid and cell redox status in wheat grown in excess copper. *Plant Physiol. Biochem.* **2002**, *40*, 591–597. [CrossRef]
32. Álvarez, C.; Ángeles Bermúdez, M.; Romero, L.C.; Gotor, C.; García, I. Cysteine homeostasis plays an essential role in plant immunity. *New Phytol.* **2012**, *193*, 165–177. [CrossRef] [PubMed]
33. Genisel, M.; Erdal, S.; Kizilkaya, M. The mitigating effect of cysteine on growth inhibition in salt-stressed barley seeds is related to its own reducing capacity rather than its effects on antioxidant system. *Plant Growth Regul.* **2015**, *75*, 187–197. [CrossRef]
34. Rausch, T.; Wachter, A. Sulfur metabolism: A versatile platform for launching defence operations. *Trends Plant Sci.* **2005**, *10*, 503–509. [CrossRef]
35. Takahashi, H.; Kopriva, S.; Giordano, M.; Saito, K.; Hell, R. Sulfur assimilation in photosynthetic organisms: Molecular functions and regulations of transporters and assimilatory enzymes. *Annu. Rev. Plant Biol.* **2011**, *62*, 157–184. [CrossRef] [PubMed]
36. Terzi, H.; Yıldız, M. Proteomic analysis reveals the role of exogenous cysteine in alleviating chromium stress in maize seedlings. *Ecotoxicol. Environ. Saf.* **2021**, *209*, 111784. [CrossRef] [PubMed]
37. Sauter, M.; Moffatt, B.; Saechao, M.C.; Hell, R.; Wirtz, M. Methionine salvage and S-adenosylmethionine: Essential links between sulfur, ethylene and polyamine biosynthesis. *Biochem. J.* **2013**, *451*, 145–154. [CrossRef]
38. Yang, S.F.; Hoffman, N.E. Ethylene biosynthesis and its regulation in higher plants. *Annu. Rev. Plant Physiol.* **1984**, *35*, 155–189. [CrossRef]
39. Al Ubeed, H.; Wills, R.; Bowyer, M.; Golding, J. Inhibition of postharvest senescence of green leafy vegetables by exogenous D-cysteine and L-cysteine as precursors of hydrogen sulphide. *J. Hortic. Sci. Biotechnol.* **2019**, *94*, 620–626. [CrossRef]
40. Allen, R.G.; Pereira, L.S.; Raes, D.; Smith, M. Crop evapotranspiration-Guidelines for computing crop water requirements-FAO Irrigation and drainage paper 56. *Fao Rome* **1998**, *300*, D05109.

41. Costache, M.A.; Campeanu, G.; Neata, G. Studies concerning the extraction of chlorophyll and total carotenoids from vegetables. *Rom. Biotechnol. Lett.* **2012**, *17*, 7702–7708.
42. Bates, L.S.; Waldren, R.P.; Teare, I. Rapid determination of free proline for water-stress studies. *Plant Soil* **1973**, *39*, 205–207. [CrossRef]
43. Chow, P.S.; Landhäusser, S.M. A method for routine measurements of total sugar and starch content in woody plant tissues. *Tree Physiol.* **2004**, *24*, 1129–1136. [CrossRef] [PubMed]
44. Ünyayar, S.; Keleþ, Y.; Ünal, E. Proline and ABA levels in two sunflower genotypes subjected to water stress. *Bulg. J. Plant Physiol.* **2004**, *30*, 34–47.
45. Heath, R.L.; Packer, L. Photoperoxidation in isolated chloroplasts: I. Kinetics and stoichiometry of fatty acid peroxidation. *Arch. Biochem. Biophys.* **1968**, *125*, 189–198. [CrossRef]
46. Velikova, V.; Yordanov, I.; Edreva, A. Oxidative stress and some antioxidant systems in acid rain-treated bean plants: Protective role of exogenous polyamines. *Plant Sci.* **2000**, *151*, 59–66. [CrossRef]
47. Huang, D.; Ou, B.; Prior, R.L. The chemistry behind antioxidant capacity assays. *J. Agric. Food Chem.* **2005**, *53*, 1841–1856. [CrossRef]
48. De Carvalho, L.M.J.; Gomes, P.B.; de Oliveira Godoy, R.L.; Pacheco, S.; do Monte, P.H.F.; de Carvalho, J.L.V.; Nutti, M.R.; Neves, A.C.L.; Vieira, A.C.R.A.; Ramos, S.R.R. Total carotenoid content, α-carotene and β-carotene, of landrace pumpkins (*Cucurbita moschata* Duch): A preliminary study. *Food Res. Int.* **2012**, *47*, 337–340. [CrossRef]
49. Chang, C.-C.; Yang, M.-H.; Wen, H.-M.; Chern, J.-C. Estimation of total flavonoid content in propolis by two complementary colorimetric methods. *J. Food Drug Anal.* **2002**, *10*, 3. [CrossRef]
50. Shahidi, F.; Naczk, M.; Griffiths, W. Food phenolics: Sources, chemistry, effects, applications. *Trends Food Sci. Technol.* **1996**, *7*, 243. [CrossRef]
51. Association of Official Analytical Chemists. *Official Methods of Analysis. Association of Official Analytical Chemists. Official Method 985.33. Vitamin C, (Reduced Ascorbic Acid) in Ready-to-Feed Milk Based Infant Formula 2, 6-Dichloroindophenol Titrimetric Method*; Association of Official Analytical Chemists International: Washington, DC, USA, 1990; pp. 1108–1109.
52. Bradford, M.M. A rapid and sensitive method for the quantitation of microgram quantities of protein utilizing the principle of protein-dye binding. *Anal. Biochem.* **1976**, *72*, 248–254. [CrossRef]
53. Beyer, F.W., Jr.; Fridovich, I. Assaying for superoxide dismutase activity: Some large consequences of minor changes in conditions. *Anal. Biochem.* **1987**, *161*, 559–566. [CrossRef]
54. Cakmak, I.; Strbac, D.; Marschner, H. Activities of hydrogen peroxide-scavenging enzymes in germinating wheat seeds. *J. Exp. Bot.* **1993**, *44*, 127–132. [CrossRef]
55. Nakano, Y.; Asada, K. Hydrogen peroxide is scavenged by ascorbate-specific peroxidase in spinach chloroplasts. *Plant Cell Physiol.* **1981**, *22*, 867–880. [CrossRef]
56. Dias, M.A.; Costa, M.M. Effect of low salt concentrations on nitrate reductase and peroxidase of sugar beet leaves. *J. Exp. Bot.* **1983**, *34*, 537–543. [CrossRef]
57. American Association of Cereal Chemists. *International Methods Approved of the American Association of Cereal Chemists*, 11th ed.; American Association of Cereal Chemists Inc.: St. Paul, MN, USA, 2012.
58. Association of Official Analytical Chemists. *Official Methods of the Association of Official Analytical Chemists*, 19th ed.; Association of Official Analytical Chemists: Arlington, VA, USA, 2012.
59. Snedecor, G.W.; Cochran, W.G. *Statistical Methods*; Iowa State University Press: Iowa City, IA, USA, 1967.
60. Rousseeuw, P.J. Silhouettes: A graphical aid to the interpretation and validation of cluster analysis. *J. Comput. Appl. Math.* **1987**, *20*, 53–65. [CrossRef]
61. Wickelmaier, F. *An Introduction to MDS*; Sound Quality Research Unit, Aalborg University: Aalborg, Denmark, 2003.
62. Mihailović, N.; Lazarević, M.; Dželetović, Z.; Vučković, M.; Đurđević, M. Chlorophyllase activity in wheat, *Triticum aestivum* L. leaves during drought and its dependence on the nitrogen ion form applied. *Plant Sci.* **1997**, *129*, 141–146. [CrossRef]
63. Hassan, N.; Ebeed, H.; Aljaarany, A. Exogenous application of spermine and putrescine mitigate adversities of drought stress in wheat by protecting membranes and chloroplast ultra-structure. *Physiol. Mol. Biol. Plants* **2020**, *26*, 233–245. [CrossRef]
64. Ibrahim, M.F.M.; Ibrahim, H.A.; Abd El-Gawad, H.G. Folic acid as a protective agent in snap bean plants under water deficit conditions. *J. Hortic. Sci. Biotechnol.* **2021**, *96*, 94–109. [CrossRef]
65. Xu, H.; Vavilin, D.; Vermaas, W. Chlorophyll b can serve as the major pigment in functional photosystem II complexes of cyanobacteria. *Proc. Natl. Acad. Sci. USA* **2001**, *98*, 14168–14173. [CrossRef] [PubMed]
66. Hoober, J.K. The Role of Chlorophyll b in Photosynthesis: Hypothesis. In *Photosynthesis*; Apple Academic Press: Point Pleasant, NJ, USA, 2016; pp. 104–118. [CrossRef]
67. Perveen, S.; Iqbal, M.; Saeed, M.; Iqbal, N.; Zafar, S.; Mumtaz, T. Cysteine-induced alterations in physicochemical parameters of oat (*Avena sativa* L. var. Scott and F-411) under drought stress. *Biol. Futur.* **2019**, *70*, 16–24. [CrossRef] [PubMed]
68. Sadak, M.S.; Abd El-Hameid, A.R.; Zaki, F.S.; Dawood, M.G.; El-Awadi, M.E. Physiological and biochemical responses of soybean (*Glycine max* L.) to cysteine application under sea salt stress. *Bull. Natl. Res. Cent.* **2020**, *44*, 1–10. [CrossRef]
69. Ibrahim, M. Induced drought resistance in common bean (*Phaseolus vulgaris* L.) by exogenous application with active yeast suspension. *Middle East J. Appl. Sci.* **2014**, *4*, 806–815.

70. Abd El-Gawad, H.G.; Mukherjee, S.; Farag, R.; Abd Elbar, O.H.; Hikal, M.S.; Abou El-Yazied, A.; Abd Elhady, S.A.; Helal, N.A.; Elkelish, A.; El Nahhas, N.; et al. Exogenous γ-aminobutyric acid (GABA)-induced signaling events and field performance associated with mitigation of drought stress in *Phaseolus vulgaris* L. *Plant Signal. Behav.* **2020**, *16*. [CrossRef]
71. Abd Elhady, S.A.; El-Gawad, H.G.A.; Ibrahim, M.F.; Mukherjee, S.; Elkelish, A.; Azab, E.; Gobouri, A.A.; Farag, R.; Ibrahim, H.A.; El-Azm, N.A. Hydrogen peroxide supplementation in irrigation water alleviates drought stress and boosts growth and productivity of potato plants. *Sustainability* **2021**, *13*, 899. [CrossRef]
72. Hasan, M.M.; Rahman, M.A.; Skalicky, M.; Alabdallah, N.M.; Waseem, M.; Jahan, M.S.; Ahammed, G.J.; El-Mogy, M.M.; El-Yazied, A.A.; Ibrahim, M.F.M.; et al. Ozone Induced Stomatal Regulations, MAPK and Phytohormone Signaling in Plants. *Int. J. Mol. Sci.* **2021**, *22*, 6304. [CrossRef] [PubMed]
73. Turk, H.; Erdal, S.; Karayel, U.; Dumlupinar, R. Attenuation of lead toxicity by promotion of tolerance mechanism in wheat roots by lipoic acid. *Cereal Res. Commun.* **2018**, *46*, 424–435. [CrossRef]
74. Parry, A.D.; Horgan, R. Carotenoids and abscisic acid (ABA) biosynthesis in higher plants. *Physiol. Plant.* **1991**, *82*, 320–326. [CrossRef]
75. Rolczynski, B.S.; Navotnaya, P.; Sussman, H.R.; Engel, G.S. Cysteine-mediated mechanism disrupts energy transfer to prevent photooxidation. *Proc. Natl. Acad. Sci. USA* **2016**, *113*, 8562–8564. [CrossRef]
76. Akula, R.; Ravishankar, G.A. Influence of abiotic stress signals on secondary metabolites in plants. *Plant Signal. Behav.* **2011**, *6*, 1720–1731. [CrossRef]
77. Nakabayashi, R.; Yonekura-Sakakibara, K.; Urano, K.; Suzuki, M.; Yamada, Y.; Nishizawa, T.; Matsuda, F.; Kojima, M.; Sakakibara, H.; Shinozaki, K. Enhancement of oxidative and drought tolerance in Arabidopsis by overaccumulation of antioxidant flavonoids. *Plant J.* **2014**, *77*, 367–379. [CrossRef]
78. Weidner, S.; Karolak, M.; Karamac, M.; Kosinska, A.; Amarowicz, R. Phenolic compounds and properties of antioxidants in grapevine roots [*Vitis vinifera* L.] under drought stress followed by recovery. *Acta Soc. Bot. Pol.* **2009**, *78*, 97–103. [CrossRef]
79. Falcone Ferreyra, M.L.; Rius, S.; Casati, P. Flavonoids: Biosynthesis, biological functions, and biotechnological applications. *Front. Plant Sci.* **2012**, *3*, 222. [CrossRef] [PubMed]
80. Baier, M.; Dietz, K.-J. Protective function of chloroplast 2-cysteine peroxiredoxin in photosynthesis. Evidence from transgenic Arabidopsis. *Plant Physiol.* **1999**, *119*, 1407–1414. [CrossRef]
81. Abuelsoud, W.; Hirschmann, F.; Papenbrock, J. Sulfur metabolism and drought stress tolerance in plants. In *Drought Stress Tolerance in Plants*; Springer: Berlin/Heidelberg, Germany, 2016; Volume 1, pp. 227–249. [CrossRef]
82. Alnusairi, G.S.; Mazrou, Y.S.; Qari, S.H.; Elkelish, A.A.; Soliman, M.H.; Eweis, M.; Abdelaal, K.; El-Samad, G.A.; Ibrahim, M.F.M.; ElNahhas, N. Exogenous Nitric Oxide Reinforces Photosynthetic Efficiency, Osmolyte, Mineral Uptake, Antioxidant, Expression of Stress-Responsive Genes and Ameliorates the Effects of Salinity Stress in Wheat. *Plants* **2021**, *10*, 1693. [CrossRef]
83. Nahhas, N.E.; Abdelaal, K.A.; AlKahtani, M.D.; al Husnain, L.; AlGwaiz, H.I.; Hafez, Y.M.; Attia, K.A.; El-Esawi, M.A.; Ibrahim, M.F.; Elkelish, A. Biochar and jasmonic acid application attenuate antioxidative systems and improves growth, physiology, nutrient uptake and productivity of faba bean (*Vicia faba* L.) irrigated with saline water. *Plant Physiol. Biochem.* **2021**, *166*, 807–817. [CrossRef]
84. Gupta, N.; Gupta, S.; Kumar, A. Effect of water stress on physiological attributes and their relationship with growth and yield of wheat cultivars at different stages. *J. Agron. Crop Sci.* **2001**, *186*, 55–62. [CrossRef]
85. Zhao, W.; Liu, L.; Shen, Q.; Yang, J.; Han, X.; Tian, F.; Wu, J. Effects of water stress on photosynthesis, yield, and water use efficiency in winter wheat. *Water* **2020**, *12*, 2127. [CrossRef]
86. Li, G.; Gao, H.; Zhao, B.; Dong, S.; Zhang, J.; Yang, J.; Wang, J.; Liu, P. Effects of drought stress on activity of photosystems in leaves of maize at grain filling stage. *Acta Agron. Sin.* **2009**, *35*, 1916–1922. [CrossRef]
87. Barutçular, C.; Dizlek, H.; El-Sabagh, A.; Sahin, T.; EL-Sabagh, M.; Islam, M.S. Nutritional quality of maize in response to drought stress during grain-filling stages in Mediterranean climate condition. *J. Exp. Biol. Agric. Sci.* **2016**, *4*, 644–652. [CrossRef]
88. Hammad, S.A.; Ali, O.A. Physiological and biochemical studies on drought tolerance of wheat plants by application of amino acids and yeast extract. *Ann. Agric. Sci.* **2014**, *59*, 133–145. [CrossRef]
89. Xiao, R.; Wang, X.; Jiang, L.; Tang, H. Research and Application of Lipoic Acid in Plants. *IOP Conf. Ser. Earth Environ. Sci.* **2018**, *108*, 042100. [CrossRef]
90. Guan, X.; Okazaki, Y.; Zhang, R.; Saito, K.; Nikolau, B.J. Dual-localized enzymatic components constitute the fatty acid synthase systems in mitochondria and plastids. *Plant Physiol.* **2020**, *183*, 517–529. [CrossRef]
91. Mukwevho, E.; Ferreira, Z.; Ayeleso, A. Potential role of sulfur-containing antioxidant systems in highly oxidative environments. *Molecules* **2014**, *19*, 19376–19389. [CrossRef] [PubMed]
92. Hasanuzzaman, M.; Bhuyan, M.; Mahmud, J.; Nahar, K.; Mohsin, S.; Parvin, K.; Fujita, M. Interaction of sulfur with phytohormones and signaling molecules in conferring abiotic stress tolerance to plants. *Plant Signal. Behav.* **2018**, *13*, e1477905. [CrossRef] [PubMed]
93. Alzahrani, O.; Abouseadaa, H.; Abdelmoneim, T.K.; Alshehri, M.A.; El-beltagi, H.S.; El-Mogy, M.M.; Atia, M.A.M. Agronomical, physiological and molecular evaluation reveals superior salt-tolerance in bread wheat through salt-induced priming approach. *Not. Bot. Horti Agrobot. Cluj Napoca* **2021**, *49*, 12310. [CrossRef]
94. Hruskova, M.; Smejda, P. Wheat flour dough alveograph characteristics predicted by NIR Systems 6500. *Czech J. Food Sci.* **2003**, *21*, 28–33. [CrossRef]

plants

MDPI

Article

Seed Priming with Silicon as a Potential to Increase Salt Stress Tolerance in *Lathyrus odoratus*

Rasha S. El-Serafy [1], Abdel-Nasser A. El-Sheshtawy [2,*], Amira K.G. Atteya [3], Abdulrahman Al-Hashimi [4], Arshad Mehmood Abbasi [5,6] and Ibrahim Al-Ashkar [7,*]

1. Horticulture Department, Faculty of Agriculture, Tanta University, Tanta 31527, Egypt; rasha.elserafi@agr.tanta.edu.eg
2. Environment and Bio-Agriculture Department, Faculty of Agriculture, Al-Azhar University, Cairo 11651, Egypt
3. Horticulture Department, Faculty of Agriculture, Damanhour University, Damanhour 22516, Egypt; amira.khames@agr.dmu.edu.eg
4. Department of Botany and Microbiology, College of Science, King Saud University, P.O. Box 2455, Riyadh 11451, Saudi Arabia; aalhashimi@ksu.edu.sa
5. University of Gastronomic Sciences, 12042 Pollenzo, Italy; amabbasi@cuiatd.edu.pk
6. Department of Environmental Sciences, COMSATS University, Abbottabad Campus, Islamabad 22060, Pakistan
7. Department of Plant Production, College of Food and Agriculture Sciences, King Saud University, Riyadh 11451, Saudi Arabia

* Correspondence: abdel_nasser2007@azhar.edu.eg (A.-N.A.E.-S.); ialashkar@ksu.edu.sa (I.A.-A.)

Citation: El-Serafy, R.S.; El-Sheshtawy, A.-N.A.; Atteya, A.K.; Al-Hashimi, A.; Abbasi, A.M.; Al-Ashkar, I. Seed Priming with Silicon as a Potential to Increase Salt Stress Tolerance in *Lathyrus odoratus*. *Plants* **2021**, *10*, 2140. https://doi.org/10.3390/plants10102140

Academic Editors: Mukhtar Ahmed, M. Iftikhar Hussain and Adele Muscolo

Received: 9 September 2021
Accepted: 6 October 2021
Published: 9 October 2021

Abstract: Water shortage is a major problem limiting the expansion of green areas and landscapes. Using seawater as an alternative source of potable water is not a novel idea, but the issue of salt stress needs to be resolved. Salinity has a negative impact on growth and the aesthetic value of ornamental plants. In order to overcome these challenges, *Lathyrus odoratus* seeds were hydro-primed and halo-primed with silicon (Si) and silicon nanoparticles (SiNPs), and exposed to seawater levels. Seawater markedly reduced seed germination and growth of *Lathyrus* seedlings, but halo-priming was shown to significantly alleviate its negative effects. Broadly, SiNPs increased the germination percentage, reduced photosynthetic pigments and carbohydrates decrease, and enhanced water relations, despite having a negative effect on germination speed. Halo-priming significantly increased the proline content and the activities of certain enzymatic (SOD, APX and CAT) and nonenzymatic (phenolic and flavonoids) compounds, that positively influenced oxidative stress (lower MDA and H_2O_2 accumulation), resulting in seedlings with more salt stress tolerance. Halo-priming with Si or SiNPs enhanced the Si and K^+ contents, and K^+/Na^+ ratio, associated with a reduction in Na^+ accumulation. Generally, halo-priming with Si or SiNPs increased *Lathyrus* seedlings salt stress tolerance, which was confirmed using seawater treatments via improving germination percentage, seedlings growth and activation of the antioxidant machinery, which detoxifies reactive oxygen species (ROS).

Keywords: *Lathyrus odoratus*; seed priming; seawater; antioxidant; proline; SiNPs

1. Introduction

Sweet pea (*Lathyrus odoratus*), an annual herb, belongs to the Fabaceae family, and is an important ornamental plant in temperate regions. *Lathyrus* is a climbing plant that reaches up to 2 m in height using tendrils. It is cultivated for its attractive, strongly fragrant, and decorative flowers. It has a range of colors, including pure whites, pinks, oranges, reds, blues, and lavenders. *Lathyrus* is widely used as a bed plant in landscapes and gardens, and is cultivated for the floral industry. As a member of the legume family, these plants are toxic and should not be eaten.

Currently, many countries are attempting to increase the amount of green areas and landscapes in urban environments, due to their vital role in mitigating climate change, and the impact of heat [1,2], air pollution [3], and water pollution [4]. In addition, they have a positive effect in terms of reducing respiratory illness, allergies, and improving public health [3,5,6]. Expanding green areas is often challenging as a result of potable water deficiency for irrigation; thus, seeking an alternative to clean water is crucial. Using seawater as an alternative to potable water in agriculture was first attempted after the Second World War. The major challenge for using seawater to irrigate landscapes is the elevated salinity. Leaf necrosis and burns are common symptoms of foliage injury in plants irrigated with low-quality or saline water [7,8]. Moreover, it causes a reduction in shoot and roots growth. Despite this, landscape plants are highly variable in terms of salinity tolerance, which can depend on various factors including climatic conditions, soil or substrate type, irrigation method, plant species, and/or growth stage [9,10]. The impacts of salinity on the growth and appearance of landscapes have attracted much attention. This is in part due to its negative influence on the aesthetics and ornamental appearance. Much effort has been focused on inducing salt tolerance, and seed priming is among the most promising methods for achieving this.

Seed priming is a pre-sowing technique that leaves the seed partially hydrated. This hydration only allows the pre-germinative physiological and biochemical metabolic processes, without radical protrusion [11]. Seed priming can accelerate and homogenize seed emergence, enhance the growth and vigor of seedlings [12]. Halo-priming and hydro-priming are types of seed priming. Hydro-priming and halo-priming are defined as soaking seeds in water and salt solutions respectively [13]. Seed priming has been successfully demonstrated to elevate the germination percentage and speed, and enhance seedling vigor under normal and stressed environments [14,15]. Hydro and halo-priming applications enhanced plant growth and performance of wheat under salt stress conditions [16]. Silicon (Si) is a beneficial element commonly used in the halo priming technique, both in bulk size (sodium silicate) and nanoparticle size (SiNPs) [17,18]. Silicon application is an eco-friendly strategy to improve plants' salinity stress response [19]. It boosts plant resistance to salinity and drought stress [20,21], reduces the negative impact of salt stress on chlorophyll content and biomass production [22], boosts adaptive responses, such as phenolic compound production, mineral uptake, and antioxidant activity [23,24]. Seed priming with sodium silicate enhanced germination characteristics and seedling vigor of wheat plants under drought stress [16]. Halo-priming with SiNPs improved seed germination and seedling growth under salinity stress and normal conditions [18,25].

As a result of the importance of sweet pea cultivars as bedding plants in gardens and landscapes, and the scarceness of potable water for landscapes irrigation, the current study aimed to evaluate the effects of seed priming with (Si) and silicon nanoparticle (SiNPs) treatments on plant growth, leaf water status, and the biochemical and physiological traits in *Lathyrus odoratus* under seawater treatments.

2. Results

2.1. Germination Characteristics

The results in Figure 1 present the effect of seed priming application on germination characteristics of *Lathyrus* seeds exposed to seawater treatments. Halo-priming application with Si and SiNPs significantly improved the germination percentage (GP) of *Lathyrus* seeds as compared with hydro-priming application. Concerning seawater levels, a gradual decrease was observed in the GP with increasing seawater levels, as the highest salinity level (30%) significantly reduced GP by 21.72%, as compared with unsalted treatments (0%). The highest GP values of 80.02 and 80% were recorded with the treatments Si and SiNPS-primed seeds under non-stressed condition, respectively. Meanwhile, the lowest GP value was obtained by hydro-primed seeds under 30% saline condition (57.48%).

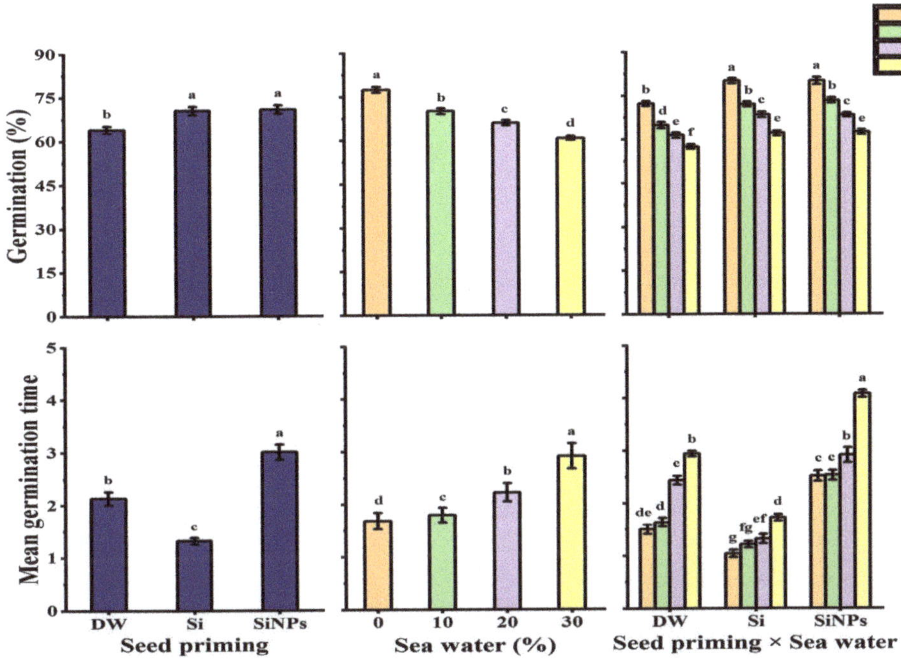

Figure 1. Effect of seed priming on germination (%) and mean germination time of *Lathyrus odoratus* seeds irrigated with seawater levels (0, 10, 20, and 30%). Data are mean value ± SE. Bars with different letters are significantly different at $p \leq 0.05$ level.

Halo-primed seeds with Si had the lowest mean germination time (MGT), which produced 56.03 and 38.32% less than SiNPs and DW, respectively. A significant and growing increase in the MGT value was observed with increased seawater levels, since the highest salinity level (30%) increased MGT by 42.12%, as compared with unsalted treatment (0%). In terms of seed priming and seawater interaction, the lowest MGT was given by Si-primed seeds under non-stressed condition (1.04), while the highest MGT was obtained by SiNPS-primed seeds under 30% saline condition (4.08).

2.2. Plant Growth

The height of *Lathyrus* seedlings significantly increased following halo-priming application (Si or SiNPs) as compared with hydro-priming application (Figure 2). Seawater treatments negatively impacted *Lathyrus* seedling height, as the lowest height of 16.44 cm was observed with 30% salinity, while unsalted treatment (0%) significantly produced the greatest seedling height (23.5 cm). Regarding the interaction, the tallest seedlings were obtained by SiNPS-primed seeds under non-stressed condition (25.43 cm). However, the shortest seedlings were given by hydro-primed seed under 30% saline condition.

When it comes to the seedling's fresh and dry weights, halo-priming application significantly enhanced seedling fresh and dry weights as compared to hydro-priming application. SiNPs-application significantly produced the highest fresh and dry weights (Figure 2). Seawater treatments led to a significant and gradual decrease in fresh and dry weights with increased seawater levels, as the lowest weights were produced by the highest seawater level (30%). In terms of the interaction, the SiNPS-primed seeds under non-stressed condition significantly produced the heaviest fresh (0.47 g) and dry (0.043 g) weights. However, the lowest fresh (0.17 g) and dry (0.018 g) weights were obtained by hydro-primed seed under 30% saline condition.

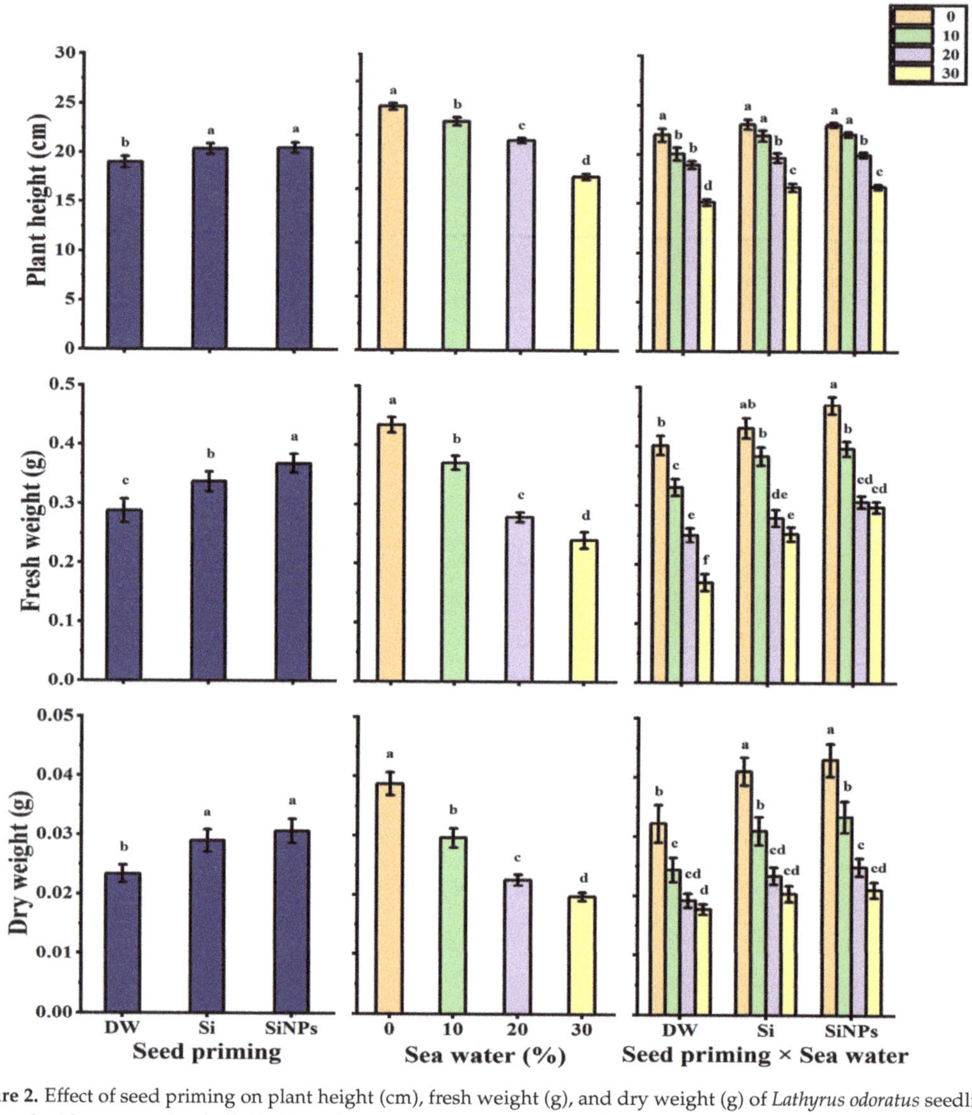

Figure 2. Effect of seed priming on plant height (cm), fresh weight (g), and dry weight (g) of *Lathyrus odoratus* seedlings irrigated with seawater levels (0, 10, 20, and 30%). Data are mean value ± SE. Bars with different letters are significantly different at $p \leq 0.05$ level.

2.3. Leaf Water Status

The water status of *Lathyrus* leaves (leaf water content; LWC) and relative water content (RWC) was significantly enhanced by halo-priming application as compared with the hydro-priming application (Figure 3). Concerning seawater treatments, increasing seawater levels led to a significant and gradual decrease in the LWC and RWC values. Leaf water status was higher with SiNPS-primed seeds under non-stressed condition, which recorded 87.51 and 87.16% for LWC and RWC, respectively. Meanwhile, a lower leaf water status was reported with hydro-primed seeds under 30% saline condition, with respective LWC and RWC of 81.53 and 80.05%.

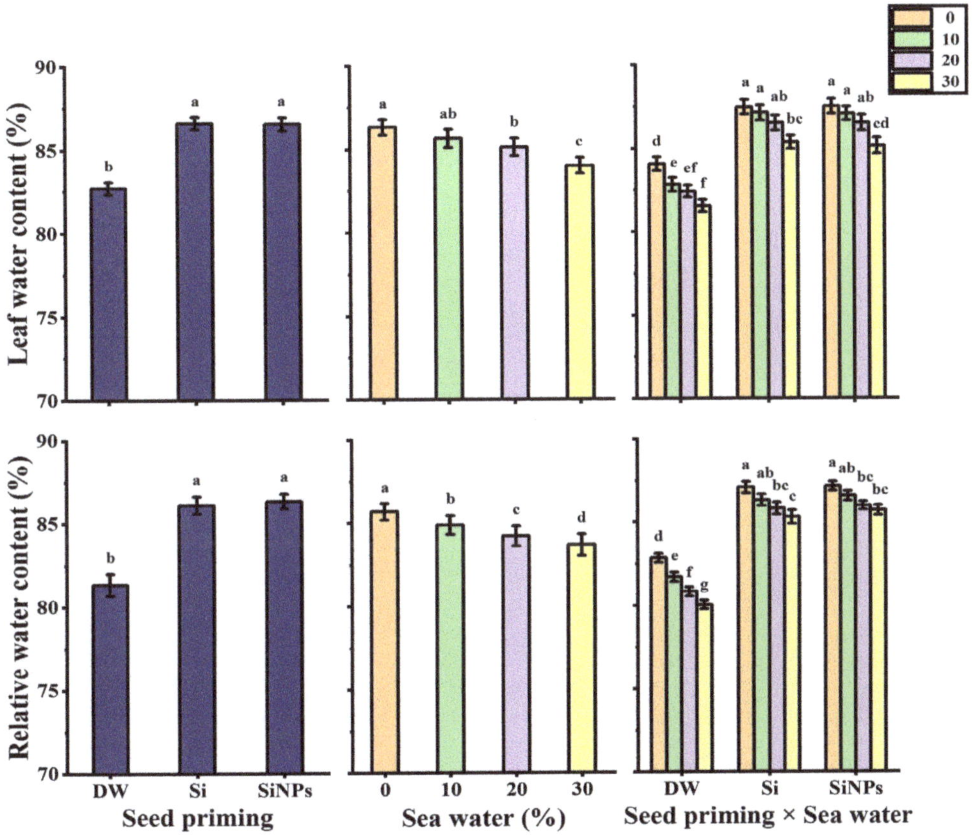

Figure 3. Effect of seed priming on leaf water content (%) and relative water content (%) of *Lathyrus odoratus* leaves irrigated with seawater levels (0, 10, 20, and 30%). Data are mean value ± SE. Bars with different letters are significantly different at $p \leq 0.05$ level.

2.4. Photosynthetic Pigments

Leaf photosynthetic pigments (total chlorophyll and carotenoids) were significantly impacted by seed priming and seawater treatments (Figure 4). In this regard, SiNPs produced the highest chlorophyll and carotenoids levels, while the lowest pigment concentrations were obtained by hydro-priming application. Leaf photosynthetic pigments were inversely proportional to seawater treatments, since total chlorophyll and carotenoids content decreased significantly as seawater level increased, reaching its lowest values under 30% salinity (1.36 and 0.367 mg g^{-1} FW for total chlorophyll and carotenoids, respectively). The treatment of SiNPS-primed seeds under non-stressed conditions significantly produced the highest chlorophyll (2.834 mg g^{-1} FW) and carotenoids (0.481 mg g^{-1} FW) concentrations, with the least chlorophyll (1.214 mg g^{-1} FW) and carotenoids (0.364 mg g^{-1} FW) concentrations observed with the treatment of hydro-primed seeds under 30% saline conditions.

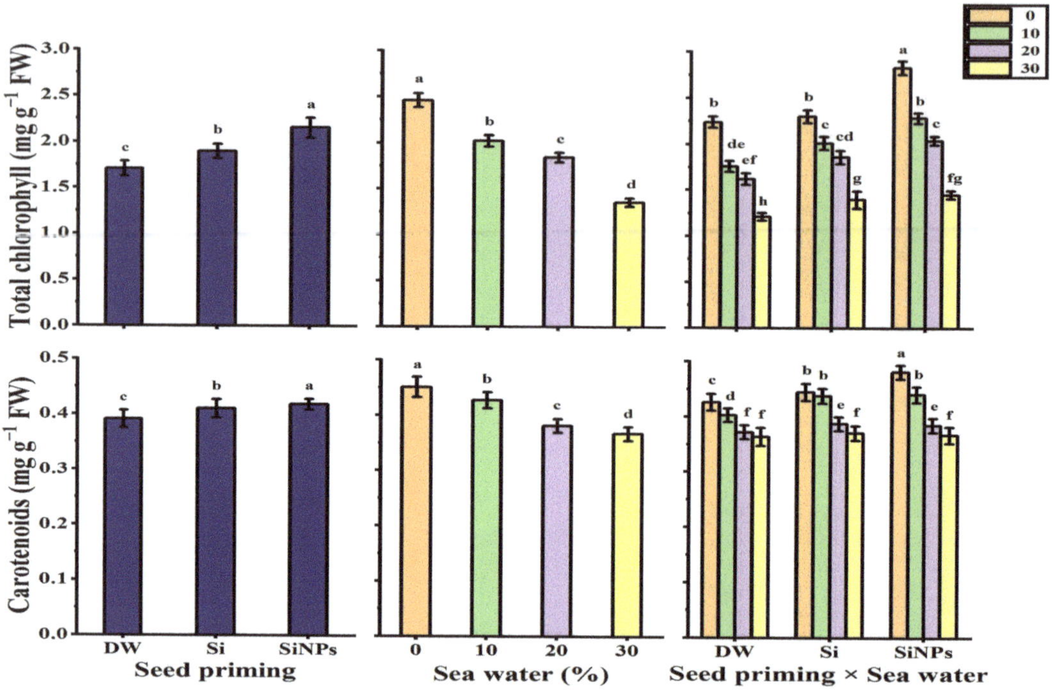

Figure 4. Effect of seed priming on total chlorophyll (mg g^{-1} FW) and carotenoids (mg g^{-1} FW) contents of *Lathyrus odoratus* leaves irrigated with seawater levels (0, 10, 20, and 30%). Data are mean value ± SE. Bars with different letters are significantly different at $p \leq 0.05$ level.

2.5. Biochemical Parameters

2.5.1. Total Carbohydrates

Seed priming application significantly enhanced total carbohydrate content in *Lathyrus* leaves (Figure 5). Halo-priming application with Si or SiNPs significantly produced higher carbohydrates than hydro-priming application, although there was a non-significant difference between both materials (Si or SiNPs). Total carbohydrates decreased gradually as seawater level increased, reaching its lowest record under 30% salinity, which was roughly 25.35% less than unsalted seedlings (0%). The treatment of SiNPS-primed seeds under non-stressed conditions significantly presented the highest carbohydrate value (27.36%), while the least carbohydrate value was observed with hydro-primed seeds under 30% saline conditions (16.91%).

2.5.2. Proline Content

The results illustrated in Figure 5 show the effect of different seed priming applications on the proline content produced in *Lathyrus* leaves subjected to seawater treatments. Proline content has been decreased by the applications of halo-priming, as the lowest value was noticed with SiNPs-application (3.29 μmol g^{-1} FW), while it was increased by 20.6% when a hydro-priming application was used. Seawater levels caused remarkable variations in the proline values. A significant and gradual increase was observed with increasing seawater levels, reaching the highest proline content under 30% salinity (5.09 μmol g^{-1} FW) against the lowest proline content found in unsalted *Lathyrus* leaves (1.98 μmol g^{-1} FW). The maximum proline content was recorded with the treatment hydro-primed seeds under 30% salinity (5.20 μmol g^{-1} FW), while the lowest proline content was given by SiNPS-primed seeds under non-stressed conditions (1.78 μmol g^{-1} FW).

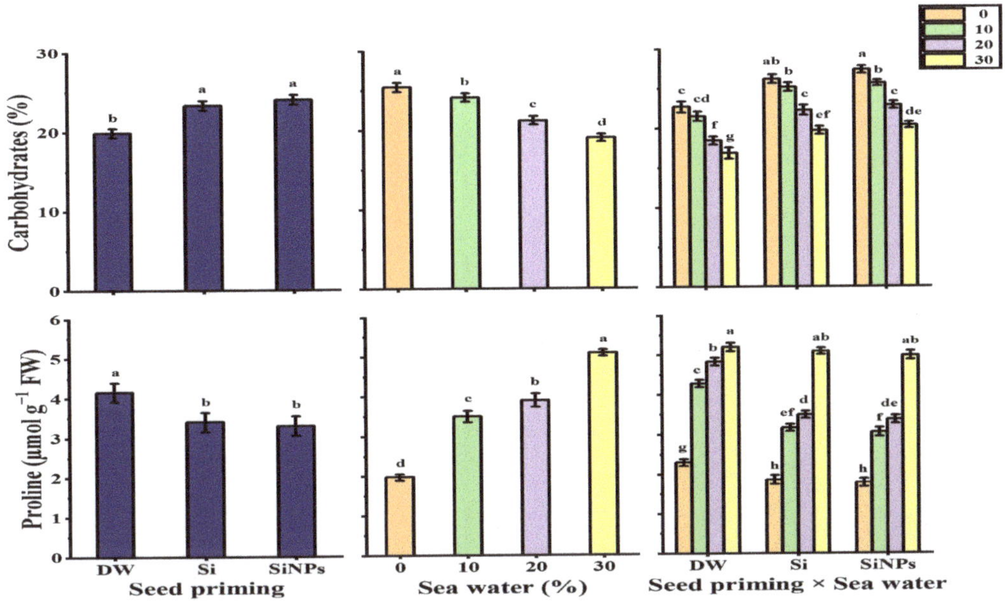

Figure 5. Effect of seed priming on carbohydrates (%) and proline (µmol g^{-1} FW) contents of *Lathyrus odoratus* leaves irrigated with seawater levels (0, 10, 20, and 30%). Data are mean value ± SE. Bars with different letters are significantly different at $p \leq 0.05$ level.

2.5.3. Total Phenols and Flavonoids Contents

The total phenols and flavonoids content significantly increased after halo-priming applications (with Si or SiNPs) as compared with hydro-priming applications (Figure 6). SiNPs significantly recorded the highest levels in this respect. Increasing seawater levels caused a significant and growing increase, reaching the highest values when plants were irrigated with 30% seawater. The highest levels of phenols (6.544 mg GAE g^{-1} DW) and flavonoids (4.170 mg CAE g^{-1} DW) were detected in *Lathyrus* leaves that had been treated with SiNPS-primed seeds under 30% salinity. Meanwhile, the hydro-primed seeds under non-stressed conditions identified the lowest values in this regard (4.670 mg GAE g^{-1} DW and 2.035 mg CAE g^{-1} DW for phenols and flavonoids, respectively).

2.5.4. Oxidative Damage Induced

Lipid peroxidation (MDA) and H_2O_2 content decreased significantly more with halo-priming applications than with hydro-priming applications (Figure 7), as SiNPs-application significantly produced the lowest values (34.2% for MDA and 70.9% for H_2O_2 lower as compared to hydro-priming applications) in this respect. The MDA and H_2O_2 values gradually increased with increasing seawater levels, with the maximum values being obtained under 30% seawater. The lowest levels of MDA and H_2O_2 were presented by SiNPS-primed seeds under non-stressed conditions, but hydro-primed seeds under 30% saline conditions significantly caused the maximum values in this respect.

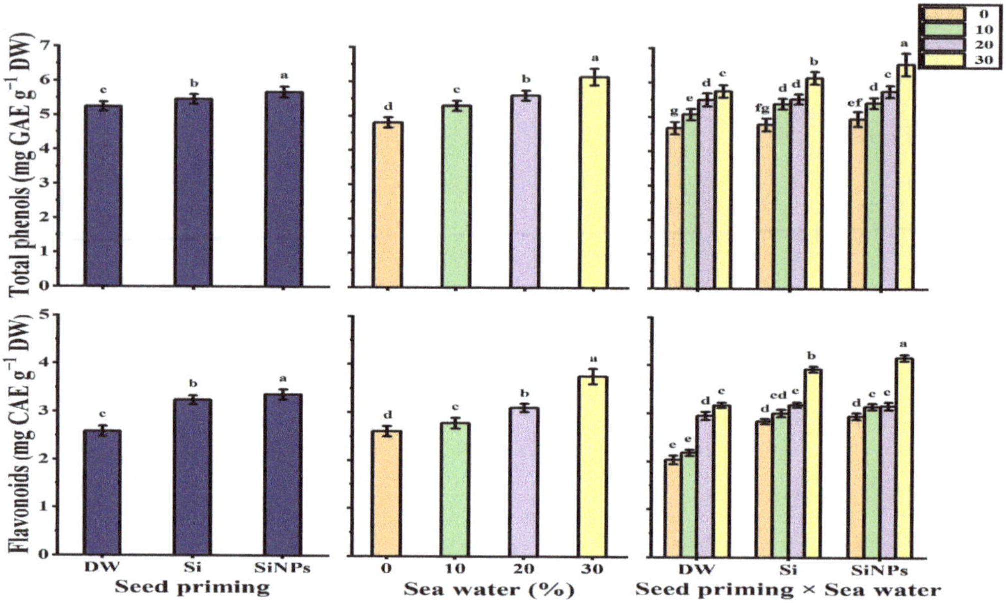

Figure 6. Effect of seed priming on total phenols (mg GAE g^{-1} DW) and flavonoids (mg CAE g^{-1} DW) contents of *Lathyrus odoratus* leaves irrigated with seawater levels (0, 10, 20, and 30%). Data are mean value ± SE. Bars with different letters are significantly different at $p \leq 0.05$ level.

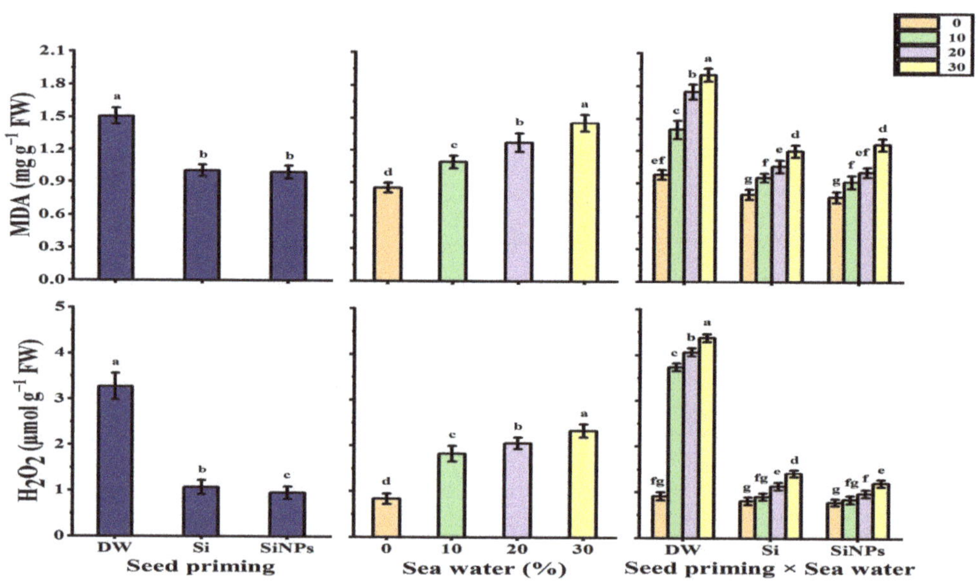

Figure 7. Effect of seed priming on MDA (mg g^{-1} FW) and H_2O_2 (μmol g^{-1} FW) contents of *Lathyrus odoratus* leaves irrigated with seawater levels (0, 10, 20, and 30%). Data are mean value ± SE. Bars with different letters are significantly different at $p \leq 0.05$ level.

2.5.5. Antioxidant Enzyme Activities

Applications of halo-priming significantly increased the activities of superoxide dismutase (SOD), catalase (CAT), and ascorbate peroxidase (APX) enzymes in *Lathyrus* leaves relative to hydro-priming applications, as the highest values were produced by the SiNPs-application (Figure 8). The maximum levels of SOD and APX activities were observed by 30% salinity, while CAT activity decreased with increasing seawater levels. In terms of the interaction effect, the highest levels of SOD and APX were detected by SiNPS-primed seeds under 30% salinity, while the highest CAT value was given by the treatment of SiNPs-primed seeds under non-stressed conditions. On the other hand, the lowest SOD and APX activities were obtained by hydro-primed seeds under non-stressed conditions and by hydro-primed seeds under 30% saline conditions for CAT activity.

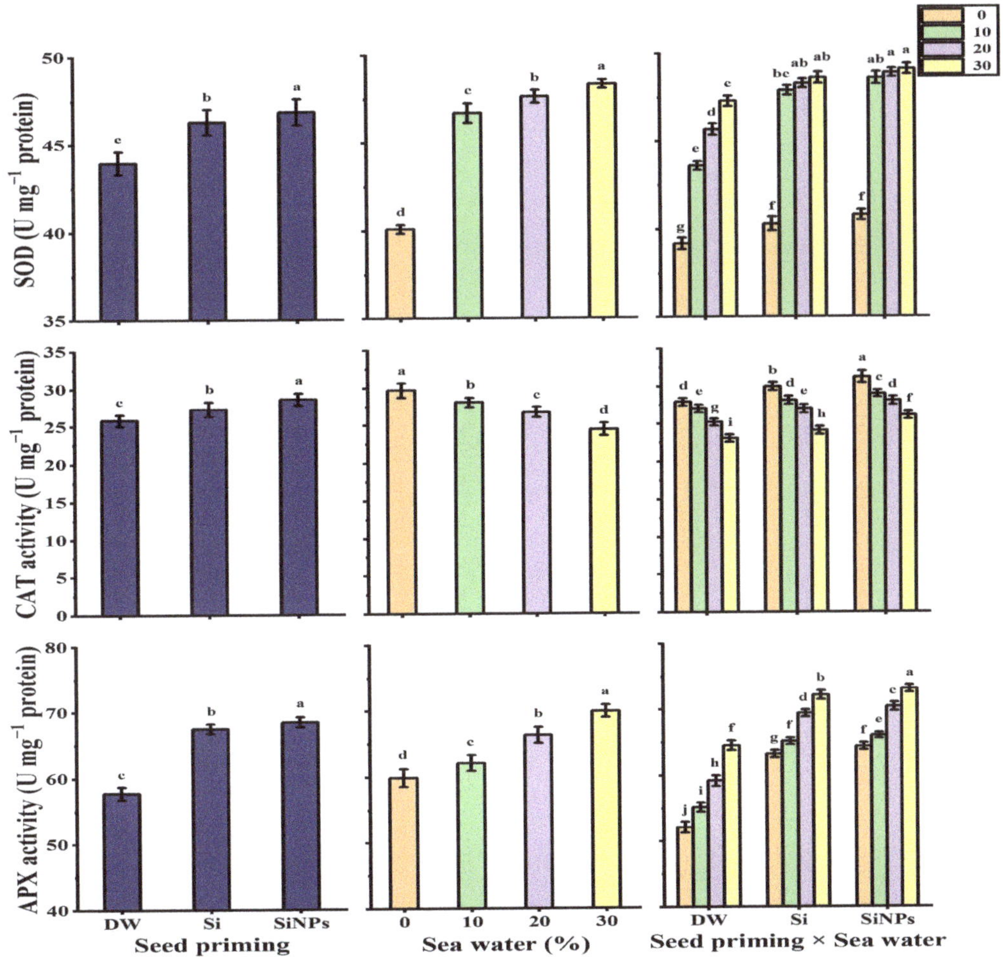

Figure 8. Effect of seed priming on SOD (U mg^{-1} protein), CAT (U mg^{-1} protein), and APX (U mg^{-1} protein) activities of *Lathyrus odoratus* leaves irrigated with seawater levels (0, 10, 20, and 30%). Data are mean value ± SE. Bars with different letters are significantly different at $p \leq 0.05$ level.

2.6. Ion Contents

The results depicted in Figure 9 indicate the content of Si, K^+, and Na^+ accumulated in *Lathyrus* leaves, and the K^+/Na^+ ratio in response to priming applications under seawater treatments. Leaf Si content was significantly elevated following halo-priming applications, as compared with the hydro-priming application. Increasing seawater levels caused an increase in Si content reaching its greatest value at 20% salinity, and then decreased after that. Regarding the interaction effect, both Si and SiNPS-primed seeds under 20% saline conditions significantly showed the highest values of Si in *Lathyrus* leaves. The hydro-priming application significantly exhibited the highest Na^+ content, but SiNPs-application significantly recorded the lowest Na^+ value (30.7% less than hydro-priming application). A gradual and significant increase was noticed with increasing salinity levels. Concerning the interaction, the maximum Na^+ content was obtained by hydro-primed seeds under 30% salinity. Meanwhile, the Si and SiNPS-primed seeds under non-stressed conditions had the lowest Na^+ content.

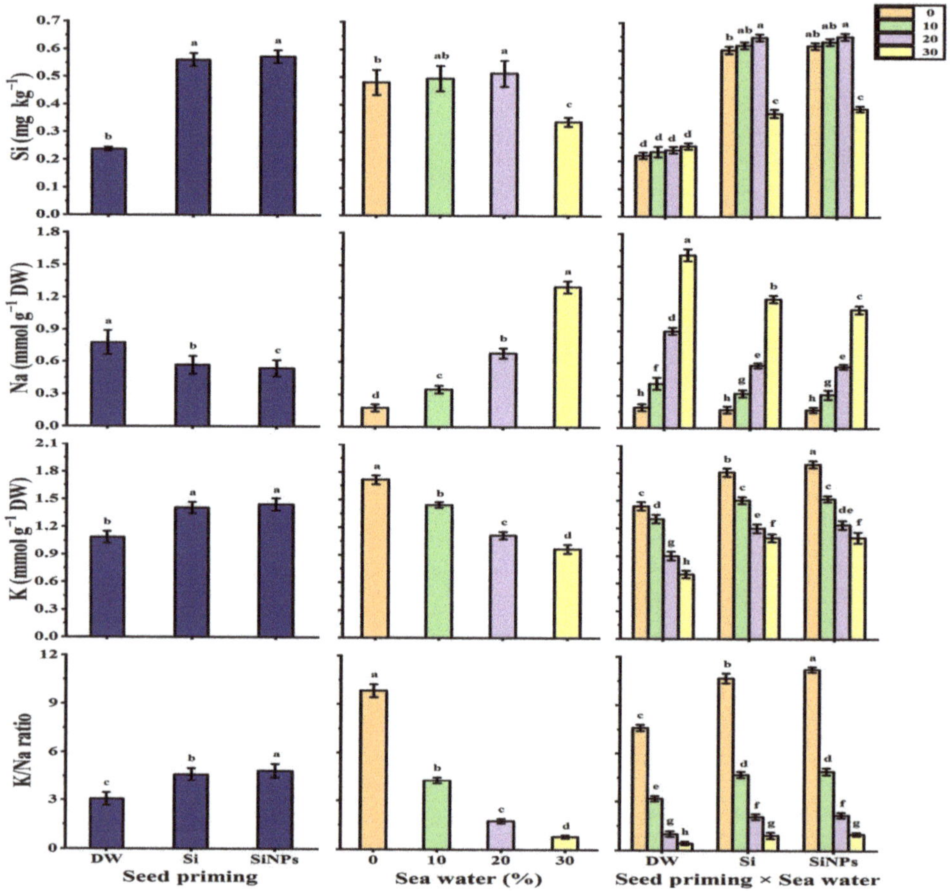

Figure 9. Effect of seed priming on Si (mg Kg^{-1}), Na (mmol g^{-1} DW), K (mmol g^{-1} DW) contents, and K/Na ratio (g) of *Lathyrus odoratus* leaves irrigated with seawater levels (0, 10, 20, and 30%). Data are mean value ± SE. Bars with different letters are significantly different at $p \leq 0.05$ level.

The K^+ content and K^+/Na^+ ratio significantly increased with halo-priming applications as compared with hydro-priming applications, which significantly recorded the lowest K^+ value and K^+/Na^+ ratio. A gradual and significant decrease was observed with increasing seawater concentration. The treatment of SiNPS-primed seeds under non-stressed conditions significantly exhibited the highest K^+ value (1.9 mmol g^{-1} DW) and K^+/Na^+ ratio (11.23). On the other hand, the treatment of hydro-primed seeds under 30% saline conditions significantly exhibited the lowest K^+ (0.70 mmol g^{-1} DW) and K^+/Na^+ ratio (0.44). The mechanisms involved in seed priming with Si or SiNPs for gaining salt stress tolerance in *Lathyrus* seedlings are presented in Figure 10.

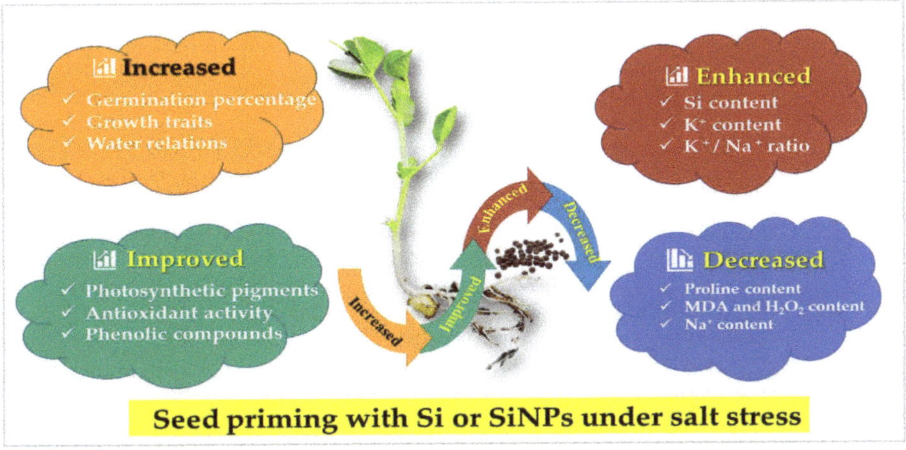

Figure 10. The mechanisms involved in seed primed *Lathyrus* (with Si or SiNPs) for gaining salt stress tolerance.

3. Discussion

Plant growth and development, as well as productivity, are predicted by seed germination. In the current research study, salt stress decreased GP and prolonged MGT, as well as causing a pronounced inhibition of *Lathyrus* seedling growth. Salinity stress decreased seed germination as a result of limited water absorption, slowed the breakdown of seed storage material, and inhibited the production of storage proteins [26,27]. Basically, salinity decreases cell expansion and division, and inhibits physiological and biochemical processes [28,29], which causes a reduction in the photosynthetic rate, dry matter accumulation, and total carbohydrates content. Furthermore, salt stress reduces chlorophyll synthesis and activates degraded chlorophyllase enzymes [30,31]. In accordance with our results, the negative effects of salt on plant growth features have been previously reported by Al-Yasi et al.; Attia et al., [32,33].

In contrast, Si and SiNPs treatments improved the seed germination of *Lathyrus* seeds. Si was found to be directly linked to the physiological process of seed germination in *Glycyrrhiza uralensis* under saline conditions [29]. Halo-priming with SiNPs exhibited the highest MGT value. Similar results were obtained by Siddiqui and Al-Whaibi [34].

The increase in mean germination time following SiNPs seed priming may be due to the ability of silica to absorb moisture from the surroundings [35]. Furthermore, 1 g of SiO_2 nanoparticles (7 nm) has a surface adsorption of 400 m^2 [36]. Increasing the seawater levels caused a reduction in the GP and an increase in the MGT. Salts negatively impact the levels of endogenous phytohormones, which essentially inhibit seed germination and plant growth [37].

In our study, halo-priming applications enhanced plant height, shoot FW and DW, leaf water status, leaf pigments, and total carbohydrate under salinity stress, more than the hydro-priming application. Silicon promotes cell elongation and division, leading to elevated plant height [38,39].

Under salinity conditions, Si enhances photosynthetic activity by decreasing ion toxicity and ROS content, preserving the structure and function of the organelles responsible for photosynthesis [37], maintaining stomatal conductance, transpiration, membrane permeability, net photosynthesis, and chlorophyll levels [29]. Moreover, Si decreases the leaf curve angle and increases leaf flatness, allowing for increasing light interception and more photosynthetic pigments [39], and thus more carbohydrates and dry matter accumulation. Silicon enhances growth performance directly, through blocking Na^+ transport, and indirectly, through different physiological processes that alleviate the negative effects of salinity [37]. Increased biomass production is the main indicator of plant resistance [40].

Leaf water status is a good indicator of water relations in plants. Preserving a good water status in plant cells helps to maintain osmotic adjustments and the activity of metabolic processes, and increases plant resistance under salinity stress [41,42]. Under seawater treatments, the LWC and RWC of *Lathyrus* leaves decreased gradually with increasing seawater levels. Salts decrease the osmotic pressure in plant cells, so they have a negative effect on water uptake by plant roots [43]. Halo-priming treatments exhibited better leaf water status. Silicon decreases plant transpiration [29] due to Si accumulation as external layers above cell walls of leaves and stems, leading to thicker leaves and stem cuticle. In addition, Si improves stem hydraulic conductance [39]. Silicon alters the osmotic pressure, which increases plant tolerance under salinity stress conditions [44,45]. Higher water content in Si-plants grown under saline conditions is mainly associated with salt dilution inside the plant, leading to plant growth improvements [44]. Therefore, it can be concluded that Si improves the leaf water status and mitigates the osmotic stress induced by seawater treatments in *Lathyrus* leaves.

Under seawater treatments, proline content gradually increased with increased seawater levels. Proline is normally produced in high amounts under salinity stress [46]. Proline plays a vital role in osmotic adjustment, sub-cellular structure protection, enzyme activities, and can also increase the cellular turgor pressure, which is responsible for cell expansion under salinity conditions [47,48]. Halo-priming treatments presented lower levels of proline. Si decreases proline accumulation under salinity stress, which indicates the role of Si in alleviating damage caused by salts [49]. Total phenols and flavonoids contents of the metabolic *Lathyrus* leaf extract are gradually increased with higher seawater concentrations, because salts stimulate phenolic compound synthesis.

Phenolic compounds, including flavonoids, are among the many sources of antioxidants in plants and are considered to be a response to protect against the oxidative damage caused by salts [50]. The chemical structure of phenols enables them to deactivate singlet oxygen and act as hydrogen donors, allowing them to scavenge ROS [51,52]. Halo-priming treatments demonstrated a significant increase in the total phenolic compound, and SiNPs were superior in this respect which may be due to the nanoscale size of the insoluble SiNPs accumulated in the epidermis. This may allow constitutional phenols to be produced on the large adsorption surfaces of epidermal cells [53,54].

Plants alleviate the oxidative damage that occurs under saline conditions through nonenzymatic (phenolic compounds) generation [55]. These processes play a vital role in protecting plant cells from the oxidative damage [56] that occurs at the cell membrane [28,57], and in ion balance and water status [58].

In our study, plants subjected to salt stress exhibited higher H_2O_2 and MDA levels; however, halo-priming treatments revealed a significant decrease in this regard. H_2O_2 negatively impacts cell membrane lipids and causes oxidative damage, which was evidenced in the increased MDA accumulation (the indicator of lipid peroxidation) [59]. Si significantly increased antioxidant enzyme activities (CAT, APX, and SOD) and reduced ROS accumulation (H_2O_2) in *Lathyrus* leaves. In the case of saline conditions, the overproduction of reactive oxygen species (ROS) exposes plant cells at risk by inducing lipid peroxidation, protein oxidation, nucleic acid damage, enzyme inhibition, and the initiation of the programmed cell death process [56]. The antioxidant enzymes CAT, SOD, and APX, as well as non-enzymatic antioxidant substances (phenolic and flavonoids compounds, proline, and carotenoids), had the ability to scavenge ROS compounds [55,56]. Si mitigates the negative effects of salt stress by enforcing the antioxidant defense capability, which reduces lipid peroxidation and plasma membrane permeability [49].

Silicon increases plant tolerance by regulating stress-related phytohormone biosynthesis [60]. Using seawater to irrigate *Lathyrus* seedlings leads to increasing Si and Na^+ in the leaves and reduces the amount of K^+, and the K^+/Na^+ ratio. Elevated K^+/Na^+ ratio stimulates plant tolerance to salinity stress [61]. Under salinity stress, the K^+/Na^+ ratio decreased due to Na^+ toxicity which inhibits K^+ uptake. This was also caused by Na^+ and K^+ competition on binding sites [62]. Halo-priming treatments showed less Na^+ and more K^+ accumulation in *Lathyrus* leaves than hydro-priming applications. Silicon increased K^+ concentration in plant cells under saline conditions [49]. Under salinity stress, Si reduces the net rate of Na^+ uptake and accumulation in plants [63]. The deposited Si beneath the cell walls of the roots, binds with Na^+, causing an increase in K^+ uptake and a reduction in Na^+ transported to the plant shoots [49]. Application of Si substantially increased K^+ and decreased Na^+ content in the cytoplasm because of the activity of H+-ATPase in the tonoplasts and plasma membrane, as well as H+-PPase activation in tonoplasts under salt-stress conditions [64]. The effects of Si on Na^+ transport resulted from the blockage of the apoplastic pathway [65], which alleviated Na^+ toxicity under salt stress.

4. Materials and Methods

4.1. Location and Plant Materials

This pot study was undertaken at a greenhouse at the Faculty of Agriculture farm, Tanta University, Tanta, Egypt, (latitude of 30°47′ N: and longitude 31°0′ E), during the 2019 and 2020 winter seasons. Mature and uniform *Lathyrus* seeds were surface-sterilized for 5 min using sodium hypochlorite (10%), and then washed with distilled water. The sterilized seeds were divided into three groups for priming with varying solutions; the first group was hydo-primed with distilled water; the second group was halo-primed with sodium silicate (Si) solution at 50 mg/L; and the third group was halo-primed with 20 mg/L of silicon nanoparticle (SiNPs) solution. The seeds were primed for 9 h, and then the primed seeds were naturally air-dried. Each seed group was divided into four groups, the first seed group was irrigated with tap water (1.52 dS m^{-1}); the second, third, and fourth groups were subjected to saltinity stress using seawater levels of 10% seawater + 90% tap water (9.33 dS m^{-1}), 20% seawater + 80% tap water (14.87 dS m^{-1}), and 30% seawater + 70% tap water (21.60 dS m^{-1}), respectively. Primed seeds were sown in plastic pots of 25 cm diameter containing 9 kg soil, (10 seeds/pot) on September 15th of both seasons. After sowing, the pots were irrigated three times for the week, in order to reach the saturation percentage with seawater levels of 0, 5, 10, and 15% for the second, third, and fourth groups, respectively, to avoid osmotic shock. From the second week, seawater levels increased to the planned levels of 0, 10, 20, and 30% seawater for the second, third, and fourth group respectively; all pots were irrigated every two days with the seawater levels to reach the field capacity. Some physical and chemical properties of the soil were determined as follows: sand, 67.24%; silt, 11.14%; and clay, 21.62%; EC, 1.66 dS m^{-1}; pH, 7.34; Ca^{2+}, 8.45 meq L^{-1}; total N^+, 0.26%; $PO_4{}^{3-}$, 0.041%, and K^+, 0.06%. Data of growth performance and biochemical analysis were estimated 45 days after sowing.

4.2. Seawater Source and Chemicals

Seawater was collected from Marsa Matrouh beach, Mediterranean Sea, Egypt. The water analysis is presented in Table 1. Sodium silicate (Na_2SiO_3, 99% purity) was bought from Sigma-Aldrich Chemie (St. Louis, MO, USA), and silicon nanoparticles (15–45 nm, 99.5% purity) were bought from Nano-technology Laboratory, Faculty of Science, Tanta University, Tanta, Egypt).

Table 1. A chemical analysis of seawater.

Component	EC (dS/m)	pH	HCO_3 (meq/L)	CL (meq/L)	SO_4^{-2} (meq/L)	Ca^{+2} (meq/L)	Mg^{+2} (meq/L)	Na^{+} (meq/L)	K^{+} (meq/L)
Concentration	40.51	6.81	5.59	415	74.35	42.10	14.57	435.35	1.34

4.3. Experiment Layout

The current study was performed in a factorial randomized complete design. Primed solutions were the first factor, while seawater concentrations were the second factor. The experiment consisted of 12 treatments, with six replicates for each treatment; each replicate consisted of three pots, and the pot contained 10 seeds.

4.4. Germination and Growth Characteristics

In order to calculate germinated seeds, the seeds were checked and counted on a daily basis. Seeds were considered germinated when cotyledon appeared above the soil surface. Germination percentage (GP) was estimated on day 18 and calculated according to the following equation:

$$GP\ (\%) = (\text{number of germinated seeds on day } 18/\text{total seeds number}) \times 100.$$

Germination speed is expressed as mean germination time (MGT) according to the equation of Ellis and Roberts [66].

$$MGT = \sum Dn/\sum n.$$

where, n refers to the number of germinated seeds on day D0, and D refers to the number of days from the beginning of the germination experiment. The seedlings were left to grow for 27 days, and order to plant height, shoot fresh weight (FW), shoot dry weight (DW) were determined.

4.5. Leaf Water Status

To determine the leaf water status, the relative water content (RWC) and leaf water content (LWC) were estimated according to Clarke and Mccaig [67] and Barrs [68] as follows: two leaf samples were separated, and immediately weighed, and the fresh weight (FW) was recorded. After that, the same leaf samples were saturated in distilled water at 4 °C for 24 h and their turgid weight was recorded (TW). Then, the same leaf samples were oven-dried at 70 °C for 48 h until reaching a constant and the dry weight (DW) was recorded. RWC and LWC were determined using the following formulae:

$$LWC = ((FW - DW) \div FW) \times 100$$

$$RWC = (FW - DW) \div (TW - DW) \times 100$$

4.6. Photosynthetic Pigments

Leaf pigments of sweet pea were estimated using methanol, as previously described by El-Serafy [69] according to Dere, et al. [70]'s protocol. Samples of fresh leaves (0.2 g) were homogenized in 96% methanol (10 mL) for 1 min. The homogenate was filtrated, and centrifuged for 10 min at 2500 rpm. The supernatant was used for chlorophyll determination using an UVVIS spectrophotometer at wavelengths of 666 nm, 653 nm, and 470 nm for chlorophyll a, b, and total carotenoids, respectively, and its contents are presented in mg g^{-1} FW.

4.7. Biochemical Parameters

4.7.1. Total Carbohydrates

Total carbohydrates were estimated as described by Weinmann [71]. In brief, 0.5 g of dried leaves was mixed with 1 N sulfuric acid (10 mL) in a glass tube. The tube was bolted and heated at 100 °C in an oven over night. The total carbohydrates content was estimated colorimeterically following the method of Dubois, et al. [72]. A total of 1 mL of sugar solution was added to phenol solution 5% (1 mL) followed by 5.0 mL sulfuric acid. The mixture was shaken thoroughly and maintained at 23–30 °C in a water bath for 20 min. The developed color was determined at 490 nm wavelength throughout the UVVIS spectrophotometer analysis, and total carbohydrates content is expressed as percentage.

4.7.2. Proline Content

The proline content was determined as explained by Bates et al. [73] protocol. Briefly, proline extract, ninhydrin acid and glacial acetic acid at volumes of 2, 2, 2 mL were mixed and incubated in a boiling water bath for 1 h, and then were placed in an ice bath. The 520 nm wavelength was used for absorbance determination. A standard curve was constructed at certain levels of authentic proline.

4.7.3. Phenols and Flavonoids Determination

Total phenol content in leaves was determined with the Folin–Ciocalteu procedure using the standard of gallic acid, as previously described by El-Serafy and El-Sheshtawy [74], according to Boateng, et al. [75], with some modifications. Dried leaf samples (1 g) were mixed with 50 mL of methanol 80% and macerated at room temperature for two days. The extract was maintained below 4 °C for total phenol estimation after being fully solvent removed. A total of 1 mL of leaf extract was mixed with Folin–Ciocalteau reagent at a volume of 1 mL, and left to stand for incubation (5 min). Then, a 2 mL of Na_2CO_3 solution (70 g/L) was supplemented. Again, it was left for incubation at 25 °C for 2 h. Thereafter, the absorbance was estimated at a wavelength of 750 nm. Phenolic content is expressed as mg GAE g^{-1} DW. The methods of Boateng et al. [75] and Talukdar [76] were used for total flavonoids estimation based on the aluminum chloride procedure. The mixture of 0.5 mL of the extract and 0.5 mL of aluminum chloride (2%) was left for incubation at room temperature for 45 min. Then, the absorbance was determined at 420 nm wavelength for the resulting mixture. Catechin (CAE) was used to calculate the standard curve, and the flavonoids content is expressed in mg CAE g^{-1} DW.

4.7.4. Lipid Peroxidation Estimation

The content of MDA was determined and utilized as an indicator of lipid peroxidation in *Lathyrus* leaves under tested treatments. MDA was estimated as described by Heath [77] with little modification. A sample of 0.5 g of fresh leaves was centrifuged for 10 min at 12,000× *g* after mixing with 5.0 mL of TCA 5% (*w*/*v*). A total of 2 mL of the extract was added to 2 mL of TBA (0.6%), and then heated for 10 min in a water bath (95 °C). The absorbance was estimated at 532 and 600 nm wavelengths. MDA content (mg g^{-1} FW) was calculated using the following formula

$$\text{MDA content} = 6.45 \times (\text{A532} - \text{A600}) - 0.56 \times \text{A450}.$$

4.7.5. H_2O_2 Content

The content of H_2O_2 in *Lathyrus* leaves was estimated as described by Patterson [78]. A total of 0.5 g of each sample was homogenized in 6 mL of chilled acetone (100%) and centrifuged for 10 min at 4 °C, at 12,000 g. Then, 1 mL of the final extract was added to 5% Ti $(SO_4)_2$ (0.1 mL), and 0.2 mL of NH_4OH solution. Thereafter, the mixture was centrifuged for 10 min at 3000 *g*. A total of 4 mL of H_2SO_4 (2 M) was used for dissolving. The optical density was measured at 412 nm wavelength. For calibration, a standard curve was formulated using various levels of H_2O_2, and the obtained data were recorded as μmol g^{-1} FW.

4.7.6. Antioxidant Enzyme Activities

Fresh leaves were used for enzyme extraction as described by Murkherje and Choudhuri [79], with simple modifications. A total of 0.3 g of fresh leaves was ground with 0.1 mM potassium phosphate buffer (PBS) solution (pH 7.8) and made into a homogenate under ice conditions, which was centrifuged at 10,000 *g* for 20 min at 4 °C. The obtained supernatant was retained at 4 °C for enzyme activity determination.

For superoxide dismutase (SOD, EC 1.15.1.1) determination, the nitro blue tetrazolium procedure of Giannopolitis and Ries [80] was utilized as follows: 0.1 mL of enzyme extract was mixed with 100 mM PBS (pH 7.8), Na_2CO_3 1.5 mM, NBT 2.25 mM, methionine 200 mM, EDTA 3 mM, riboflavin 0.06 mM, in addition to distilled water. The reactions tubes with or without enzymes extract (control) were illuminated for 10 min with a 15 W fluorescent lamp; the blank tubes were not illuminated. The absorbance was spectrophotometrically estimated at the 560 nm wavelength. SOD unit was expressed as the amount of enzyme required to inhibit the rate of NBT reduction by 50% in the controls tubes.

Catalase (CAT, EC 1.11.1.6) activity determination was run according to Aebi [81]. In brief, 3 mL of the reaction solution was mixed with 50 mmol L^{-1} PBS (pH 7.0) and 10 mmol L^{-1} of H_2O_2 solution. Then, the enzyme activity was estimated by calculating the amount of H_2O_2 which was consumed at 240 nm for 2 min.

Ascorbate peroxidase (APX, EC 1.11.1.11) determination was conducted as described by Nakano and Asada [82]. A fresh leaf sample (0.1 g) was mixed with 0.2 mL of extraction buffer, which consisted of 3.0 mM EDTA and 0.1 M Na-phosphate adjusted to pH 7.0, and mixed with 1.0% Triton X-100 and 1.0% polyvinylpyrrolidone. Thereafter, the mixture was centrifuged at 10,000 *g* for 20 min. The absorbance estimation was determined at 290 nm wavelength. The reaction buffer consisted of 0.1 mM H_2O_2, 0.5 mM ascorbate, 0.05 mL of extract containing enzyme, and 0.1 mM EDTA mL^{-1}; the reaction was performed at 25 °C for 5 min. The coefficient of absorbance of 2.8 mM^{-1} cm^{-1} was used to calculate the activity of APX.

4.7.7. Ion Estimation

Si was estimated according to Snyder [83] using ICP-OES. Briefly, 0.1 g of ground leaf was mixed with 3.0 mL of NaOH (18.5 M) in 55 mL TeflonH vessels. Then, the mixture was heated up to 200 °C for 15 min in a microwave, and maintained at this temperature for 15 min. A total amount of 2 mL H_2O_2 was added to the mixture after cooling to the room temperature. Then, the mixture was re-heated to 200 °C for 15 min, and left for 5 min at 200 °C. The mixture was filtered after cooling. A total of 9 mL of deionized water was added to 1 mL of the filtrate and injected into the ICP-OES for determination. Dried leaf samples (0.5 g) were digested with 0.5% HNO_3 according to Deal [84]. Sodium (Na^+) and potassium (K^+) contents were determined using flame photometry, and are expressed as mmol g^{-1} DW.

4.8. Statistical Analysis

The present investigation was designed in a complete randomized layout in factorial arrangement with two factors. The data sets of both tested seasons were collected and subjected to ANOVA using the SPSS program Base 9, SPSS Inc. USA. A combined analysis was performed. Duncan multiple rang test was used at $p \leq 0.05$ probability level to compare mean differences according to Waller and Duncan [85]. The results were presented as means ± SE.

5. Conclusions

This research study investigated the potential of seed priming with Si and SiNPs to enhance salt stress tolerance in *Lathyrus*. Halo-priming application with SiNPs effectively exhibited improved salt tolerance against seawater treatments in *Lathyrus* seedlings. Despite SiNPs application increasing MGT, their seedlings showed similar characteristics to the seeds primed with Si, in terms of growth characteristics and salt stress tolerance. Halo-priming with SiNPs improved growth traits, carbohydrates accumulation, photosynthetic pigments content, K^+/Na^+ ratio, and enzymatic (SOD, APX and CAT) and nonenzymatic (phenolic compounds) generation in salted seedlings. The decrease in MDA and H_2O_2 contents in halo-priming treatments protected the cell membrane from deterioration. Using the seed priming technique for salt stress tolerance in *Lathyrus* seeds not only significantly influenced plant growth and resistance, but also enhanced their aesthetic value and ornamental appearance. This is because the seed priming technique helped maintain the dry matter and carbohydrate accumulation values, and preserved the physiological processes under seawater treatments.

Author Contributions: Conceived and designed the experiments, A.-N.A.E.-S., R.S.E.-S. and A.K.G.A.; performed the experiments, R.S.E.-S., A.-N.A.E.-S. and A.K.G.A.; agro-physiologic measurements, R.S.E.-S. and A.-N.A.E.-S.; ion concentrations analysis: R.S.E.-S., A.-N.A.E.-S. and A.K.G.A.; analyzed the data: A.-N.A.E.-S., I.A.-A. and A.-N.A.E.-S.; biochemical parameters measurements, R.S.E.-S. and A.K.G.A.; edited the manuscript: R.S.E.-S., A.-N.A.E.-S., I.A.-A., A.K.G.A. and A.M.A.; final approval of the version to be published: R.S.E.-S., A.-N.A.E.-S., A.K.G.A., I.A.-A., A.A.-H. and A.M.A. All authors have read and agreed to the published version of the manuscript.

Funding: Project number (RSP-2021/298), King Saud University, Riyadh, Saudi Arabia.

Acknowledgments: The authors extend their appreciation to the researchers supporting project number (RSP-2021/298), King Saud University, Riyadh, Saudi Arabia.

Conflicts of Interest: The authors declare no conflict of interest.

References

1. Al-Gretawee, H.; Rayburg, S.; Neave, M. The cooling effect of a medium sized park on an urban environment. *Int. J. GEOMATE* **2016**, *11*, 2541–2546. [CrossRef]
2. Gunawardena, K.R.; Wells, M.J.; Kershaw, T. Utilising green and bluespace to mitigate urban heat island intensity. *Sci. Total Environ.* **2017**, *584–585*, 1040–1055. [CrossRef] [PubMed]
3. WHO (World Health Organization). *Urban Green Space Interventions and Health: A Review of Impacts and Effectiveness*; World Health Organization Regional Office for Europe: Copenhagen, Denmark, 2017. Available online: http://www.euro.who.int/__data/assets/pdf_file/0010/337690/FULL-REPORT-for-LLP.pdf?ua=1 (accessed on 15 May 2017).
4. Zupancic, T.; Westmacott, C.; Bulthuis, M. *The Impact of Green Space on Heat and Air Pollution in Urban Communities: A Meta-Narrative Systematic Review*; David Suzuki Foundation: Vancouver, Canada, 2015.
5. Lovasi, G.S.; Quinn, J.W.; Neckerman, K.M.; Perzanowski, M.S.; Rundle, A. Children living in areas with more street trees have lower prevalence of asthma. *J. Epidemiol. Community Health* **2008**, *62*, 647–649. [CrossRef]
6. Hanski, I.; von Hertzen, L.; Fyhrquist, N.; Koskinen, K.; Torppa, K.; Laatikainen, T.; Karisola, P.; Auvinen, P.; Paulin, L.; Mäkelä, M.; et al. Environmental biodiversity, human microbiota, and allergy are interrelated. *Proc. Natl. Acad. Sci. USA* **2012**, *109*, 8334–8339. [CrossRef] [PubMed]
7. Jordan, L.A.; Davitt, D.A.; Morris, R.L.; Neuman, D.S. Foliar damage to ornamental trees sprinkler-irrigated with reuse water. *Irrig. Sci.* **2001**, *21*, 17–25.
8. Wu, L.; Guo, X.; Harivandi, A. Salt tolerance and salt accumulation of landscape plants irrigated by sprinkler and drip irrigation systems. *J. Plant Nutr.* **2001**, *24*, 1473–1490. [CrossRef]
9. Fox, L.J.; Grose, J.N.; Appleton, B.L.; Donohue, S.J. Evaluation of treated effluent as an irrigation source for landscape plants. *J. Environ. Hortic.* **2005**, *23*, 174–178. [CrossRef]
10. Niu, G.; Cabrera, R.I. Growth and physiological responses of landscape plants to saline water irrigation: A review. *HortScience* **2010**, *45*, 1605–1609. [CrossRef]
11. Bradford, K.J. Manipulation of seed water relations via osmotic priming to improve germination under stress conditions. *HortScience* **1986**, *21*, 1105–1112.
12. McDonald, M.B. Seed priming. In *Seed Technology and Its Biological Basis*; Black, M., Bewley, J.D., Eds.; CRC Press LLC: Boca Raton, FL, USA, 2000; pp. 287–325.

13. Ghassemi-Golezani, K.; Aliloo, A.A.; Valizadeh, M.; Moghaddam, M. Effects of hydro and osmo-priming on seed germination and field emergence of lentil (*Lens culinaris* Medik.). *Not. Bot. Hort. Agrobot. Cluj* **2008**, *36*, 29–33.
14. Esmaielpour, B.; Ghassemi-Golezani, K.; Khoei, F.R.; Gregoorian, V.; Toorchi, M. The effect of NaCl priming on cucumber seed-ling growth under salinity stress. *J. Food Agric. Environ.* **2006**, *4*, 347–349.
15. Sivritepe, N.; Sivritepe, H.O.; Eris, A. The effects of NaCl priming on salt tolerance in melon seedlings grown under saline con-ditions. *Sci. Hortic.* **2003**, *97*, 229–237. [CrossRef]
16. Hameed, A.; Afzal, I.; Iqbal, N. Seed priming and salinity induced variations in wheat (*Triticum aestivum* L.) leaf protein profile. *Seed Sci. Technol.* **2010**, *38*, 236–241. [CrossRef]
17. Hameed, A.; Sheikh, M.A.; Jamil, A.; Basra, S.M.A. Seed priming with sodium silicate enhances seed germination and seedling growth in wheat (*Triticum aestivum* L.) under water deficit stress induced by polyethylene glycol. *Pak. J. Life Soc. Sci.* **2013**, *11*, 19–24.
18. Bao-Shan, L.; Chun-hui, L.; Li-jun, F.; Shu-chun, Q.; Min, Y. Effect of TMS (nanostructured silicon dioxide) on growth of Changbai larch seedlings. *J. For. Res.* **2004**, *15*, 138–140. [CrossRef]
19. Almeida, D.M.; Oliveira, M.M.; Saibo, N.J. Regulation of Na+ and K+ homeostasis in plants: Towards improved salt stress tolerance in crop plants. *Genet. Mol. Biol.* **2017**, *40*, 326–345. [CrossRef]
20. Zhu, Y.; Gong, H. Beneficial effects of silicon on salt and drought tolerance in plants. *Agron. Sustain. Dev.* **2014**, *34*, 455–472. [CrossRef]
21. Rizwan, M.; Ali, S.; Ibrahim, M.; Farid, M.; Adrees, M.; Bharwana, S.A.; Zia-Ur-Rehman, M.; Qayyum, M.F.; Abbas, F. Mechanisms of silicon-mediated alleviation of drought and salt stress in plants: A review. *Environ. Sci. Pollut. Res.* **2015**, *22*, 15416–15431. [CrossRef] [PubMed]
22. Seal, P.; Das, P.; Biswas, A.K. Versatile potentiality of silicon in mitigation of biotic and abiotic stresses in plants: A review. *Am. J. Plant Sci.* **2018**, *9*, 1433–1454. [CrossRef]
23. Ahanger, M.A.; Alyemeni, M.N.; Wijaya, L.; Alamri, S.A.; Alam, P.; Ashraf, M.; Ahmad, P. Potential of exogenously sourced kinetin in protecting Solanum lycopersicum from NaCl-induced oxidative stress through up-regulation of the antioxidant system, ascorbate-glutathione cycle and glyoxalase system. *PLoS ONE* **2018**, *13*, e0202175. [CrossRef]
24. Ahmad, P.; Ahanger, M.A.; Alam, P.; Alyemeni, M.N.; Wijaya, L.; Ali, S.; Ashraf, M. Silicon (Si) Supplementation Alleviates NaCl Toxicity in Mung Bean [*Vigna radiata* (L.) Wilczek] Through the modifications of physio-biochemical attributes and key antioxidant enzymes. *J. Plant Growth Regul.* **2019**, *38*, 70–82. [CrossRef]
25. Sabaghnia, N.; Janmohammadi, M. Effect of nano-silicon particles application on salinity tolerance in early growth of some lentil genotypes. *Ann. Univ. Mariae Curie-Sklodowska Sectio C Biol.* **2014**, *69*, 39–55.
26. Ashraf, M.; Foolad, M.R. Pre-sowing seed treatment. A shotgun approach to improve germination, plant growth, and crop yield under saline and non-saline conditions. *Adv. Agronomy* **2005**, *88*, 223–271.
27. Eslami, V.; Behdani, M.A.; Ali, S. Effect of salinity on germination and early seedling growth of canola cultivars. *Environ. Stress Agric. Sci.* **2009**, *1*, 39–46.
28. Debez, A.; Chaibi, W.; Bouzid, S. Effect of NaCl and growth regulators on germination of *Atriplex halimus* L. *Cah. Agric.* **2001**, *10*, 135–138.
29. Zhang, X.H.; Zhou, D.; Cui, J.J.; Ma, H.L.; Lang, D.Y.; Wu, X.L.; Wang, Z.S.; Qiu, H.Y.; Li, M. Effect of silicon on seed germi-nation and the physiological characteristics of *Glycyrrhiza uralensis* under different levels of salinity. *J. Hortic. Sci. Biotechnol.* **2015**, *90*, 439–443. [CrossRef]
30. Santos, C.V. Regulation of chlorophyll biosynthesis and degradation by salt stress in sunflower leaves. *Sci. Hortic.* **2004**, *103*, 93–99. [CrossRef]
31. Al-Ashkar, I.; Alderfasi, A.; Ben Romdhane, W.; Seleiman, M.F.; El-Said, R.A.; Al-Doss, A. Morphological and genetic diversity within salt tolerance detection in eighteen wheat genotypes. *Plants* **2020**, *9*, 287. [CrossRef]
32. Al-Yasi, H.; Attia, H.; Alamer, K.; Hassan, F.; Ali, E.; Elshazly, S.; Siddique, K.H.M.; Hessini, K. Impact of drought on growth, photosynthesis, osmotic adjustment, and cell wall elasticity in Damask rose. *Plant Physiol. Biochem.* **2020**, *150*, 133–139. [CrossRef]
33. Attia, H.; Al-Yasi, H.; Alamer, K.; Esmat, F.; Hassan, F.; Elshazly, S.; Hessini, K. Induced anti-oxidation efficiency and others by salt stress in Rosa damascena Miller. *Sci. Hortic.* **2020**, *274*, 109681. [CrossRef]
34. Siddiqui, M.H.; Al-Whaibi, M.H. Role of nano-SiO_2 in germination of tomato (*Lycopersicum esculentum* seeds Mill.). *Saudi J. Biol. Sci.* **2014**, *21*, 13–17. [CrossRef]
35. Pesaran, A.A. Moisture Transport in Silica Gel Particle Beds. Ph.D. Thesis, University of California, Los Angeles, CA, USA, 1983.
36. Zhu, Z.; Wei, G.; Li, J.; Qian, Q.; Yu, J. Silicon alleviates salt stress and increases antioxidant enzymes activity in leaves of salt-stressed cucumber (*Cucumis sativus* L.). *Plant Sci.* **2004**, *167*, 527–533. [CrossRef]
37. Munns, R.; Tester, M. Mechanisms of salinity tolerance. *Annu. Rev. Plant Biol.* **2008**, *59*, 651–681. [CrossRef]
38. Hossain, M.T.; Mori, R.; Soga, K.; Wakabayashi, K.; Kamisaka, S.; Fujii, S.; Yamamoto, R.; Hoson, T. Growth promotion and an increase in cell wall extensibility by silicon in rice and some other poaceae seedlings. *J. Plant Res.* **2002**, *115*, 23–27. [CrossRef]
39. El-Serafy, R.S. Effect of Silicon and Calcium on Productivity and Flower Quality of Carnation. Ph.D. Thesis, Tanta University, Tanta, Eqypt, 2015.
40. Wang, J.; Naser, N. Improved performance of carbon paste amperometric biosensors through the incorporation of fumed silica. *Electroanalytical* **1994**, *6*, 571–575. [CrossRef]

41. Meng, Y.; Yin, Q.; Yan, Z.; Wang, Y.; Niu, J.; Zhang, J.; Fan, K. Exogenous silicon enhanced salt resistance by maintaining K+/Na+ homeostasis and antioxidant performance in Alfalfa leaves. *Front. Plant Sci.* **2020**, *11*, 11. [CrossRef]
42. Al-Ashkar, I.; Romdhane, W.B.; El-Said, R.A.; Ghazy, A.; Attia, K.; Al-Doss, A. Agro-physiologic responses and stress-related gene expression of four doubled haploid wheat lines under salinity stress conditions. *Biology* **2021**, *10*, 56. [CrossRef] [PubMed]
43. Hoffmann, J.; Berni, R.; Hausman, J.-F.; Guerriero, G. A review on the beneficial role of silicon against salinity in non-accumulator crops: Tomato as a model. *Biomolecules* **2020**, *10*, 1284. [CrossRef] [PubMed]
44. Romero-Aranda, M.R.; Jurado, O.; Cuartero, J. Silicon alleviates the deleterious salt effect on tomato plant growth by improving plant water status. *J. Plant Physiol.* **2006**, *163*, 847–855. [CrossRef] [PubMed]
45. Liu, P.; Yin, L.; Wang, S.; Zhang, M.; Deng, X.; Zhang, S.; Tanaka, K. Enhanced root hydraulic conductance by aquaporin reg-ulation accounts for silicon alleviated salt-induced osmotic stress in *Sorghum bicolor* L. *Environ. Exp. Bot.* **2015**, *111*, 42–51. [CrossRef]
46. Ashraf, M.; Foolad, M. Roles of glycine betaine and proline in improving plant abiotic stress resistance. *Environ. Exp. Bot.* **2007**, *59*, 206–216. [CrossRef]
47. Sairam, R.K.; Tyagi, A. Physiology and molecular biology of salinity stress tolerance in plants. *Curr. Sci. India* **2004**, *86*, 407–421.
48. Kishor, P.K.; Sangam, S.; Amrutha, R.N.; Laxmi, P.S.; Naidu, K.R.; Rao, K.R.S.S.; Rao, S.; Reddy, K.J.; Sreenivasulu, N. Regulation of proline biosynthesis, degradation, uptake and transport in higher plants: Its implications in plant growth and abiotic stress tolerance. *Curr. Sci. India* **2005**, *88*, 424–438.
49. Khan, A.; Khan, A.L.; Muneer, S.; Kim, Y.-H.; Al-Rawahi, A.; Al-Harrasi, A. Silicon and salinity: Crosstalk in crop-mediated stress tolerance mechanisms. *Front. Plant Sci.* **2019**, *10*, 1429. [CrossRef] [PubMed]
50. Berni, R.; Luyckx, M.; Xu, X.; Legay, S.; Sergeant, K.; Hausman, J.-F.; Lutts, S.; Cai, G.; Guerriero, G. Reactive oxygen species and heavy metal stress in plants: Impact on the cell wall and secondary metabolism. *Environ. Exp. Bot.* **2019**, *161*, 98–106. [CrossRef]
51. Abdel-Hameed, E.-S.S. Total phenolic contents and free radical scavenging activity of certain Egyptian Ficus species leaf samples. *Food Chem.* **2009**, *114*, 1271–1277. [CrossRef]
52. Di Meo, F.; Lemaur, V.; Cornil, J.; Lazzaroni, R.; Duroux, J.-L.; Olivier, Y.; Trouillas, P. Free radical scavenging by natural polyphenols: Atom versus electron transfer. *J. Phys. Chem. A* **2013**, *117*, 2082–2092. [CrossRef] [PubMed]
53. Li, W.B.; Shi, X.H.; Wang, H.; Zhang, F.S. Effects of silicon on rice leaves resistance to ultraviolet-B. *Acta Bot. Sin.* **2004**, *46*, 691–697.
54. El-Serafy, R.S. Silica Nanoparticles enhances physio-biochemical characters and postharvest quality of *Rosa hybrida* L. cut flowers. *J. Hortic. Res.* **2019**, *27*, 47–54. [CrossRef]
55. AAl-Ashkar, I.; Alderfasi, A.; El-Hendawy, S.; Al-Suhaibani, N.; El-Kafafi, S.; Seleiman, M.F. Detecting salt tolerance in doubled haploid wheat lines. *Agronomy* **2019**, *9*, 211. [CrossRef]
56. Gill, S.S.; Tuteja, N. Reactive oxygen species and antioxidant machinery in abiotic stress tolerance in crop plants. *Plant Physiol. Biochem.* **2010**, *48*, 909–930. [CrossRef]
57. Munns, R. Comparative physiology of salt and water stress. *Plant Cell Environ.* **2002**, *25*, 239–250. [CrossRef]
58. Coskun, D.; Britto, D.T.; Huynh, W.Q.; Kronzucker, H.J. The Role of silicon in higher plants under salinity and drought stress. *Front. Plant Sci.* **2016**, *7*, 1072. [CrossRef]
59. Guo, Q.; Liu, L.; Barkla, B.J. Membrane Lipid Remodeling in Response to Salinity. *Int. J. Mol. Sci.* **2019**, *20*, 4264. [CrossRef]
60. Hamayun, M.; Sohn, E.-Y.; Khan, S.A.; Shinwari, Z.K.; Khan, A.L.; Lee, I.-J. Silicon alleviates the adverse effects of salinity and drought stress on growth and endogenous plant growth hormones of soybean (*Glycine max* L.). *Pak. J. Bot.* **2010**, *42*, 1713–1722.
61. Himabindu, Y.; Chakradhar, T.; Reddy, M.C.; Kanygin, A.; Redding, K.E.; Chandrasekhar, T. Salt-tolerant genes from halo-phytes are potential key players of salt tolerance in glycophytes. *Environ. Exp. Bot.* **2016**, *124*, 39–63. [CrossRef]
62. Azooz, M.M.; Metwally, A.; Abou-Elhamd, M.F. Jasmonate-induced tolerance of Hassawi okra seedlings to salinity in brackish water. *Acta Physiol. Plant.* **2015**, *37*, 77. [CrossRef]
63. Yan, G.; Fan, X.; Peng, M.; Yin, C.; Xiao, Z.; Liang, Y. Silicon improves rice salinity resistance by alleviating ionic toxicity and osmotic constraint in an organ-specific pattern. *Front. Plant Sci.* **2020**, *11*, 260. [CrossRef] [PubMed]
64. Soleimannejad, Z.; Abdolzadeh, A.; Sadeghipour, H.R. Beneficial effects of silicon application in alleviating salinity stress in halophytic puccinellia distans plants. *Silicon* **2019**, *11*, 1001–1010. [CrossRef]
65. Flam-Shepherd, R.; Huynh, W.Q.; Coskun, D.; Hamam, A.M.; Britto, D.T.; Kronzucker, H.J. Membrane fluxes, bypass flows, and sodium stress in rice: The influence of silicon. *J. Exp. Bot.* **2018**, *69*, 1679–1692. [CrossRef] [PubMed]
66. Ellis, R.H.; Roberts, E.H. The quantification of aging and survival in orthodox seeds. *Seed Sci. Technol.* **1981**, *9*, 373–409.
67. Clarke, J.M.; McCaig, T.N. Excised-leaf water retention capability as an indicator of drought resistance of triticum genotypes. *Can. J. Plant Sci.* **1982**, *62*, 571–578. [CrossRef]
68. Barrs, H.D. Determination of water deficits in plant tissues. *Water Defic. Plant Growth* **1968**, *1*, 235–368.
69. El-Serafy, R.S. Phenotypic plasticity, biomass allocation, and biochemical analysis of cordyline seedlings in response to oligo-chitosan foliar spray. *J. Soil Sci. Plant Nutr.* **2020**, *20*, 1503–1514. [CrossRef]
70. Dere, Ş.; Güneş, T.; Sivaci, R. Spectrophotometric determination of chlorophyll-A, B and total carotenoid contents of some algae species using different solvents. *Tr. J. Botany* **1998**, *22*, 13–17.
71. Weinmann, H. Determination of total available carbohydrates in plants. *Plant Physiol.* **1947**, *22*, 279–290. [CrossRef]
72. Dubois, M.; Gilles, K.A.; Hamilton, J.K.; Rebers, P.A.; Smith, F. Colorimetric method for determination of sugars and related substances. *Anal. Chem.* **1956**, *28*, 350–356. [CrossRef]

73. Bates, L.S.; Waldren, R.P.; Teare, I.D. Rapid determination of free proline for water-stress studies. *Plant Soil* **1973**, *39*, 205–207. [CrossRef]
74. El-Serafy, R.S.; El-Sheshtawy, A.A. Effect of nitrogen fixing bacteria and moringa leaf extract on fruit yield, estragole content and total phenols of organic fennel. *Sci. Hortic.* **2020**, *265*, 109209. [CrossRef]
75. Boateng, J.; Verghese, M.; Walker, L.T.; Ogutu, S. Effect of processing on antioxidant contents in selected dry beans (*Phaseolus* spp. L.). *LWT* **2008**, *41*, 1541–1547. [CrossRef]
76. Talukdar, D. Flavonoid-Deficient Mutants in Grass Pea (*Lathyrus sativus* L.): Genetic control, linkage relationships, and mapping with aconitase and s-nitrosoglutathione reductase isozyme loci. *Sci. World J.* **2012**, *2012*, 345983. [CrossRef]
77. Heath, R.L.; Packer, L. Photoperoxidation in isolated chloroplasts. I. Kinetics and stoichiometry of fatty acid peroxidation. *Arch. Biochem. Biophys.* **1968**, *125*, 189–198. [CrossRef]
78. Patterson, B.D.; Macrae, E.A.; Ferguson, I.B. Estimation of hydrogen peroxide in plant extracts using titanium (IV). *Anal. Biochem.* **1984**, *139*, 487–492. [CrossRef]
79. Murkherje, S.; Choudhuri, M. Implication of water stress-induced changes in the levels of endogenous ascorbic acid and hydrogenperoxide in Vigna seedlings. *Physiol. Plant* **1983**, *58*, 166–170. [CrossRef]
80. Giannopolitis, C.N.; Ries, S.K. Superoxide dismutases: I. Occurrence in higher plants. *Plant Physiol.* **1977**, *59*, 309–314. [CrossRef] [PubMed]
81. Aebi, H. Catalase in vitro. *Methods Enzymol.* **1984**, *105*, 121–126. [PubMed]
82. Nakano, Y.; Asada, K. Hydrogen Peroxide is Scavenged by Ascorbate-specific Peroxidase in spinach chloroplasts. *Plant Cell Physiol.* **1981**, *22*, 867–880.
83. Snyder, G.H. Methods for silicon analysis in plants, soils, and fertilizers. In *Studies in Plant Science*; Elsevier: Amsterdam, The Netherlands, 2001; Volume 8, pp. 185–196.
84. Deal, S.B. Flame photometric determination of sodium and potassium in zinc cadmium sulfide phosphors. *Anal. Chem.* **1954**, *26*, 598–599. [CrossRef]
85. Waller, R.A.; Duncan, D.B. A bayes rule for the symmetric multiple comparisons problem. *J. Am. Stat. Associ.* **1969**, *64*, 1484–1503.

plants

MDPI

Review

Unraveling Sorghum Allelopathy in Agriculture: Concepts and Implications

M. Iftikhar Hussain [1,2,*], Subhan Danish [3], Adela M. Sánchez-Moreiras [1,2], Óscar Vicente [4], Khawar Jabran [5], Usman Khalid Chaudhry [6], Ferdinando Branca [7] and Manuel J. Reigosa [1]

1 Department of Plant Biology and Soil Science, Universidad de Vigo, Campus Lagoas-Marcosende, 36310 Vigo, Spain; adela@uvigo.es (A.M.S.-M.); mreigosa@uvigo.es (M.J.R.)
2 CITACA, Agri-Food Research and Transfer Cluster, Campus da Auga, Universidad de Vigo, 32004 Ourense, Spain
3 Department of Soil Science, Faculty of Agricultural Sciences and Technology, Bahauddin Zakariya University, Multan 60800, Pakistan; sd96850@gmail.com
4 Institute for the Conservation and Improvement of Valencian Agrodiversity (COMAV), Universitat Politècnica de València, Camino de Vera s/n, 46022 Valencia, Spain; ovicente@upvnet.upv.es
5 Department of Plant Production and Technologies, Faculty of Agricultural Sciences and Technologies, Niğde Ömer Halisdemir University, Niğde 51240, Turkey; khawarjabran@gmail.com
6 Department of Agricultural Genetic Engineering, Faculty of Agricultural Sciences and Technologies, Niğde Ömer Halisdemir University, Niğde 51240, Turkey; ukojla0455@gmail.com
7 Department of Agriculture, Food and Environment, Università degli Studi di Catania - Piazza Università, 95131 Catania, Italy; fbranca@unict.it
* Correspondence: mih786@gmail.com; Tel.: +34-986-812-616

Citation: Hussain, M.I.; Danish, S.; Sánchez-Moreiras, A.M.; Vicente, Ó.; Jabran, K.; Chaudhry, U.K.; Branca, F.; Reigosa, M.J. Unraveling Sorghum Allelopathy in Agriculture: Concepts and Implications. *Plants* **2021**, *10*, 1795. https://doi.org/10.3390/plants10091795

Academic Editor: Roberta Masin

Received: 5 July 2021
Accepted: 23 August 2021
Published: 28 August 2021

Abstract: Allelopathy is an ecological phenomenon that involves the production and release of biomolecules from different crops, cultivated plants, and bacteria or fungi into the soil rhizosphere and impacts other organisms in the vicinity. Sorghum possesses vital allelopathic characteristics due to which it produces and releases different biomolecules from its root hairs, stems, and grains. Several studies have reported that sorghum acts as an allelopathic crop, decreasing the growth and eco-physiological attributes of surrounding plants and weeds growing simultaneously or subsequently in the field. Sorghum allelopathy has been exploited in the context of green manure, crop rotations, cover crops, and intercropping or mulching, whereas plant aqueous extracts or powder might be an alternate method of weed control. A diverse group of allelochemicals, including benzoic acid, p-hydroxybenzoic acid, vanillic acid, ferulic acid, chlorogenic acid, m-coumaric acid, p-coumaric acid, gallic acid, caffeic acid, p-hydroxibenzaldehyde, dhurrin, sorgoleone, m-hydroxybenzoic acid and protocatechuic acid, have been isolated and identified from different plant tissues of sorghum and root exudates. These allelochemicals, especially sorgoleone, have been investigated in terms of their mode(s) of action, specific activity and selectivity, release in the rhizosphere and uptake and translocation in sensitive species. The present review describes the importance of sorghum allelopathy as an ecological tool in managing weeds, highlighting the most recent advances in the allelochemicals present in sorghum, their modes of action, and their fate in the ecosystem. Further research should focus on the evaluation and selection of sorghum cultivars with high allelopathic potential, so that sorghum allelopathy can be better utilized for weed control and yield enhancement.

Keywords: weed suppression; allelochemicals; sorgoleone; benzoquinone; phenolics; cropping systems

1. Introduction

1.1. Weeds and Challenges to Modern Crop Production

The presence of weeds in agricultural fields decreases the quantity as well as the quality of the agricultural products, resulting in enormous financial losses for farmers [1]. Weeds are considered undesirable and detrimental plants that have harmful effects on the growth of desired plants and reduces the production potential of those desired plants.

Weeds compete with crops for resources such as light, nutrients, space, and water, causing plant yields to suffer [2]. The presence of weeds is very common in crop plant fields, decreasing crop yields and increasing production costs, and consequently making crop production less cost-efficient [3]. Weeds cause a reduction of crop yields because of disturbances of plant growth due to allelopathy, competition, or both [4]. In recent decades, the use of herbicides to control weeds is causing severe problems and danger to the ecosystem, plants, and human beings. Meanwhile, long-term application of herbicides is the cause of generating resistance in weeds, which is currently becoming a serious problem in the development of sustainable agriculture worldwide. For example, triazines were one of the first popular groups of herbicides, which were applied widely due to their significant inhibition of the photosynthesis of various weeds [5–7]. The costs and unsustainability of current weed management are becoming increasingly apparent to farmers, to the public and to policymakers. This is evidenced by increasing demand for organic produce and recent discussions around banning widely used herbicides such as glyphosate [8]. Reduced reliance on chemical herbicides has led to searching for alternate natural products, such as diverse secondary metabolites, which could serve as lead compounds for weed management in the future [9]. The use of allelopathy may help improve plant and environmental productivity through the ecological management of weeds, pests, and plant diseases. In the last two decades, there has been a clear focus on plant-based natural products with the potential to replace chemical herbicides [10–12].

Allelopathy refers to the beneficial or harmful impact of one plant on its neighboring plants with the release of allelochemicals that influence their growth. Allelochemicals are a less toxic, safer, range of chemicals released by plants via volatilization, exudation, leaching, or residue decomposition [13]. Crop plants, such as soybean [14], sunflower [15], wheat [16], alfalfa [17], maize [18], sesame [19], rice [20], sorghum [21] and many others, have demonstrated allelopathic impacts on certain weed species. However, sorghum (*Sorghum bicolor* L.) is a well-known allelopathic crop that has the potential to suppress the weed growth due to the synthesis of sorgoleone [22,23]. It contains a range of allelochemicals, i.e., benzoic, p-hydroxybenzoic, vanillic, ferulic, chlorogenic, m-coumaric, p-coumaric, gallic, and caffeic acids [24], p-hydroxybenzaldehyde [25], dhurrin, sorgoleone [26], m-hydroxybenzoic acid and protocatechuic acid [27] with the potential to reduce weed growth. A comprehensive list of allelochemicals and secondary metabolites present in different plant parts of sorghum (roots, stems, foliage and panicle) is documented in Table 1. In sorghum, these allelochemicals are synthesized at greater concentrations in the adult stage of the plants [28]. Their toxicity can persist up to 22 to 28 weeks [29]. The allelochemicals are released into the soil rhizosphere during the plant life-cycle [30] or by the incorporation of crop debris, i.e., stubble [31] or stalk [32] into the soil.

The mode of action of natural products includes multiple mechanisms, such as the reduction in percent and rate of germination together with reductions in root and shoot growth [33,34], interference with photosystem-II through electron transport [12], [35,36] and primary action on ATP production. In addition, this includes the inhibition of chloroplast oxygen evolution, a strong effect on mitochondrial function, alteration of nutrient absorption, chlorophyll pigments, carbon isotope discrimination [11,12], or water use efficiency [12,37]. The biochemical and physiological action mode of some common phenolic compounds on target plant species is shown in Table 2.

Table 1. A comprehensive list of allelochemicals and secondary metabolites present in different parts (roots, stems, foliage and panicle) of *Sorghum bicolor* L.

Plant Species	Plant Parts	Secondary Metabolites	References
Sorghum bicolor	stems, leaves, roots	ferulic, p-coumaric, syringic, vanillic and p-hydroxybenzoic acids	[38]
Sorghum bicolor	roots	p-coumaric acid, m-hydroxybenzoic acid and protocatechuic acid	[27]

Table 1. *Cont.*

Plant Species	Plant Parts	Secondary Metabolites	References
Sorghum bicolor	whole plant	benzoic acid, p-hydroxybenozoic acid, vanillic acid, m-coumaric acid, p-coumaric acid, gallic acid, caffeic acid, ferulic acid and chlorogenic acid	[39]
Sorghum bicolor	whole plant	vanillic acid, p-hydroxybenzaldehyde, p-coumaric acid and ferulic acid	[25]
Sorghum bicolor	roots	sorgoleone	[40,41]
Sorghum bicolor	stems	methyl-1-(2-proponyl)-hydrozine, 1-aziridineethanol, 5-chloro-2-pentanone and 2-(methylseleno)-ethanamine	[42]

Table 2. Biochemical and physiological mode of action of some of the common phenolic compounds on the target plant species, as reported in the literature.

Compounds	Mechanisms	Target Species	References
Ferulic and p-hydroxybenzoic acids	Inhibition of photosynthetic attributes	*Rumex acetosa*	[36]
Ferulic and p-hydroxybenzoic acids	Inhibition of relative water content, photosynthetic performance and carbon isotope discrimination	*Lolium perenne*	[12]
Ferulic, p-coumaric, o-hydroxyphenyl acetic acid	Stimulation of chlorophyll degradation mechanism	*Oryza sativa*	[43]
P-hydroxybenzoic acid	Inhibits seedling growth, induces water stress, stomatal closure		[44]
Hydroxyamic acid	Mitotic interference, inhibits seedling growth	*Lactuca sativa*	[45]
Caffeine	Inhibits cell division, abnormal root growth	*Zea mays*	[46]
Caffiec acid	Inhibits seed germination, plant growth, disruption of plant–water relationship, reduce chlorophyll contents	*Euporbia esula*	[47]
2-Benzoxazolinone (BOA)	Inhibits plasma membrane bound H^{+}-ATPase in roots	*Avena fatua*	[48]
//	Inhibits germination, seedling growth, induces oxidative stress	*Lactuca sativa*	[49]
//	Disruption of plant–water relationship, adverse effect on transpiration and photosynthesis	*Lactuca sativa*	[45]
			[50]
Caffeic, p-coumaric, ferulic, salicylic acids	Induces water stress Glycine max,	*Sorghum bicolor*	[51]
Benzoic acid and cinnamic acid	Disruption of membrane or alter membrane permeability, efflux of ions, reduce chlorophyll content by damage of thylakoid membrane		[52]
Ferulic and p-hydroxybenzoic acids	Inhibition of photosynthesis, growth and carbon isotope discrimination	*Lactuca sativa*	[53]
Benzoxazolin-2(3*H*)-one (BOA) and cinnamic acid	Inhibition of leaf water content, photosystem-II efficiency, photon energy, photochemical quenching	*Dactylis glomerata, Lolium perenne, Rumex acetosa*	[35]
Cinnamic acid	Decrease of photochemical efficiency of PSII, quantum yield, fluorescence quenching, non-photochemical quenching, portion of absorbed photon energy thermally dissipated, photon energy absorbed by PSII antennae and trapped by "closed" PSII reaction centers, and carbon isotope composition	*Lactuca sativa*	[54]
Phenolic compounds	Reduction in hydraulic conductivity, net nutrient uptake	*Glycine max*	[52]
DIMBOA, MBOA	Inhibits seed germination	*Avena fatua*	[55]
p-coumaric, vanillic, ferulic acids	Inhibit photosynthesis and protein synthesis		[56]
Benzoxazolin-2(3*H*)-one, cinnamic acid	Reduction in leaf water relations, carbon isotopes discrimination, intrinsic water use efficiency	*Dactylis glomerata, Lolium perenne, Rumex acetosa*	[37]
p-hydroxybenzoic acid	Biochemical, physiological and isotopic traits inhibition	*Dactylis glomerata*	[57]

1.2. Weed Management Strategies and Sorghum Allelopathy

Weeds are serious pests of plant species, and cause huge biological and economic crop losses, disrupt functioning, and suppress growth, development and yield of crops. The development of sustainable weed control strategies is urgently needed because of environmental pollution and evolution of herbicide resistance in weeds. Indeed, allelopathy is very important in natural, sustainable, and integrated weed management programs [58]. Sorgoleone, an allelopathic chemical secreted from Sorghum bicolor as root exudates in dryland, constitutes an excellent example of a natural herbicide [59]. At the juvenile stage, sorghum plants secrete significant concentrations of sorgoleone, reaching high concentrations in the root hairs (0.5 mg g−1 of root fresh weight) [22,60]. The potential of this allelopathic chemical is high in the suppression and inhibition of weed growth without disturbing the crop species [60]. It also offers an auspicious platform to spot its potential as a natural herbicide. Most broadleaf and grass weeds are susceptible to the herbicidal potential of sorgoleone. The persistence of sorgoleone is high in soil due to its hydrophobic nature and that it is absorbed by soil; thus, it possesses a long-term herbicidal activity effect that lasts for seven weeks after incorporation [61]. Sorgoleone directly affects the photosynthetic apparatus by disturbing the minerals and water uptake, especially in lower plants [62]. In addition to the above, it also inhibits electron transport in mitochondria and chloroplasts. The effectiveness of sorgoleone as an herbicide is comparable to synthetic herbicides for commercial use [63]. Allelochemicals released from sorghum plants have a direct influence on plant growth under laboratory, greenhouse and field experiments [10,64].

Allelochemicals secreted by sorghum plants directly influenced the growth of cultivated plants (such as rice and maize) in laboratory, greenhouse and field experiments [65,66]. Sorghum phytotoxicity differs with the plant organ, age, environmental factors, genotype and targeted weed species. Sorghum can be utilized in various ways to affect weeds, e.g., as surface mulch [67], by assimilation in soil [68], in aqueous extracts sprays [21], or by rotation [69], smothering [70] or mix cropping [71]. Figure 1 illustrates how sorghum can impact different weeds through several biological control practices. Suppressive effects on purple nutsedge density by incorporation of sorghum roots, stems and leaves in the soil have been reported by [72]. Similarly, foliar addition of a sorgaab (sorghum water extract) decreased the density and dry weight of purple nutsedge up to 44 and 67%, respectively, with an increase in maize grain yield of 44% [73]. Allelopathic effects of sorghum depend upon the genotype, age, location, environmental conditions and cropping system.

The only study about the formulation of sorgoleone available is by Uddin et al. [74]. According to this study, it was wettable powder formulation with 4.6% active ingredient, i.e., sorgoleone; the formulation was prepared by blending methanol dissolved active ingredient with different carriers (e.g., kaolin 79.2%, SiO2 9.2%) and surfactant polyoxymethylene monooctadecyl ether. These authors reported that the germination process and seedling growth of *Setaria viridis* and *Aeschynomene indica* was decreased. Sorgoleone (0.2 g active ingredient (a.i.) L^{-1}) completely reduced germination and seedling growth of broadleaf weeds (*Galium spurium, Rumex japonicus, Aeschynomene indica*, and *Amaranthus retroflexus*). A 20–25% inhibition was observed in weeds after application of sorgoleone as a post-emergence herbicide. Meanwhile, it was observed that sorgoleone 4.6 wettable powder (WP) is more effective in inhibiting the weed plant biomass and growth [74]. In an independent experiment, [64], the joint action of *Sorghum bicolor* (root exudate) and *Fagopyrum* spp. (root extract) on grasses (*Setaria viridis*) and broadleaf weeds (*Galium spurium, Rumex japonicus, Aeschynomene indica*, and *Amaranthus retroflexus*) under greenhouse conditions was observed A mixture of the two extracts (150 $\mu g\ ml^{-1}$ of sorgoleone and 7.5 mg ml^{-1} of hairy root extract alone) significantly decreased germination and growth of target seedlings; among them, the broadleaf weeds, *Galium spurium, Aeschynomene indica, Rumex japonicus,* and *Amaranthus retroflexus* were the most susceptible.

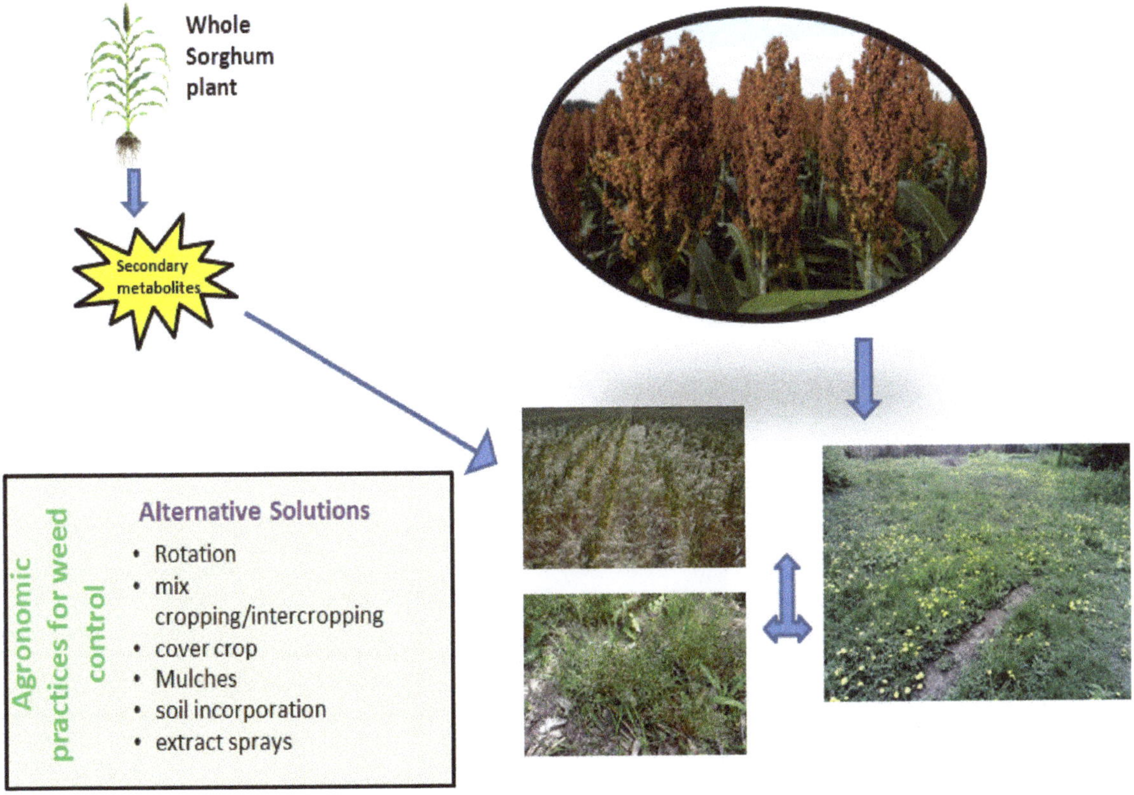

Figure 1. An illustration to demonstrate how sorghum can impact through several biological control practices on different weeds.

Allelopathic potential of root exudates of *Sorghum bicolor* on physiological traits of *Triticum aestivum* L., *Triticum durum* Desf., *Hordeum spontaneum* K. Koch., *Avena fatua* L. and *Phalaris minor* Retz. were studied [75]. They showed that *Phalaris minor* Desf. was the most sensitive in terms of reduced length, dry weight, and chlorophyll content as compared to untreated control. In another study, seedling growth of broadleaf weed species was suppressed more than grass weeds [76]. Weston et al. [77] published a comprehensive review on allelopathic potential and phytotoxicity of sorghum under laboratory, greenhouse and field conditions. They argued that allelochemicals in sorghum tissues vary depending on the plant parts, cultivars and age. The use of sorghum residues as green manure also induced adverse effects on weeds when incorporated as organic matter [78]. A variable class of polyphenols, such as dhurrin and sorgoleone, was also documented from sorghum roots, shoot and exudates [79].

The allelopathic potential of two sorghum varieties (Enkath and Rabeh) against different weeds was evaluated at 26.6 plant m^{-2} planting densities, assessing their effects on common purslane growth during 2009–2010 [80]. They found a significant reduction in weed root and shoot biomass (46–57%) compared to the control, following the treatment with sorghum. Sorghum cv. Enkath was more phytotoxic than cv. Rabeh. The main mechanism responsible for weed growth inhibition included extensive root growth of sorghum and allelochemicals released into the surrounding soil rhizosphere [80]. According to another study [81], sorghum accessions (353) from selected African countries (Botswana, Malawi, Mozambique, Namibia, Tanzania, South Africa, Zambia and Zimbabwe), showed significant variation of 334.62–584.69 μg mg^{-1} root fresh weight in production of sorgoleone. Among all the tested accessions, the South African landrace IS9456 produced the highest amount of sorgoleone (584.69 μg mg^{-1} root fresh weight), followed by an accession from Botswana and a wild sorghum accession from Zimbabwe. The authors concluded that wild sorghum varieties were superior in sorgoleone production compared to improved varieties and hence possess more phytotoxic potential against weeds [81]. The seeds were sown in pots, and sor1 gene expression was measured through RNA sampling from roots collected at 5, 10, 15, 20, 25, and 30 days after seedling emergence (DAE). In the inhibition test, cotton and three weeds were examined during single planting or planting with S. bicolor. The result showed an early expression of sor1 genes in several S. bicolor accessions by 5 days after emergence (DAE). Just one accession demonstrated the expression of sor1 up to 30 DAE. The plant biomass (roots and shoots dry weight) of spiny sandbur (*Cenchrus echinatus*) and Bermuda grass (Cynodon dactylon) was highly decreased. However, it is important to mention that the cotton intercropping with S. bicolor did not show any negative effects [82].

The allelopathic potential of sorghum has been demonstrated by several researchers in both laboratory and field studies [83,84], [21]. Three sorghum varieties (Hybrid sorghum IS41245 and GDLP 34-5-5-3) were evaluated to check their phytotoxicity and production of secondary metabolites such as sorgoleone [85]. Sorgoleone production and release of biological nitrification inhibition (BNI) activity by roots are strongly correlated (1 μg of sorgoleone is equivalent to 1 allylthiourea (ATU) activity). Soil nitrification was significantly inhibited by sorgoleone, and it was variety dependent. In this context, GDLP 34-5-5-3 and Hybrid sorgo exhibited greater production and release of sorgoleone and BNI than the variety IS41245 [85]. Sorgoleone is a hydrophobic molecule from the root hairs that exudates into the soil environment and affects the growth of weeds competing with sorghum [28]. The biosynthetic pathways of this molecule are relatively well known, except for some unknown enzymes. GC-MS analysis showed that the suppression of CYP71AM1 (P450 enzyme) in *S. bicolor* was mediated through RNAi and caused a decrease in sorgoleone production [86]. The authors concluded that CYP71AM1 contributes to the biosynthetic pathway of the allelochemical sorgoleone. Additionally, [87] also documented nitrification inhibition due to the release of allelochemicals from sorghum root hairs in the soil rhizosphere. The allelopathic potential of aqueous extract of two sorghum hybrids (Medovyi and Dovista) and the variety Sylosne 42, was evaluated against germination and seedling growth of *Beta vulgaris* L. and hybrid Ukrainian MS 97. Morphological traits, such as bud number, leaf length and plant height were highly reduced after 14 days of treatment. The results showed that the aqueous extract of Medovyi seeds was less phytotoxic than that of Sylosne 42 [88].

2. Role of Sorghum Allelopathy in Agro-Ecosystem

The use of allelopathy in agricultural practices has been identified as a traditional means to control weeds and has become an important field of study [13]. One approach to utilize this development is to screen numerous crops and their cultivars for their allelopathic properties. Injurious after-effects of sorghum on subsequent crops have long been known to farmers without knowing the actual cause [89]. Experiments were conducted to evaluate the allelopathic effects of different crops, including sorghum [90]. They tested/screened various crops/plant species for their allelopathic effects. They found that sorghum was a highly allelopathic crop because its residues (allowed to decompose in the field) reduced the weed population up to 95%. Based on these studies, several scientific workers [91,92] proposed that crop residue of winter planted sorghum could be utilized for natural weed control.

Previously, Cheema [39] has worked on the allelopathic potential of sorghum in the field, and its possible use to control the weeds. He found that sorghum is a highly allelopathic crop, which exhibits effects on the subsequent crops in rotation, and it also influences weeds selectively. It was also observed that sorghum root residues, incorporated with soil, suppressed the growth (dry weight) of weeds such as *Chenopodium album*, *Phalaris minor*, *Avena fatua*, *Rumex dentatus*, *Senebiera didyma*, *Polygonum bellardi* and *Anagalis arvensis* by 20–48%, while purple nutsedge growth was decreased by 28 to 92%. On the contrary, the growth of *Melilotus parviflora* was promoted by the sorghum residues. It was noted that the amount of the material (sorghum) incorporated into the soil determined the observed effects, so that the greater the quantity, the stronger the allelopathic effect.

Sorghum showed significant quantity of allelochemicals in stem, leaves and roots [38]. The chemical composition of sorghum residues showed significant concentration of phenolic acids, especially, p-coumaric acid, along with ferulic, syringic, vanillic and p-hydroxybenzoic acids. Subsequently, it was revealed that sorghum residues were significantly more toxic at the time of harvest and that it requires approximately 22–28 weeks to decompose [93]. Several phenolic compounds were identified from sorghum, including p-coumaric acid, m-hydroxybenzoic acid and protocatechuic acid as the principal inhibitors in sorghum roots [27], whereas dhurrin and sorgoleone were more important allelochemicals present in sorghum shoots [26]. Sorgoleone, which is released from the roots of living sorghum, is phytotoxic to several weeds, even at low concentrations [94]. Following these studies, Cheema [24] identified nine allelochemicals in sorghum herbage, namely benzoic, p-hydroxybenzoic, vanillic, m-coumaric, p-coumaric, gallic, caffeic, ferulic and chlorogenic acids, while some unknown compounds were also present in residues. Similarly, vanillic acid, p-hydroxybenzaldehyde, p-coumaric acid and ferulic acid were also detected in four sorghum hybrids, with p-coumaric acid present at a significantly higher concentration (7618 μg per g of plant dry weight) than ferulic acid [25].

Sorghum allelochemicals are produced either in the early seedling stage or near maturity. It was reviewed that phenolic acid concentration was higher at each growth stage [77]; even upon harvest a considerable amount of phenolic acid was observed [95]. The concentration of phenolic acid in young plants was again increased at the time of heading. Cheema [24] observed that whole-crop sorghum incorporated at the pre-flowering stage showed no allelopathic effects on wheat and weeds. However, the incorporation of mature sorghum roots, leaves and stems exhibited very strong allelopathic effects on the weeds and the wheat crop. In a later study, [95] found that the total phenol pool size of sorghum differed from 4 to 156 kg/ha in above-ground parts of the plant and from 1 to 16 kg/ha in roots.

Allelopathic compounds of sorghum are species-specific and discriminatory in their action, i.e., they inhibit the growth of some species, but might not affect certain species and may have stimulatory effects on others [39]. The allelochemicals can also inhibit the sprouting and growth of seedlings [68]. Previous studies documented a primary action on ATP by sorgoleone, which then inhibits chloroplast and mitochondrial functions. Sorgoleone has the potential to block chloroplast function at the photosystem-II complex, whereas benzoic acid alters mineral uptake, chlorophyll content, photosynthesis, carbon flow and phytohormone activities. Inhibition of weeds through sorghum allelopathy resulted from the joint activity of various allelochemicals on various cell target sites was proposed by [40]. Gonzalez [41] also proved that sorgoleone is a strikingly intense inhibitor of electron transport in photosystem-II in both confined chloroplasts and PS-II layers. It is clear from the above information that sorghum allelochemicals affected most processes directly or indirectly related to growth. However, their effects were species-specific and concentration-dependent.

Eco-Physiological Impact of Sorghum Allelochemicals

Allelopathy phenomena include examples when one crop may destroy or encourage the germination, growth and yield of the associated crop(s) growing with it (crop mixtures or intercropping) or of the following crop (monoculture or crop rotations) through the release of leachates or washings from germinating seeds or decomposing crop residues [96]. Figure 2 shows an overall view of sorghum allelopathy, including the sources of allelochemicals production from plant parts, i.e., leachates from the aerial parts, surface mulch, soil incorporation, the spray of aqueous extracts, rotation, smothering, root exudates, or mixed cropping. Moreover, factors affecting allelopathy are also depicted.

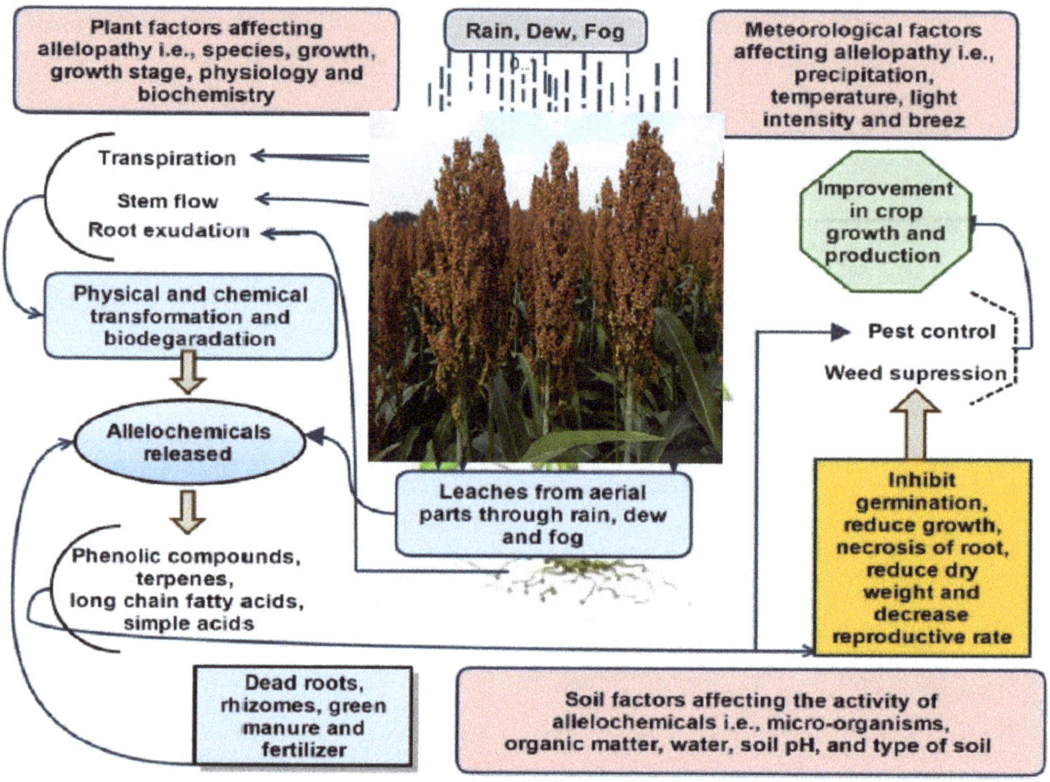

Figure 2. Overall processes of sorghum allelopathy in the soil environment and factors affecting allelopathy. The figure describes the way of allelochemical production from the plant parts, leachates from aerial parts, surface mulch, soil incorporation, spray of aqueous extracts, rotation, smothering, root exudates, and mix cropping.

The germination and seedling growth of *Sphenostylis sternocarpa* (African Yam beans) was evaluated under the treatment of sorghum stem and maize roots aqueous extracts [97]. They reported that sorghum stem aqueous extracts had significant effect on radicle growth of both plants while degree of inhibition was increased with the increase in the concentrations of the extracts. Matos et al. [98] evaluated the bioherbicidal potential of sorghum carried out on *Cyperus rotundus* L. young seedlings with four types of sorghum extract: root extraction in alcohol, leaf extraction in alcohol, root extraction in water and leaf extraction in water, and five concentrations (0%, 20%, 40%, 80% and 100%). The results demonstrated that sorghum leaf extract had a significant impact on *C. rotundus* by interfering in plant growth attributes. The alcohol and aqueous extract showed significant growth retardation in *C. rotundus*, while leaf had more promising effects than roots [98].

Both extracts inhibited the tomato seed germination. In clay soil, *B. napus* extracts increased the bacterial population; however, *S. halepense* extracts restricted bacterial growth but stimulated fungal populations. Kim [99] explored the efficacy of allelochemicals from sorghum residues and water extracts and revealed that seed germination and the development of shoots and roots of crops such as radish, wheat, and rice were inhibited, while maize was less sensitive. The allelochemicals were extracted as fractions of chemical compounds such as methylene chloride, ethyl ether, hexane, and ethyl acetate. In another study, Ben-Hammouda [100] determined the variability of allelopathic effects among sorghum hybrids. Extracts obtained from different parts of the sorghum plant indicated considerable contrasts in phytotoxicity to wheat seedlings. Each extract exhibited a different level of phytotoxicity, which differed depending on the assayed hybrid and the performed measurement. The sorghum and sunflower water extracts (100% and 50% concentration), applied directly to wheat leaves at 30 days after sowing (DAS), increased the wheat grain yield by 5–14% over the control [101]. The maximum upturn in wheat yield (14%) was obtained in plots where 100 per cent sorghum water extract was sprayed, which was attributed to increased weed destruction and translocation of assimilates to the grain resulting from reduced competition. The extract treatments enhanced 1000-grain weight and the number of grains per spikelet, while the number of fertile tillers and spikelet length was reduced. The efficiency of sorgaab as a natural weed inhibitor was evaluated in Raya (Mustard) (*Brassica nigra*) [102]. They reported that the yield of the Raya crop was considerably increased (33–58%) over the control by applying one to three sprays of sorgaab. A significant effect of sorgaab treatment was also observed for plant height, the number of pods per plant and 1000-seed weight. Cheema and Khaliq [67] conducted experiments to explore the allelopathic effect of sorghum on mungbean. Applications of three sprays of sorghum water extract (at 15, 30 and 45 days after sowing (DAS)) and sorghum mulching at 10 and 15 t ha^{-1} increased the grain yield by 18.8, 7.2 and 12.8%, respectively, over the corresponding controls. This improvement was mainly attributed to a better weed control and enlarged leaf area, number of pods per plant and number of grains per pod. In another study, Cheema et al. [73] demonstrated that sorgaab foliar spray enhanced maize grain yields up to 13–44%, whereas the yield was increased by 36–40% when mature and chopped sorghum herbage was applied on the soil surface at the time of sowing. Likewise, three foliar sprays of sorgaab (sorghum water extract), applied after 15, 30 and 45 days of sowing were found to be effective in control of *Cyperus rotundus* L. in maize, as contrasted to hand weeding. Similarly, Cheema et al. [21] checked one and two foliar sprays of sorgaab against different varieties of wheat. The results showed that wheat grain yield was increased by 10–22%, and that leaf area, productive tillers, grain number, 1000-grain weight and harvest index were also improved. The cultivar Parwaz-94 was observed to be the most receptive to sorgaab, showing the largest increment in grain yield.

3. Sorghum Allelopathy and Sustainable Weed Management

Herbicide-resistant weeds are becoming increasingly competitive in agriculture systems, reducing the yield of most of the crops, particularly cereals and food grain crops [103–107]. Meanwhile, efforts are on the top agenda to include allelopathic crop cultivars, e.g., wheat, rice and sunflower that are yield stable and can also demonstrate phytotoxic influence on weeds [108]. Sorghum is an important grain crop with significant potential to suppress weeds under laboratory and field settings [84]. In less developed agriculture, weeds provide stiff competition to crops, thereby limiting crop growth, yield and economic profit [109].

Any inexpensive weed control measure would be helpful to farmers, hence, many plant species have been tested for their weed management potential, as they provide effective control by the suppression of the weed germination in agro-ecosystems. The effect of allelochemicals in *Sorghum bicolor* was previously reported [92]. Subsequently, numerous studies have been carried out to investigate the allelopathic potential of sorghum water extracts, sorghum mulch, and sorghum as cover crops on different weeds, and the reported results indicated that mature sorghum expressed selective, species-specific and concentration-dependent allelopathic effects [67,72,101,110]. The allelopathic impacts of sorghum on different weed species are documented in Table 3.

Table 3. Allelopathic effect of *Sorghum bicolor* L. alone and in association with other crops and their phytotoxic potential.

Plant Part	Phytotoxicity (% Reduction over Control)	References
	(Sorghum Crop Residues)	
Sorghum root residues	25–50	[31]
Shoot extract	35–90	[100], [67,101] [111]
Shoot residues	42–98	[67]
Green manure, sorghum mulch	23–90	[31,92,112], [112]
Living roots	62–78	
Crop residue	29	[21]
Crop residue	35, 38, 49, 36	[32]
	23, 44	[113]
Sorghum	32, 35, 40	[32]
Sorghum	59	[114,115]
	Joint action of Sorghum + other crop residues	
Sorghum herbage	23–41; 21–41	[116]
Sorghum +Eucalyptus	13–18; 28–32	//
Sorghum + Sunflower	30–35; 24–39	//
		//
Sorghum + Sesamum	21–24; 19–24	//
Sorghum + Tobacco	10–14; 14	//
Sorghum + Brassica	21–27; 28–24	//
Sorghum + Sunflower	36–55; 42–63	//
		//
Sorghum + Sunflower + Rice	18, 10, 17	[117]
Sorghum herbage	40	[118]
	Sorghum bicolor × *Sorghum sudanese*	
Sorghum root residues	20–60	[29] [119]
Shoot extract	85–20	[120]
Shoot residues	25	[119] [92,112]
Green manure, sorghum mulch	0–30	[29,112]
Living roots	50–90	[119] [29]

3.1. Use of Sorghum Water Extracts for Weed Suppression

Sorghum allelopathy has been used to control weeds in crop rotations [31] and intercropping systems [110] and by the use of sorghum mulches [32]. Similarly, the use of sorghum water extracts has shown significant suppression of weeds [67]. Allelopathic potential of water extracts were evaluated from different sorghum parts on weeds and crops in laboratory and greenhouse experiments [99]. They revealed that the allelopathic potential of sorghum was species-specific and relied upon source and concentration. Aqueous extract of sorghum leaves stems and roots significantly decreased the germination and seedling development of *Echinochloa colona* and radishes. They concluded that stem extract induced the most prominent inhibitory impact on *E. colona*, while each of the three extracts produced a similar reaction in radishes. In another study, [99] isolated toxic compounds of sorghum, and its chemical composition was resolved as far as their hindrance of germination and seedling development of *E. colona* and radishes. All hexane, ethyl ether, methylene chloride, ethyl acetic acid derivation and aqueous fractions were checked individually, and results showed that the ethyl ether fraction had the maximum inhibitory activity on *E. colona*. Of the eight fractions separated by rapid chromatography, the fraction with the dissolvable mixes of butanol: acetic acid: water (8:1:1) had the greatest lethality to plant species, *E. colona* and radish. Liquid chromatography coupled to mass spectrometry was used to identify the toxic compounds 1 methyl-1-(2-proponyl)-hydrazine, 1-aziridineethanol, 5-chloro-2-pentanone and 2-(methylseleno)-ethanamine.

The feasibility of using aqueous extracts of allelopathic crops viz. sorghum and sunflower were investigated for weed control in wheat [101]. Spraying 100% water extracts of sorghum and sunflower after 30 days following wheat sowing diminished aggregate weed thickness up to 48% and 32% and whole weeds dry weight up to 51% and 51%, respectively. The weed biomass of *Rumex dentatus*, *Chenopodium album*, *Coronopus didymus*, and *Fumaria parviflora*, was reduced by 74%, 38%, 62% and 40%, respectively. Souza et al. (1999) evaluated the allelopathic impact of sorgoleone from sorghum root exudates upon *Phaseolus vulgaris* and *Amaranthus retroflexus*. Based on visual symptoms, *P. vulgaris* and *A. retroflexus* were the least and most susceptible species to sorgoleone, respectively. Root and shoot dry weights of *P. vulgaris* displayed an inversely proportional relationship with sorgoleone concentration. Khaliq et al. [114] sprayed sorgaab sorghum water extract that is obtained after soaking mature sorghum herbage in water for a period of one to two days for its weed control activity on soybean. Spraying of sorgaab at 25 and 45 DAS reduced the dry weight of all weeds by 20 to 42%, approximately, except that of *Trianthema portulacastrum*, which showed a yield increase of 9% over the control. Pendimethalin spray was also very effective in weed control but was more costly than sorgaab spray [114]. In another study, sorghum phytotoxicity was evaluated against various weeds in field-planted mungbean [121]. Plant dry biomass of target weeds (*Convolvulus arvensis* and *Portulaca oleracea*) decreased by about 60% and 75%, respectively, when treated with sorgaab foliar spray at 15, 30 and 45 DAS, while *Trianthema portulacastrum* remained unaffected [121]. Sorgaab reduced the weed thickness and dry weight by 32–62% and 47–75%, respectively, compared to the control, in raya crop [102]. Cheema et al. [121] conducted a field trial to observe the feasibility of sorghum allelopathy against the weed in traditional cotton. Sorgaab sprays decreased the total weed density by 13–54% and biomass by 87%. Cheema et al. [67] compared the concentration and frequency of sorgaab applications with hand weeding and chemical herbicide for controlling weeds in flooded wheat in a semi-arid district of Punjab. The dry weight and thickness of weeds were controlled by using sorgaab up to 35–49% and 22–46%, respectively, corresponding an increase in grain yield by 10–21%. Two foliar sprays of 10% sorgaab at 30 and 60 DAS were used to control the weeds in wheat with maximum yield. Chemical weedicides and the hand weeding technique were found to be wasteful for weed control because of higher costs in both cases.

Ahmad et al. [122] assessed the allelopathic potential of sorgaab as natural weed control in maize. Spraying of sorgaab suppressed the total weed density by 34–57% and horse purslane (*Trianthema portulacastrum*) density by 24–40%; the total dry weight reduction ranged from 13 to 34%, and that of horse purslane from 12 to 34%. In an independent study, Cheema et al. [21] used sorghum aqueous extracts as a foliar treatment against some winter weeds in four wheat varieties. One (30 DAS) and two (30 and 60 DAS) foliar applications of SWE impacted negatively the thickness and biomass of many weed species, such as *Chenopodium album*, *Phalaris minor*, *Avena fatua*, *Convolvulus arvensis*, and *Rumex dentatus*. On the other hand, the growth and density of *Melilotus parviflora* were improved. The obtained results showed that total biomass and weed thickness were significantly decreased. The Parwaz-94 variety was the most receptive to the aqueous sorghum extracts, showing the greatest increase in grain yield. The compound substances discharged by the plant deposits left on the dirt surface act uniquely in contrast to those released by the fused plant buildups.

3.2. Use of Sorghum Residues/Mulches for Weed Suppression

Weed growth could be suppressed by growing sorghum crops because the sorghum residues present on the soil surface release different allelochemicals which suppress the weed germination and seedling development [96]. The chemical substances released by the plant residues left on the soil surface respond differently than those released by plant residues incorporated into the soil. In the former case, they might be concentrated on the soil surface while, in the latter, the allelochemicals were diluted into the soil, following soil incorporation. Since the intensity of the allelopathic effect depends on the concentration of allelochemicals, their action is more intense on the soil surface under mulch [96]. On the other hand, when the release of these products is slower, the effects can be noticed for a more extended period. The higher the amount of plant material used for mulch, the greater is the total amount of allelochemicals present in the mulch and released, leading to a higher concentration of allelochemicals into the soil [123].

Allelopathic cover crops have been extensively used to inhibit weeds in organic agriculture. In this context, sorghum crop mulch and crop residues could contribute exceptionally to weed control [91]. Sorghum and sudan grass used as mulch resulted in reductions of weed biomass by approximately 90% and 85%, respectively. These authors concluded that the sorghum residues or mulches were allelopathic and could provide excellent suppression of several annual weeds. In another study, it was revealed that wheat, barley, oat, rye, sorghum and sudan grass mulch were very effective in the suppression of several weed species [92]. Seedling growth and biomass of purslane and smooth crab grass significantly decreased by 70% and 80%, respectively, following treatment with sorghum mulch. The residues of sorghum and sudan grass completely inhibited smooth grass seed development for 60 days, whereas wheat, oat, barley and rye residues likewise reduced the aggregate weed biomass up to 75%, and also the early season weed development. In a field trial, Cheema and Ahmad [72] demonstrated that the combination of whole sorghum plants or different sorghum parts, separated or blended, generally suppressed the growth of weeds, except for *Melilotus parviflora*, which was promoted. In situ integration of sorghum roots reduced the dry weight of other weeds by 26 – 49%. The sorghum's allelopathic effects relied upon the phase of sorghum integration, the quantity of sorghum mass incorporated into the soil and its developmental stage. These experiments showed that sorghum residues could be adequately used to manage some of the weeds in wheat fields.

The sorghum residues incorporation into the soil as surface mulch at 0, 0.5, 1.0, 2.0 and 3.0% w/v, showed that the efficacy of sorghum allelochemicals was species-specific and depended upon the source and concentration [99]. The sorghum stem residue considerably restricted the seed germination of *E. colona* and radishes, but not that of rice. The crop residues (maize, proso millet, safflower, grain sorghum and winter wheat) incorporation into soil inhibited the seedling growth in goat grass (*Aegilops triuncialis*), and to a lesser extent in winter wheat [124]. For instance, the residue of sorghum grain decreased seedling development of goat grass by 78% and that of winter wheat by 50%. The sorghum stem deposits impressively limited seed germination of *E. colona* and radishes, yet not that of rice. The chopped residues of four crops (sunflower, sorghum, rice and wheat) was incorporated in cotton fields at 5, 7.5 and 10 t ha^{-1} each [68]. Maximum reduction in weed population (ca. 52%) was observed in plots where wheat residue was applied at 5.0 t ha^{-1}. This was followed by wheat (7.5 t ha^{-1}), rice (7.5 t ha^{-1}) and sorghum (10.0 t ha^{-1}), with a reduction in the weed population with respect to the controls of about 40%, in all cases. Regarding dry weed biomass, the maximum reduction was observed in plots receiving sorghum crop residues at 10.0 t ha^{-1}, amounting to 45.3% less than in the control.

Narwal et al. [70] observed the following order of weed suppression: pearl millet > maize > sorghum > cluster bean > cowpea. The residual suppression effect on weeds even persisted in the next crop. The sorghum herbage (applied as surface mulch at 10 and 15 t ha^{-1}) in mungbean fields showed a significant reduction in the dry weight of *Trianthema partulacastrum* by 14 to 20% and 18 to 45%, respectively [32]; on the other hand, the reduction in thickness and dry weight of other weeds (*Cyperus rotundus*, *C. arvensis* and

P. oleracea) was in the range of 52–68% and 60–77%, respectively. Cheema and Khaliq [67] studied the efficacy of allelochemicals of sorghum stalk integrated into the soil on rabi weeds and wheat crop. Mature sorghum chopped herbage (2 to 6 Mg ha^{-1}) caused the reduction of weed dry weight by 20–41% and 42–56%, respectively, and an increase in wheat grain yield by 6 to 17%. In another study, [32] conducted a field trial to check the potential of sorghum allelochemicals to control the weeds in desi cotton, showing that sorghum mulching (3.5, 7.0, 10.5 t ha^{-1}) suppressed the cumulative density of weeds by 23–62%, whereas 52–70% and 54–64% reductions were noted by using chemical treatment and hand weeding, respectively. The reduction in weed biomass under sorghum mulching was up to 56%.

3.3. *Effect of Sorghum in Crop Rotation*

Inclusion of sorghum in a rotation can help to control weeds through secretion of allelochemicals, which ultimately suppress the weeds. In a field trial in Nebraska, grain sorghum reduced the weed density, biomass and seedling growth in soybeans or maize [125]. In areas where sorghum has been included in the cropping system, weed infestation was constantly lower after few years with arrangements of four lines of grain sorghum with soybeans or maize [125]. Sorghum residues regularly delayed the growth of wheat crop; however, they did not influence yields, most likely due to the degradation of the allelochemicals in the soil over time [31]. No-till sorghum stover had little impact on stand establishment, yet every row decreased the yields of wheat grains, potentially on the grounds that allelochemicals drained gradually. In the rice-wheat crop rotation system, grain sorghum was cultivated before the rice planting. It was observed that this rotation with sorghum reduced the weed density in the succeeding rice crop with less herbicide application [69]. Likewise, the winter weeds may be controlled due to wheat replacement by oat and berseem clover (*Trifolium alexandrinum* L.).

3.4. *Intercropping of Sorghum*

Intercropping is a typical cultivation framework amongst livestock farmers of the emerging world. The main aim behind mixing harvests or planting in an adjacent grouping is to amplify, utilize and lessen the danger of crop disappointment. Intercropping maintains soil ripeness, reduces disintegration and may decrease insect harms. It has been guaranteed that one purpose behind this is the destruction of weeds [126]. Intercropping efficiency for weed control relies on the species consolidated, their relative extents and plant geometry in the field [127]. The output of intercropping frameworks can be decreased or improved, relying upon the inhibitory or stimulatory impacts of crops but ensuring that the other resources, such as light, nutrients, water and space are not limited [128]. In intercropping frameworks, the development and yield of segment crops increase because of more prominent supplement retention or better weed control than in harvests, yet the underlying mechanisms are not completely understood. Root exudates play a noteworthy part in the efficiency of crop mixtures as they may enhance crop development and yield of component crops through enhanced ion exchange, greater nutrient uptake and partial weed control, compared with pure crops [129,130].

3.4.1. Allelochemicals Biosynthesis and Abiotic Stress Resistance

Plants as being sessile grow under natural environmental conditions where so many factors are involved for their nurturing. Therefore, any deviation from their required growth conditions at different growth stages exerts pressure [131]. Abiotic stresses are environmental adversities that negatively influence the plant growth and cellular functioning [132]. Abiotic stresses are the major hurdles in sustainable agriculture development. Currently, it is the main challenge for maintaining plant growth and crop productivity under such stress scenario for sustainable agriculture. All these environmental factors alone or in combination disrupt plant functions. Abiotic stresses are the chief cause of deprived yield and crop failure of sorghum [133]. Drought is one of the main abiotic stresses that is increasing at a rapid rate. A plant experiences drought once during its growth stage or throughout its life in certain regions. While living in the same biota, plants compete within their species and with species of other plant communities for nutrients and space. Allelochemicals are produced as tools for survival under these conditions. Survival of sorghum is difficult due to decline in available water resources, and there is a great need to adapt new strategies to grapple these stress factors. Plants produce phenolic acids in response to stress that work as osmoprotectants and antioxidants to scavenge oxidative stress [134]. Alteration in phenolic concentration is indispensable for plant survival. The exogenous application of phenolic acids helps plants in coping with harsh environmental conditions [135]. Moreover, phenolic acid is naturally a part of the allelochemicals that plants produce in high concentration with fluctuating environmental conditions [136]. Currently, studies are being carried out to observe the beneficial concentration of allelochemicals (phenolic acids) for the survival of plants and to protect them from environmental adversities. The residues of sorghum crop were used to extract water that resulted in inhibition of the germination and growth of the surrounding plants. This reduction in growth was due to phenolic acids, which are the characteristic feature of sorghum allelopathy [113]. Additionally, allelopathic sorghum was manipulated for the suppression of weed growth in wheat. The allelopathic plant extract (sorgaab) from sorghum was analyzed and it revealed higher concentrations of phenolic acids [137]. It has been reported that these phenolic compounds are among the plant secondary metabolites that are effective for abiotic stress tolerance in plants. Multifarious strategies can be adopted to cope with abiotic stresses, but sorgaab extraction from sorghum leaves proved to be efficient for minimizing the influence of adverse environmental factors [115].

3.4.2. Production of Allelochemicals in Response to Abiotic Stresses

Allelochemicals have the potential to suppress the growth of weeds by disrupting water relations of plants in the root cell membrane. Additionally, they also result in biochemical changes for the alleviation of oxidative stress after their exposure to abiotic stresses [138]. Regardless of their benefits to cope with abiotic stresses, allelochemicals have not been given proper attention to explore their benefits to cope with environmental stresses [139]. Previous studies have explored potential groups of allelochemicals that confer stress tolerance. These studies assisted in bridging the gap of the positive role of allelochemicals that can exploited for stress resistance in sorghum [140]. However, the concentration of allelochemicals generally varies, as they are produced differentially during different growth stages, likewise the sensitivity of the plant against abiotic stresses also varies [141]. Allelochemicals that are produced in high concentrations in response to abiotic stresses include terpenoids and phenolic acids [142]. The synthesis of sorgoleone, dhurrin, and kinetin occurs in root, stem and leaves of sorghum and work as a first line of defense to alleviate abiotic stresses [143,144]. Higher accumulation of phenolic acids is positively correlated for abiotic stress tolerance of sorghum [137,145]. Allelochemicals are known to alleviate abiotic stresses. Numerous stress conditions alter the levels and synthesis of allelochemicals [146]. Fluctuations in temperature, decreased availability of water, and nutrient stress are the main environmental factors influencing the allelopathy. Additionally, herbicidal applications and heavy metals are also reported for differential regulation of allelochemicals [147]. Meanwhile, climatic factors also influence the synthesis of allelochemicals. It was reported that root growth of sorghum was influenced due to fluctuation in temperature, as optimum root growth goes along with sorgoleone production. The increase in temperature causes heat stress conditions that ultimately suppress the sorgoleone production in sorghum. The plants growing near the sorghum exert competition stresses that intensify the influence of abiotic stresses, resulting in decreased sorgoleone production [10].

3.4.3. Stress Signaling by Allelochemicals in Sorghum

Abiotic stresses influence the transcriptional regulation of the allelochemicals. Sorghum growing under natural growth conditions is directly influenced by environmental stresses. Plants sense stress conditions and send signals to activate various molecular mechanisms in cells that resultantly cause physio-biochemical changes in plants to adapt to changed environmental conditions [148]. Plants even have the potential to send signals to neighboring plants with excessive production of allelochemicals under certain conditions [149]. Plants respond to stress signals by perceiving external harsh conditions and transmit between plant cells. The release of various type of allelochemicals such as soluble chemicals or volatile organic compounds helps in the regulation of soil microbes that confers a beneficial role by changing physio-chemical properties of the surroundings in the soil, which assist in inhibiting the growth of the competitor plants. As plants send signals to the neighboring plants, likewise, they also perceive beneficial signals from neighbors, which includes plant volatiles [150]. Abiotic stress signals in plants are perceived by increased levels of abscisic acid (ABA), calcium, and reactive oxygen species (ROS) that are commonly involved for some other pathways as well. Thereby, the response of allelopathic chemicals toward environmental pressure is likely to be related to elevated levels of ABA, calcium or ROS in plants [151].

Sorghum perceives environmental stresses and transmits a signal to the nucleus through complex cellular signaling networks that involves secondary messengers, i.e., calcium-associated proteins, reactive oxygen intermediates (ROIs), and mitogen-activated protein kinase (MAPK) cascades. The signaling network activates several transcriptional pathways that results in regulation of stress related genes resulting in physio-biochemical changes to protect the cellular membrane of plants [151,152].

Allelochemicals modify the mitogen-activated protein kinases (MAPK), which is the main enzyme for ethylene production [153]. The synthesis of allelochemicals occurs with the intervention of antioxidant enzymes that further assist in scavenging oxidative stress [154]. Moreover, these allelochemicals can trigger the gene expression pattern of root meristematic tissues that eventually assist root growth functions under stressful environments [83]. The literature shows great work has been done to understand the abiotic response of sorghum at molecular levels, but less work has been done to reveal the molecular basis of allelochemicals for conferring stress tolerance in sorghum.

3.4.4. Genetic Factors Responsible for Sorgoleone Production

Natural products from plants offer a broad array of molecules with great diversity in their structure, biological activity and toxicologically, that can be used for managing weeds. The sorgoleone has been studied thoroughly [154–156]. Firstly, it was discovered during studying secondary metabolites that influenced the germination of witchweed [155]. It was noticed that allelochemicals can be absorbed by growing seedlings via hypocotyl and cotyledon, resulting in hindering the photosynthesis process. The sorgoleone sustain in soil for longer period than herbicides. Currently, studies are being done to identify the QTLs to enhance the production of sorgoleone in sorghum. Numerous studies have explored the biosynthetic pathway involved to produce sorgoleone [156]. Identification of genes controlling the production of allelochemicals would help in improving our knowledge regarding their synthesis pathways, release mechanisms into the soil rhizosphere, and corresponding phytotoxicity against different weeds. Genetic mechanisms responsible for the allelopathic effect of sorghum as a biological weed control are a new challenge, and fewer studies have focused on genetic factors. Recently, one of the studies by Shehzad et al. [157] highlighted that sorgoleone is not only a phenolic compound that contains allelochemical characteristics, but it also synthesizes other chemicals for the inhibition of the growth of neighboring plants. The SOR1 gene is responsible for sorgoleone production; it was reported that its higher transcript levels were observed from different root, stem and leaves of sorghum [43]. It was further confirmed from another study that showed that the higher expression of SOR1 resulted in weed suppression, and additionally the intercropping of sorghum and wheat exhibited no deleterious effect on cotton [82].

4. Conclusions and Future Perspectives

The study of the regulation of sorgoleone production by sorghum root hairs can increase the possibilities of employing sorghum as mulch or cover crop for effective management of germinating weed seedlings. The effect of sorgoleone resembles pre-emergent soil herbicides such as pendimethalin. Several researchers have proposed using a systematic approach employing candidate crops with better secondary metabolite profiles, and different agronomic techniques for better weed management under field settings. The phytotoxicity of sorghum and allelopathic interference has been elaborated under laboratory, greenhouse and in field trials. The present review also highlights the allelochemicals production under abiotic stresses, stress signaling by allelochemicals, and genetic factors responsible for sorgoleone production in sorghum. Different multidisciplinary approaches that incorporate sorghum crops for strategic weed control might be an alternative with great potential, using secondary metabolites that can also serve as lead compounds for herbicide discovery programs. These approaches should ideally have to be focused on weed control by employing agro-ecological and agronomic practices for better suppressing weeds at pre- and post-emergence stages, representing an alternative to genetically modified crops, which are considered by many (at least in the EU) as possibly harmful to the ecosystem and environment.

Author Contributions: All authors have contributed equally to the preparation of this review article. All authors have read and agreed to the published version of the manuscript.

Funding: This research received no external funding.

Data Availability Statement: It was a review article. All the data has been mentioned in the manuscript.

Acknowledgments: This work was not supported by any external funding.

Conflicts of Interest: The authors declare that they have no conflict of interest.

References

1. Sarić-Krsmanović, M.; Gajić Umiljendić, J.; Radivojević, L.; Šantrić, L.; Potočnik, I.; Đurović-Pejčev, R. Bio-herbicidal effects of five essential oils on germination and early seedling growth of velvetleaf (*Abutilon theophrasti* Medik.). *J. Environ. Sci. Health B* **2019**, *54*, 247–251. [CrossRef] [PubMed]
2. Macías, A.F.; Mejías, F.J.R.; Molinillo, J.M.G. Recent advances in allelopathy for weed control: From knowledge to applications *Pest. Manag. Sci.* **2019**, *75*, 2413–2436.
3. Ali, H.H.; Peerzada, A.M.; Hanif, Z.; Hashim, S.; Chauhan, B.S. Weed management using crop competition in Pakistan: A review. *Crop. Prot.* **2017**, *95*, 22–30. [CrossRef]
4. Farooq, N.; Abbas, T.; Tanveer, A.; Jabran, K. Allelopathy for weed management. In *Coevolution of Secondary Metabolites*; Springer: Berlin/Heidleberg, Germany, 2020; pp. 505–519.
5. Bo, A.B.; Khaitov, B.; Umurzokov, M.; Cho, K.M.; Park, K.W.; Choi, J.S. Biological control using plant pathogens in weed management. *Weed Turfgrass Sci.* **2020**, *9*, 11–19.
6. Hicks, H.L.; Comont, D.; Coutts, S.R. The factors driving evolved herbicide resistance at a national scale. *Nat. Ecol. Evol.* **2018**, *2*, 529–536. [CrossRef] [PubMed]
7. Bàrberi, P. Ecological weed management in sub-Saharan Africa: Prospects and implications on other agroecosystem services. *Adv. Agron.* **2019**, *156*, 219–264.
8. Reganold, J.P.; Wachter, J.M. Organic agriculture in the twenty-first century. *Nat. Plants* **2016**, *2*, 15221. [CrossRef] [PubMed]
9. Farooq, M.; Jabran, K.; Cheema, Z.A.; Wahid, A.; Siddique, K.H.M. The role of allelopathy in agricultural pest management. *Pest. Manag. Sci.* **2011**, *67*, 493–506. [CrossRef]
10. Dayan, F.E.; Howell, J.L.; Weidenhamer, J.D. Dynamic root exudation of sorgoleone and its in planta mechanism of action. *J. Exp. Bot.* **2009**, *60*, 2107–2117. [CrossRef]
11. Hussain, M.I.; Reigosa, M.J. Evaluation of herbicide potential of sesquiterpene lactone and flavonoid: Impact on germination, seedling growth indices and root length in *Arabidopsis thaliana*. *Pak. J. Bot.* **2014**, *46*, 995–1000.
12. Hussain, M.I.; Reigosa, M.J. Evaluation of photosynthetic performance and carbon isotope discrimination in perennial ryegrass (*Lolium perenne* L.) under allelochemicals stress. *Ecotoxicology* **2017**, *26*, 613–624. [CrossRef] [PubMed]
13. Rice, E.L. *Allelopathy*, 2nd ed.; Academic Press: New York, NY, USA, 1984.

14. Rose, S.J.; Burnside, O.C.; Specht, J.E.; Swisher, B.A. Competition and allelopathy between soybeans and weeds. *Agron. J.* **1984**, *76*, 523–528. [CrossRef]
15. Leather, G.R. Weed control using allelopathic sunflowers and herbicide. *Plant Soil* **1987**, *98*, 17–23. [CrossRef]
16. Cheema, Z.A.; Ahmad, S.; Gill, M.A.; Amin, M. Effect of different weed control practices on cotton weeds and its yield. *Pak. J. Agric. Sci.* **1990**, *27*, 386–393.
17. Chung, I.-M.; Miller, D.A. Allelopathic influence of nine forage grass extracts on germination and seedling growth of alfalfa. *Agron. J.* **1995**, *87*, 767–772. [CrossRef]
18. Buhler, D.D.; Mester, T.C.; Kohler, K.A. The effect of maize residues and tillage on emergence of *Setaria faberi, Abutilon theophrasti, Amaranthus retroflexus* and *Chenopodium album*. *Weed Res.* **1996**, *36*, 153–165. [CrossRef]
19. Chandrasekhar, C.N.; Manian, K.; Kandasamy, O.S. Studies on allelopathic potential of sesame haulm on physiology and growth of purple nutsedge (*Cyperus rotundus*) and bermuda grass (*Cynodon dactylon*). *Ind. J. Weed Sci.* **1998**, *30*, 94–95.
20. Ahn, J.K.; Chung, I.M. Allelopathic potential of rice hulls on germination and seedling growth of barnyard grass. *Agron. J.* **2000**, *11*, 62–67.
21. Cheema, Z.A.; Iqbal, M.; Ahmad, R. Response of wheat varieties and some rabi weeds to allelopathic effects of sorghum water extract. *Int. J. Agri. Biol.* **2002**, *4*, 52–55.
22. Uddin, M.R.; Park, K.W.; Kim, Y.K.; Park, S.U.; Pyon, J.Y. Enhancing sorgoleone levels in grain sorghum root exudates. *J. Chem. Ecol.* **2010**, *36*, 914–922. [CrossRef]
23. Razzaq, A.; Cheema, Z.A.; Jabran, K.; Hussain, M.; Farooq, M.; Zafar, M. Reduced herbicide doses used together with allelopathic sorghum and sunflower water extracts for weed control in wheat. *J. Plant. Prot. Res.* **2012**, *52*, 281–285. [CrossRef]
24. Cheema, Z.A. Weed Control in Wheat through Sorghum Allelochemicals. Ph.D. Thesis, Department of Agronomy, University of Agriculture, Faisalabad, Pakistan, 1988.
25. Cherney, D.J.; Patterson, J.A.; Cherney, J.H.; Axtell, J.D. Fibro and soluble phenolic monomer composition of morphological components of sorghum stover. *J. Sci. Food Agric.* **1991**, *54*, 645–649. [CrossRef]
26. Nicollier, G.F.; Pope, D.F.; Thompson, A.C. Biological activity of dhurrin and other compounds from johnson grass (*Sorghum helepense*). *J. Agri. Food Chem.* **1983**, *31*, 748–751. [CrossRef]
27. Burgos-Leon, W.; Gaury, F.; Nicou, R.; Chopart, T.L.; Dommergues, Y. Etudes et travaux:un cas de fatigue des sols induite par la culture du sorgho. *Agron. Trop.* **1980**, *35*, 319–334.
28. Dayan, F.E. Factors modulating the levels of the allelochemical sorgoleone in *Sorghum bicolor*. *Planta* **2006**, *224*, 339–346. [CrossRef] [PubMed]
29. Weston, L.A.; Czarnota, M.A. Activity and persistence of sorgoleone, a long-chain hydroquinone produced by *Sorghum bicolor*. *J. Crop. Prod.* **2001**, *4*, 363–377. [CrossRef]
30. Jabran, K. *Manipulation of Allelopathic Crops for Weed Control*; Springer International Publishing: Cham, Switzerland, 2017.
31. Roth, C.M.; James, P.S.; Gary, M.P. Allelopathy of sorghum on wheat under several tillage systems. *Agron. J.* **2000**, *92*, 855–860. [CrossRef]
32. Cheema, Z.A.; Asim, M.; Khaliq, A. Sorghum allelopathy for weed control in cotton (*Gossypium arboreum* L.). *Int. J. Agri. Biol.* **2000**, *2*, 37–41.
33. Hussain, M.I.; González, L.; Reigosa, M.J. Germination and growth response of four plant species towards different allelochemicals and herbicides. *Allelop. J.* **2008**, *22*, 101–110.
34. Hussain, M.I.; Reigosa, M.J. Higher peroxidase activity, leaf nutrient contents and carbon isotope composition changes in *Arabidopsis thaliana* are related to rutin stress. *J. Plant Physiol.* **2014**, *171*, 1325–1333. [CrossRef]
35. Hussain, M.I.; González, L.; Reigosa, M.J. Allelopathic potential of *Acacia melanoxylon* R. Br. on the germination and root growth of native species. *Weed Biol. Manag.* **2011**, *11*, 18–28. [CrossRef]
36. Hussain, M.I.; Reigosa, M.J. Secondary metabolites, ferulic acid and *p*-hydroxybenzoic acid induced toxic effects on photosynthetic process in *Rumex acetosa* L. *Biomolecules* **2021**, *11*, 233. [CrossRef]
37. Hussain, M.I.; Reigosa, M.J. Seedling growth, leaf water status and signature of stable carbon isotopes in C_3 perennials exposed to natural phytochemicals. *Aust. J. Bot.* **2012**, *60*, 676–684. [CrossRef]
38. Guenzi, W.D.; McCalla, T.M. Phenolic acids in oats, wheat, sorghum and corn residues and their phytotoxicity. *Agron. J.* **1966**, *58*, 303–304. [CrossRef]
39. Cheema, Z.A. Sorghum allelopathy-A new weed control technology for enhancing wheat productivity. *J. Anim. Plant. Sci.* **1998**, *8*, 19–21.
40. Einhellig, F.A. Mechanism of action of allelochemicals in allelopathy. In *Allelopathy: Organisms, Processes and Applications*; Inderjit, Dakshini, K.M.M., Einhellig, F.A., Eds.; American Chemical Society: Washington, DC, USA, 1995; pp. 96–116.
41. Gonzalez, V.M.J.; Kazimir, C.I.; Nimbal, L.A.; Weston, G.M. Cheniae. Inhibition of a photosystem II electron transfer reaction by the natural product sogoleone. *J. Agric. Food Chem.* **1997**, *45*, 1415–1421. [CrossRef]
42. Kim, S.Y.; DeDatta, S.K.; Robles, R.P.; Kim, K.U.; Lee, S.C.; Shin, D.H. Allelopathic effects of sorghum extract and residues on selected crops and weeds. *Korean J. Weed Sci.* **1994**, *14*, 34–41.
43. Yang, X.; Scheffler, B.E.; Weston, L.A. SOR1, a gene associated with bioherbicide production in sorghum root hairs. *J. Exp. Bot.* **2004**, *55*, 2251–2259. [CrossRef] [PubMed]

44. Barkosky, R.R.; Einhellig, F.A. Allelopathic interference of plant-water relationships by para-hydroxybenzoic acid. *Bot. Bull. Acad. Sin.* **2003**, *44*, 53–58.
45. Sanchez-Moreiras, A.M.; Martinez, A.; Gonzalez, L.; Pellisier, F.; Regiosa, M.J. *Mode of Action of the Hydoxamic Acid BOA and Other Related Compounds*; CRC Press: Boca Raton, FL, USA, 2004; pp. 239–252.
46. Anaya, A.L.; Waller, G.R.; Owuor, P.O.; Friedman, J.; Chou, C.H.; Suzuki, T.; Arroyo-Estrada, J.F.; Cruz-Ortega, R. The role of caffeine in the production decline due to autotoxicity in coffee and tea plantations. In *Allelopathy from molecules to ecosystems*; Reigosa, M., Pedrol, N., Eds.; Science Publishers: Enfield, UK, 2002; pp. 71–92.
47. Barkosky, R.R.; Einhellig, F.A.; Butler, J.L. Caffeic acid-induced changes in plant–water relationships and photosynthesis in leafy spurge *Euphorbia esula*. *J. Chem. Ecol.* **2000**, *26*, 2095–2109. [CrossRef]
48. Friebe, A.; Roth, U.; Kück, P.; Schnabl, H.; Schulz, M. Effects of 2, 4-dihydroxy-1, 4-benzoxazin-3-ones on the activity of plasma membrane H+-ATPase. *Phytochemistry* **1997**, *44*, 979–983. [CrossRef]
49. Singh, H.P.; Batish, D.R.; Setia, N.; Kohli, R.K. Herbicidal activity of volatile oils from *Eucalyptus citriodora* against *Parthenium hysterophorus*. *Ann. App. Biol.* **2005**, *146*, 89–94. [CrossRef]
50. Kato-Noguchi, H.; Ino, T. Possible involvement of momilactone B in rice allelopathy. *J. Plant. Physiol.* **2005**, *162*, 718–721. [CrossRef] [PubMed]
51. Barkosky, R.R.; Einhellig, F.A. Effects of salicylic acid on plant-water relationships. *J. Chem. Ecol.* **1993**, *19*, 237–247. [CrossRef]
52. Baziramakenga, R.; Leroux, G.D.; Simard, R.R. Effects of benzoic and cinnamic acids on membrane permeability of soybean roots. *J. Chem. Ecol.* **1995**, *21*, 1271–1285. [CrossRef]
53. Hussain, M.I.; González, L.; Reigosa, M.J. Phytotoxic effect of allelochemicals and herbicides on photosynthesis, growth and carbon isotope discrimination in *Lactuca sativa*. *Allelop. J.* **2010**, *26*, 157–174.
54. Hussain, M.I.; González, L.; Souto, C.; Reigosa, M.J. Ecophysiological responses of native plants to phytotoxic effect of *Acacia melanoxylon* R. Br. *Agrofor. Syst.* **2011**, *83*, 149–166. [CrossRef]
55. Perez, F.J. Allelopathic effect of hydroxamic acids from cereals on *Avena sativa* and *A. fatua*. *Phytochemistry* **1990**, *29*, 773–776. [CrossRef]
56. Hussain, M.I.; Reigosa, M.J. Characterization of xanthophyll pigments, photosystem II photochemistry, heat energy dissipation, reactive oxygen species generation and carbon isotope discrimination during artemisinin-induced stress in *Arabidopsis thaliana*. *PLoS ONE* **2015**, *10*, e0114826. [CrossRef]
57. Mersie, W.; Singh, M. Phenolic acids affect photosynthesis and protein synthesis by isolated leaf cells of velvet-leaf. *J. Chem. Ecol.* **1993**, *19*, 1293–1301. [CrossRef] [PubMed]
58. Jabran, K.; Mahajan, G.; Sardana, V.; Chauhan, B.S. Allelopathy for weed control in agricultural systems. *Crop. Prot.* **2015**, *72*, 57–65. [CrossRef]
59. Tibugari, H.; Chiduza, C.; Mashingaidze, A.B.; Mabasa, S. High sorgoleone autotoxicity in sorghum (*Sorghum bicolor* (L.) Moench) varieties that produce high sorgoleone content. *S. Afr. J. Plant Soil.* **2020**, *37*, 160–167. [CrossRef]
60. Besançon, T.E.; Dayan, F.E.; Gannon, T.W.; Everman, W.J. Conservation and divergence in sorgoleone production of sorghum species. *J. Environ. Qual.* **2020**, *49*, 368–377. [CrossRef]
61. Dayan, F.E.; Rimando, A.M.; Pan, Z.; Baerson, S.R.; Gimsing, A.L.; Duke, S.O. Sorgoleone. *Phytochemistry* **2010**, *71*, 1032–1039. [CrossRef] [PubMed]
62. Ashraf, R.; Sultana, B.; Yaqoob, S.; Iqbal, M. Allelochemicals and crop management: A review. *Curr. Sci.* **2017**, *3*, 1–13.
63. Jesudas, P.A.; Kingsley, S.J.; Ignacimuthu, S. Sorgoleone from *Sorghum bicolor* as a potent bioherbicide. *Res. J. Recent Sci.* **2014**, *3*, 32–36.
64. Uddin, M.R.; Park, K.W.; Pyon, J.Y.; Park, S.U. Combined herbicidal effect of two natural products (sorgoleone and hairy root extract of tartary buckwheat) on crops and weeds. *Aust. J. Crop. Sci.* **2013**, *7*, 227.
65. Farooq, M.; Nawaz, A.; Ahmad, E.; Nadeem, F.; Hussain, M.; Siddique, K.H. Using sorghum to suppress weeds in dry seeded aerobic and puddled transplanted rice. *Field Crop. Res.* **2017**, *214*, 211–218. [CrossRef]
66. Farooq, M.; Khan, I.; Nawaz, A.; Cheema, M.A.; Siddique, K.H. Using sorghum to suppress weeds in autumn planted maize. *Crop. Prot.* **2020**, *133*, 105162. [CrossRef]
67. Cheema, Z.A.; Khaliq, A. Use of sorghum allelopathic properties to control weeds in irrigated wheat in a semi-arid region of Punjab. *Agri. Ecosyst. Environ.* **2000**, *79*, 105–112. [CrossRef]
68. Ahmad, S.; Rehman, A.; Cheema, Z.A.; Tanveer, A.; Khaliq, A. Evaluation of some crop residues for their allelopathic effects on germination and growth of cotton and cotton weeds. In Proceedings of the 4th Pakistan Weed Science Conference Faisalabad, Faisalabad, Pakistan, 26–27 March 1994; 1995; pp. 63–71.
69. Narwal, S.S. Weed management in rice: Wheat rotation by allelopathy. *Crit. Rev. Plant Sci.* **2000**, *19*, 249–266. [CrossRef]
70. Narwal, S.S.; Sarmah, M.K. Effect of wheat residues and forage crops on the germination and growth of weeds. *Allelop. J.* **1996**, *3*, 229–240.
71. Kondap, S.M.; Rao, A.R.; Reddy, G.V. Studies on the effect of planting patterns and weeding intervals in sorghum based intercropping system on weed infestation and yield. *Madras Agri. J.* **1990**, *77*, 64–69.
72. Cheema, Z.A.; Ahmad, S. Allelopathy: A potential tool for weed management. In Proceedings of the National Seminar on Role of Plant Health and Care in Agriculture Production, Faisalabad, Pakistan, 28–29 December 1988.

73. Cheema, Z.A.; Khaliq, A.; Saeed, S. Weed control in maize (*Zea mays* L.) through sorghum allelopathy. *J. Sust. Agric.* **2004**, *23*, 73–86. [CrossRef]
74. Uddin, M.R.; Park, S.U.; Dayan, F.E.; Pyon, J.Y. Herbicidal activity of formulated sorgoleone, a natural product of sorghum root exudate. *Pest. Manag. Sci.* **2014**, *70*, 252–257. [CrossRef] [PubMed]
75. Naby, K.Y.; Ali, K.A. Allelopathic potential of *Sorghum bicolor* L. root exudates on growth and chlorophyll content of wheat and some grassy weeds. In Proceedings of the International Collaborative Conference of Modern Agricultural Technologies, Erbil, Iraq, 24–25 March 2021; Volume 761, p. 012085.
76. Uddin, M.R.; Thwe, A.A.; Kim, Y.B.; Park, W.T.; Chae, S.C.; Park, S.U. Effects of jasmonates on sorgoleone accumulation and expression of genes for sorgoleone biosynthesis in sorghum roots. *J. Chem. Ecol.* **2013**, *39*, 712–722. [CrossRef] [PubMed]
77. Weston, L.A.; Alsaadawi, I.S.; Baerson, S.R. Sorghum allelopathy from ecosystem to molecule. *J. Chem. Ecol.* **2013**, *39*, 142–153. [CrossRef]
78. Tani, E.; Abraham, E.; Chachalis, D.; Travlos, I. Molecular, genetic and agronomic approaches to utilizing pulses as cover crops and green manure into cropping systems. *Int. J. Mol. Sci.* **2017**, *18*, 1202. [CrossRef]
79. Lucas, S.T. Sustainable Suppression of Weeds through Ecological Use of Cover Crops. In *Cover Crops and Sustainable Agriculture*; CRC Press: Boca Raton, FL, USA, 2021; pp. 69–83.
80. Al-Bedairy, N.R.; Alsaadawi, I.S.; Shati, R.K. Combining effect of allelopathic *Sorghum bicolor* L. (Moench) cultivars with planting densities on companion weeds. *Arch. Agron. Soil Sci.* **2013**, *59*, 955–961. [CrossRef]
81. Tibugari, H.; Chiduza, C.; Mashingaidze, A.B.; Mabasa, S. Quantification of sorgoleone in sorghum accessions from eight southern African countries. *S. Afr. J. Plant Soil* **2019**, *36*, 41–50. [CrossRef]
82. Santos, R.C.D.; Ferraz, G.D.M.G.; Albuquerque, M.B.D.; Lima, L.M.D.; Melo Filho, P.D.A.; Ramos, A.D.R. Temporal expression of the sor1 gene and inhibitory effects of *Sorghum bicolor* L. Moench on three weed species. *Acta Bot. Bras.* **2014**, *28*, 361–366. [CrossRef]
83. Alsaadawi, I.S.; Dayan, F.E. Potentials and prospects of sorghum allelopathy in agroecosystems. *Allelop. J.* **2009**, *2*, 24.
84. Głąb, L.; Sowiński, J.; Bough, R.; Dayan, F.E. Allelopathic potential of sorghum (*Sorghum bicolor* (L.) Moench) in weed control: A comprehensive review. *Adv. Agron.* **2017**, *145*, 43–95.
85. Tesfamariam, T.; Yoshinaga, H.; Deshpande, S.P.; Rao, P.S.; Sahrawat, K.L.; Ando, Y.; Nakahara, K.; Hash, C.T.; Subbarao, G.V. Biological nitrification inhibition in sorghum: The role of sorgoleone production. *Plant Soil* **2014**, *379*, 325–335. [CrossRef]
86. Pan, Z.; Baerson, S.R.; Wang, M.; Bajs Hirschel, J.; Rimando, A.M.; Wang, X.; Nanayakkara, N.D.; Noonan, B.P.; Fromm, M.E.; Dayan, F.E.; et al. A cytochrome P450 CYP 71 enzyme expressed in Sorghum bicolor root hair cells participates in the biosynthesis of the benzoquinone allelochemical sorgoleone. *New Phytol.* **2018**, *218*, 616–629. [CrossRef] [PubMed]
87. Watanabe, T.; Venkata, S.P.; Sahrawat, K.L.; Wani, S.P.; Ito, O. Acidification in rhizospheric soil of field-grown sorghum decreases nitrification activity. *Jpn. Agric. Res. Q* **2015**, *49*, 245–253. [CrossRef]
88. Duke, O.S. Proving Allelopathy in crop-weed interactions. *Weed Sci.* **2015**, *63*, 121–132. [CrossRef]
89. Breazeale, J.F. The injurious after-effects of sorghum. *Agron. J.* **1924**, *16*, 689–701. [CrossRef]
90. Putnam, A.R.; Duke, W.A. Biological suppression of weeds: Evidence for allelopathy in accessions of cucumber. *Science* **1974**, *185*, 370–372. [CrossRef]
91. Putnam, A.R.; DeFrank, J. Use of allelopathic cover crops to inhibit weeds. In Proceedings of the IX International Congress on Plant Protection, Washington, DC, USA, 5–11 August 1979; pp. 580–582.
92. Putnam, A.R.; Defrank, J.; Barnes, J.P. Exploitation of allelopathy for weed control in annual and perennial cropping systems. *J. Chem. Ecol.* **1983**, *9*, 1001–1010. [CrossRef]
93. Guenzi, W.D.; McCalla, T.M.; Norstadt, F.A. Presence and persistence of phytotoxic substances in wheat, oat, corn and sorghum residues. *Agron. J.* **1967**, *59*, 163–165. [CrossRef]
94. Netzly, D.H.; Butler, L.G. Roots of sorghum exude hydrophobic droplets containing biologically active components. *Crop. Sci.* **1986**, *26*, 775–778. [CrossRef]
95. Sene, M.; Dore, T.; Gallet, C. Relationship between biomass and phenolic production in grain sorghum grown under different conditions. *Agron. J.* **2001**, *93*, 49–54. [CrossRef]
96. Narwal, S.S.; Tauro, P. Allelopathic problems in Indian agriculture and prospects of research. In *Allelopathy in Agriculture and Forestry*; Scientific Publishers: Jodhpur, India, 1994; pp. 37–57.
97. Ayeni, J.M.; Kyode, J. The effects of aqueous extracts from maize roots and sorghum stem on the germination and radicle growth of *Sphenosylis sternocarpa* Hochst ex. Rich.(African Yam beans). *Afr. J. General Agri.* **2021**, *5*, 117–121.
98. Matos, F.S.; Furtado, B.N.; dos Santos, M.R.; Amorim, V.A.; Borges, L.P. Biorational agriculture: Herbicidal activity of sorghum extract in control of *Cyperus rotundus* L. *Magistra* **2021**, *31*, 675–682.
99. Kim, S.Y.; DeDatta, S.K.; Robles, R.P.; Kim, K.U.; Lee, S.C.; Shin, D.H. Partial characterization of allelopathic substances in sorghum stem by different organic solvents and pH. *Korean J. Weed Sci.* **1993**, *14*, 42–48.
100. Ben-Hammouda, M.; Robert, J.K.; Harry, C.M. Phytotoxicity of extracts from sorghum plant components on wheat seedlings. *Crop. Sci.* **1995**, *35*, 1652–1656. [CrossRef]
101. Cheema, Z.A.; Luqman, M.; Khaliq, A. Use of allelopathic extracts of sorghum and sunflower herbage for weed control in wheat. *J. Anim. Plant. Sci.* **1997**, *7*, 91–93.

102. Bhatti, M.Q.L.; Cheema, Z.A.; Mahmood, T. Efficacy of sorgaab as natural weed inhibitor in raya. *Pak. J. Biol. Sci.* **2000**, *3*, 1128–1130. [CrossRef]
103. Xuan, T.D.; Tsuzuki, E.; Tawata, S.; Khanh, T.D. Methods to determine allelopathic potential of crop plants for weed control. *Allelop. J.* **2004**, *13*, 149–164.
104. Friesen, L.F.; Jones, T.L.; Acker, R.C.V.; Morrison, I.N. Identification of *Avena fatua* populations resistant to imazamethabenz, flamprop and fenoxaprop-P. *Weed Sci.* **2000**, *48*, 532–540. [CrossRef]
105. Busi, R.; Vila-Aiub, M.M.; Beckie, H.J.; Gaines, T.A.; Goggin, D.E.; Kaundun, S.S.; Lacoste, M.; Neve, P.; Nissen, S.J.; Norsworthy, J.K.; et al. Herbicide-resistant weeds: From research and knowledge to future needs. *Evol. Appl.* **2013**, *6*, 1218–1221. [CrossRef] [PubMed]
106. Duke, S.O.; Dayan, F.E.; Romagni, J.G.; Rimando, A.M. Natural products as sources of herbicides: Current status and future trends. *Weed Res.* **2000**, *40*, 99–111. [CrossRef]
107. Pimentel, D.; Lach, L.; Zuniga, R.; Morrison, D. Environmental and economic costs of nonindigenous species in the United States. *BioScience* **2000**, *50*, 53–65. [CrossRef]
108. Worthington, M.; Reberg-Horton, C. Breeding cereal crops for enhanced weed suppression: Optimizing allelopathy and competitive ability. *J. Chem. Ecol.* **2013**, *39*, 213–231. [CrossRef] [PubMed]
109. Zimdahl, R.L. *Weed Crop Competition. A Review*; International Plant Protection Center: Corvallis, OR, USA, 1980.
110. Mahmood, A.; Cheema, Z.A.; Mushtaq, M.N.; Farooq, M. Maize–sorghum intercropping systems for purple nutsedge management. *Archives Agron. Soil Sci.* **2013**, *59*, 1279–1288. [CrossRef]
111. Kim, S.Y.; DeDatta, S.K.; Robles, R.P.; Kim, K.U.; Lee, S.C.; Shin, D.H. Isolation and characterization of allelopathic substances from sorghum stem. *Korean J. Weed Sci.* **1993**, *14*, 156–162.
112. Weston, L.A. Utilization of allelopathy for weed management in agro-ecosystems. *Agron. J.* **1996**, *88*, 860–866. [CrossRef]
113. Cheema, Z.A.; Khaliq, A.; Akhtar, S. Use of sorgaab (sorghum water extract) as a natural weed inhibitor in spring mungbean. *Int. J. Agric. Biol.* **2001**, *3*, 515–518.
114. Khaliq, A.; Cheema, Z.A.; Mukhtar, M.A.; Ahmad, S.M. Evaluation of sorghum (*Sorghum bicolor*) water extract for weed control in soybean. *Int. J. Agri. Biol.* **1999**, *1*, 23–26.
115. Khaliq, A.; Shakeel, M.; Matloob, A.; Hussain, S.; Tanveer, A.; Murtaza, G. Influence of tillage and weed control practices on growth and yield of wheat. *Philipp. J. Crop. Sci.* **2013**, *38*, 12–20.
116. Jamil, M.; Cheema, Z.A.; Mushtaq, M.N.; Farooq, M.; Cheema, M.A. Alternative control of wild oat and canary grass in wheat fields by allelopathic plant water extracts. *Agron. Sustain. Dev.* **2009**, *29*, 475–482. [CrossRef]
117. Rehman, A.; Cheema, Z.A.; Khaliq, A.; Arshad, M.; Mohsan, S. Application of sorghum, sunflower and rice water extract combinations helps in reducing herbicide dose for weed management in rice. *Int. J. Agric. Biol.* **2010**, *12*, 901–906.
118. Wazir, I.; Sadiq, M.; Baloch, M.S.; Awan, I.U.; Khan, E.A.; Shah, I.H.; Nadim, M.A.; Khakwani, A.A.; Bakhsh, I. Application of bio-herbicide alternatives for chemical weed control in rice. *Pak. J. Weed Sci. Res.* **2011**, *17*, 245–252.
119. Geneve, R.L.; Weston, L.A. Growth reduction of eastern redbud (*Cercis canadensis* L.) seedlings caused by interaction with a sorghum-sudan grass hybrid (Sudex). *J. Environ. Hort.* **1988**, *6*, 24–26.
120. Weston, L.A.; Harmon, R.; Mueller, S. Allelopathic potential of sorghum-sudangrass hybrid (sudex). *J. Chem. Ecol.* **1989**, *15*, 1855–1865. [CrossRef] [PubMed]
121. Cheema, Z.A.; Rakha, A.; Khaliq, A. Use of sorgaab and sorghum mulch for weed management in mungbean. *Pak. J. Agri. Sci.* **2000**, *37*, 140–144.
122. Ahmad, A.; Cheema, Z.A.; Ahmad, R. Evaluation of sorgaab as natural weed inhibitor in maize. *J. Anim. Plant. Sci.* **2000**, *10*, 141–146.
123. Rodrigues, B.N.; Passini, T.; Ferreira, A.G. *Research on allelopathy in Brazil*; Allelopathy update; Narwal, S.S., Ed.; Science Publishers: Enfield, NH, USA, 1999; Volume 1, pp. 307–323.
124. Anderson, R.L. Crop residue reduces jointed goat gras (*Aegilops cylindrica*) seedling growth. *Weed Tech.* **1993**, *7*, 717–722. [CrossRef]
125. Einhellig, F.A.; Rasmussen, T.A. Prior cropping with grain sorghum inhibits weeds. *J. Chem. Ecol.* **1989**, *15*, 951–960. [CrossRef]
126. Shi, X.; Zhao, X.; Ren, J.; Dong, J.; Zhang, H.; Dong, Q.; Jiang, C.; Zhong, C.; Zhou, Y.; Yu, H. Influence of peanut, sorghum, and soil salinity on microbial community composition in interspecific interaction zone. *Front. Microbiol.* **2021**, *12*, 1306. [CrossRef] [PubMed]
127. Satheeshkumar, N.; Thukkaiyannan, P.; Ponnuswamy, K.; Santhi, P. Effect of sowing and weed management methods and intercrops on weed control and grain yield of sorghum under intercropping situation. *Crop. Res.* **2011**, *41*, 46–51.
128. Kaur, S.; Kaur, R.; Chauhan, B.S. Understanding crop-weed-fertilizer-water interactions and their implications for weed management in agricultural systems. *Crop. Prot.* **2018**, *103*, 65–72. [CrossRef]
129. Cobb, A.B.; Wilson, G.W.; Goad, C.L.; Bean, S.R.; Tesso, T.T.; Wilson, J.D. Assessing the influence of farm fertility amendments, field management, and sorghum genotypes on soil microbial communities and grain quality. *App. Soil Ecol.* **2017**, *119*, 367–374. [CrossRef]
130. Khashi u Rahman, M.; Zhou, X.; Wu, F. The role of root exudates, CMNs, and VOCs in plant–plant interaction. *J. Plant. Interact.* **2019**, *14*, 630–636. [CrossRef]
131. Chaudhry, U.K.; Gökçe, Z.N.Ö.; Gökçe, A.F. Drought and salt stress effects on biochemical changes and gene expression of photosystem II and catalase genes in selected onion cultivars. *Biologia.* **2021**, 1–15.

132. Chaudhry, U.K.; Gökçe, Z.N.Ö.; Gökçe, A.F. Effects of salinity and drought stresses on the physio-morphological attributes of onion cultivars at bulbification stage. *Int. J. Agric. Biol.* **2020**, *24*, 1681–1689.
133. Junaid, M.D.; Chaudhry, U.K.; Gökçe, A.F. *Climate Change and Plant Growth–South Asian Perspective*, 1st ed.; CRC Press: Boca Raton, FL, USA, 2021; Volume 37.
134. Chaudhry, U.K.; Junaid, M.D.; Gökçe, A.F. Influence of Environmental Adversities on Physiological Changes in Plants. In *Developing Climate-Resilient Crops*; CRC Press: Boca Raton, FL, USA, 2021; pp. 85–110.
135. Hasanuzzaman, M.; Bhuyan, M.H.M.; Zulfiqar, F.; Raza, A.; Mohsin, S.M.; Mahmud, J.A.; Fujita, M.; Fotopoulos, V. Reactive oxygen species and antioxidant defense in plants under abiotic stress: Revisiting the crucial role of a universal defense regulator. *Antioxidants* **2020**, *9*, 681. [CrossRef]
136. Haig, T. Allelochemicals in plants. In *Allelopathy in Sustainable Agriculture and Forestry*; Springer: New York, NY, USA, 2008; pp. 63–104.
137. Cheema, Z.A.; Khaliq, A.; Abbas, M.; Farooq, M. Allelopathic potential of sorghum (*Sorghum bicolor* L. Moench) cultivars for weed management. *Allelop. J.* **2007**, *20*, 167.
138. Farooq, M.; Mazhar, H.; Abdul, W.; Rushna, M. Employing aqueous allelopathic extracts of sunflower in improving salinity tolerance of rice. *J. Agric. Soc. Sci.* **2011**, *7*, 75–80.
139. Pedrol, N.; González, L.; Reigosa, M.J. Allelopathy and abiotic stress. In *Allelopathy*; Springer: Dordrecht, The Netherlands, 2006; pp. 171–209.
140. Dicko, M.H.; Gruppen, H.; Barro, C.; Traoré, A.S.; van Berkel, W.J.; Voragen, A.G. Impact of phenolic compounds and related enzymes in sorghum varieties for resistance and susceptibility to biotic and abiotic stresses. *J. Chem. Ecol.* **2005**, *31*, 2671–2688. [CrossRef] [PubMed]
141. Patra, B.; Schluttenhofer, C.; Wu, Y.; Pattanaik, S.; Yuan, L. Transcriptional regulation of secondary metabolite biosynthesis in plants. *Biochim. Biophys. Acta Gene Regul. Mech.* **2013**, *1829*, 1236–1247. [CrossRef] [PubMed]
142. Marchiosi, R.; dos Santos, W.D.; Constantin, R.P.; de Lima, R.B.; Soares, A.R.; Finger-Teixeira, A.; Mota, T.R.; de Oliveira, D.M.; Foletto-Felipe, M.D.P.; Abrahão, J.; et al. Biosynthesis and metabolic actions of simple phenolic acids in plants. *Phytochem. Rev.* **2020**, *19*, 865–906. [CrossRef]
143. Asao, T.; Kitawaza, H.; Tomita, K.; Suyama, K.; Yamamoto, H.; Hosoki, T.; Pramanik, M.H.R. Mitigation of cucumber autotoxicity in hydroponic culture using microbial strain. *Sci. Hort.* **2004**, *99*, 207–214. [CrossRef]
144. Al-Tawaha, A.R.M.; Odat, N. Use of sorghum and maize allelopathic properties to inhibit germination and growth of wild barley (*Hordeum spontaneum*). *Not. Bot. Hort. Agrobot. Cluj* **2010**, *38*, 124–127.
145. Weir, T.L.; Park, S.W.; Vivanco, J.M. Biochemical and physiological mechanisms mediated by allelochemicals. *Curr. Opin. Plant. Biol.* **2004**, *7*, 472–479. [CrossRef] [PubMed]
146. Maqbool, N.; Wahid, A.; Farooq, M.; Cheema, Z.A.; Siddique, K.H.M. Allelopathy and abiotic stress interaction in crop plants. In *Allelopathy*; Springer: Berlin/Heidelberg, Germany, 2013; pp. 451–468.
147. Meiners, S.J.; Kong, C.H.; Ladwig, L.M.; Pisula, N.L.; Lang, K.A. Developing an ecological context for allelopathy. *Plant Ecol.* **2012**, *213*, 1221–1227. [CrossRef]
148. Roberts, J.A.; Hussain, A.; Taylor, I.B.; Black, C.R. Use of mutants to study long-distance signalling in response to compacted soil. *J. Exp. Bot.* **2002**, *53*, 45–50. [CrossRef]
149. Pickett, J.A. Chemical ecology and sustainable food production. *J. Chem. Ecol.* **2012**, *38*, 1071. [CrossRef]
150. Caruso, C.M.; Parachnowitsch, A.L. Do plants eavesdrop on floral scent signals? *Trends Plant Sci.* **2016**, *21*, 9–15. [CrossRef]
151. Knight, H.; Knight, M.R. Abiotic stress signaling pathways: Specificity and cross-talk. *Trends Plant Sci.* **2001**, *6*, 262–267. [CrossRef]
152. Johnson, S.M.; Lim, F.L.; Finkler, A.; Fromm, H.; Slabas, A.R.; Knight, M.R. Transcriptomic analysis of *Sorghum bicolor* responding to combined heat and drought stress. *BMC Genom.* **2014**, *15*, 456. [CrossRef]
153. Pitzschke, A.; Hirt, H. Mitogen-activated protein kinases and reactive oxygen species signaling in plants. *Plant Physiol.* **2006**, *141*, 351–356. [CrossRef]
154. Taiz, L.; Zeiger, E.; Møller, I.M.; Murphy, A. *Plant Physiology and Development*, 6th ed.; Sinauer Associates Incorporated: Sunderland, MA, USA, 2015.
155. Chang, M.; Netzly, D.H.; Butler, L.G.; Lynn, D.G. Chemical regulation of distance. Characterization of the first natural host germination stimulant for *Striga asiatica*. *J. Am. Chem. Soc.* **1986**, *108*, 7858–7860. [CrossRef] [PubMed]
156. Duke, S.O.; Scheffler, B.E.; Dayan, F.E.; Weston, L.A.; Ota, E. Strategies for using transgenes to produce allelopathic crops. *Weed Technol.* **2001**, *15*, 826–834. [CrossRef]
157. Shehzad, T.; Okuno, K. Genetic analysis of QTLs controlling allelopathic characteristics in sorghum. *PLoS ONE* **2020**, *15*, 235896. [CrossRef]

plants

MDPI

Article

Exogenous Nitric Oxide Reinforces Photosynthetic Efficiency, Osmolyte, Mineral Uptake, Antioxidant, Expression of Stress-Responsive Genes and Ameliorates the Effects of Salinity Stress in Wheat

Ghalia S. H. Alnusairi [1], Yasser S. A. Mazrou [2,3], Sameer H. Qari [4], Amr A. Elkelish [5], Mona H. Soliman [6,*], Mohamed Eweis [6], Khaled Abdelaal [7], Gomaa Abd El-Samad [8], Mohamed F. M. Ibrahim [9] and Nihal ElNahhas [10]

1 Department of Biology, College of Science, Jouf University, Sakaka 72388, Saudi Arabia; gshalnusairi@ju.edu.sa
2 Business Administration Department, Community College, King Khalid University, Guraiger, Abha 62529, Saudi Arabia; ymazrou@kku.edu.sa
3 Faculty of Agriculture, Tanta University, Tanta 31512, Egypt
4 Biology Department, Al-Jumum University College, Umm Al-Qura University, Mecca 21955, Saudi Arabia; shqari@uqu.edu.sa
5 Botany Department, Faculty of Science, Suez Canal University Ismailia, Ismailia 41522, Egypt; amr.elkelish@science.suez.edu.eg
6 Botany and Microbiology Department, Faculty of Science, Cairo University, Giza 12613, Egypt; amradel807080@googelmail.com
7 Plant Pathology and Biotechnology Laboratory, Excellence Center (EPCRS), Faculty of Agriculture, Kafrelsheikh University, Kafr Elsheikh 33516, Egypt; khaled.elhaies@gmail.com
8 Department of Agronomy, Faculty of Agriculture, Ain Shams University, Cairo 11566, Egypt; Gomaa_abdelsamad@agr.asu.edu.eg
9 Department of Agricultural Botany, Faculty of Agriculture, Ain Shams University, Cairo 11566, Egypt; Ibrahim_mfm@agr.asu.edu.eg
10 Department of Botany and Microbiology, Faculty of Science, Alexandria University, Alexandria 21526, Egypt; nihal.elnahhas@alexu.edu.eg
* Correspondence: monahsh1@gmail.com; Tel.: +20-1005-145-454

Citation: Alnusairi, G.S.H.; Mazrou, Y.S.A.; Qari, S.H.; Elkelish, A.A.; Soliman, M.H.; Eweis, M.; Abdelaal, K.; El-Samad, G.A.; Ibrahim, M.F.M.; ElNahhas, N. Exogenous Nitric Oxide Reinforces Photosynthetic Efficiency, Osmolyte, Mineral Uptake, Antioxidant, Expression of Stress-Responsive Genes and Ameliorates the Effects of Salinity Stress in Wheat. *Plants* **2021**, *10*, 1693. https://doi.org/10.3390/plants10081693

Academic Editor: Juan Barceló

Received: 10 July 2021
Accepted: 12 August 2021
Published: 18 August 2021
Corrected: 22 February 2022

Abstract: Salinity stress is one of the major environmental constraints responsible for a reduction in agricultural productivity. This study investigated the effect of exogenously applied nitric oxide (NO) (50 μM and 100 μM) in protecting wheat plants from NaCl-induced oxidative damage by modulating protective mechanisms, including osmolyte accumulation and the antioxidant system. Exogenously sourced NO proved effective in ameliorating the deleterious effects of salinity on the growth parameters studied. NO was beneficial in improving the photosynthetic efficiency, stomatal conductance, and chlorophyll content in normal and NaCl-treated wheat plants. Moreover, NO-treated plants maintained a greater accumulation of proline and soluble sugars, leading to higher relative water content maintenance. Exogenous-sourced NO at both concentrations up-regulated the antioxidant system for averting the NaCl-mediated oxidative damage on membranes. The activity of antioxidant enzymes increased the protection of membrane structural and functional integrity and photosynthetic efficiency. NO application imparted a marked effect on uptake of key mineral elements such as nitrogen (N), potassium (K), and calcium (Ca) with a concomitant reduction in the deleterious ions such as Na^+. Greater K and reduced Na uptake in NO-treated plants lead to a considerable decline in the Na/K ratio. Enhancing of salt tolerance by NO was concomitant with an obvious down-regulation in the relative expression of SOS1, NHX1, AQP, and OSM-34, while D2-protein was up-regulated.

Keywords: nitric oxide; salinity stress; antioxidant system; osmolytes; photosystem II; Na^+/H^+ antiporters; *Triticum aestivum* L.

1. Introduction

Being sessile, plants are frequently confronted with various environmental stresses, resulting in considerable alternation in metabolism, which leads to a serious threat within the yield and crop production [1,2]. Among the key stress factors, salinity seriously affects the growth and development of plants [3–5]. Nowadays, at global levels, the problem of salinity, particularly in the arid and semiarid regions, is increasing, and the problem has been intensified due to the continuous usage of salt-rich water for irrigation purposes [6,7]. Such agricultural malpractices have to lead to the continuous addition of excess toxic salts to soil, rendering the productive lands a saline wasteland. It has been estimated that 5–7% of the global land and approximately 20% of irrigated land areas are affected by high salinity [8–10]. Salinity stress proved to be detrimental to the growth and development of existing crops through induction of aberrations in physiological and biochemical processes, including chlorophyll synthesis, photosynthesis, respiration, and ion homeostasis [11–13]. Moreover, salinity stress negatively affects the metabolism, particularly the nitrogen or carbon assimilatory pathway, which in turn reflects reduced growth and yield [14–16].

It has been reported that the availability of excess toxic salts at the subcellular level triggers the excessive generation of reactive oxygen species (ROS), resulting in the induction of oxidative stress [1,17,18]. These ROS include radicals such as superoxide, hydrogen peroxide, hydroxyl, and peroxide, which are injurious to plant metabolism and growth at higher concentrations [19–21]. The accumulation of ROS in plant tissue can result in the oxidation of lipids [22,23], proteins [20,24], nucleic acids [25,26], and chlorophyll [27,28], thereby disturbing the structural and functional integrity of cells. To avoid the deleterious impact of accumulated ROS and also prevent their further generation, several defensive mechanisms are being evolved to improve cellular functioning [29,30].

These mechanisms include the selective absorption of mineral ions for osmotic adjustment and up-regulation of antioxidant and osmoprotective agents accumulated in the cytoplasm and organelles [11,31]. Additionally, osmolytes such as proline, glycine betaine, and sugars accumulate to maintain cellular functioning by making the cells osmotically stable. This takes place through the maintenance of the cell water content and assists in scavenging ROS, hence maintaining the enzyme structure and functioning [32,33]. Furthermore, the antioxidant defense system contributes to the neutralization of toxic ROS for preventing oxidative damage effects [19,34].

Nitric oxide (NO) is an essential radical molecule implicated in several physiological and biochemical functions of plants [13,35]. NO is involved in growth and development, as well as in defense responses to a variety of abiotic stresses, including salinity [36,37]. Working with different plant species, researchers have considered NO as an endogenous signaling molecule implicated in the regulation and coordination of the signaling network [38,39]. NO is the leading molecule for several physiological and adaptive biochemical changes [40]. The resultant effects, whether beneficial or deleterious, have been shown depending on the concentration and the site of production [41,42].

Wheat (*Triticum aestivum* L.) is one of the most important and strategic cereal crops in temperate zones. It is one of the preferred meals that are being used by approximately 36% of the whole population. Worldwide, wheat supplies almost 55% of the carbohydrates and 20% of the food calories consumed globally [43]. During the last 20 centuries, the demand for wheat has doubled due to urbanization and industrialization century. Salinity stress directly affects wheat phenological aspects, root growth rate, root/shoot ratio, and total dry matter yield [44]. Therefore, exogenous application of NO can improve salinity tolerance by maintaining the antioxidant and osmolyte metabolism for better growth and yield in salt-exposed seedlings. With this hypothesis, the present investigation was aimed to analyze the possible involvement of exogenous NO in the regulation of growth of wheat through the regulation of physiological, biochemical, and molecular attributes under salt stress.

2. Results

2.1. Growth Parameters

Results depicting the influence of NaCl stress and NO application on growth parameters such as length and fresh and dry biomass in wheat are shown in (Table 1). Wheat seedlings treated with 100 mM of NaCl showed a significant decline in shoot height and fresh and dry weight, which was, however, mitigated by exogenous application of NO. Relative to control, the observed decline in length and fresh and dry weight of wheat was 36.70%, 38.88% and 41.13% due to 100 mM NaCl. NO application at 50 µM caused an enhancement of 14.25%, 4.82% and 12.98% in length and fresh and dry weight, respectively, while 100 µM NO caused a maximum increase of 22.52%, 24.63% and 40.70% over the control. Application of NO (50 and 100 µM) mitigated the effect of NaCl by 12.26% and 21.59% in shoot length, 16.30% and 32.29% in fresh weight, and 25.09% and 35.89% in dry weight over the NaCl-stressed plants (Table 1).

Table 1. Effect of salinity stress (100 mM NaCl) on shoot length (cm) and shoot fresh and dry weight (g) in wheat (*Triticum aestivum* L.) with and without exogenous application of NO. Data presented are mean (±SE) of three replicates and different letters denote significant difference at $p \leq 0.05$.

	Shoot Length (cm)	Shoot Fresh Weight (gm)	Shoot Dry Weight (gm)
Control	22.26 ± 1.006c	4.573 ± 0.417c	1.206 ± 0.103bc
NaCl (100 mM)	14.09 ± 0.899e	2.725 ± 0.453f	0.7071 ± 0.012e
NO 50 µM	25.96 ± 2.65b	4.805 ± 0.116b	1.386 ± 0.055b
NO 100 µM	28.73 ± 1.08a	6.068 ± 0.073a	2.034 ± 0.06a
NaCl + NO 50 µM	16.06 ± 0.125d	3.256 ± 0.169e	0.944 ± 0.05d
NaCl + NO 100 µM	17.97 ± 0.047d	4.025 ± 0.18d	1.103 ± 0.1c

2.2. No Protects Chlorophyll and Photosynthetic Attributes in Salt-Stressed Wheat Plants

Salinity stress resulted in a considerable decline in the synthesis of chlorophyll a (Chl a), chlorophyll b (Chl b), carotenoid, and total pigments reflecting in decreased photosynthetic rate and stomatal conductance (Figure 1A–F). Relative to control, NaCl treatment reduced the Chl 'a' by 31.1%, Chl 'b' by 32%, carotenoids by 26.56%, and total pigments by 38.7%. Application of NO at both concentrations improved the chlorophyll synthesis significantly over the control and also ameliorated the negative effects of salinity by 3.62%, 4.07%, 7.23% and 13.33% with 50 µM NO and by 5.42%, 15.2%, 13.23% and 28.1% with 100 µM NO over the NaCl-treated counterparts. At 100 µM, application of NO (100 mM NaCl + 100 µM NO) ameliorated the negative effect of NaCl on stomatal conductance and photosynthetic efficiency by 28.39% and 36.51%, respectively, over the NaCl-treated plants (Figure 1E,F).

2.3. No Maintains Leaf RWC by Improving the Proline and Sugar Content

Results showing the effect of NaCl and NO on the synthesis of proline, sugars, and RWC are depicted in (Figure 2A–C). NO application improved the proline and soluble sugar accumulation in both normal and NaCl-stressed wheat plants conditions, reflecting in increased RWC in them over the controls. RWC was declined in NaCl-stressed plants by 24.12% over control. However, an enhancement of 2.26% and 5.99% was observed with 50 µM and 100 µM NO, respectively. Application of NO to NaCl-stressed plants mitigated the negative effects on RWC by 3.25% at NaCl + 50 µM NO and by 8.27% at NaCl + 100 µM NO over the NaCl (100 mM)-stressed plants. NO-treated plants (100 mM NaCl + 100 µM NO) exhibited maximal accumulation of proline and soluble sugars with an increase of 36.78% and 52.66% over the control plants, resulting in maximal amelioration of salinity-induced decline in RWC (Figure 2A–C).

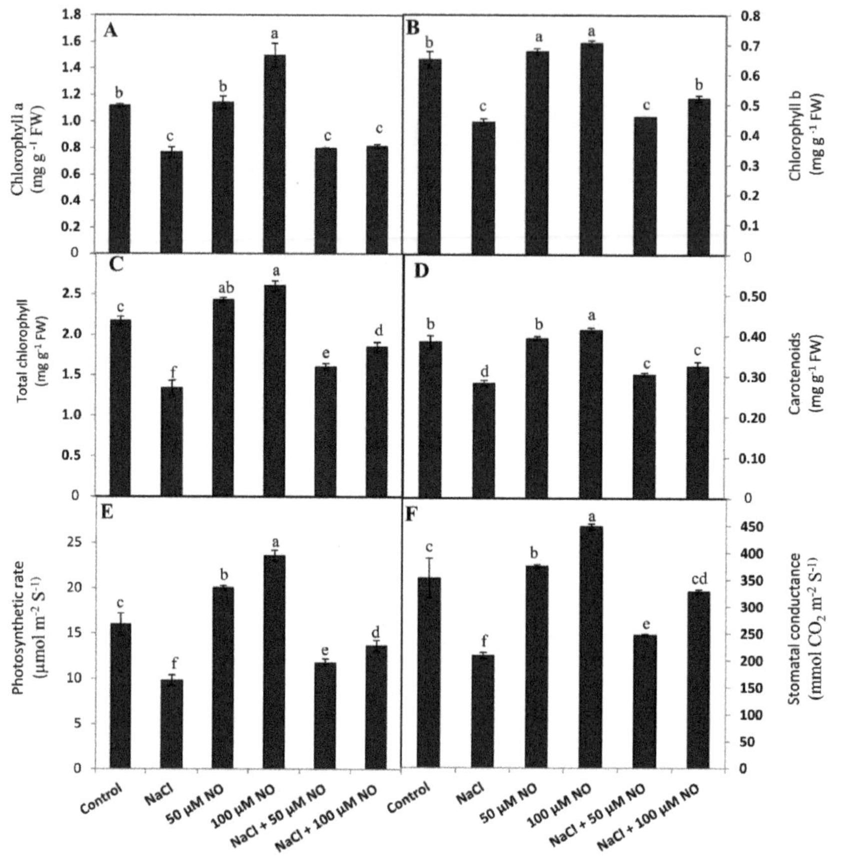

Figure 1. Effect of salinity stress (100 mM NaCl) on (**A**) chlorophyll a, (**B**) chlorophyll b, (**C**) total chlorophyll, (**D**) carotenoids, (**E**) photosynthetic rate, and (**F**) stomatal conductance in wheat (*Triticum aestivum* L.) with and without exogenous application of NO (50 and 100 μM). Data presented are mean (±SE) of three replicates, and bars with different letters denote significant difference at $p \leq 0.05$.

2.4. No Application Reduces Oxidative Damage by Preventing Lipid Peroxidation, Hydrogen Peroxide, and Improving Membrane Stability Index

Exogenous application of NO prevented the oxidative stress by reducing the formation of hydrogen peroxide (H_2O_2), and hence the lipid peroxidation resulting in an increased membrane stability index (Figure 3A–C). NaCl stress triggered the generation of H_2O_2 by 52.97% over control, causing an increase of 34.97% in lipid peroxidation and subsequently a decline of 27.27% in the membrane stability index in them. Exogenous application of NO at 50 and 100 μM proved significant in declining the NaCl-mediated generation of H_2O_2 and hence preventing lipid peroxidation and protecting membrane stability. Maximal protection to wheat seedlings was exhibited by seedlings treated with 100 μM NO, reducing the generation of H_2O_2 and lipid peroxidation by 21.35% and 17.24% over the control, and at 100 mM NaCl + 100 μM NO, H_2O_2 and lipid peroxidation was ameliorated by 21.14% and 17.19% over the NaCl-stressed plants. The naCl-mediated decline in the membrane stability index was assuaged by 4.2% and 7.62% in 50 μM and 100 μM NO-supplemented seedlings, respectively (Figure 3A–C).

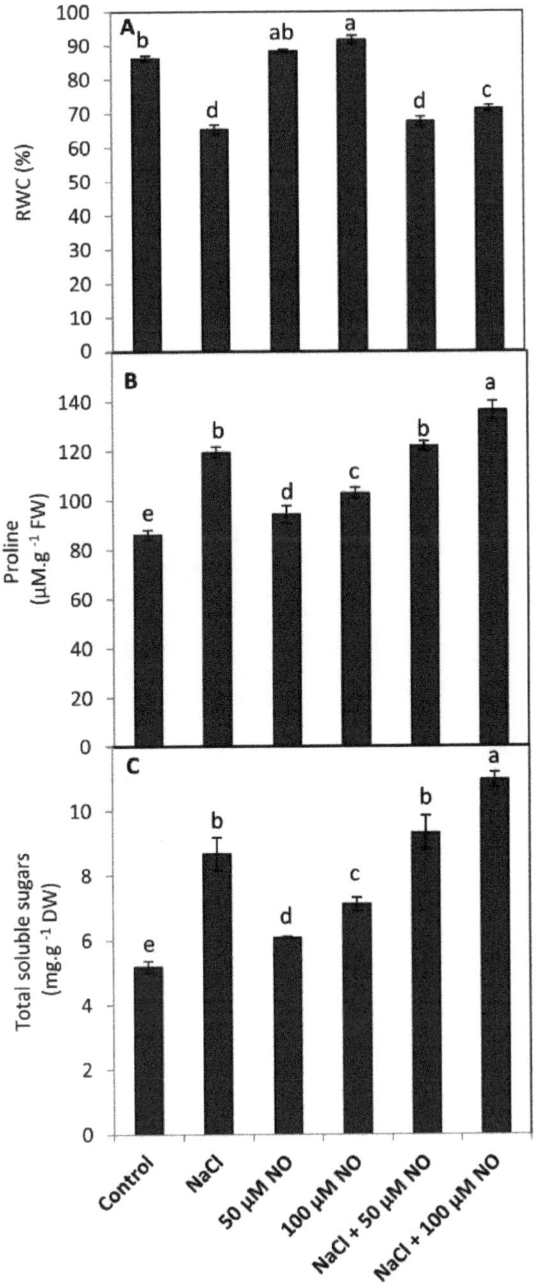

Figure 2. Effect of salinity stress (100 mM NaCl) on (**A**) RWC, (**B**) proline, and (**C**) soluble sugar content in wheat (*Triticum aestivum* L.) with and without exogenous application of NO (50 µM and 100 µM). Data presented are mean (±SE) of three replicates, and bars with different letters denote significant difference at $p \leq 0.05$.

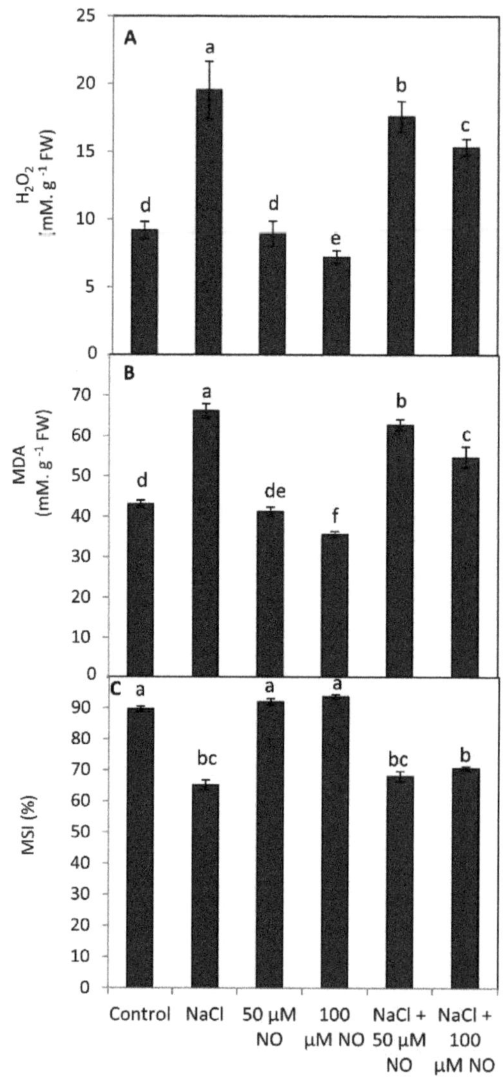

Figure 3. Effect of salinity stress (100 mM NaCl) on (**A**) hydrogen peroxide, (**B**) lipid peroxidation (MDA), and (**C**) membrane stability index (MSI) in wheat (*Triticum aestivum* L.) with and without exogenous application of NO (50 and 100 µM). Data presented are mean (±SE) of three replicates, and bars with different letters denote significant difference at $p \leq 0.05$.

2.5. Exogenous NO Up-Regulates Antioxidant System in Salinity-Stressed Wheat

In both normal and NaCl-exposed wheat seedlings, the activities of the antioxidant enzymes and the contents of non-enzymatic antioxidants studied (glutathione (GSH) and ascorbic acid (ASA)) were observed to increase with the application of NO, and obvious effects were observed with 100 µM NO (Figure 4A–F). Increase in the activities due to 100 mM NaCl was 34.86% in superoxide dismutase (SOD), 26.17% in catalase (CAT), 35.67% ascorbate peroxidase (APX), and 30.3% in glutathione reductase (GR) activity over the control plants. GSH increased by 28.04%, while ASA content decreased by 15.29% due to NaCl treatment. Application of NO increased the activity of SOD, CAT, APX, and GR by 5.54%, 8.83%, 22.22% and 7.45% at 50 µM and by 13.41%, 18.96%, 33.33% and 25.96% at

100 µM over the control plants. Relative to the NaCl-treated seedlings, NaCl + NO-treated ones showed a further increase in the activity of SOD, CAT, APX, and GR, with the increase being more obvious in 100 mM NaCl + 100 µM NO-treated plants showing a percentage increase of 8.27% for SOD, 13.39% for CAT, 5.44% for APX, and 20.9% for GR. ASA was increased by 7.77% and 17.2% by application of 50 µM and 100 µM NO, respectively, and amelioration of 5.88% and 13.6% in ASA content was observed when applied to NaCl-stressed plants. Accumulation of GSH was enhanced maximally in NO-supplemented plants with a percent increase of 29.44% and 33.33% with NaCl + 50 µM NO and NaCl + 100 µM NO, respectively, over the control (Figure 4A–F).

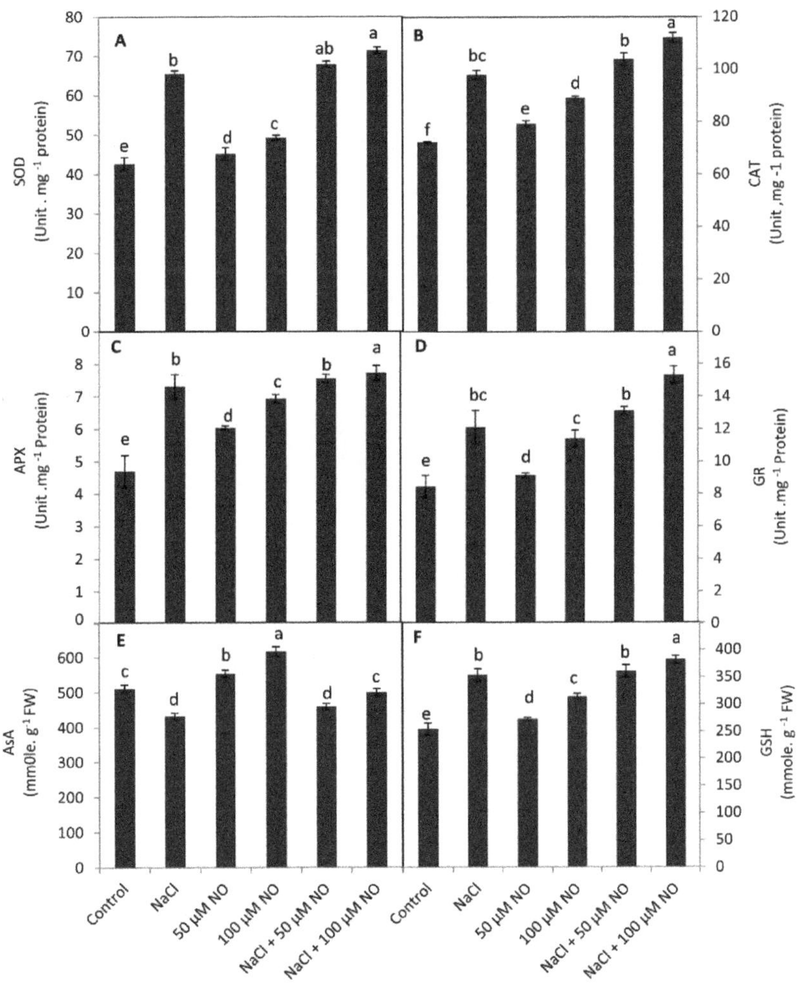

Figure 4. Effect of salinity stress (100 mM NaCl) on activity of (**A**) superoxide dismutase, (**B**) catalase, (**C**) ascorbate peroxidase, (**D**) glutathione reductase, and (**E**) ascorbic acid and (**F**) reduced glutathione content (*Triticum aestivum* L.) with and without exogenous application of NO (50 µM and 100 µM). Data presented are mean (±SE) of three replicates, and bars with different letters denote significant difference at $p \leq 0.05$.

2.6. No Improves Uptake of N, K, and Ca under Salinity Stress

Mineral elements were estimated to assess the salt tolerance mediated by exogenous-applied NO. NaCl treatment declined the uptake of nitrogen (N), potassium (K), and calcium (Ca) while increased the accumulation of sodium both in the leaf as well as root (Table 2). Relative to control, N, K, and Ca were observed to decrease by 27.31%, 45.07% and 40.7% in leaf and 25.2%, 32.5% and 34.62% in root tissues, respectively. In leaf tissues, application of NO increased N, K, and Ca by 8.34%, 10.92% and 7.75% at 50 µM and by 19.04%, 25.33% and 28.63% at 100 µM concentrations, respectively. Such a positive influence of NO was also observed when applied to NaCl-treated plants, resulting in significant amelioration of decline in their uptake. Relative to salinity-stressed plants, uptake of N, K, and Ca increased by 23.2%, 33.76% and 32.2%, respectively, with NaCl + 100 µM reflecting in the mitigation of ill effects of salinity stress. In NaCl-treated plants, sodium accumulation increased by 54.06% and 43.58% in leaf and root over the control plants. However, a decline of 7.35% and 31.25% was observed in NO-treated plants at 50 µM and 100 µM, respectively, in leaf tissues, and a similar trend was maintained in root tissues. Application of NO to NaCl-treated plants limited the uptake of Na by 14.35% and 31.12% at NaCl + 50 µM and NaCl + 100 µM, respectively (Table 2).

Table 2. Effect of salinity stress (100 mM NaCl) on nitrogen, potassium calcium, sodium, and Na/K ratio in wheat (*Triticum aestivum* L.) with and without exogenous application of NO. Data presented are mean (±SE) of three replicates, and different letters denote significant difference at $p \leq 0.05$.

	N (mg g^{-1})		Na (mg g^{-1})		K (mg g^{-1})		Ca (mg g^{-1})		Na/K %	
	Leaf	**Root**	**Leaf**	**Root**	**Leaf**	**Root**	**Leaf**	**Root**	**Leaf**	**Root**
Control	35.37 ± 1.05c	17.46 ± 1.02c	7.072 ± 0.51d	9.98 ± 0.87d	45.64 ± 0.74c	26.55 ± 1.44c	6.78 ± 0.502c	4.91 ± 0.54c	0.154	0.375
NaCl (100 mM)	25.71 ± 1.05f	13.06 ± 0.96d	15.39 ± 1.15a	17.69 ± 0.6a	25.07 ± 0.27f	17.92 ± 0.89f	4.02 ± 0.327ef	3.21 ± 0.29ef	0.613	0.987
NO 50 µM	38.59 ± 0.61b	19.3 ± 1.12b	6.55 ± 0.37d	7.66 ± 0.42e	51.24 ± 1.22b	30.02 ± 1.04b	7.35 ± 0.168b	5.56 ± 0.38b	0.127	0.255
NO 100 µM	43.69 ± 1.13a	25.31 ± 0.97a	4.86 ± 0.08e	5.51 ± 0.55f	61.13 ± 1.12a	34.75 ± 1.03a	9.5 ± 0.465a	6.93 ± 0.14a	0.079	0.158
NaCl + NO 50 µM	27.95 ± 0.11e	14.92 ± 0.97d	13.18 ± 0.61b	15.29 ± 1.01b	31.41 ± 0.6e	19.43 ± 0.65e	4.58 ± 0.302e	3.6 ± 0.51e	0.419	0.786
NaCl + NO 100 µM	33.48 ± 0.44cd	16.68 ± 1.08c	10.6 ± 0.53c	12.71 ± 0.54c	37.85 ± 0.9d	24.25 ± 0.58cd	5.93 ± 0.153d	4.56 ± 0.4cd	0.28	0.524

2.7. NO Regulates the Expression of SOS1, NHX1, AQP, OSM-34, and D2-Protein

To further explore the protective effect of exogenous NO on the wheat plants under saline conditions, the relative gene expression of SOS1, NHX1, AQP, OSM-34, and D2 proteins was investigated by real-time (RT)-qPCR (Figure 5). Plants subjected to salt stress (100 mM NaCl) dramatically exhibited an obvious and significant ($p \leq 0.05$) increase in the relative expression of SOS1, NHX1, AQP, and OSM-34 compared to the non-saline conditions (control). However, an opposite and significant trend was observed with respect to D2-protein.

On the other hand, under salt-stress conditions, NO-treated plants (50 or 100 µM) showed a significant decrease in the relative gene expression of SOS1, NHX1, AQP, and OSM-34 compared to the salt-affected NO-untreated plants (100 mM NaCl), while the relative gene expression of D2-protein revealed an obvious and significant improvement with the treatments of NO at 50 or 100 µM compared to the NO-untreated plants under saline conditions. In this context, the treatment of 100 µM NO was more potent than the other treatment of 50 µM.

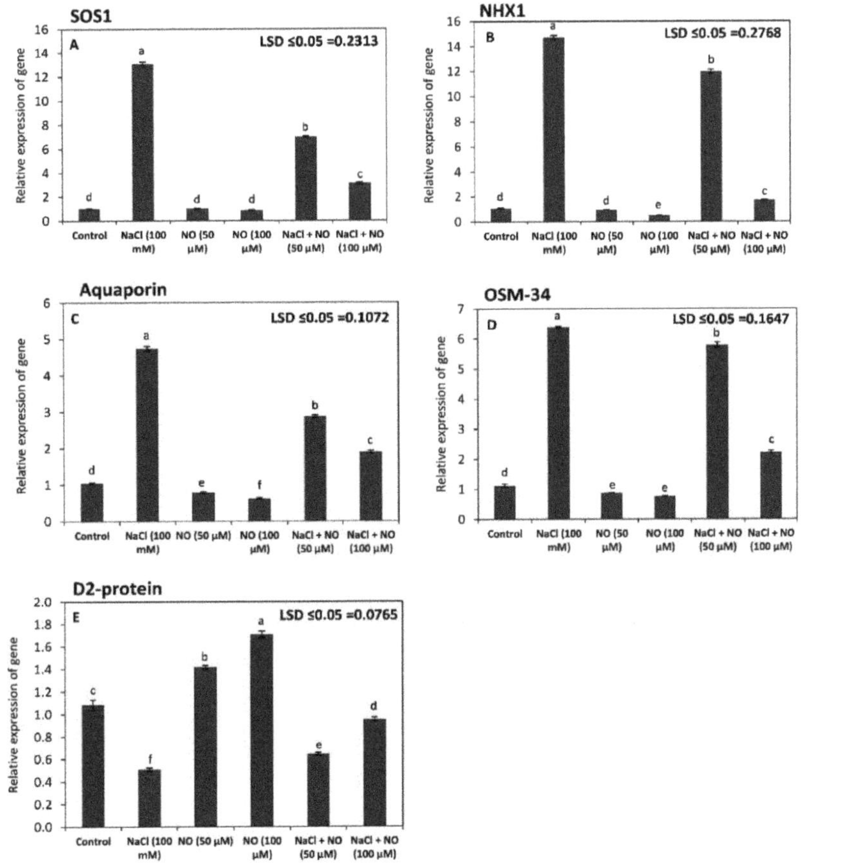

Figure 5. Effect of salinity stress (100 mM NaCl) on relative gene expression of (**A**) SOS1, (**B**) NHX1, (**C**) Aquaporin, (**D**) OSM-34, and (**E**) D2 protein (*Triticum aestivum* L.) with and without exogenous application of NO (50 µM and 100 µM). Data presented are mean (±SE) of three replicates, and bars with different letters denote significant difference at $p \leq 0.05$.

3. Discussion

Increased salinity levels in soils due to agricultural malpractices have led to the failure of net agricultural productivity and crop survival [45]. Salinity stress has resulted in a considerable decline in crop yields such as soybean [46] and tomato [47]. Hence, the need for the prevention of crop losses to meet the demands of ever-increasing human populations [48]. Accordingly, research investigations are needed to provide some better alternatives for improving the protection mechanisms against stress to enhance the yield. In this connection, we analyzed the efficiency of exogenously supplied NO in enhancing growth through its involvement in the regulation of key physiological and biochemical attributes. Increased salinity concentrations in soil solution potentially reduce the morphological parameters by restricting the cell division, leading to declined cellular elongation and growth [49]. NO has the potential to ameliorate such adverse effects and enhance the stress resilience of plants, which can be ascribed to the key role in stress mitigation [50,51]. Growth parameters showed an increase with increasing concentration of NO from 50 µM to 100 µM and also ameliorated the decline triggered by salinity (100 mM NaCl) to a considerable extent (Table 1). Our results are in corroboration with Fatma and Khan (2014) [41] for mustard, Ahmad et al. (2016) [52] for chickpea, and Fan et al. (2013) [53] for cucumber.

The ameliorative impact of exogenous NO on the growth parameters is believed to be due to its cumulative effect on the uptake and accumulation of key mineral nutrients [54]. In the present study, exogenous application of NO prevented excess accumulation of Na in upper plant parts, preventing the intensity of NaCl-mediated generation of oxidative stress in wheat plants (Table 2). Exogenously sourced NO triggers the vacuolar H^+-ATPase and increased Na^+/H^+ antiport activity, thereby reflecting inefficient compartmentalization of Na^+ [55].

Similar to our observation, the positive implication of NO in improving the K^+, Ca^{2+}, and Mg^{2+} content and preventing salinity-triggered decline has been demonstrated in *Gossypium hirsutum* [56]. Exogenously sourced NO increased uptake of N, K, and Ca, reaching maximal values with 100 µM NO under normal conditions and also maintaining its salinity stress amelioration potential. Earlier, it has been demonstrated that NO improves the uptake of N, P, and K in wheat seedlings subjected to cadmium stress [57], reflecting improved photosynthesis. Exogenously applied NO may have led to the maintenance of cytosolic Na concentrations by improving the expression of Na^+ transporters and H^+ pumps [58]. In addition, NO application potentiates the Ca-mediated stress amelioration by improving its uptake and partitioning to cellular compartments, and in the present investigation, NO proved beneficial in improving the Ca and K uptake, causing a significant reduction in Na/K ratio [59,60]. Lowering the Na/K ratio strengthens the cellular metabolism and protects enzyme activity, synthesis of metabolites, and yields performance of crop plants [19,61]. Exogenous NO application restricted the uptake of deleterious sodium ions, resulting in a significant decrease in the Na/K ratio by way of increasing mineral uptake. In the present study, NO-mediated reduction in Na/K ratio contributes to reduced susceptibility of wheat plants by mediating the exclusion of the deleterious ions for the maintenance of the cellular osmotic potential [37,60]. NO application was advantageous in improving the chlorophyll and carotenoid pigments at both the concentrations and imparted a positive impact on the photosynthetic efficiency and stomatal conductance. Earlier reports detect the reduction in chlorophyll contents in NaCl-stressed plants [62,63]. Enhancement of photosynthetic capacity following NO treatment was observed in salt-stressed wheat plants by assisting in the synthesis of other pigments protecting components such as cysteine and reduced glutathione that protect the chlorophyll breakdown as well as up-regulation of the activity of enzymes involved in chlorophyll biosynthesis [64,65]. The optimal presence of NO modulates photosynthetic functioning by improving CO_2 assimilation, photosynthetic rate, and chlorophyll fluorescence characteristic under normal and stressful conditions [66].

Similar to our finding, exposure to salinity stress triggers the accumulation of osmolytes for better protection of cellular functioning [19,67]. Moreover, NO application leads to the maintenance of relative water content by improving the accumulation of proline and soluble sugars to maintain the water potential below the external solution [68]. Overall physiologically, these osmoprotectants reduce cellular osmotic potential [69], and a reduction in the hydraulic conductivity of the membranes occurs, possibly by decreasing the number of water channels (aquaporins) [70]. NO application may have improved the expression of proteins involved in the biosynthesis of osmolytes, mainly proline, and be associated with a decrease in its catabolizing enzymes [71].

Salinity treatment was observed to increase the generation of free radical H_2O_2, causing more significant damage to membrane lipids and loss of membrane stability index [72]. For preventing the oxidative damage that is triggered by ROS, plants intensify the ROS-scavenging mechanisms [73]. In the present study, 100 mM NaCl stress raised the hydrogen peroxide and lipid peroxidation level, and NO supplementation in salt-stressed plants maintained higher activity of an antioxidant system (SOD, CAT, APX, and GR) than untreated plants. These results were in accordance with Elkahoui et al. (2005) [74] in *Catharanthus roseus*, Hernandez et al. (2010) [75] in *Brassica oleracea*, and Carrasco-Ríos and Pinto (2014) [76] in *Zea mays*. Earlier evidence from Lamattina et al. (2003) and Corpas and Barroso (2015) [77,78] showed that NO mediated better growth and significant amelioration of the oxidative damage due to increased stabilization of macromolecules such as lipids, proteins, and nucleic acids. Moreover, exogenously supplied NO controlled the production of antioxidants and accumulation of ROS and directly interacted with lipid alkoxyl and peroxyl radicals, leading to preventing the propagation of radical-mediated lipid oxidation [79]. NO-treated plants significantly up-regulated the antioxidant system in imparted quick elimination of excess accumulated free radicals, resulting in the stability of structural and functional aspects of cellular membranes [52,76]. Greater activity of SOD, CAT, APX, and GR has been correlated with improved stress tolerance, and NO-mediated enhancement in their activities will lead to a further reduction in the intensity of oxidative stress [54].

Under saline conditions, achieving ion homeostasis and conserving water status and photosynthesis are the most important challenges for plant growth and development. Therefore, different plant species develop a wide array of defensive mechanisms that can protect them from the destructive effect of salt stress. In the current study, activation of the signaling pathways of the plasma membrane (*SOS1*) and vacuolar (*NHX1*) Na^+/H^+ antiporters were observed by increasing the relative gene expression of the salt-stressed plants compared to the unstressed ones (Figure 5). These two proteins have been found to be responsible for excluding Na^+ ions from the cytosol to outside plasma membrane or inside vacuole, respectively. These responses enable plants to survive under salt stress by avoiding Na^+ toxicity on different plant metabolisms and maintain ion homeostasis in the cytoplasmic matrix [80]. Aquaporins (AQPs) are well-known membrane channel proteins that are responsible for water, metal ions, gasses, and small neutral solutes transport during biotic and abiotic stresses [81]. Similarly, osmotin (*OSM-34*) is a cysteine-rich protein synthesized in vacuoles to function as an osmoregulator under low water potential [82]. It can also control the oxidative damage induced by ROS, specifically H_2O_2, and isolate Na^+ in the vacuoles during salt stress [83]. Furthermore, overexpression of *OSM-34* has been found to reduce lipid peroxidation and increase the proline content under different stresses [83]. In this study, the overexpression of aquaporin and *OSM-34* in the salt-affected plants reveals the importance of both proteins in the regulation of osmotic potential and keeping plant–water relations under saline conditions. Conversely, exposing plants to salt stress led to diminishing the relative expression of D2-protein, which is considered one of the core proteins in the photosystem II center reaction. This protein is vulnerable to the oxidative damage and photoinhibition process during stress conditions [84].

Applied NO has been found to counteract the detrimental effects of osmotic stressors, i.e., drought [13] and salinity [85,86]. These effects may be due to enhancing the antioxidant capacity, osmotic potential, nutrient homeostasis, and gas exchange [13,86]. In the present investigation, exogenous NO, particularly at 100 μM, led to a significant downregulation in the relative expression of *SOS1*, *NHX1*, *AQP*, and *OSM-34* of the salt-affected plants compared to the salt-stressed NO-untreated plants, while an obvious improvement in the expression of D2- protein was observed by the treatments of NO (50 μM and 100 μM) in the salt-stressed plants compared to the NO-untreated plants. These responses may imply that applied NO can preserve photosynthesis, osmotic potential and minimize Na^+ toxicity in the salt-stressed plants, as there is no need to activate the ionic homeostasis (*SOS1/NHX1*) or osmotic (*AQP/OSM-34*)-related proteins with enhancing the photosynthetic efficiency (D2-protein) of the NO-treated plants under salt stress conditions.

4. Materials and Methods

4.1. Experimental Design and Treatment

Seeds of *Triticum aestivum* L. were surface-sterilized using 0.5% sodium hypochlorite for 3 min and were repeatedly washed with distilled water and sown in earthen pots having a diameter of 27 cm filled with peat, compost, and sand (4:1:1). At the time of sowing, pots were supplied with 250 mL of full-strength Hoagland's solution [87]. After seedling growth for 10 days, pots were divided into two groups, and one set was supplied with modified Hoagland's solution containing 100 mM NaCl, and another set was given normal Hoagland's solution every alternate day. NO (in the form of sodium nitroprusside as NO donor) at the concentration of 50 μM and 100 μM (10 mL per pot) was applied foliarly to both normal and NaCl-treated sets, and the control was supplied with an equal amount of distilled water and was also maintained. Seedlings were allowed to grow for another 20 days. After 1 month, plants were analyzed for different parameters such as chlorophyll pigments, photosynthetic functioning, oxidative damage attributes, osmolytes, and antioxidant system. The pot was laid in a complete randomized block design with five replications.

4.2. Estimation of Photosynthetic Pigments and Measurement of Stomatal Conductance and Photosynthetic Efficiency

Photosynthetic pigments were quantified in fresh leaves after extracting in dimethyl sulfoxide (DMSO), and the optical density of the supernatant was measured by spectrophotometer at 480, 510, 645, 663 nm against DMSO [88]. Photosynthetic efficiency and stomatal conductance were measured in upper fully expanded leaves at 13:00 by using the infrared gas analyzer (CID-340, Photosynthesis System, Bio-Science, Pullman, WA USA).

4.3. Determination of Leaf Water Content, Proline, and Soluble Sugars

Relative water content (RWC) was determined by punching leaf discs from fresh treated and normal plants, and their fresh weights were determined. After that, the same leaf discs were kept in Petri dishes containing distilled water for 1 h to gain turgidity. After recording turgid weight, the leaf discs were oven-dried at 80 °C for 24 h to record the dry weight [89]. Calculations were completed using the following formula:

$$\text{RWC (\%)} = \text{Fresh weight} - \text{Dry weight}/\text{Turgid weight} - \text{dry weight} \times 100 \quad (1)$$

For proline, 0.5 g of leaf sample was extracted in 3% (*w/v*) sulphosalicylic acid and was subjected to centrifugation at 3000× *g* for 20 min. A total of 2 mL of the supernatant was reacted with 2 mL of glacial acetic acid and 2 mL of ninhydrin reagent at boiling temperature for 1 h. Subsequently, the samples were kept on an ice bath, and the content of proline was separated using toluene, and absorbance was read spectrophotometrically at 520 nm [90]. Using the data obtained from the standard curve prepared at known concentrations, linear regression is completed (comparing absorbance vs. proline concentration).

For estimation of sugar content, dry plant sample was extracted in boiling ethanol (80 *v*/*v*) and centrifuged for 20 min at 5000× *g*. The concentration of soluble sugars was measured by reacting the extract with anthrone reagent, and optical density was recorded at 585 nm. A standard curve of glucose was used to determine the soluble sugar content [91].

4.4. Measurement of Membrane Stability Index, Lipid Peroxidation, and Hydrogen Peroxide

Membrane stability index (MSI) was determined by chopping 0.1 g fresh leaf tissue in test tubes containing 10 mL distilled water. After that, tubes were boiled at 40 °C for recording the electric conductivity (EC_1), and same tubes were boiled at 100 °C, and again EC (EC_2) was recorded [92]. Percent MSI was calculated using the formula:

$$\text{(MSI)} = [1 - (EC_1/EC_2)] \times 100 \quad (2)$$

Lipid peroxidation was measured by estimating the formation of malonaldehyde (MDA) content. For lipid peroxidation determination, fresh leaves were extracted in trichloroacetic acid (1%, *w/v*, TCA). After centrifugation at 10,000× *g* for 5 min, 1.0 mL supernatant was mixed with 0.5% thiobarbituric acid, and mixture was boiled at 95 °C for half an hour. After that, tubes were kept on ice bath followed by centrifugation for 5 min at 5000× *g* for clarification, and optical density was read at 532 nm and 600 nm [93]. The MDA concentration was determined by dividing the difference in absorbance (A532–A600) by its molar extinction coefficient (155 mM^{-1} cm^{-1}), and results expressed as mmol g^{-1} fresh weight.

The concentration of hydrogen peroxide (H_2O_2) was estimated by extracting fresh leaf samples in 0.1% (*w/v*) TCA using pestle mortar. After centrifugation at 12,000× *g* for 15 min, a known volume of the supernatant was mixed with 0.5 mL of 10 mM potassium phosphate buffer (pH 7.0) and 1 M potassium iodide (1 mL). Subsequently, the optical density of the mixture was taken at 390 nm [94], and computation was completed using a standard curve of H_2O_2.

4.5. Assay of Antioxidant Enzymes

Antioxidant enzymes were extracted by homogenizing 5.0 g fresh leaves in chilled pestle and mortar using 50 mM sodium phosphate buffer (pH 7.0) containing 1% (*w/v*) polyvinyl pyrrolidine. The resulting homogenate was used as the enzyme source after centrifugation at 15,000× *g* for 20 min at 4 °C, and the protein content was determined by following [95].

The activity of superoxide dismutase (SOD, EC 1.15.1.1) was determined by adopting the method of [96]. Briefly, enzyme aliquot was incubated under light and dark to monitor the photoreduction of nitroblue tetrazolium (NBT) at 560 nm, and the activity of SOD was expressed as enzyme unit (EU) mg^{-1} protein. One unit of enzyme activity represents the amount of enzyme required for 50% inhibition of NBT reduction at 560 nm. For assaying activity of catalase (CAT, EC 1.11.1.6), [97]'s method was adopted, and change in optical density was recorded at 240 nm. An extinction coefficient of 36×10^3 mM^{-1} cm^{-1} was used for calculation and expressed as EU mg^{-1} protein. For determination of ascorbic peroxidase (APX) activity, 0.1 mL enzyme was added to 1 mL potassium phosphate buffer (100 mM, pH 7.0), 0.1 mM EDTA, 0.5 mM ascorbate, and 0.1 mM H_2O_2. The disappearance of H_2O_2 was observed as a change in absorbance at 290 nm [98]. The reaction was initiated by addition of hydrogen peroxide, and oxidation of ascorbate was followed by the decrease in absorbance at 290 nm at 30 s interval for 5 min. One unit of APX activity is defined as the amount of enzyme that oxidizes 1 μM of ascorbate per min at room temperature. For determination of glutathione reductase (GR, EC 1.6.4.2), activity change in absorbance was recorded at 340 nm for 3 min following [99]. For calculation of activity, an extinction coefficient of 6.2 mM^{-1} cm^{-1} was used.

4.6. Determination of Ascorbate and Reduced Glutathione

For determination of ascorbic acid, fresh leaf samples were macerated in 5% (*w/v*) TCA, and to the extract, 2% (*w/v*) dinitrophenyl-hydrazine and 10% (*w/v*) thiourea were added. The resultant mixture was kept in a boiling-water bath for 15 min and brought to room temperature followed by centrifugation at 1000× *g* for 10 min. For dissolving the resulting pellet, 80% (*v/v*) H_2SO_4 was added, and optical density was taken at 530 nm [100]. A standard curve of ascorbic acid was used for calculation.

Reduced glutathione (GSH) was estimated by homogenizing 500 mg fresh leaf tissue in a phosphate buffer. After centrifugation at 3000× *g* for 15 min, 40 μL of 5, 5-dithiobis-2-nitrobenzoic acid was added to 500 μL supernatant and allowed to stand for 10 min. Absorbance was taken at 412 nm [101]. A standard curve of GSH was used for calculation.

4.7. Estimation of Mineral Ions

Estimation of mineral ions, including Na, K, and Ca, was completed using the flame photometer after acid digesting the dried tissue [102]. For determining the nitrogen content in treated and untreated tissues, the method of Subbiah and Asija [103] was adopted.

4.8. Gene Expression

Total mRNA was isolated from 0.5 g shoot parts of wheat plant of all treatments after 2 weeks of salinity and NO foliar application using Total RNA extraction kit (Sigma-Aldrich) according to the manufacturer's protocol. The purified RNA was quantitated spectrophotometrically and analyzed on 1% (*w/v*) agarose gel. Reverse transcription of RNA was performed. The reaction mixture contained 10 as oligo dT primer (10 pml/μL), 2.5 μL

5 × buffer, 2.5 µL $MgCl_2$, 2.5 µL 2.5 mM dNTPs, 4 µL oligo (dT), 0.2 µL (5 Unit/µL) reverse transcriptase (Promega, Walldorf, Germany), and 2.5 µL RNA. RT-PCR amplification was performed in a thermal cycler PCR, programmed at 42 °C for 1 h and 72 °C for 20 min. Quantitative real-time PCR for Gene expression analysis used a SYBR® Green-based method. Primers of 5 specific genes and housekeeping gene (reference gene) were used in real-time analysis using (Rotor-Gene, Düsseldorf, Germany). A total reaction volume of 20 µL was used. Reactions included 2 µL of template, 10 µL of SYBR Green Master Mix, 2 µL of reverse primer, 2 µL of forward primer, and sterile distilled water for a total volume of 20 µL. PCR assays were performed using the following conditions: 95 °C for 15 min followed by 40 cycles of 95 °C for 30 s and 58 °C for 30 s. The CT of each sample was used to calculate ΔCT values (target gene CT subtracted from β-Actin and tubulin gene CT). The relative gene expression was determined using the $2^{-\Delta\Delta Ct}$ method [104].

4.9. Statistical Analysis

Data are mean (±SE) of three replicates, and for testing significance of data, Duncan's Multiple Range Test was performed using One-Way ANOVA, and the least significant difference (LSD) was calculated at $p < 0.05$.

5. Conclusions

Conclusively, it can be said that salinity reduces the growth of wheat through alterations in the physiological and biochemical parameters studied. NaCl treatment increased the lipid peroxidation inducing membrane dysfunction, hence leading to impeding the uptake of important mineral elements. However, the application of NO effectively lessened the negative impact of salinity on growth and physio-biochemical parameters by improving osmolytes and antioxidant metabolism. NO treatment dispelled the salt-stress-mediated ravage by restricting the excess accumulation of Na^+ and better scavenging of ROS through the up-regulated antioxidant system (enzymatic and non-enzymatic). NO-mediated osmoregulation in wheat plants directly affected the ROS scavenging and the mineral nutrients uptake in wheat, leading to significantly alleviated NaCl stress. This protective effect of NO extended to the molecular level by affecting the relative gene expression of the ionic homeostasis (*SOS1/NHX1*), osmotic (*AQP/OSM-34*), and photosystem II (D2-protein)-related proteins. We recommend that future works focus on how the responses of wheat genotypes contrast in salt tolerance to NO treatment under both normal and salt-stress conditions (in different stages of wheat ontogenesis).

Author Contributions: Conceptualization, M.H.S. and G.S.H.A.; methodology, S.H.Q., A.A.E., M.H.S., G.A.E.-S., M.F.M.I., M.E. and N.E.; software, S.H.Q., A.A.E., M.H.S., G.A.E.-S., M.F.M.I., M.E., Y.S.A.M., K.A. and N.E.; validation, S.H.Q., A.A.E., M.H.S., G.A.E.-S., M.F.M.I., M.E. and N.E.; formal analysis, S.H.Q., A.A.E., M.H.S., G.A.E.-S., M.F.M.I., M.E. and N.E.; investigation, G.S.H.A., S.H.Q., A.A.E., M.H.S., G.A.E.-S., M.F.M.I., Y.S.A.M., K.A., M.E. and N.E.; resources, G.S.H.A., A.A.E., M.H.S., G.A.E.-S., M.F.M.I., M.E., Y.S.A.M., K.A. and N.E.; data curation, A.A.E., M.H.S., M.F.M.I. and N.E.; writing—original draft preparation A.A.E., M.H.S., M.F.M.I. and N.E.; writing—review and editing, S.H.Q., A.A.E., M.H.S., Y.S.A.M., K.A., G.A.E.-S., M.F.M.I., M.E. and N.E.; visualization, S.H.Q., M.E., Y.S.A.M., K.A. and N.E. All authors have read and agreed to the published version of the manuscript.

Funding: The authors acknowledge the Deanship of Scientific Research at King Khalid University for funding this work through the Program of Research Groups under grant number (RGP 01/258/42).

Data Availability Statement: Not applicable.

Acknowledgments: The authors strongly acknowledge the Deanship of Scientific Research at King Khalid University for funding this work through the Program of Research Groups under grant number (RGP 01/258/42). Y.S.A.M. and K.A. would like to extend their appreciation to the Graduate Studies and Research Affairs Sector of Kafrelsheikh University and Tanta University, Egypt. We also thank all staff members of our laboratories for their helpful discussions and comments.

Conflicts of Interest: The authors declare no conflict of interest.

References

1. Ibrahim, M.F.M.; Elbar, O.H.A.; Farag, R.; Hikal, M.; El-Kelish, A.; El-Yazied, A.A.; Alkahtani, J.; El-Gawad, H.G.A. Melatonin Counteracts Drought Induced Oxidative Damage and Stimulates Growth, Productivity and Fruit Quality Properties of Tomato Plants. *Plants* **2020**, *9*, 1276. [CrossRef]
2. Abd Elhady, S.A.; El-Gawad, H.G.A.; Ibrahim, M.F.M.; Mukherjee, S.; Elkelish, A.; Azab, E.; Gobouri, A.A.; Farag, R.; Ibrahim, H.A.; El-Azm, N.A. Hydrogen Peroxide Supplementation in Irrigation Water Alleviates Drought Stress and Boosts Growth and Productivity of Potato Plants. *Sustainability* **2021**, *13*, 899. [CrossRef]
3. Eid, A.M.; Fouda, A.; Abdel-Rahman, M.A.; Salem, S.S.; Elsaied, A.; Oelmüller, R.; Hijri, M.; Bhowmik, A.; Elkelish, A.; Hassan, S.E.-D. Harnessing Bacterial Endophytes for Promotion of Plant Growth and Biotechnological Applications: An Overview. *Plants* **2021**, *10*, 935. [CrossRef] [PubMed]
4. Ismail, M.A.; Amin, M.A.; Eid, A.M.; Hassan, S.E.-D.; Mahgoub, H.A.; Lashin, I.; Abdelwahab, A.T.; Azab, E.; Gobouri, A.A.; Elkelish, A.; et al. Comparative Study between Exogenously Applied Plant Growth Hormones versus Metabolites of Microbial Endophytes as Plant Growth-Promoting for *Phaseolus Vulgaris* L. *Cells* **2021**, *10*, 1059. [CrossRef] [PubMed]
5. Nahhas, N.E.; Abdelaal, K.A.; AlKahtani, M.D.; Al Husnain, L.; AlGwaiz, H.I.; Hafez, Y.M.; Attia, K.A.; El-Esawi, M.A.; Ibrahim, M.F.; Elkelish, A. Biochar and Jasmonic Acid Application Attenuate Antioxidative Systems and Improves Growth, Physiology, Nutrient Uptake and Productivity of Faba Bean (*Vicia Faba* L.) Irrigated with Saline Water. *Plant Physiol. Biochem.* **2021**, *166*, 807–817. [CrossRef] [PubMed]
6. Khalil, A.M.A.; Hassan, S.E.-D.; Alsharif, S.M.; Eid, A.M.; Ewais, E.E.-D.; Azab, E.; Gobouri, A.A.; Elkelish, A.; Fouda, A. Isolation and Characterization of Fungal Endophytes Isolated from Medicinal Plant Ephedra Pachyclada as Plant Growth-Promoting. *Biomolecules* **2021**, *11*, 140. [CrossRef] [PubMed]
7. Elsaeed, S.M.; Zaki, E.G.; Ibrahim, T.M.; Ibrahim Talha, N.; Saad, H.A.; Gobouri, A.A.; Elkelish, A.; Mohamed el-kousy, S. Biochar Grafted on CMC-Terpolymer by Green Microwave Route for Sustainable Agriculture. *Agriculture* **2021**, *11*, 350. [CrossRef]
8. Kibria, M.G.; Hossain, M.; Murata, Y.; Hoque, M.A. Antioxidant Defense Mechanisms of Salinity Tolerance in Rice Genotypes. *Rice Sci.* **2017**, *24*, 155–162. [CrossRef]
9. Ibrahim, M.F.M.; Ibrahim, H.A.; Abd El-Gawad, H.G. Folic acid as a protective agent in snap bean plants under water deficit conditions. *J. Hortic. Sci. Biotechnol.* **2021**, *96*, 94–109. [CrossRef]
10. Abd El-Gawad, H.G.; Mukherjee, S.; Farag, R.; Abd Elbar, O.H.; Hikal, M.; Abou El-Yazied, A.; Abd Elhady, S.A.; Helal, N.; ElKelish, A.; El Nahhas, N.; et al. Exogenous γ-Aminobutyric Acid (GABA)-Induced Signaling Events and Field Performance Associated with Mitigation of Drought Stress in *Phaseolus Vulgaris* L. *Plant Signal. Behav.* **2021**, *16*, 1853384. [CrossRef]
11. Yan, N.; Marschner, P.; Cao, W.; Zuo, C.; Qin, W. Influence of Salinity and Water Content on Soil Microorganisms. *Int. Soil Water Conserv. Res.* **2015**, *3*, 316–323. [CrossRef]
12. Hashim, A.M.; Elkelish, A.; Alhaithloul, H.A.; El-hadidy, S.M.; Farouk, H. Environmental Monitoring and Prediction of Land Use and Land Cover Spatio-Temporal Changes: A Case Study from El-Omayed Biosphere Reserve, Egypt. *Environ. Sci. Pollut. Res.* **2020**, *27*, 42881–42897. [CrossRef] [PubMed]
13. Bin-Jumah, M.; Abdel-Fattah, A.-F.M.; Saied, E.M.; El-Seedi, H.R.; Abdel-Daim, M.M. Acrylamide-Induced Peripheral Neuropathy: Manifestations, Mechanisms, and Potential Treatment Modalities. *Environ. Sci. Pollut. Res.* **2021**, *28*, 13031–13046. [CrossRef]
14. Ashraf, M.; Harris, P.J.C. Photosynthesis under Stressful Environments: An Overview. *Photosynthetica* **2013**, *51*, 163–190. [CrossRef]
15. Azab, E.; Hegazy, A.K.; Gobouri, A.A.; Elkelish, A. Impact of Transgenic Arabidopsis Thaliana Plants on Herbicide Isoproturon Phytoremediation through Expressing Human Cytochrome P450-1A2. *Biology* **2020**, *9*, 362. [CrossRef]
16. Din, A.F.; Ibrahim, M.F.; Farag, R.; El-Gawad, H.G.A.; El-Banhawy, A.; Alaraidh, I.A.; Rashad, Y.M.; Lashin, I.; El-Yazied, A.A.; Elkelish, A.; et al. Influence of Polyethylene Glycol on Leaf Anatomy, Stomatal Behavior, Water Loss, and some Physiological Traits of Date Palm Plantlets Grown In Vitro and Ex Vitro. *Plants* **2020**, *9*, 1440. [CrossRef]
17. Acosta-Motos, J.R.; Ortuño, M.F.; Bernal-Vicente, A.; Diaz-Vivancos, P.; Sanchez-Blanco, M.J.; Hernandez, J.A. Plant Responses to Salt Stress: Adaptive Mechanisms. *Agronomy* **2017**, *7*, 18. [CrossRef]
18. Salem, S.S.; El-Belely, E.F.; Niedbała, G.; Alnoman, M.M.; Hassan, S.E.-D.; Eid, A.M.; Shaheen, T.I.; Elkelish, A.; Fouda, A. Bactericidal and In-Vitro Cytotoxic Efficacy of Silver Nanoparticles (Ag-NPs) Fabricated by Endophytic Actinomycetes and Their Use as Coating for the Textile Fabrics. *Nanomaterials* **2020**, *10*, 2082. [CrossRef]
19. Ahanger, M.A.; Agarwal, R.M. Salinity Stress Induced Alterations in Antioxidant Metabolism and Nitrogen Assimilation in Wheat (*Triticum Aestivum* L.) as Influenced by Potassium Supplementation. *Plant Physiol. Biochem.* **2017**, *115*, 449–460. [CrossRef]
20. Ibrahim, M.F.; El-Samad, A.; Ashour, H.; El-Sawy, A.M.; Hikal, M.; Elkelish, A.; El-Gawad, H.A.; El-Yazied, A.A.; Hozzein, W.N.; Farag, R. Regulation of Agronomic Traits, Nutrient Uptake, Osmolytes and Antioxidants of Maize as Influenced by Exogenous Potassium Silicate under Deficit Irrigation and Semiarid Conditions. *Agronomy* **2020**, *10*, 1212. [CrossRef]
21. Hashim, A.M.; Alharbi, B.M.; Abdulmajeed, A.M.; Elkelish, A.; Hozzein, W.N.; Hassan, H.M. Oxidative Stress Responses of Some Endemic Plants to High Altitudes by Intensifying Antioxidants and Secondary Metabolites Content. *Plants* **2020**, *9*, 869. [CrossRef]
22. Farmer, E.E.; Mueller, M.J. ROS-Mediated Lipid Peroxidation and RES-Activated Signaling. *Annu. Rev. Plant Biol.* **2013**, *64*, 429–450. [CrossRef]

23. Soliman, M.; Qari, S.H.; Abu-Elsaoud, A.; El-Esawi, M.; Alhaithloul, H.; Elkelish, A. Rapid Green Synthesis of Silver Nanoparticles from Blue Gum Augment Growth and Performance of Maize, Fenugreek, and Onion by Modulating Plants Cellular Antioxidant Machinery and Genes Expression. *Acta Physiol. Plant.* **2020**, *42*, 148. [CrossRef]
24. Krishnamurthy, A.; Rathinasabapathi, B. Oxidative Stress Tolerance in Plants. *Plant Signal. Behav.* **2013**, *8*, e25761. [CrossRef] [PubMed]
25. Sharma, P.; Jha, A.B.; Dubey, R.S.; Pessarakli, M. Reactive Oxygen Species, Oxidative Damage, and Antioxidative Defense Mechanism in Plants under Stressful Conditions. *J. Bot.* **2012**, *2012*, e217037. [CrossRef]
26. Rashad, Y.; Aseel, D.; Hammad, S.; Elkelish, A. Rhizophagus Irregularis and Rhizoctonia Solani Differentially Elicit Systemic Transcriptional Expression of Polyphenol Biosynthetic Pathways Genes in Sunflower. *Biomolecules* **2020**, *10*, 379. [CrossRef]
27. Tripathy, B.C.; Oelmüller, R. Reactive Oxygen Species Generation and Signaling in Plants. *Plant Signal. Behav.* **2012**, *7*, 1621–1633. [CrossRef] [PubMed]
28. Soliman, M.H.; Abdulmajeed, A.M.; Alhaithloul, H.; Alharbi, B.M.; El-Esawi, M.A.; Hasanuzzaman, M.; Elkelish, A. Saponin Biopriming Positively Stimulates Antioxidants Defense, Osmolytes Metabolism and Ionic Status to Confer Salt Stress Tolerance in Soybean. *Acta Physiol. Plant.* **2020**, *42*, 114. [CrossRef]
29. Panieri, E.; Santoro, M.M. ROS Homeostasis and Metabolism: A Dangerous Liason in Cancer Cells. *Cell Death Dis.* **2016**, *7*, e2253. [CrossRef]
30. Soliman, M.; Elkelish, A.; Souad, T.; Alhaithloul, H.; Farooq, M. Brassinosteroid Seed Priming with Nitrogen Supplementation Improves Salt Tolerance in Soybean. *Physiol. Mol. Biol. Plants* **2020**, *26*, 501–511. [CrossRef]
31. Elkelish, A.A.; Alhaithloul, H.A.S.; Qari, S.H.; Soliman, M.H.; Hasanuzzaman, M. Pretreatment with Trichoderma Harzianum Alleviates Waterlogging-Induced Growth Alterations in Tomato Seedlings by Modulating Physiological, Biochemical, and Molecular Mechanisms. *Environ. Exp. Bot.* **2020**, *171*, 103946. [CrossRef]
32. Soliman, M.H.; Alayafi, A.A.M.; El Kelish, A.A.; Abu-Elsaoud, A.M. Acetylsalicylic Acid Enhance Tolerance of Phaseolus Vulgaris, L. to Chilling Stress, Improving Photosynthesis, Antioxidants and Expression of Cold Stress Responsive Genes. *Bot. Stud.* **2018**, *59*, 6. [CrossRef] [PubMed]
33. Baghour, M.; Gálvez, F.J.; Sánchez, M.E.; Aranda, M.N.; Venema, K.; Rodríguez-Rosales, M.P. Overexpression of LeNHX2 and SlSOS2 Increases Salt Tolerance and Fruit Production in Double Transgenic Tomato Plants. *Plant Physiol. Biochem.* **2019**, *135*, 77–86. [CrossRef]
34. Ismail, H.; Maksimović, J.D.; Maksimović, V.; Shabala, L.; Živanović, B.D.; Tian, Y.; Jacobsen, S.-E.; Shabala, S.; Ismail, H.; Maksimović, J.D.; et al. Rutin, a Flavonoid with Antioxidant Activity, Improves Plant Salinity Tolerance by Regulating K^+ Retention and Na^+ Exclusion from Leaf Mesophyll in Quinoa and Broad Beans. *Funct. Plant Biol.* **2015**, *43*, 75–86. [CrossRef] [PubMed]
35. Falahi, H.; Sharifi, M.; Ahmadian Chashmi, N.; Zare Maivan, H. Water Stress Alleviation by Polyamines and Phenolic Compounds in Scrophularia Striata Is Mediated by NO and H_2O_2. *Plant Physiol. Biochem.* **2018**, *130*, 139–147. [CrossRef] [PubMed]
36. Du, S.-T.; Liu, Y.; Zhang, P.; Liu, H.-J.; Zhang, X.-Q.; Zhang, R.-R. Atmospheric Application of Trace Amounts of Nitric Oxide Enhances Tolerance to Salt Stress and Improves Nutritional Quality in Spinach (*Spinacia oleracea* L.). *Food Chem.* **2015**, *173*, 905–911. [CrossRef]
37. Floryszak-Wieczorek, J.; Arasimowicz-Jelonek, M.; Izbiańska, K. The Combined Nitrate Reductase and Nitrite-Dependent Route of NO Synthesis in Potato Immunity to Phytophthora Infestans. *Plant Physiol. Biochem.* **2016**, *108*, 468–477. [CrossRef]
38. Freschi, L. Nitric Oxide and Phytohormone Interactions: Current Status and Perspectives. *Front. Plant Sci.* **2013**, *4*, 398. [CrossRef]
39. Gaber, A.; Alsanie, W.F.; Kumar, D.N.; Refat, M.S.; Saied, E.M. Novel Papaverine Metal Complexes with Potential Anticancer Activities. *Molecules* **2020**, *25*, 5447. [CrossRef]
40. Ali, Q.; Daud, M.K.; Haider, M.Z.; Ali, S.; Rizwan, M.; Aslam, N.; Noman, A.; Iqbal, N.; Shahzad, F.; Deeba, F.; et al. Seed Priming by Sodium Nitroprusside Improves Salt Tolerance in Wheat (*Triticum Aestivum* L.) by Enhancing Physiological and Biochemical Parameters. *Plant Physiol. Biochem.* **2017**, *119*, 50–58. [CrossRef]
41. Fatma, M.; Khan, N.A. Nitric Oxide Protects Photosynthetic Capacity Inhibition by Salinity in Indian Mustard. *J. Func. Environ. Bot.* **2014**, *4*, 106. [CrossRef]
42. Melo, N.K.G.; Bianchetti, R.E.; Lira, B.S.; Oliveira, P.M.R.; Zuccarelli, R.; Dias, D.L.O.; Demarco, D.; Peres, L.E.P.; Rossi, M.; Freschi, L. Nitric Oxide, Ethylene, and Auxin Cross Talk Mediates Greening and Plastid Development in Deetiolating Tomato Seedlings. *Plant Physiol.* **2016**, *170*, 2278–2294. [CrossRef]
43. Shewry, P.R.; Hey, S.J. The Contribution of Wheat to Human Diet and Health. *Food Energy Secur.* **2015**, *4*, 178–202. [CrossRef]
44. Elkelish, A.A.; Soliman, M.H.; Alhaithloul, H.A.; El-Esawi, M.A. Selenium Protects Wheat Seedlings against Salt Stress-Mediated Oxidative Damage by up-Regulating Antioxidants and Osmolytes Metabolism. *Plant Physiol. Biochem.* **2019**, *137*, 144–153. [CrossRef]
45. Ke, Q.; Wang, Z.; Ji, C.Y.; Jeong, J.C.; Lee, H.-S.; Li, H.; Xu, B.; Deng, X.; Kwak, S.-S. Transgenic Poplar Expressing CodA Exhibits Enhanced Growth and Abiotic Stress Tolerance. *Plant Physiol. Biochem.* **2016**, *100*, 75–84. [CrossRef] [PubMed]
46. Egbichi, I.; Keyster, M.; Ludidi, N. Effect of Exogenous Application of Nitric Oxide on Salt Stress Responses of Soybean. *South Afr. J. Bot.* **2014**, *90*, 131–136. [CrossRef]
47. Shokat, S.; Großkinsky, D.K. Tackling Salinity in Sustainable Agriculture—What Developing Countries May Learn from Approaches of the Developed World. *Sustainability* **2019**, *11*, 4558. [CrossRef]

48. Wani, S.H.; Dutta, T.; Neelapu, N.R.R.; Surekha, C. Transgenic Approaches to Enhance Salt and Drought Tolerance in Plants. *Plant Gene* **2017**, *11*, 219–231. [CrossRef]
49. Shi, Q.; Ding, F.; Wang, X.; Wei, M. Exogenous Nitric Oxide Protect Cucumber Roots against Oxidative Stress Induced by Salt Stress. *Plant Physiol. Biochem.* **2007**, *45*, 542–550. [CrossRef]
50. Mostofa, M.G.; Fujita, M.; Tran, L.S.P. Nitric Oxide Mediates Hydrogen Peroxide- and Salicylic Acid-Induced Salt Tolerance in Rice (*Oryza Sativa* L.) Seedlings. *Plant Growth Regul.* **2015**, *77*, 265–277. [CrossRef]
51. Asgher, M.; Per, T.S.; Masood, A.; Fatma, M.; Freschi, L.; Corpas, F.J.; Khan, N.A. Nitric Oxide Signaling and Its Crosstalk with Other Plant Growth Regulators in Plant Responses to Abiotic Stress. *Environ. Sci. Pollut. Res. Int.* **2017**, *24*, 2273–2285. [CrossRef]
52. Ahmad, N.; Malagoli, M.; Wirtz, M.; Hell, R. Drought Stress in Maize Causes Differential Acclimation Responses of Glutathione and Sulfur Metabolism in Leaves and Roots. *BMC Plant Biol.* **2016**, *16*, 247. [CrossRef] [PubMed]
53. Fan, H.-F.; Du, C.-X.; Ding, L.; Xu, Y.-L. Effects of Nitric Oxide on the Germination of Cucumber Seeds and Antioxidant Enzymes under Salinity Stress. *Acta Physiol. Plant.* **2013**, *35*, 2707–2719. [CrossRef]
54. Ahmad, P.; Ahanger, M.A.; Alyemeni, M.N.; Wijaya, L.; Alam, P. Exogenous Application of Nitric Oxide Modulates Osmolyte Metabolism, Antioxidants, Enzymes of Ascorbate-Glutathione Cycle and Promotes Growth under Cadmium Stress in Tomato. *Protoplasma* **2018**, *255*, 79–93. [CrossRef]
55. Marvasi, M. Potential Use and Perspectives of Nitric Oxide Donors in Agriculture. *J. Sci. Food Agric.* **2017**, *97*, 1065–1072. [CrossRef] [PubMed]
56. Dong, Y.J.; Jinc, S.S.; Liu, S.; Xu, L.L.; Kong, J. Effects of Exogenous Nitric Oxide on Growth of Cotton Seedlings under NaCl Stress. *J. Soil Sci. Plant Nutr.* **2014**, *14*, 1–13. [CrossRef]
57. Basalah, M.; Ali, H.; Al-whaibi, M.; Siddiqui, D.M.H.; Sakran, A.; Alsahli, A. Nitric Oxide and Salicylic Acid Mitigate Cadmium Stress in Wheat Seedlings. *J. Pure Appl. Microbiol.* **2013**, *7*, 139–148.
58. Simontacchi, M.; Galatro, A.; Ramos-Artuso, F.; Santa-María, G.E. Plant Survival in a Changing Environment: The Role of Nitric Oxide in Plant Responses to Abiotic Stress. *Front. Plant Sci.* **2015**, *6*, 977. [CrossRef]
59. Lu, S.; Zhuo, C.; Wang, X.; Guo, Z. Nitrate Reductase (NR)-Dependent NO Production Mediates ABA- and H_2O_2-Induced Antioxidant Enzymes. *Plant Physiol. Biochem.* **2014**, *74*, 9–15. [CrossRef]
60. Kolbert, Z. Implication of Nitric Oxide (NO) in Excess Element-Induced Morphogenic Responses of the Root System. *Plant Physiol. Biochem.* **2016**, *101*, 149–161. [CrossRef]
61. Shabala, S.; Pottosin, I. Regulation of Potassium Transport in Plants under Hostile Conditions: Implications for Abiotic and Biotic Stress Tolerance. *Physiol. Plant.* **2014**, *151*, 257–279. [CrossRef]
62. Jabborova, D.; Annapurna, K.; Paul, S.; Kumar, S.; Saad, H.A.; Desouky, S.; Ibrahim, M.F.M.; Elkelish, A. Beneficial Features of Biochar and Arbuscular Mycorrhiza for Improving Spinach Plant Growth, Root Morphological Traits, Physiological Properties, and Soil Enzymatic Activities. *JoF* **2021**, *7*, 571. [CrossRef]
63. Elkelish, A.; Ibrahim, M.F.M.; Ashour, H.; Bondok, A.; Mukherjee, S.; Aftab, T.; Hikal, M.; El-Yazied, A.A.; Azab, E.; Gobouri, A.A.; et al. Exogenous Application of Nitric Oxide Mitigates Water Stress and Reduces Natural Viral Disease Incidence of Tomato Plants Subjected to Deficit Irrigation. *Agronomy* **2021**, *11*, 87. [CrossRef]
64. Turan, S.; Tripathy, B.C. Salt-Stress Induced Modulation of Chlorophyll Biosynthesis during de-Etiolation of Rice Seedlings. *Physiol. Plant.* **2015**, *153*, 477–491. [CrossRef]
65. Fatma, M.; Masood, A.; Per, T.S.; Khan, N.A. Nitric Oxide Alleviates Salt Stress Inhibited Photosynthetic Performance by Interacting with Sulfur Assimilation in Mustard. *Front. Plant Sci.* **2016**, *7*, 521. [CrossRef]
66. Procházková, D.; Haisel, D.; Wilhelmová, N.; Pavlíková, D.; Száková, J. Effects of Exogenous Nitric Oxide on Photosynthesis. *Photosynthetica* **2013**, *51*, 483–489. [CrossRef]
67. Azooz, M.; Youssef, A.; Ahmad, P. Evaluation of Salicylic Acid (SA) Application on Growth, Osmotic Solutes and Antioxidant Enzyme Activities on Broad Bean Seedlings Grown under Diluted Seawater. *Inter. J. Plant Physiol. Biochem.* **2011**, *3*, 253–264. [CrossRef]
68. Bahrami-Rad, S.; Hajiboland, R. Effect of Potassium Application in Drought-Stressed Tobacco (*Nicotiana Rustica* L.) Plants: Comparison of Root with Foliar Application. *Ann. Agric. Sci.* **2017**, *62*, 121–130. [CrossRef]
69. Munns, R. Comparative Physiology of Salt and Water Stress. *Plant Cell Environ.* **2002**, *25*, 239–250. [CrossRef]
70. Kaldenhoff, R.; Ribas-Carbo, M.; Sans, J.F.; Lovisolo, C.; Heckwolf, M.; Uehlein, N. Aquaporins and Plant Water Balance. *Plant Cell Environ.* **2008**, *31*, 658–666. [CrossRef]
71. Dawood, M.G.; El-Awadi, M.E. Alleviation of Salinity Stress on *Vicia Faba* L. Plants via Seed Priming with Melatonin. *Acta Biológica Colomb.* **2015**, *20*, 223–235. [CrossRef]
72. Abdelgawad, H.; Zinta, G.; Hegab, M.M.; Pandey, R. High Salinity Induces Different Oxidative Stress and Antioxidant Responses in Maize Seedlings Organs. *Front. Plant Sci.* **2016**, *7*, 276. [CrossRef] [PubMed]
73. Saleh, A.; Abu-Elsaoud, A.; El Kelish, A.; Sahadad, M.; Mohamed, E. Role of External Proline on Enhancing Defence Mechanisms of *Vicia Faba* L. against Ultraviolet Radiation. *Am. Eurasian J. Sustain. Agric.* **2015**, *9*, 22–34.
74. Elkahoui, S.; Hernández, J.A.; Abdelly, C.; Ghrir, R.; Limam, F. Effects of Salt on Lipid Peroxidation and Antioxidant Enzyme Activities of Catharanthus Roseus Suspension Cells. *Plant Sci.* **2005**, *168*, 607–613. [CrossRef]
75. Hernandez, M.; Fernandez-Garcia, N.; Diaz-Vivancos, P.; Olmos, E. A Different Role for Hydrogen Peroxide and the Antioxidative System under Short and Long Salt Stress in Brassica Oleracea Roots. *J. Exp. Bot.* **2010**, *61*, 521–535. [CrossRef] [PubMed]

76. Carrasco-Ríos, L.; Pinto, M. Effect of Salt Stress on Antioxidant Enzymes and Lipid Peroxidation in Leaves in Two Contrasting Corn, 'Lluteno' and 'Jubilee'. *Chil. J. Agric. Res.* **2014**, *74*, 89–95. [CrossRef]
77. Lamattina, L.; García-Mata, C.; Graziano, M.; Pagnussat, G. Nitric Oxide: The Versatility of an Extensive Signal Molecule. *Annu. Rev. Plant Biol.* **2003**, *54*, 109–136. [CrossRef]
78. Corpas, F.J.; Barroso, J.B. Functions of Nitric Oxide (NO) in Roots during Development and under Adverse Stress Conditions. *Plants* **2015**, *4*, 240–252. [CrossRef]
79. Ayala, A.; Muñoz, M.F.; Argüelles, S. Lipid Peroxidation: Production, Metabolism, and Signaling Mechanisms of Malondialdehyde and 4-Hydroxy-2-Nonenal. *Oxid. Med. Cell Longev.* **2014**, *2014*, 360438. [CrossRef]
80. Dutta, D.; Esmaili, M.; Overduin, M.; Fliegel, L. Expression and Detergent Free Purification and Reconstitution of the Plant Plasma Membrane Na^+/H^+ Antiporter SOS1 Overexpressed in Pichia Pastoris. *Biochim. Biophys. Acta Biomembr.* **2020**, *1862*, 183111. [CrossRef]
81. Afzal, Z.; Howton, T.C.; Sun, Y.; Mukhtar, M.S. The Roles of Aquaporins in Plant Stress Responses. *J. Dev. Biol.* **2016**, *4*, 9. [CrossRef] [PubMed]
82. Hakim; Ullah, A.; Hussain, A.; Shaban, M.; Khan, A.H.; Alariqi, M.; Gul, S.; Jun, Z.; Lin, S.; Li, J.; et al. Osmotin: A Plant Defense Tool against Biotic and Abiotic Stresses. *Plant Physiol. Biochem.* **2018**, *123*, 149–159. [CrossRef] [PubMed]
83. Bashir, M.A.; Silvestri, C.; Ahmad, T.; Hafiz, I.A.; Abbasi, N.A.; Manzoor, A.; Cristofori, V.; Rugini, E. Osmotin: A Cationic Protein Leads to Improve Biotic and Abiotic Stress Tolerance in Plants. *Plants* **2020**, *9*, 992. [CrossRef] [PubMed]
84. Kale, R.; Hebert, A.E.; Frankel, L.K.; Sallans, L.; Bricker, T.M.; Pospíšil, P. Amino Acid Oxidation of the D1 and D2 Proteins by Oxygen Radicals during Photoinhibition of Photosystem II. *Proc. Natl. Acad. Sci. USA* **2017**, *114*, 2988–2993. [CrossRef] [PubMed]
85. Goyal, V.; Jhanghel, D.; Mehrotra, S. Emerging Warriors against Salinity in Plants: Nitric Oxide and Hydrogen Sulphide. *Physiol. Plant.* **2021**, *171*, 896–908. [CrossRef]
86. Hasanuzzaman, M.; Inafuku, M.; Nahar, K.; Fujita, M.; Oku, H. Nitric Oxide Regulates Plant Growth, Physiology, Antioxidant Defense, and Ion Homeostasis to Confer Salt Tolerance in the Mangrove Species, Kandelia Obovata. *Antioxidants* **2021**, *10*, 611. [CrossRef]
87. Hoagland, D.R.; Arnon, D.I. The Water-Culture Method for Growing Plants without Soil. *Circ. Calif. Agric. Exp. Stn.* **1950**, *347*, 32.
88. Hiscox, J.D.; Israelstam, G.F. Erratum: A Method for the Extraction of Chlorophyll from Leaf Tissue without Maceration. *Can. J. Bot.* **2011**, *58*, 403. [CrossRef]
89. Smart, R.E.; Bingham, G.E. Rapid Estimates of Relative Water Content. *Plant Physiol.* **1974**, *53*, 258–260. [CrossRef]
90. Sadasivam, S.; Manickam, A. *Biochemical Methods*; New Age International: New Delhi, India, 1996; ISBN 978-81-224-0976-5.
91. Shields, R.; Burnett, W. Determination of Protein-Bound Carbohydrate in Serum by Modified Anthrone Method. *Anal. Chem.* **1960**, *32*, 885–886. [CrossRef]
92. Ibrahim, M.; Bondok, A.; Al-Senosy, N.K.; Younis, R. Stimulation some of defense mechanisms in tomato plants under water deficit and Tobacco mosaic virus (TMV). *World J. Agric. Sci.* **2015**, *11*, 289–302.
93. Heath, R.L.; Packer, L. Photoperoxidation in Isolated Chloroplasts: I. Kinetics and Stoichiometry of Fatty Acid Peroxidation. *Arch. Biochem. Biophys.* **1968**, *125*, 189–198. [CrossRef]
94. Cheeseman, J.M. Hydrogen Peroxide Concentrations in Leaves under Natural Conditions. *J. Exp. Bot.* **2006**, *57*, 2435–2444. [CrossRef]
95. Lowry, O.H.; Rosebrough, N.J.; Farr, A.L.; Randall, R.J. Protein measurement with the folin phenol reagent. *J. Biol. Chem.* **1951**, *193*, 265–275. [CrossRef]
96. Beyer, W.F.; Fridovich, I. Assaying for Superoxide Dismutase Activity: Some Large Consequences of Minor Changes in Conditions. *Anal. Biochem.* **1987**, *161*, 559–566. [CrossRef]
97. Aebi, H. Catalase in vitro. In *Methods in Enzymology*; Oxygen Radicals in Biological Systems; Academic Press: Cambridge, MA, USA, 1984; Volume 105, pp. 121–126.
98. Nakano, Y.; Asada, K. Hydrogen Peroxide Is Scavenged by Ascorbate-Specific Peroxidase in Spinach Chloroplasts. *Plant. Cell Physiol.* **1981**, *22*, 867–880. [CrossRef]
99. Carlberg, I.; Mannervik, B. Glutathione reductase. In *Methods in Enzymology*; Glutamate, Glutamine, Glutathione, and Related Compounds; Academic Press: Cambridge, MA, USA, 1985; Volume 113, pp. 484–490.
100. Mukherjee, S.P.; Choudhuri, M.A. Implications of Water Stress-Induced Changes in the Levels of Endogenous Ascorbic Acid and Hydrogen Peroxide in Vigna Seedlings. *Physiol. Plant.* **1983**, *58*, 166–170. [CrossRef]
101. Ellman, G.L. Tissue Sulfhydryl Groups. *Arch. Biochem. Biophys.* **1959**, *82*, 70–77. [CrossRef]
102. Wolf, B. A Comprehensive System of Leaf Analyses and Its Use for Diagnosing Crop Nutrient Status. *Commun. Soil Sci. Plant. Anal.* **1982**, *13*, 1035–1059. [CrossRef]
103. Subbiah, B.V.; Asija, G.L. A Rapid Procedure for the Estimation of Available Nitrogen in Soils. *Curr. Sci.* **1956**, *25*, 259–260.
104. Togawa, T.; Dunn, W.A.; Emmons, A.C.; Nagao, J.; Willis, J.H. Developmental Expression Patterns of Cuticular Protein Genes with the R&R Consensus from Anopheles Gambiae. *Insect Biochem. Mol. Biol.* **2008**, *38*, 508–519. [CrossRef] [PubMed]

plants

MDPI

Article

Unraveling the Influence of Land-Use Change on $\delta^{13}C$, $\delta^{15}N$, and Soil Nutritional Status in Coniferous, Broadleaved, and Mixed Forests in Southern China: A Field Investigation

Taimoor Hassan Farooq [1,2,3,*], Xiaoyong Chen [2,4], Awais Shakoor [5], Yong Li [2,3], Jun Wang [2,3], Muhammad Haroon U. Rashid [6], Uttam Kumar [7] and Wende Yan [2,3,*]

1 Bangor College China, a Joint Unit of Bangor University, Wales, UK and Central South University of Forestry and Technology, Changsha 410004, China
2 National Engineering Laboratory for Applied Technology in Forestry and Ecology in South China, Central South University of Forestry and Technology, Changsha 410004, China
3 College of Life Science and Technology, Central South University of Forestry and Technology, Changsha 410004, China
4 College of Arts and Sciences, Governors State University, University Park, IL 60484, USA
5 Department of Environment and Soil Sciences, University of Lleida, Avinguda Alcalde Rovira Roure 191, 25198 Lleida, Spain
6 College of Forestry, Fujian Agriculture and Forestry University, Fuzhou 350002, China
7 College of Plant Protection, Fujian Agriculture and Forestry University, Fuzhou 350002, China
* Correspondence: taimoorhassan2055@gmail.com (T.H.F.); csfuywd@hotmail.com (W.Y.)

Citation: Farooq, T.H.; Chen, X.; Shakoor, A.; Li, Y.; Wang, J.; Rashid, M.H.U.; Kumar, U.; Yan, W. Unraveling the Influence of Land-Use Change on $\delta^{13}C$, $\delta^{15}N$, and Soil Nutritional Status in Coniferous, Broadleaved, and Mixed Forests in Southern China: A Field Investigation. *Plants* **2021**, *10*, 1499. https://doi.org/10.3390/plants10081499

Academic Editors: M. Iftikhar Hussain, Adele Muscolo and Mukhtar Ahmed

Received: 25 June 2021
Accepted: 17 July 2021
Published: 21 July 2021

Abstract: Natural isotopic abundance in soil and foliar can provide integrated information related to the long-term alterations of carbon (C) and nitrogen (N) cycles in forest ecosystems. We evaluated total carbon (TC), total nitrogen (TN), and isotopic natural abundance of C ($\delta^{13}C$) and N ($\delta^{15}N$) in soil and foliar of coniferous plantation (CPF), natural broadleaved forest (NBF), and mixed forest stands at three different soil depths (i.e., 0–10, 10–20, and 20–40 cm). This study also explored how soil available nutrients are affected by different forest types. Lutou forest research station, located in Hunan Province, central China, was used as the study area. Results demonstrated that the topsoil layer had higher TC and TN content in the mixed forest stand, resulting in a better quality of organic materials in the topsoil layer in the mixed forest than NBF and CPF. In general, soil TC, TN, and $\delta^{15}N$ varied significantly in different soil depths and forest types. However, the forest type did not exhibit any significant effect on $\delta^{13}C$. Overall, soil $\delta^{13}C$ was significantly enriched in CPF, and $\delta^{15}N$ values were enriched in mixed forest. Foliar C content varied significantly among forest types, whereas foliar N content was not significantly different. No big differences were observed for foliar $\delta^{15}N$ and $\delta^{13}C$ across forest types. However, foliar $\delta^{13}C$ and $\delta^{15}N$ were positively related to soil $\delta^{13}C$ and $\delta^{15}N$, respectively. Foliar N, soil and foliar C:N ratio, soil moisture content (SMC), and forest type were observed as the major influential factors affecting isotopic natural abundance, whereas soil pH was not significantly correlated. In addition, forest type change and soil depth increment had a significant effect on soil nutrient availability. In general, soil nutrient availability was higher in mixed forest. Our findings implied that forest type and soil depth alter TC, TN, and soil $\delta^{15}N$, whereas $\delta^{13}C$ was only driven by soil depth. Moreover, plantations led to a decline in soil available nutrient content compared with NBF and mixed forest stands.

Keywords: stable isotope; isotopic composition; C and N cycling; vegetation type; soil health

1. Introduction

Human activities and different natural environmental factors produce abrupt, large scale, irreversible changes and alter forest structure and composition, consequently resulting in the changes of biogeochemical cycles [1,2]. Species composition significantly affects the quality and quantity of carbon (C) and nitrogen (N) input by controlling surface soil and vegetation layer C and N contents and their turnover rates [3–5]. N is the essential element in the natural ecosystem, and is a vital limiting resource for plant growth [5]. The C cycle is crucial as it influences soil respiration and the plant photosynthetic process. The ^{13}C and ^{15}N natural abundance in the soils is a dynamic function of the rate and isotopic composition and the C and N transformations in the forest ecosystem as it provides cohesive insights into C and N cycles [6–8]. Over the years, the isotopic composition of C and N in soil has been used to evaluate soil C turnover rates [9]. Despite its significance, the increase in $\delta^{13}C$ and $\delta^{15}N$ values in soils remains challenging, especially given the various factors that may discriminate against ^{13}C and ^{15}N natural abundance within soil profiles [9]. Therefore, ecologists seek numerous pathways to better understand how changes in the forest type impact the ecosystem functioning.

Land is an indispensable natural resource for ecosystem functioning. In general, soil organic carbon usually changes with human activities and land-use change. Litter inputs usually lower the soil $\delta^{13}C$ and $\delta^{15}N$, whereas a higher decomposition rate enhances $\delta^{13}C$ and $\delta^{15}N$ [10,11]. However, in stable forest vegetation, soil organic carbon turnover rates do not considerably alter with soil depth profiles [7,12]. In addition, the N cycle processes in forests are impacted by vegetation type change, thus influencing soil $\delta^{15}N$ [13–16]. These biogeochemical cycles play an essential role in maintaining the overall ecosystem and soil fertility [17,18].

Plantations are usually fast-growing species with shorter rotation cycles compared to broadleaved natural forests and mixed-species forests. Ever-increasing forest product demand has led to an upsurge in plantation forests worldwide [19]. Since the 1980s, different large-scale afforestation forestry programs have been introduced in China to meet the ever-increasing demand for good quality timber and other forest products, resulting in the conversion of many natural and mixed forests into coniferous monoculture plantations [19,20]. These plantations are man-made and managed differently than natural forest stands. Natural broadleaf forest species are generally more N-rich than coniferous plantations and have less N retention capacity [21]. Moreover, natural forests have typically stable vegetation with natural rotation cycles and less human interference than plantations. Previous studies have described that soil quality indicators and different silvicultural practices such as intercropping/inter cultivation, fallow period, irrigation application techniques, use of organic and inorganic fertilizers, along with human activities, affect greenhouse gas emissions [21–24], soil C and N storage, loss of inorganic N, soil fertility, and soil structure [25,26]. However, in global analyses, details pertaining to fluctuations in soil and foliar ^{13}C and ^{15}N natural abundance and their relation with influential factors is scarce regarding the different sub-tropical forest types of eastern Asia. Soil nutrient availability is a crucial indicator of soil fertility [27]. Soil fertility in forest ecosystems can be influenced by numerous factors such as canopy dynamics, litter quality, parent material, land-use history, species composition, environmental factors, and atmospheric deposition [28–30]. Thus, this can directly affect the isotopic abundance, forest productivity, and sustainability [31–33]. Therefore, studying the $\delta^{15}N$, $\delta^{13}C$, TC, and TN content in leaves and soil, and variability in soil fertility is expected to provide insights into the species' resource utilization efficiency in an ecosystem and the consequences of different silvicultural methods for these biogeochemical cycles.

The objectives of this study were: (i) to characterize the variability of soil and foliar total C (TC), total N (TN) content, and ^{13}C and ^{15}N natural abundance in vertical depth soil profiles in different forest types; and (ii) to analyze the effect of forest type change (different forest types) on soil nutrient availability. Soil and foliar stable isotopic ratios are influenced by atmospheric N_2 inputs and processes that regulate N cycling and soil organic matter (SOM) transformation. Moreover, different forest management regimes influence the C and N isotopic abundance, for example, clear-cutting practices in plantations increase soil N availability, which can enrich $\delta^{15}N$. Therefore, we hypothesize that: (a) TC and TN content and $\delta^{13}C$ and $\delta^{15}N$ in soil and foliar would significantly vary in different forest types and at different soil depths; moreover, (b) forest structure and composition would also considerably influence the soil nutrient availability.

2. Materials and Methods

2.1. Study Area

The study was conducted at the Lutou forest ecosystem observation and research Station, Pingjiang County, Yueyang, Hunan Province, China. The geographic coordinates of the research station are 113°51′52″~113°58′24″ E, 28°31′17″~28°38′00″ N (Figure 1). The total area of the research station is 4762 ha. The soil type is the same in all three forest types (Lateritic red soil). The study area is located in the humid subtropical zone. The climate of the study area is warm and humid with abundant rainfall, sufficient sunshine, and four distinct seasons. The mean annual temperature is 17.07 °C, and the mean annual precipitation is 1312 mm, with an average humidity of 82%.

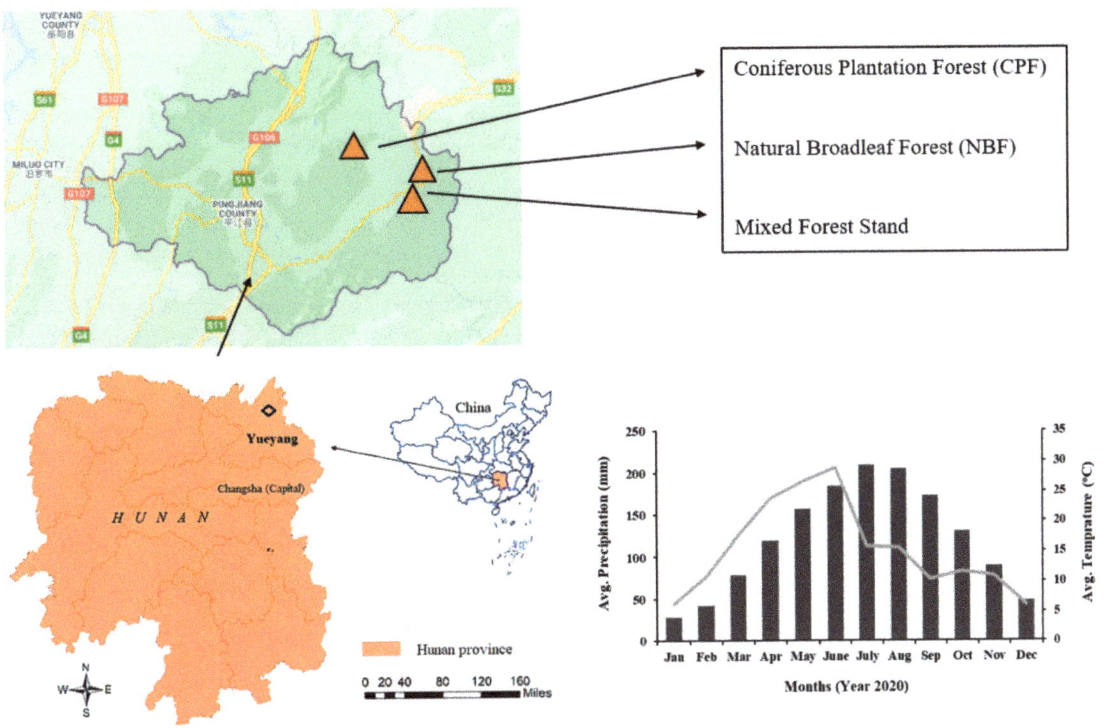

Figure 1. Location and climatic conditions of the study area.

2.2. Forest Types and Species

Natural broadleaved forest (NBF), coniferous plantation forest (CPF), and mixed forest stands were used in the study. NBF and mixed forest stands were adjacent to each other, and the CPF stand was at a 300–400 m (0.19–0.25 mile) distance from the first two forest stands. All three forest stands were present in the same research station. There were no water channels present on the ground in any forest type and watering was natural because of the abundant rainfall. Litterfall biomass and understory vegetation were apparently more in the mixed forest stand. We selected those trees where shrub presence around the canopy projection area of trees was significantly less than the usual shrub presence in the stands. Therefore, the direct influence of the shrub layer would be minimal. Still, these are field conditions; hence indirect effects might be possible. The characteristics of the study area and forest types are shown in Table 1.

Table 1. Characteristics of the study area and selected forest types. Values are mean ± SE. Different small letters represent the significant difference among forest types at $p \leq 0.05$.

	Mixed Forest	NBF [a]	CPF
Species	*C. eyeri* + *P. massoniana* [b]	*C. eyeri*	*C. lanceolata*
Elevation	777 m (a.s.l)	800 m (a.s.l)	427 m (a.s.l)
Soil pH (0–40 cm)	4.60 ± 0.06^{b}	4.40 ± 0.15^{c}	4.81 ± 0.02^{a}
Soil BD (0–40 cm) [c]	1.28 ± 0.03^{a}	1.12 ± 0.08^{c}	1.25 ± 0.03^{b}
SMC (%) (0–40 cm)	16.3^{a}	11.63^{b}	10.48^{b}
Forest age	30–60 years	60 years	30 years

[a] NBF is natural broadleaf forest; CPF is coniferous plantations forest. [b] *C. eyeri*, *Castanopsis eyeri*; *C. lanceolata*, *Cunninghamia lanceolata*; *P. massoniana* is *Pinus massoniana*. [c] BD, soil bulk density; SMC, soil moisture content.; a.s.l, above sea level.

Detailed information about (1) NBF, (2) CPF, and (3) mixed forest stands is mentioned below.

1. The mid-subtropical zone of China is covered with natural forests dominated by broadleaf evergreen species, and *Castanopsis eyeri* (Fagaceae) is one of them [34]. *C. eyeri* is a vital part of forest ecosystems and provides services such as water conservation, biodiversity protection, biomass maintenance, and local climate regulation [35]. The *C. eyeri* (NBF) stand in the research station is the largest and most complete community of Castonpsis, and therefore has crucial protection, scientific research, and landscape values. The forest is neat, and the forest canopy undulating. At the Lutou forest farm, it is mainly distributed on mountain slopes. The average slope gradient is 20–25°. The shrub layer includes species such as *Rhododendron simsii, Ilex formosana, Daphniphyllum oldhami, Toxicodendron vernicifluum, Cyclobalanopsis glauca, Eurya tetragonoclada, Symplocos sumuntia, Rhododendron simiarum, Lindera aggregate, Eurya muricat, Dendropanax dentiger*, and *Myrica rubra*.
2. Chinese fir (*Cunninghamia lanceolata*) is a typical evergreen coniferous timber tree species. The pure natural forest of *C. lanceolata* is rare to find, and mainly presents as an artificial forest plantation. The CPF stand at Lutou forest station are present on both flat soil and mountain slopes. The average slope gradient is 18–23°. Besides timber production, climate regulation, and soil and water conservation, it also provides biomass energy [21,36]. The CPF stand was established 30 years ago under the Chinese afforestation program for forest area enhancement in 1980–1990. It was established as a plantation forest after clearing the site. The shrub layers' primary species include *Litsea cubeba, Eurya mauricata, Rhus chinensis, Sapium discolor, Rhododendron mariesii, Diplospora dubia, Ilex chinensis*, and *Rubus lambertianus*.
3. The *P. massoniana* + *C. eyeri* mixed forest is an important forest type at Lutou forest station. In Lutou, it is distributed in the ridges, hillside, and mountain slopes. The average slope gradient was 20–25°. *P. massoniana* is an evergreen coniferous species that has been widely planted in the red soils of southern China since 1980, mainly for soil conservation purposes [37,38]. At the Lutou forest ecosystem observation and research station, the mixed forest stand is formed by planting *P. massoniana* in the natural regenerated *C. eyeri*. The shrub layer primarily comprises *Loropetalum chinensis, Rhododendron longipetalon, Ilex latifolia, Camellia oleifera, Phobe hupehensis, Ilex ficifolia, Eugenia glabra, Rubus cranbergii, Rata thunbergia*, etc.

2.3. Soil and Foliar Sampling

Sampling was carried out in the first week of October 2020. Soil samples were collected from nine selected locations (three locations from each forest type). Three 20 × 20 m plots were established in each forest stand. The soil samples were obtained with a steel soil auger (3.5 cm diameter) up to a depth of 0–10, 10–20, and 20–40 cm soil. After collection, soil samples were appropriately cleaned, and roots/stones were sorted out. Three composite samples per forest type per soil depth were prepared for the analysis. A sufficient number of mature and healthy foliage (5–8 per tree) of each forest type (three plots in each stand) was collected from four to six different individuals (at least 5 m apart from each other) using a long-handled pruner and pooled three composite samples for each forest type. Leaves were placed in a clean, perforated plastic bag [15]. We tried to select the plots with uniform topography to minimize the local terrain impact on trees/vegetation.

Field moist soils were air-dried at room temperature and sieved through a 2 mm mesh size to remove stones, roots, and plant residues. The foliage samples were gently rinsed with distilled water and oven-dried for 72 h at 65 °C. Later, soil and foliar samples were finely ground with a ball mill (JXFSTPRP-64, Jingxin Co. Ltd., China) and used to measure $\delta^{13}C$ and $\delta^{15}N$, C and N content, SOC, and soil nutrient availability analysis [9,15].

2.4. Isotopic ^{13}C, ^{15}N Abundance, and C, N Content Analysis

$\delta^{13}C$, $\delta^{15}N$, TC, and TN content were measured using an isotope ratio mass spectrometer (IRMS) (IsoPrime 100, Isoprime Ltd., Cheadle, UK), connected to a CN elemental analyzer (Vario MICRO cube, Elementar, Germany). C and N isotopic abundances were calculated as $\delta^{13}C$ and $\delta^{15}N$ (‰) using the following formula:

$$\delta^{13}C \text{ and } \delta^{15}N\ (‰) = (R_{sample}/R_{standard} - 1) \times 1000$$

where R_{sample} is the stable isotopic ratio in the samples and $R_{standard}$ is the ratio in the standard. The Vienna Pee Dee Belemnite (VPDB) was used as a standard for $\delta^{13}C$ and atmospheric N_2 was used as the standard for $\delta^{15}N$. The precision of isotopic composition was checked using internal standards (i.e., acetanilide, L-histidine, D-glutamic, and glycine) and positive and negative values were observed. Positive delta values indicate that there is a greater percentage of isotope presence relative to the standard whereas negative values indicate a lesser percentage of isotope presence relative to the standard. In general, the analytical precision for $\delta^{13}C$ and $\delta^{15}N$ was better than 0.2‰ [9,15].

2.5. Determination of Soil Available Nutrients

Available nitrogen (AN) was determined by the Kjeldahl method [39]. Available phosphorus (AP) was determined by the diacid extraction spectrophotometric colorimetry method [40], and available potassium (AK) was determined using a flame photometer method by ammonium acetate extraction [41]. Moreover, soil moisture content (SMC) was calculated based on wet and dry weight, whereas soil pH was determined using a potentiometric method (1:2.5 soil:water). Details about the soil laboratory analysis are also available in our published papers [42,43].

2.6. Data Analysis

Two-way analysis of variance (ANOVA) at a 5% probability level was performed to determine the effects of forest types and different soil depth profiles on the natural abundance of C and N isotopes, soil TC and TN content, and soil nutrient availability. One-way ANOVA at 5% probability level was performed to determine the differences between foliar $\delta^{13}C$, $\delta^{15}N$ and, foliar TC and TN content among forest types. A regression analysis was conducted to analyze the foliar and soil isotopic abundance relationship. The Pearson correlation test was performed to observe the relationship between isotopic dynamics and influential factors. Means that exhibited significant differences were compared using post-hoc Tukey's HSD significance test. All statistical analyses were performed using the SPSS Statistical Package (SPSS 17.0, Chicago, IL, USA).

3. Results

3.1. Soil and Foliar TC, TN Content, and C:N Ratio

Two-way ANOVA revealed that soil TC, TN content, and C:N ratio varied significantly among forest types ($p < 0.05$) and soil depths ($p < 0.001$) (Tables 2 and A1). Soil TN content ranged from 0.56 g·kg^{-1} to 3.33 g·kg^{-1}, while the soil TC content varied from 1.69 g·kg^{-1} to 46.88 g·kg^{-1}. Moreover, the soil C:N ratio ranged from 3.01 to 14.06 (Table 2). Soil TN, TC, and C:N ratio decreased as the soil depth increased among all forest types. In the topmost soil layer, TC, TN, and soil C:N ratio was higher in the mixed forest stand, while for the remaining two soil layers, it was greater in CPF (Table 2).

Table 2. Soil and foliar TC, TN content, and C:N ratio in different forest types at different soil depths.

	Elements	Depth	Mix Forest	NBF	CPF
Soil	TN [a] (g·kg^{-1})	0–10	3.33 ± 0.03 ^{aA}	1.56 ± 0.01 ^{cA}	2.10 ± 0.02 ^{bA}
		10–20	1.12 ± 0.03 ^{bB}	0.77 ± 0.01 ^{cB}	1.83 ± 0.01 ^{aB}
		20–40	0.79 ± 0.04 ^{bC}	0.56 ± 0.01 ^{cC}	1.19 ± 0.01 ^{aC}
	TC (g·kg^{-1})	0–10	46.88 ± 0.44 ^{aA}	16.63 ± 0.24 ^{cA}	26.02 ± 0.28 ^{bA}
		10–20	10.19 ± 0.07 ^{bB}	4.17 ± 0.09 ^{cB}	19.40 ± 0.06 ^{aB}
		20–40	5.33 ± 0.06 ^{bC}	1.69 ± 0.03 ^{cC}	8.91 ± 0.09 ^{aC}
	C:N	0–10	14.06 ± 0.10 ^{aA}	10.65 ± 0.23 ^{bA}	12.39 ± 0.12 ^{aA}
		10–20	9.07 ± 0.25 ^{bB}	5.40 ± 0.06 ^{cB}	10.62 ± 0.12 ^{cB}
		20–40	6.73 ± 0.32 ^{aC}	3.01 ± 0.07 ^{bC}	7.48 ± 0.15 ^{aC}
Foliar	TN (g·kg^{-1})	-	27.28 ± 1.96 ^{ab}	28.92 ± 1.43 ^{a}	26.40 ± 0.68 ^{b}
	TC (g·kg^{-1})	-	482.23 ± 3.77 ^{b}	495.72 ± 0.27^{a}	475.30 ± 0.42 ^{c}
	C:N	-	15.33 ± 1.56 ^{a}	17.21 ± 0.83 ^{a}	18.02 ± 0.48 ^{a}

Note: Values are means ±SE. The different small letter represents the significant differences between forest types and different capital letters between different vertical soil depths at $p \leq 0.05$. [a] TN, total nitrogen; TC, total carbon.

In terms of foliage values, one-way ANOVA showed that foliar C content varied significantly among forest types ($p < 0.05$), whereas foliar N content and foliar C:N ratio were not significantly different ($p = 0.242$ and $p = 0.155$, respectively) (Table 2).

3.2. C^{13} and ^{15}N Natural Abundance

Two-way ANOVA depicted that soil δ^{15}N varied significantly among forest types ($p > 0.001$) and soil depths ($p = 0.048$). However, forest type ($p = 0.078$) and interaction of forest type × soil depth ($p = 0.329$) had no significant effect on soil δ^{13}C, whereas soil δ^{13}C varied significantly with soil depth ($p = 0.028$) (Figure 2 and Table A1). A positive correlation was present between soil δ^{15}N-soil depth ($r = 0.606$) and soil δ^{13}C-soil depth ($r = 0.470$). Overall, soil δ^{15}N values ranged from −1.61‰ to 5.19‰, and soil δ^{13}C values ranged from −28.23% to −25.00% (Figure 2).

One-way ANOVA illustrated that forest type had a considerable effect on foliar δ^{13}C ($p = 0.03$) and foliar δ^{15}N ($p = 0.047$). Foliar δ^{15}N values ranged from 6.39‰ to 8.90‰, and foliar δ^{13}C values ranged from −25.4‰ to −29.7‰. Foliar δ^{15}N was enriched in the NBF stand, and δ^{13}C was less negative in the mixed forest stand (Figure 2).

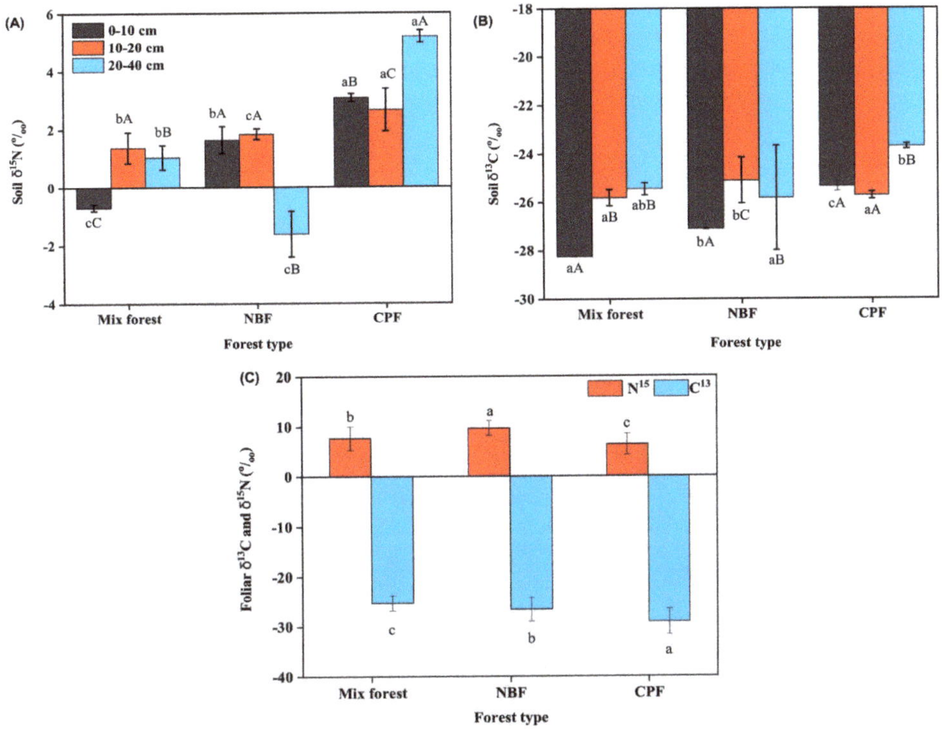

Figure 2. The natural abundance of (**A**) soil ^{15}N (‰) and (**B**) soil ^{13}C (‰) at different soil depths and (**C**) foliar ^{15}N (‰) and ^{13}C (‰) in different forest types. Values are means ±SE. The different small letters represent the significant differences between forest types, and different capital letters represent the significant difference between different soil depths $p \leq 0.05$.

3.3. Relationship between Soil, Foliar Isotopic Abundance, and C:N Ratio

A strong positive linear relationship was observed between soil and foliar δ^{13}C (r = 0.78, p = 0.04) and soil and foliar δ^{15}N (r = 0.56, p= 0.03) across forest types and soil depths (Figure 3A,B).

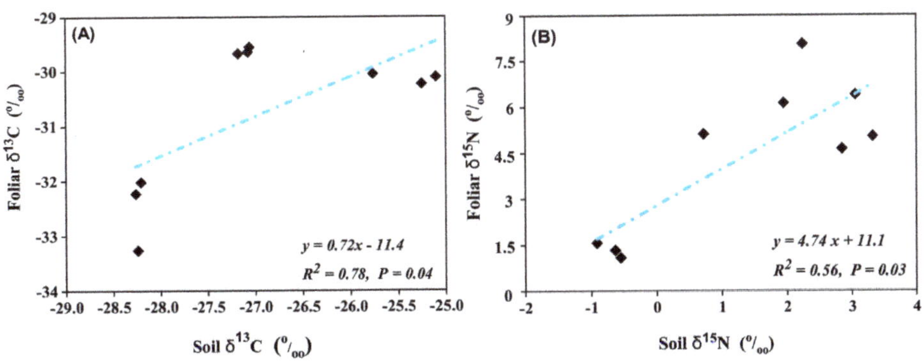

Figure 3. Linear relationships between (**A**) soil and foliar δ^{13}C and (**B**) soil and foliar δ^{15}N (‰) across three forest types at a 0–40 cm soil depth.

3.4. Correlation between Isotopic Abundance and Potentially Influential Factors

Soil $\delta^{13}C$ was positively correlated to SMC and foliar C:N ratio, while negatively correlated to soil and foliar TN and soil TC whereas soil $\delta^{15}N$ was positively related to soil C:N ratio, foliar C:N ratio, and negatively correlated to SMC, foliar TC, and TN content (Table 3).

Table 3. Pearson correlation between soil and foliar $\delta^{13}C$ and $\delta^{15}N$ and potentially influential factors across forest types and soil depths.

	Foliar $\delta^{13}C$		**Soil $\delta^{13}C$**			**Foliar $\delta^{15}N$**		**Soil $\delta^{15}N$**	
	r	*p*	*r*	*p*		*r*	*p*	*r*	*p*
Foliar TN [a]	0.220	0.263	−0.526 **	0.001	Foliar TN	0.923 **	0.001	−0.414 *	0.032
Soil TN	0.118	0.551	−0.470 *	0.015	Soil TN	−0.152	0.448	−0.058	0.774
Foliar TC	0.047	0.81	−0.190	0.346	Foliar TC	0.395 *	0.041	−0.526 **	0.004
Soil TC	0.167	0.403	−0.504 **	0.012	Soil TC	−0.124	0.536	−0.126	0.531
Foliar C:N	−0.724 **	0.001	0.551 **	0.018	Foliar C:N	−0.780 **	0.001	0.404 *	0.036
Soil C:N	0.088	0.665	−0.406 *	0.038	Soil C:N	−0.194 *	0.033	0.141 *	0.048
Soil pH	−0.198	0.324	0.107	0.591	Soil pH	−0.285	0.149	0.194	0.332
BD	0.567 **	0.002	−0.043	0.862	BD	−0.098	0.626	0.076	0.707
SMC	−0.431	0.130	0.359 **	0.015	SMC	−0.321 **	0.001	−0.478 **	0.001
Forest type	−0.261	0.186	0.267	0.179	Forest type	−0.463 *	0.015	0.616 **	0.001
Soil depth	0.005	1.02	0.470 *	0.013	Soil depth	0.005	1.01	0.040 *	0.842

Note: Correlation was significant at the 0.05 * and 0.01 ** level (2-tailed). [a] TN, total nitrogen; TC, total carbon; BD, soil bulk density; SMC, soil moisture content.

Foliar $\delta^{13}C$ was positively and negatively correlated to soil BD and foliar C:N ratio, respectively, whereas foliar $\delta^{15}N$ was positively related to foliar TC and TN content, and negatively correlated to foliar C:N ratio and SMC (Table 3).

3.5. Soil Nutrient Availability

Two-way ANOVA illustrated that AN, AP, and AK varied significantly in different forest types and soil depths ($p < 0.05$) (Tables 4 and A1). In the 0–10 cm and 20–40 cm soil layers, AN content (177 $mg \cdot kg^{-1}$ and 109 $mg \cdot kg^{-1}$) was observed to be higher in the mixed forest; in 10–20 cm, it was higher in the NBF (155 $mg \cdot kg^{-1}$). AN content decreased as the soil depth increased in the mixed forest and NBF. AP content ranged from 29 $mg \cdot kg^{-1}$ to 82 $mg \cdot kg^{-1}$. In the 0–10 cm soil layer, the highest AP content was observed in CPF (71 $mg \cdot kg^{-1}$), while in the remaining two layers (10–20 and 20–40 cm), the maximum content was measured in the mixed forest. Among all forest types and soil depths, AK ranged from 12 $g \cdot kg^{-1}$ to 147 $g \cdot kg^{-1}$. In the topmost soil layer, it was higher in the mixed forest stand. In both the 10–20 cm and 20–40 cm layers, it was higher in NBF. Generally, soil nutrient content decreased as the soil depth increased across all forest types (Table 4).

Table 4. Available soil nutrients in different forest types at different soil depths.

Elements	Depth	Mix Forest	NBF	CPF
Available N ($mg \cdot kg^{-1}$)	0–10	177 ± 2.06 aA	167 ± 2.04 bA	155 ± 2.2 cA
	10–20	142 ± 2.49 bB	155 ± 2.45 aB	137 ± 1.58 cB
	20–40	109 ± 1.0 aC	105 ± 3.54 bC	98 ± 1.15 cC
Available P ($mg \cdot kg^{-1}$)	0–10	62 ± 2.42 bB	55 ± 0.75 cA	71 ± 0.69 aA
	10–20	82 ± 2.46 aA	54 ± 0.47 bA	36 ± 1.22 cB
	20–40	60 ± 2.59 aB	38 ± 0.29 bB	29 ± 0.15 cC
Available K ($mg \cdot kg^{-1}$)	0–10	147 ± 0.37 aA	104 ± 1.94 bB	91 ± 0.81 cA
	10–20	41 ± 1.02 bB	119 ± 1.71 aA	22 ± 0.85 cB
	20–40	22 ± 0.07 bC	41 ± 0.92 aC	12 ± 0.21 cC

Note: Values are means ±SE. The different small letters represent the significant differences between forest types, and different capital letters represent the significant difference between soil depths at $p < 0.05$.

4. Discussion

Among all forest types, soil TN content decreased as the soil depth increased. In the topsoil layer, both TC and TN contents were enriched in the mixed forest. It indicates that litter input was higher in the mixed forest due to mixed species composition and more understory vegetation, which increased soil organic matter (SOM) more than both NBF and CPF stands. Since soil organic carbon changes with the vertical soil depth profiles, soil depth is widely used as a SOM decomposition rate indicator [44–46]. It also points toward the better leaf litter quality in mixed forest with higher N release potential during decomposition. This study observed a considerable difference in soil and foliar TC, TN content, and C:N ratio among the NBF, CPF, and mixed forest. Tree species vary in their nutrient transformation potential with changing forest structure and canopy dynamics. These factors influence the various biogeochemical cycles [12,26,42]. Species autecology affects the decomposition and the turnover rate of C and N dynamics [47]. Neirynck et al. [48] mentioned that different tree species such as *Acer psedoplatanus*, *Fraxinus excelsior*, and *Tilia platyphyllos* significantly affect the C:N ratio and it was related mainly to litter quality. Lovett et al. [49] also revealed that forest structure alters the C and N dynamics because species composition affects nutrient cycling under their different canopy structures, which can substantially influence the forest ecosystems' overall nutrient turnover rates [50,51].

In this study, a higher abundance of soil ^{15}N and ^{13}C was observed in CPF. Enriched soil δ^{15}N and δ^{13}C in CFP could be due to more intensive forest management practices than that in the NBF and mixed forest stands because the plantations are artificially managed forests with added activities of humans and livestock. Pardo et al. [13] and Yang et al. [52] also reported that different silvicultural techniques in plantations such as clear-cutting, slash burning, site preparation, thinning, and pruning could enrich the soil δ^{13}C and δ^{15}N in plantation forests because of higher soil C and N loss such as enhanced nitrification followed by nitrate loss, which could have a direct impact on isotopic abundance. Rehman et al. [53] reported that various forest protection activities could influence stable isotopic abundance and C:N ratios. Moreover, it is generally believed that multiple factors such as light availability, litter input, water use efficiency, soil physical, chemical, and biological properties altered by the forest types may affect natural isotopic abundance in forest ecosystems [50,51].

Forest type and soil depth significantly impacted soil δ^{15}N, whereas soil δ^{13}C was only impacted by soil depth; moreover, both the soil δ^{13}C and δ^{15}N were enriched in the topmost soil layer. Higher isotopic values in the upper soil layer can be due to higher SOC. Ellert and Janzen [54] reported higher C content in the uppermost layer, providing ample SOM that strongly influences the δ^{13}C in the topsoil, whereas lesser C content in the deeper layers determines the lesser δ^{13}C values. Ngaba et al. [9] also reported that soil depth increment significantly impacted the C and N isotopic abundance. Forest type had a significant effect on foliar C (highest in CPF) and N (highest in NBF) abundance. Usually, foliar δ^{15}N values increased under N-rich conditions; in our study, higher foliar and soil δ^{15}N were observed in CPF. On average, foliar and soil δ^{15}N patterns in sub-tropical forests are higher than other forest types because of the gaseous N losses related to microbial activities. Our study showed foliar C and N positively correlated with the soil isotopic values; this indicates that foliar isotopic values of different species are consistent concerning soil values.

When we studied the correlation of potentially influential factors concerning δ^{13}C and δ^{15}N in soil and foliage, we observed that SMC, foliar TN, and foliar C:N ratio was usually significantly related to soil foliar ^{13}C and ^{15}N natural abundance, either positively or negatively. These factors can affect biogeochemical processes such as N mineralization, nitrification/denitrification, soil respiration, root dynamics, thus altering the rate of δ^{15}N [55,56]. Moisture content is a vital factor influencing the nutrient content and δ^{13}C of the plant and soil. Philips et al. [57] stated that the δ^{13}C depends on moisture conditions, either caused by low atmospheric humidity, less precipitation, or low soil moisture. Moreover, drought can also potentially influence the C fluxes and soil $\delta^{13}C$ [58]. C isotopic abundance primarily depends on the water use efficiency; therefore, soil water resources can affect the isotopic abundance [59]. Ngaba et al. [9] also stated mean annual precipitation and land-use as the significant factor altering soil $\delta^{13}C$. While analyzing the ^{15}N abundance between the north and south slopes of forests, Chen et al. [5] also reported soil moisture content, leaf N content, and leaf C:N as the significant factors related to isotopic abundance. This inferred that water retention capacity should be the critical factor concerning the isotopic abundance of plant tissues.

In our study, a significant difference was observed in variability in macronutrient availability among all forest types. Generally, AN and AK were higher in mixed forest stand and AP in CPF. Plantation forests mostly have fast-growing species with successive short rotations; hence, the fast growth demands more extensive nutrient availability, which causes an imbalance in soils under plantations compared to mixed-species forest. Kooch et al. [47] reported the decline in soil fertility with successive planting of fast-growing monoculture species at the same site. The decline in soil available nutrients with soil depth increase could be due to a large amount of plant litter deposition in the topsoil layer as its decomposition results in a higher accumulation of soil nutrients in the uppermost soil than the lower soil layers. Farooq et al. [43] cited that species structure influences the soil properties, which is more noticeable at the upper layer of soil. Our study is also in line with Selvaraj et al. [60], who stated that both available and soil total nutrient content decreased with the soil depth, irrespective of stand age; moreover, Groppo et al. [61] and Breulmann et al. [62] also supported this phenomenon regardless of species and land use. However, the relative importance of the various soil-forming factors always remains debated [28–30,60]. Soil conditions solely do not affect directly, but with an interplay of other related factors such as the autecology of tree species, litter quality, parent material, soil community effects, human activities, and local climatic conditions [3,63–65]. These impacts are likely to have significant consequences for belowground communities [1]. Similarly, different land uses cannot directly affect soil fertility, but it can influence through indirect effects such as organic/inorganic amendments [66–68], various stresses for plant and nutrient availability [69,70], different planting materials [71,72], and atmospheric deposition levels and turnover rates [73].

5. Conclusions

This study explored how $\delta^{13}C$, $\delta^{15}N$, TC, and TN content of soil and foliage and soil nutrient availability is affected by different forest types. Forest type and soil depth significantly affected the soil $\delta^{15}N$, while forest type effect on soil $\delta^{13}C$ was not significant. Significant changes in soil $\delta^{15}N$ among different forest types indicated that ^{15}N signatures might dominate the processes of N cycling in different forest types. Across forest types and soil depths, the observed $\delta^{13}C$ and $\delta^{15}N$ variations were attributed to SMC, soil TC, soil TN, and soil C:N ratio. Variations in foliar $\delta^{13}C$ were accredited to differences in soil BD and foliar C:N ratio, whereas variation in foliar $\delta^{15}N$ was accredited to SMC, foliar TN, and TC. Our study also highlights that, on average, available forms of soil nutrients were less in monoculture Chinese fir stands (CPF) than NBF and mixed forests. This difference was possibly caused by higher litter biomass accumulation from broadleaved and mixed-species litter compared to CPF. Thus, the presence of broadleaved species within a monoculture stand along with other management practices such as the introduction of multi-layering and multi-aged plantations, avoiding clear-cutting practice, and burying the cutting leftover rather than burning could improve soil health, sustainable management, and production, which would enhance both economic and environmental aspects to the forest industry. Overall, our results highlight the importance of forest types and composition when studying biogeochemical cycling and its response in different terrestrial ecosystems. Moreover, this study will also help to further understand the role of C and N cycles concerning soil productivity.

Author Contributions: Conceptualization, T.H.F., U.K. and W.Y.; Data curation, A.S.; Formal analysis, T.H.F. and Y.L.; Funding acquisition, W.Y.; Methodology, T.H.F., X.C. and M.H.U.R.; Project administration, J.W. and W.Y.; Resources, W.Y.; Software, T.H.F. and A.S.; Validation, A.S. and W.Y.; Writing—original draft, T.H.F.; Writing—review & editing, X.C. and W.Y. All authors have read and agreed to the published version of the manuscript.

Funding: This research was financially supported by postdoctoral research funding from the Central South University of Forestry and Technology (70702-45200003), Scientific Research Fund of Hunan Provincial Education Department (18A150) and Key Research and Development Project of Hunan (2020NK2022).

Institutional Review Board Statement: Not applicable.

Informed Consent Statement: Not applicable.

Data Availability Statement: All the available data have been presented in the manuscript.

Acknowledgments: We would like to thank Xiang Zhang and Arfien Sayman for their help in conducting fieldwork.

Conflicts of Interest: The authors declare that they have no known competing interests.

Abbreviations

NBF—natural broadleaved forest; CPF—coniferous plantations forest; SMC—soil moisture content; BD—soil bulk density; SOM—soil organic matter; MAT—mean annual temperature; MAP—mean annual precipitation; AN—available nitrogen; AP—available phosphorus; AK—available potassium.

Appendix A

Table A1. Two-way ANOVA results (p values) for the studied soil variables at $p \leq 0.05$.

Variable	Forest Type	Soil Depth	Forest Type × Soil Depth
TN	<0.05	<0.001	0.02
TC	0.04	<0.001	0.04
$\delta^{15}N$	<0.001	0.048	<0.001
$\delta^{13}C$	0.078	0.028	0.329
C:N	<0.001	<0.001	0.04
SOC	0.01	0.01	0.03
AN	0.04	<0.001	0.02
AP	0.02	0.03	<0.001
AK	0.05	0.05	0.02

References

1. Ayres, E.; Dromph, K.M.; Cook, R.; Ostle, N.; Bardgett, R.D. The influence of belowground herbivory and defoliation of a legume on nitrogen transfer to neighbouring plants. *Funct. Ecol.* **2007**, *21*, 256–263. [CrossRef]
2. Farooq, T.H.; Yan, W.; Chen, X.; Shakoor, A.; Rashid, M.H.U.; Gilani, M.M.; He, Z.; Wu, P. Dynamics of canopy development of *Cunninghamia lanceolata* mid-age plantation in relation to foliar nitrogen and soil quality influenced by stand density. *Glob. Ecol. Conserv.* **2020**, *24*, e01209. [CrossRef]
3. Finzi, A.C.; Van Breemen, N.; Canham, C.D. Canopy tree-soil interactions within temperate forests: Species effects on soil carbon and nitrogen. *Ecol. Appl.* **1998**, *8*, 440–446.
4. Macinnis-Ng, C.; Schwendenmann, L. Litterfall, carbon and nitrogen cycling in a southern hemisphere conifer forest dominated by kauri (*Agathis australis*) during drought. *Plant Ecol.* **2015**, *216*, 247–262. [CrossRef]
5. Chen, C.; Jia, Y.; Chen, Y.; Mehmood, I.; Fang, Y.; Wang, G. Nitrogen isotopic composition of plants and soil in an arid mountainous terrain: South slope versus north slope. *Biogeosciences* **2018**, *15*, 369. [CrossRef]
6. Pan, Y.; Birdsey, R.A.; Fang, J.; Houghton, R.; Kauppi, P.E.; Kurz, W.A.; Phillips, O.L.; Shvidenko, A.; Lewis, S.L.; Canadell, J.G. A large and persistent carbon sink in the world's forests. *Science* **2011**, *333*, 988–993. [CrossRef] [PubMed]
7. Brunn, M.; Spielvogel, S.; Sauer, T.; Oelmann, Y. Temperature and precipitation effects on $\delta^{13}C$ depth profiles in SOM under temperate beech forests. *Geoderma* **2014**, *235*, 146–153. [CrossRef]
8. Wang, C.; Houlton, B.Z.; Liu, D.; Hou, J.; Cheng, W.; Bai, E. Stable isotopic constraints on global soil organic carbon turnover. *Biogeosciences* **2018**, *15*, 987–995. [CrossRef]

9. Ngaba, M.J.Y.; Hu, Y.-L.; Bol, R.; Ma, X.-Q.; Jin, S.-F.; Mgelwa, A.S. Effects of land use change from natural forest to plantation on C, N and natural abundance of ^{13}C and ^{15}N along a climate gradient in eastern China. *Sci. Rep.* **2019**, *9*, 1–12. [CrossRef]
10. Natelhoffer, K.J.; Fry, B. Controls on natural nitrogen-15 and carbon-13 abundances in forest soil organic matter. *Soil Sci. Soc. Am. J.* **1988**, *52*, 1633–1640. [CrossRef]
11. Ehleringer, J.R.; Buchmann, N.; Flanagan, L.B. Carbon isotope ratios in belowground carbon cycle processes. *Ecol. Appl.* **2000**, *10*, 412–422. [CrossRef]
12. Guillaume, T.; Damris, M.; Kuzyakov, Y. Losses of soil carbon by converting tropical forest to plantations: Erosion and decomposition estimated by δ^{13}C. *Glob. Chang. Biol.* **2015**, *21*, 3548–3560. [CrossRef] [PubMed]
13. Pardo, L.H.; Hemond, H.F.; Montoya, J.P.; Fahey, T.J.; Siccama, T.G. Response of the natural abundance of ^{15}N in forest soils and foliage to high nitrate loss following clear-cutting. *Can. J. For. Res.* **2002**, *32*, 1126–1136. [CrossRef]
14. Wang, A.; Fang, Y.T.; Chen, D.X.; Koba, K.; Makabe, A.; Luo, T.-S.; Yoh, M. Variations in nitrogen-15 natural abundance of plant and soil systems in four remote tropical rainforests, southern China. *Oecologia* **2014**, *174*, 567–580. [CrossRef] [PubMed]
15. Mgelwa, A.S.; Hu, Y.L.; Liu, J.-F.; Qiu, Q.; Liu, Z.; Ngaba, M.J.Y. Differential patterns of nitrogen and δ^{15}N in soil and foliar along two urbanized rivers in a subtropical coastal city of southern China. *Environ. Pollut.* **2019**, *244*, 907–914. [CrossRef]
16. Robinson, D. δ^{15}N as an integrator of the nitrogen cycle. *Trends Ecol. Evol.* **2001**, *16*, 153–162. [CrossRef]
17. Lim, S.S.; Kwak, J.H.; Lee, K.S.; Chang, S.X.; Yoon, K.-S.; Kim, H.-Y.; Choi, W.-J. Soil and plant nitrogen pools in paddy and upland ecosystems have contrasting δ^{15}N. *Biol. Fertil. Soils* **2015**, *51*, 231–239. [CrossRef]
18. Denk, T.R.A.; Mohn, J.; Decock, C.; Lewicka-Szczebak, D.; Harris, E.; Butterbach-Bahl, K.; Kiese, R.; Wolf, B. The nitrogen cycle: A review of isotope effects and isotope modeling approaches. *Soil Biol. Biochem.* **2017**, *105*, 121–137. [CrossRef]
19. Farooq, T.H.; Tigabu, M.; Ma, X.; Zou, X.; Liu, A.; Odén, P.C.; Wu, P. Nutrient uptake, allocation and biochemical changes in two Chinese fir cuttings under heterogeneous phosphorus supply. *iForest* **2018**, *11*, 411–417. [CrossRef]
20. Farooq, T.H.; Wu, W.; Tigabu, M.; Ma, X.; Zou, X.H.; Rashid, M.H.U.; Gilani, M.M.; Wu, P.F. Growth, biomass production and root development of Chinese fir in relation to initial planting density. *Forests* **2019**, *10*, 236. [CrossRef]
21. Farooq, T.H.; Yan, W.; Rashid, M.H.U.; Tigabu, M.; Gilani, M.M.; Zou, X.H.; Wu, P.F. Chinese fir (*Cunninghamia Lanceolata*) a green gold of China with continues decline in its productivity over the successive rotations: A review. *Appl. Ecol. Environ. Res.* **2019**, *17*, 11055–11067. [CrossRef]
22. Shakoor, A.; Shakoor, S.; Rehman, A.; Ashraf, F.; Abdullah, M.; Shahzad, S.M.; Farooq, T.H.; Ashraf, M.; Manzoor, M.A.; Altaf, M.M.; et al. Effect of animal manure, crop type, climate zone, and soil attributes on greenhouse gas emissions from agricultural soils—A global meta-analysis. *J. Clean. Prod.* **2020**, *278*, 124019. [CrossRef]
23. Shakoor, A.; Shahzad, S.M.; Chatterjee, N.; Arif, M.S.; Farooq, T.H.; Altaf, M.M.; Tufail, M.A.; Dar, A.A.; Mehmood, T. Nitrous oxide emission from agricultural soils: Application of animal manure or biochar? A global meta-analysis. *J. Environ. Manag.* **2021**, *285*, 112170. [CrossRef] [PubMed]
24. Shakoor, A.; Shahbaz, M.; Farooq, T.H.; Sahar, N.E.; Shahzad, S.M.; Altaf, M.M.; Ashraf, M. A global meta-analysis of greenhouse gases emission and crop yield under no-tillage as compared to conventional tillage. *Sci. Total Environ.* **2021**, *750*, 142299. [CrossRef]
25. Fang, Y.; Gundersen, P.; Mo, J.; Zhu, W. Nitrogen leaching in response to increased nitrogen inputs in subtropical monsoon forests in southern China. *For. Ecol. Manag.* **2009**, *257*, 332–342. [CrossRef]
26. Anaya, C.A.; Huber-Sannwald, E. Long-term soil organic carbon and nitrogen dynamics after conversion of tropical forest to traditional sugarcane agriculture in East Mexico. *Soil Tillage Res.* **2015**, *147*, 20–29. [CrossRef]
27. Li, Y.; Zhang, J.; Chang, S.X.; Jiang, P.; Zhou, G.; Shen, Z.; Wu, J.; Lin, L.; Wang, Z.; Shen, M. Converting native shrub forests to Chinese chestnut plantations and subsequent intensive management affected soil C and N pools. *For. Ecol. Manag.* **2014**, *312*, 161–169. [CrossRef]
28. Guo, X.; Meng, M.; Zhang, J.; Chen, H.Y.H. Vegetation change impacts on soil organic carbon chemical composition in subtropical forests. *Sci. Rep.* **2016**, *6*, 29607. [CrossRef]
29. Maes, S.L.; Blondeel, H.; Perring, M.P.; Depauw, L.; Brūmelis, G.; Brunet, J.; Decocq, G.; den Ouden, J.; Härdtle, W.; Hédl, R. Litter quality, land-use history, and nitrogen deposition effects on topsoil conditions across European temperate deciduous forests. *For. Ecol. Manag.* **2019**, *433*, 405–418. [CrossRef]
30. Cornwell, W.K.; Cornelissen, J.H.C.; Amatangelo, K.; Dorrepaal, E.; Eviner, V.T.; Godoy, O.; Hobbie, S.E.; Hoorens, B.; Kurokawa, H.; Pérez-Harguindeguy, N. Plant species traits are the predominant control on litter decomposition rates within biomes worldwide. *Ecol. Lett.* **2008**, *11*, 1065–1071. [CrossRef] [PubMed]
31. Shelton, R.E.; Jacobsen, K.L.; McCulley, R.L. Cover crops and fertilization alter nitrogen loss in organic and conventional conservation agriculture systems. *Front. Plant Sci.* **2018**, *8*, 2260. [CrossRef]
32. Thiffault, E.; Hannam, K.D.; Paré, D.; Titus, B.D.; Hazlett, P.W.; Maynard, D.G.; Brais, S. Effects of forest biomass harvesting on soil productivity in boreal and temperate forests—A review. *Environ. Rev.* **2011**, *19*, 278–309. [CrossRef]
33. Ringeval, B.; Augusto, L.; Monod, H.; Van Apeldoorn, D.; Bouwman, L.; Yang, X.; Achat, D.L.; Chini, L.P.; Van Oost, K.; Guenet, B. Phosphorus in agricultural soils: Drivers of its distribution at the global scale. *Glob. Chang. Biol.* **2017**, *23*, 3418–3432. [CrossRef]
34. Hu, Z.H.; Qian, H.Y.; Yu, M.J. The niche of dominant species populations in *Castanopsis eyrei* forest in Gutian Mountain National Nature Reserve. *Acta Ecol. Sin.* **2009**, *29*, 3670–3677.

35. Ren, Y.; Peng, D.; Pan, J.; Hong, Z.; Ye, G. Relationship between species biodiversity and biomass accumulation in a *Castanopsis eyrei* Forest in Wuyishan Mountain under different temporal and spatial scales. *Sci. Silvae Sin.* **2010**, *46*, 33–38.
36. Farooq, T.H.; Shakoor, A.; Wu, X.; Li, Y.; Rashid, M.H.U.; Zhang, X.; Gilani, M.M.; Kumar, U.; Chen, X.; Yan, W. Perspectives of plantation forests in the sustainable forest development of China. *iForest* **2021**, *14*, 166. [CrossRef]
37. Cao, L.; Liang, Y.; Wang, Y.; Lu, H. Runoff and soil loss from *Pinus massoniana* forest in southern China after simulated rainfall. *Catena* **2015**, *129*, 1–8. [CrossRef]
38. Gilani, M.M.; Tigabu, M.; Liu, B.; Farooq, T.H.; Rashid, M.H.U.; Ramzan, M.; Ma, X. Seed germination and seedling emergence of four tree species of southern China in response to acid rain. *J. For. Res.* **2020**, *32*, 471–481. [CrossRef]
39. Ryan, J.; Estefan, G.; Rashid, A. *Soil and Plant Analysis Laboratory Manual*; ICARDA: Beirut, Lebanon, 2007.
40. Bray, R.H.; Kurtz, L.T. Determination of total, organic, and available forms of phosphorus in soils. *Soil Sci.* **1945**, *59*, 39–46. [CrossRef]
41. Schollenberger, C.; Simon, R. Determination of exchange capacity and exchangeable bases in soil ammonium acetate method. *Soil Sci.* **1945**, *59*, 13–24. [CrossRef]
42. Farooq, T.H.; Ma, X.; Rashid, M.H.U.; Wu, W.; Xu, J.; Tarin, M.W.K.; He, Z.; Wu, P. Impact of stand density on soil quality in Chinese Fir (*Cunninghamia Lanceolata*) monoculture. *Appl. Ecol. Environ. Res.* **2019**, *17*, 3553–3566. [CrossRef]
43. Farooq, T.H.; Kumar, U.; Mo, J.; Shakoor, A.; Wang, J.; Rashid, M.H.U.; Tufail, M.A.; Chen, X.; Yan, W. Intercropping of Peanut–Tea Enhances Soil Enzymatic Activity and Soil Nutrient Status at Different Soil Profiles in Subtropical Southern China. *Plants* **2021**, *10*, 881. [CrossRef] [PubMed]
44. Obalum, S.E.; Chibuike, G.U.; Peth, S.; Ouyang, Y. Soil organic matter as sole indicator of soil degradation. *Environ. Monit. Assess.* **2017**, *189*, 176. [CrossRef] [PubMed]
45. Lal, R. Managing organic matter content for restoring health and ecosystem services of soils of India. *J. Indian Soc. Soil Sci.* **2020**, *68*, 1–5. [CrossRef]
46. Bhowmik, A.; Kukal, S.S.; Saha, D.; Sharma, H.; Kalia, A.; Sharma, S. Potential indicators of soil health degradation in different land use-based ecosystems in the Shiwaliks of Northwestern India. *Sustainability* **2019**, *11*, 3908. [CrossRef]
47. Kooch, Y.; Rostayee, F.; Hosseini, S.M. Effects of tree species on topsoil properties and nitrogen cycling in natural forest and tree plantations of northern Iran. *Catena* **2016**, *144*, 65–73. [CrossRef]
48. Neirynck, J.; Mirtcheva, S.; Sioen, G.; Lust, N. Impact of *Tilia platyphyllos* Scop., *Fraxinus excelsior* L., *Acer pseudoplatanus* L., *Quercus robur* L. and *Fagus sylvatica* L. on earthworm biomass and physico-chemical properties of a loamy topsoil. *For. Ecol. Manag* **2000**, *133*, 275–286. [CrossRef]
49. Lovett, G.M.; Weathers, K.C.; Arthur, M.A. Control of Nitrogen Loss from Forested Watersheds by Soil Carbon: Nitrogen Ratio andTree Species Composition. *Ecosystems* **2002**, *5*, 712–718. [CrossRef]
50. Ariz, I.; Cruz, C.; Neves, T.; Irigoyen, J.J.; García, C.; Nogués, S.; Aparicio-Tejo, P.M.; Aranjuelo, I. Leaf $\delta^{15}N$ as a physiological indicator of the responsiveness of N_2-fixing alfalfa plants to elevated [CO_2], temperature and low water availability. *Front. Plant Sci.* **2015**, *6*, 574. [CrossRef]
51. Lawson, S.; Pike, C. Stable isotope ratios and reforestation potential in *Acacia koa* populations on Hawai'i. *Ann. For. Res.* **2017**. [CrossRef]
52. Yang, Y.S.; Guo, J.; Chen, G.; Xie, J.; Gao, R.; Li, Z.; Jin, Z. Carbon and nitrogen pools in Chinese fir and evergreen broadleaved forests and changes associated with felling and burning in mid-subtropical China. *For. Ecol. Manag.* **2005**, *216*, 216–226. [CrossRef]
53. Rahman, M.M.; Zhang, X.; Ahmed, I.; Iqbal, Z.; Zeraatpisheh, M.; Kanzaki, M.; Xu, M. Remote sensing-based mapping of senescent leaf C: N ratio in the sundarbans reserved forest using machine learning techniques. *Remote Sens.* **2020**, *12*, 1375. [CrossRef]
54. Ellert, B.H.; Janzen, H.H. Long-term biogeochemical cycling in agroecosystems inferred from 13C, 14C and 15N. *J. Geochem. Explor.* **2006**, *88*, 198–201. [CrossRef]
55. Shi, H.; Wang, X.; Xu, M.; Zhang, H.; Luo, Y. Characteristics of soil C: N ratio and δ 13C in wheat-maize cropping system of the North China Plain and influences of the Yellow River. *Sci. Rep.* **2017**, *7*, 1–9. [CrossRef]
56. Aguilos, M.; Stahl, C.; Burban, B.; Hérault, B.; Courtois, E.; Coste, S.; Wagner, F.; Ziegler, C.; Takagi, K.; Bonal, D. Interannual and seasonal variations in ecosystem transpiration and water use efficiency in a tropical rainforest. *Forests* **2019**, *10*, 14. [CrossRef]
57. Phillips, C.L.; Nickerson, N.; Risk, D.; Kayler, Z.E.; Andersen, C.; Mix, A.; Bond, B.J. Soil moisture effects on the carbon isotope composition of soil respiration. *Rapid Commun. Mass Spectrom.* **2010**, *24*, 1271–1280. [CrossRef] [PubMed]
58. Palta, J.A.; Gregory, P.J. Drought affects the fluxes of carbon to roots and soil in ^{13}C pulse-labelled plants of wheat. *Soil Biol. Biochem.* **1997**, *29*, 1395–1403. [CrossRef]
59. Burke, I.C.; Yonker, C.M.; Parton, W.J.; Cole, C.V.; Flach, K.; Schimel, D.S. Texture, climate, and cultivation effects on soil organic matter content in US grassland soils. *Soil Sci. Soc. Am. J.* **1989**, *53*, 800–805. [CrossRef]
60. Selvaraj, S.; Duraisamy, V.; Huang, Z.; Guo, F.; Ma, X. Influence of long-term successive rotations and stand age of Chinese fir (*Cunninghamia lanceolata*) plantations on soil properties. *Geoderma* **2017**, *306*, 127–134. [CrossRef]
61. Groppo, J.D.; Lins, S.R.M.; de Camargo, P.B.; Assad, E.D.; Pinto, H.S.; Martins, S.C.; Salgado, P.R.; Evangelista, B.; Vasconcellos, E.; Sano, E.E. Changes in soil carbon, nitrogen, and phosphorus due to land-use changes in Brazil. *Biogeosciences* **2015**, *12*, 4765–4780. [CrossRef]

62. Breulmann, M.; Boettger, T.; Buscot, F.; Gruendling, R.; Schulz, E. Carbon storage potential in size–density fractions from semi-natural grassland ecosystems with different productivities over varying soil depths. *Sci. Total Environ.* **2016**, *545*, 30–39. [CrossRef]
63. McGrath, M.J.; Luyssaert, S.; Meyfroidt, P.; Kaplan, J.O.; Burgi, M.; Chen, Y.; Erb, K.; Gimmi, U.; McInerney, D.; Naudts, K. Reconstructing European forest management from 1600 to 2010. *Biogeosciences* **2015**, *12*, 4291–4316. [CrossRef]
64. Wu, P.; Wang, G.; Farooq, T.H.; Li, Q.; Zou, X.; Ma, X. Low phosphorus and competition affect Chinese fir cutting growth and root organic acid content: Does neighboring root activity aggravate P nutrient deficiency? *J. Soils Sediments* **2017**, *17*, 2775–2785. [CrossRef]
65. Rashid, M.H.U.; Tigabu, M.; Chen, H.; Farooq, T.H.; Ma, X.; Wu, P. Calcium-mediated adaptive responses to low phosphorus stress in Chinese fir. *Trees* **2020**, *34*, 825–834. [CrossRef]
66. Saqib, H.M.U.; Ahmad, I.; Rashid, M.H.U.; Farooq, T.H.; Asif, M.; Kashif, M.; Iqbal, A.; Nawaz, M.F. Effect of Compost Application on the Growth of *Acacia nilotica*. *Cercet. Agron. Mold.* **2019**, *177*, 66–73. [CrossRef]
67. Rashid, M.H.U.; Farooq, T.H.; Iqbal, W.; Asif, M.; Islam, W.; Lin, D.C.; Ahmad, I.; Wu, P.F. Role of indole acetic acid on growth and biomass production of athel tree (*tamarix aphylla*) by using different cutting lengths. *Appl. Ecol. Environ. Res.* **2020**, *18*, 3805–3816. [CrossRef]
68. Rehman, A.; Arif, M.S.; Tufail, M.A.; Shahzad, S.M.; Farooq, T.H.; Ahmed, W.; Mehmood, T.; Farooq, M.R.; Javed, Z.; Shakoor, A. Biochar potential to relegate metal toxicity effects is more soil driven than plant system: A global meta-analysis. *J. Clean. Prod.* **2021**, *316*, 128276. [CrossRef]
69. Yousaf, M.S.; Farooq, T.H.; Ahmad, I.; Gilani, M.M.; Rashid, M.H.U.; Gautam, N.P.; Islam, W.; Asif, M.; Wu, P. Effect of Drought Stress on the Growth and Morphological Traits of *Eucalyptus camaldulensis* and *Eucalyptus citriodora*. *PSM Biol. Res.* **2018**, *3*, 85–91.
70. Wang, J.; Farooq, T.H.; Aslam, A.; Shakoor, A.; Chen, X.; Yan, W. Non-targeted metabolomics reveal the impact of phenanthrene stress on root exudates of ten urban greening tree species. *Environ. Res.* **2021**, *196*, 110370. [CrossRef]
71. Gilani, M.M.; Irfan, A.; Farooq, T.H.; Wu, P.; Yousaf, M.S.; Khan, M.W.; Talha, Y.; Ma, X. Effects of pre-sowing treatments on seed germination and morphological growth of *Acacia nilotica* and *Faidherbia albida*. *Sci. For.* **2019**, *122*, 374–382. [CrossRef]
72. Rashid, M.H.U.; Asif, M.; Farooq, T.H.; Gautam, N.P.; Nawaz, M.F.; Ahmad, I.; Gilani, M.M.; Wu, P. Cuttings growth response of *Dalbergia sissoo* (shisham) to soil compaction stress. *Appl. Ecol. Environ. Res.* **2019**, *17*, 1049–1059. [CrossRef]
73. Lejeune, Q.; Davin, E.L.; Gudmundsson, L.; Winckler, J.; Seneviratne, S.I. Historical deforestation locally increased the intensity of hot days in northern mid-latitudes. *Nat. Clim. Chang.* **2018**, *8*, 386–390. [CrossRef]

plants MDPI

Article

Intercropping of Peanut–Tea Enhances Soil Enzymatic Activity and Soil Nutrient Status at Different Soil Profiles in Subtropical Southern China

Taimoor Hassan Farooq [1,2,†], Uttam Kumar [3,†], Jing Mo [1,2], Awais Shakoor [4], Jun Wang [1,2], Muhammad Haroon U. Rashid [5], Muhammad Aammar Tufail [6], Xiaoyong Chen [1,7,*] and Wende Yan [1,2,*]

1 National Engineering Laboratory for Applied Technology of Forestry and Ecology in South China, Central South University of Forestry and Technology, Changsha 410004, China; taimoorhassan2055@gmail.com (T.H.F.); jingmo_changsha@hotmail.com (J.M.); jwang0829@csuft.edu.cn (J.W.)
2 College of Life Science and Technology, Central South University of Forestry and Technology, Changsha 410004, China
3 Institute of Applied Ecology, College of Plant Protection, Fujian Agriculture and Forestry University, Fuzhou 350002, China; Uttam5454@gmail.com
4 Department of Environment and Soil Sciences, University of Lleida, Avinguda Alcalde Rovira Roure 191, 25198 Lleida, Spain; awais.shakoor@udl.cat
5 College of Forestry, Fujian Agriculture and Forestry University, Fuzhou 350002, China; haroonrashid3838@outlook.com
6 Department of Civil, Environmental and Mechanical Engineering, University of Trento, via Mesiano 77, 38123 Trento, Italy; muhammad.tufail@unitn.it
7 College of Arts and Sciences, Governors State University, University Park, IL 60484, USA
* Correspondence: xchen@govst.edu (X.C.); csfuywd@hotmail.com (W.Y.)
† These authors contributed equally to this work.

Citation: Farooq, T.H.; Kumar, U.; Mo, J.; Shakoor, A.; Wang, J.; Rashid, M.H.U.; Tufail, M.A.; Chen, X.; Yan, W. Intercropping of Peanut–Tea Enhances Soil Enzymatic Activity and Soil Nutrient Status at Different Soil Profiles in Subtropical Southern China. *Plants* **2021**, *10*, 881. https://doi.org/10.3390/plants10050881

Academic Editor: M. Iftikhar Hussain

Received: 2 April 2021
Accepted: 23 April 2021
Published: 27 April 2021

Publisher's Note: MDPI stays neutral with regard to jurisdictional claims in published maps and institutional affiliations.

Abstract: Intercropping is one of the most widely used agroforestry techniques, reducing the harmful impacts of external inputs such as fertilizers. It also controls soil erosion, increases soil nutrients availability, and reduces weed growth. In this study, the intercropping of peanut (*Arachishypogaea* L.) was done with tea plants (*Camellia oleifera*), and it was compared with the mono-cropping of tea and peanut. Soil health and fertility were examined by analyzing the variability in soil enzymatic activity and soil nutrients availability at different soil depths (0–10 cm, 10–20 cm, 20–30 cm, and 30–40 cm). Results showed that the peanut–tea intercropping considerably impacted the soil organic carbon (SOC), soil nutrient availability, and soil enzymatic responses at different soil depths. The activity of protease, sucrase, and acid phosphatase was higher in intercropping, while the activity of urease and catalase was higher in peanut monoculture. In intercropping, total phosphorus (TP) was 14.2%, 34.2%, 77.7%, 61.9%; total potassium (TK) was 13.4%, 20%, 27.4%, 20%; available phosphorus (AP) was 52.9%, 26.56%, 61.1%; 146.15% and available potassium (AK) was 11.1%, 43.06%, 46.79% higher than the mono-cropping of tea in respective soil layers. Additionally, available nitrogen (AN) was 51.78%, 5.92%, and 15.32% lower in the 10–20 cm, 20–30 cm, and 30–40 cm layers of the intercropping system than in the mono-cropping system of peanut. Moreover, the soil enzymatic activity was significantly correlated with SOC and total nitrogen (TN) content across all soil depths and cropping systems. The depth and path analysis effect revealed that SOC directly affected sucrase, protease, urease, and catalase enzymes in an intercropping system. It was concluded that an increase in the soil enzymatic activity in the intercropping pattern improved the reaction rate at which organic matter decomposed and released nutrients into the soil environment. Enzyme activity in the decomposition process plays a vital role in forest soil morphology and function. For efficient land use in the cropping system, it is necessary to develop coherent agroforestry practices. The results in this study revealed that intercropping certainly enhance soil nutrients status and positively impacts soil conservation.

Keywords: *Camellia oleifera*; *Arachis hypogaea*; soil nutritional status; soil quality; cropping pattern; silvicultural methods; sustainable production

1. Introduction

The enhancement and maintenance of soil productivity and sustainability through the long-term use of different silvicultural practices, such as varying planting density and spacing, introducing native and exotic beneficial species, agroforestry and intercropping, has been widely used worldwide [1–6]. Intercropping can enhance soil quality by incorporating a significant amount of topsoil and subsoil organic matter and releasing and recycling nutrients [7–9]. It is the general opinion that intercropping is more suitable than mono-cropping for the long-term maintenance of soil fertility [10]. Intercropping systems are rarely dependent on external inputs such as fertilizers [11]; this also reduces the adverse environmental effects such as soil erosion [12]. In China, trees/legume intercropping is most commonly used because it reduces crop failure risk and improves land use [13].

Tea (*Camellia oleifera*) is an evergreen shrub of the Theaceae family, and it is widely found in central and south China [14]. Moreover, it can also survive in nutrient-depleted soils. Peanut (*Arachishypogaea* L.) is a leguminous crop that can fix the atmospheric N and increase soil fertility [15]. According to the US Department of Agriculture, it produces a considerable amount of organic N, improves soil organic matter (SOM), helps in nutrient release and recycling, and improves soil structure. When peanut was intercropped with maize, it played a crucial role in changing soil health by influencing soil microbes composition [16]. Additionally, the dominant microbial species altered due to the peanut intercropping, which shows a close and significant relationship between improving available soil nutrients and soil enzyme activities [17].

Soil enzymes are continually playing dynamic roles in the maintenance of soil health. Soil enzymes are the direct mediators for the biological catabolism of soil organic and mineral components. They are often closely associated with SOM, soil physical properties, and microbial activities and biomass. They are the better indicators of soil health as changes in enzymes occurred much earlier than other soil parameters, thus providing early indications of changes in soil health. Their activities can also be used as measures of microbial activity and soil productivity [18,19]. Although they are present in a very nominal quantity, their role in soil quality can never be ignored. Likewise, a soil nutrient's total content and soil nutrient availability directly affect the plant's growth and development, reflecting soil health [20–22]. In the southern Chinese province of Hunan, the adoption of agroforestry (intercropping) in tea interests the growers because intercropping controls weed growth and soil erosion. However, many of the sites where these plantations were established are deficient in available macronutrients, albeit a high level of total phosphorus (P). The available P rapidly forms insoluble complexes with cations, particularly aluminum and iron. Less nutrient availability in forest soils is considered as one of the most important causes of productivity decline [23]. Nutrient total status and availability are essential for plant growth and development, and thus, macronutrient availability is vital for sustainable productivity. Soil enzymes play a significant role in nutrient recycling in soil ecosystems. In addition, soil enzymes play a crucial role in maintaining soil quality. They can provide essential and early detection signals for soil metabolic activity and nutrient status changes. However, few publications have focused on how legume intercropping with tea plantation, specifically related soil enzymatic activities, influences soil fertility.

Therefore, the objectives of our current study were (1) to investigate the soil total nutrient content and soil nutrient availability; (2) and the activities of soil enzymes (sucrase, protease, urease, acid phosphatase, and catalase); and (3) to explore a correlation between soil enzymatic activity and soil nutrients in different cropping systems and soil depths. Three cropping systems, including peanut–tea intercropping, tea mono-cropping, peanut mono-cropping, and four different soil depths (0–10 cm, 10–20 cm, 20–30 cm, and 30–40 cm) were used in this study. Our hypotheses were that: (i) intercropping favors better nutrient availability and total nutrient content than monoculture systems due to better litter quality and decomposition, and nutrient availability will decrease with the increase in soil depth; (ii) soil enzymatic activity will be higher in the peanut–tea intercropping system than in tea monoculture because peanut influences soil microbial composition—thus, it can enhance soil microbial activity, which has a significant relationship with the improvement of soil enzyme activities; and (iii) there will be a positive and considerable relationship between soil nutrients and soil enzymatic activity across all cropping types and soil depths.

2. Results

2.1. Soil Organic Carbon

Among all the cropping systems and soil depth, SOC content ranged from 5.61 g.kg^{-1} to 19.93 g.kg^{-1}. The topmost layer had the highest SOC content among all the cropping systems. While comparing each cropping system at the same depth, the SOC content was significantly different except at a 10–20 cm depth of CPM and CMM. SOC content decreased as the soil depth increased in the PMM and CMM. However, in CPM, it was observed to be highest at 20–30 cm. Overall, SOC was with the order of PMM = CPM > CMM (Table 1).

Table 1. Soil organic carbon (g.kg^{-1}) content at different soil depths in *C. oleifera* mono-cropping model (CMM), peanut mono-cropping model (PMM), and Camellia–peanut inter-cropping model (CPM).

Cropping Model	0–10 cm	10–20 cm	20–30 cm	30–40 cm
PMM	19.93 ± 1.5 ^{Aa}	16.61 ± 1.39 ^{Ab}	8.62 ± 2.11 ^{Ac}	6.56 ± 1.79 ^{Ac}
CPM	16.77 ± 0.77 ^{Ba}	12.56 ± 4.48 ^{Ba}	14.37 ± 6.17 ^{Bab}	7.68 ± 2.22 ^{Bb}
CMM	13.9 ± 1.65 ^{Ca}	12.15 ± 1.93 ^{Ba}	9.89 ± 3.04 ^{Ca}	5.61 ± 3.11 ^{Cb}

Note: Two-way analysis of variance (ANOVA) test was performed and the values are mean ± SD. Different uppercase letters indicate significant differences between different cropping systems at $p < 0.05$, while different lowercase letters indicate significant differences in different depths $p < 0.05$.

2.2. Fluctuations in Soil Total Nutrient Status

TN ranged from 0.72 g.kg^{-1} to 1.51 g.kg^{-1} among all cropping types and soil depths. PMM had the highest TN content among all the cropping systems at the topmost layer. TN content decreased as the soil depth increase, apart from CPM, where it was observed as the highest at 20–30 cm. There was a significant difference present in TN at each soil layer among the three cropping systems, except at a 10–20 cm depth of CPM and CMM. In terms of TP, among all the cropping patterns and soil depths, its content ranged from 0.21 g.kg^{-1} to 0.57 g.kg^{-1}. CPM had the highest TP observed at 10–20 cm.

The TK content ranged from 0.05 g.kg^{-1} to 0.07 g.kg^{-1} among all soil layers and cropping patterns. CPM had the highest TK content, and it was observed to be the highest at 10–20 cm depth. At the same depths, there was a non-significant difference observed between PMM and CMM; however, both were significantly different to PMM. Overall, the TP and TK contents were in the order of CPM > CMM > PMM (Figure 1).

2.3. Variability in Soil Nutrient Availability

Among all the soil layers and cropping patterns, AN content ranged from 5.04 mg.kg^{-1} to 9.69 mg.kg^{-1}. There was a non-significant difference observed for AN between different cropping systems at each soil layer, except 10–20 cm depth. The AP content ranged from 7.55 mg.kg^{-1} to 110.58 mg.kg^{-1} among all the soil depths and cropping systems, and it was observed to be the highest in the topsoil layer. Overall, the AP content was observed highest in CPM. Similar to AP, AK was also observed to be higher in CPM compared to both mono-cropping systems. On average, the AN content followed the order of PMM > CPM > CMM, whereas AP content followed the order of CPM > CMM > PMM, and the AK content followed the order CPM > PMM > CMM (Figure 2).

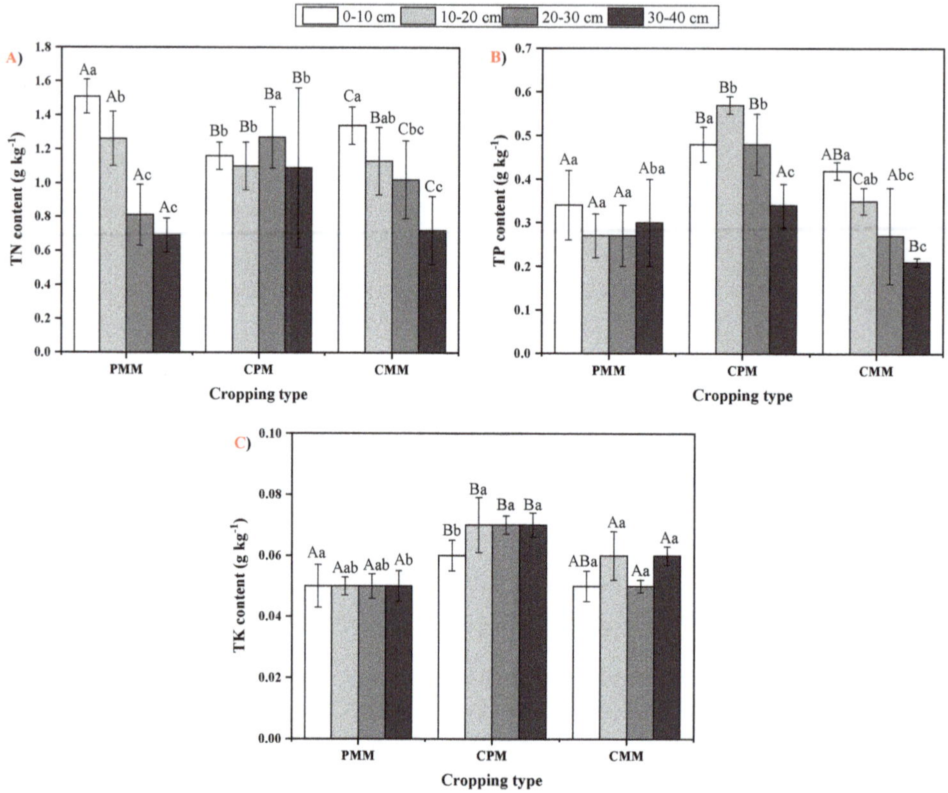

Figure 1. Variability in (**A**) soil total nitrogen (TN), (**B**) total phosphorus (TP) and (**C**) total potassium (TK) content at different soil depths in *C. oleifera* mono-cropping model (CMM), peanut mono-cropping model (PMM), and Camellia–peanut inter-cropping model (CPM). Two-way analysis of variance (ANOVA) test was conducted and the values in the columns are mean ± SD. Different uppercase letters indicate significant differences between the cropping systems at $p < 0.05$, while the different lowercase letters indicate significant differences among the soil depths $p < 0.05$.

2.4. Soil Enzymatic Responses in Different Cropping Types

All enzymes had the highest activity in 0–10 cm for each cropping system except CMM. In CMM, soil urease and protease activity were most increased at a 10–20 cm depth. The protease content varied significantly for all the cropping systems in the 0–10 cm and 30–40 cm layer and was observed to be the highest in CPM. The sucrase content followed the order PMM > CPM > CMM in all soil layers. For acid phosphatase, all layers followed the order CPM > CMM > PMM. The urease activity varied significantly among all cropping systems in the 0–10 cm, 10–20 cm, and 30–40 cm layers; only in the 20–30 cm layer was there no significant difference observed. Catalase content was not significantly different in the 0–10 cm, 20–30 cm and 30–40 cm among the cropping systems, and it was the highest in PMM. However, in the 10–20 cm layer, it was observed highest in CPM (Figure 3).

2.5. Correlation between Soil Nutrients and Soil Enzymatic Activity

(1) In PMM, sucrase, protease and catalase had no significant correlation with any nutrient. Urease was positively correlated to SOC, TN and AK, while negatively correlated to TK. Acid phosphatase was positively correlated with SOC, TN and AK. (2) In CPM, sucrase was positively related to TN; protease to SOC; urease to SOC, AN and AK; acid phosphatase to SOC; and catalase to SOC and AK. (3) In CMM, sucrase was positively

related to AP, while protease was negatively associated with AN. Urease, acid phosphatase and catalase enzymes were positively associated with SOC, TN, TP, AP and AK (Table 2).

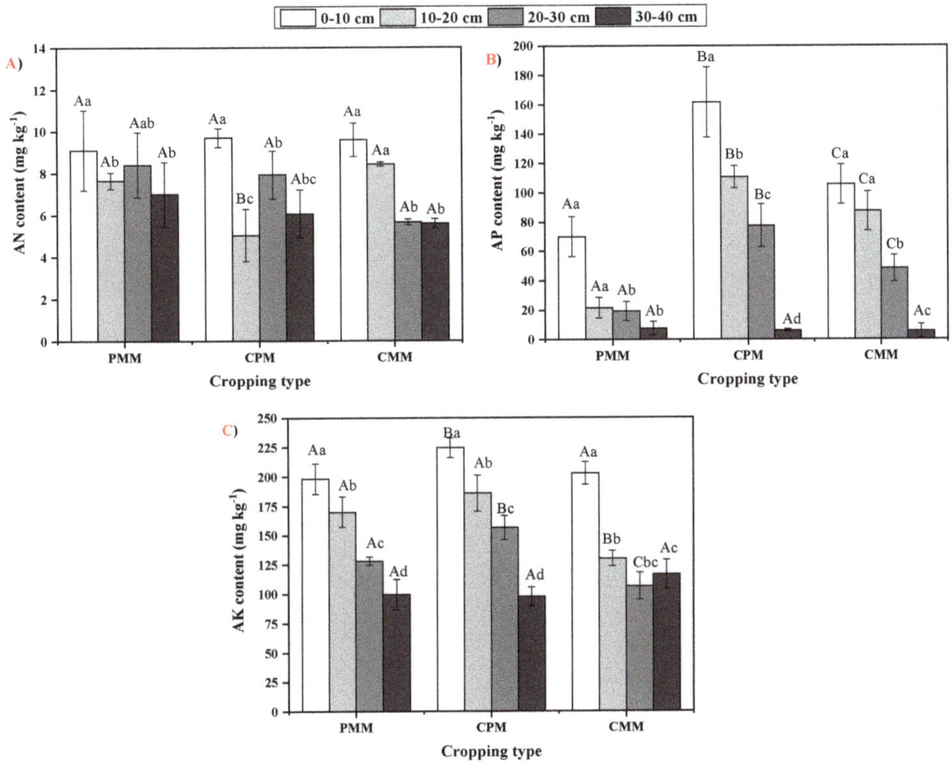

Figure 2. Variability in (**A**) soil available nitrogen (AN), (**B**) available phosphorus (AP) and (**C**) available potassium (AK) content in different soil depths in *C. oleifera* mono-cropping model (CMM), peanut mono-cropping model (PMM), and Camellia–peanut inter-cropping model (CPM). Two-way analysis of variance (ANOVA) test was conducted and the values in the columns are mean ± SD. Different uppercase letters indicate significant differences between cropping systems at $p < 0.05$, while different lowercase letters indicate significant differences among soil depths $p < 0.05$.

2.6. Path Coefficients between Soil Nutrients and Soil Enzyme Activity

In PMM, among all the soil nutrients, the direct path coefficients of soil TN (0.821) and SOC (−0.534) on soil sucrase activity were relatively large. The direct path coefficients of SOC (0.689) and AP (0.524) indicate the substantial direct effect of these nutrients on protease activity. SOC also has a robust direct path coefficient with acid phosphatase activity (1.045). The direct path coefficients of SOC (2.203) and TN (−2.069) on catalase activity indicate that both had strong positive and negative effects on catalase activity, respectively (Table A1).

In CPM, soil TN, SOC and AN had a robust positive effect on sucrase enzyme activity. However, the path coefficient was small. SOC (0.469) had a sizeable direct path coefficient with protease indicates the substantial direct effect on protease activity. Moreover, SOC (0.7074) also has a robust positive path coefficient with urease activity. The direct path coefficients of all soil nutrients with catalase activity were also positive. Compared with the other four enzymes, the direct and indirect path coefficients of soil nutrient factors with acid phosphatase were relatively small, indicating that soil nutrient factors had a more negligible effect on acid phosphatase activity (Table A2).

In CMM, soil AP (0.532) had a more significant direct path coefficient on sucrase and TP (0.905), which strongly affected protease activity shows. The effect of soil nutrients on urease activity was positive with a relatively more minor path coefficient. The direct path coefficient of AP (0.885) on acid phosphatase was rather significant. In addition, TN (0.436) had a direct effect on catalase activity (Table A3). Two-way ANOVA results of the testes soil variables are shown in Table 3.

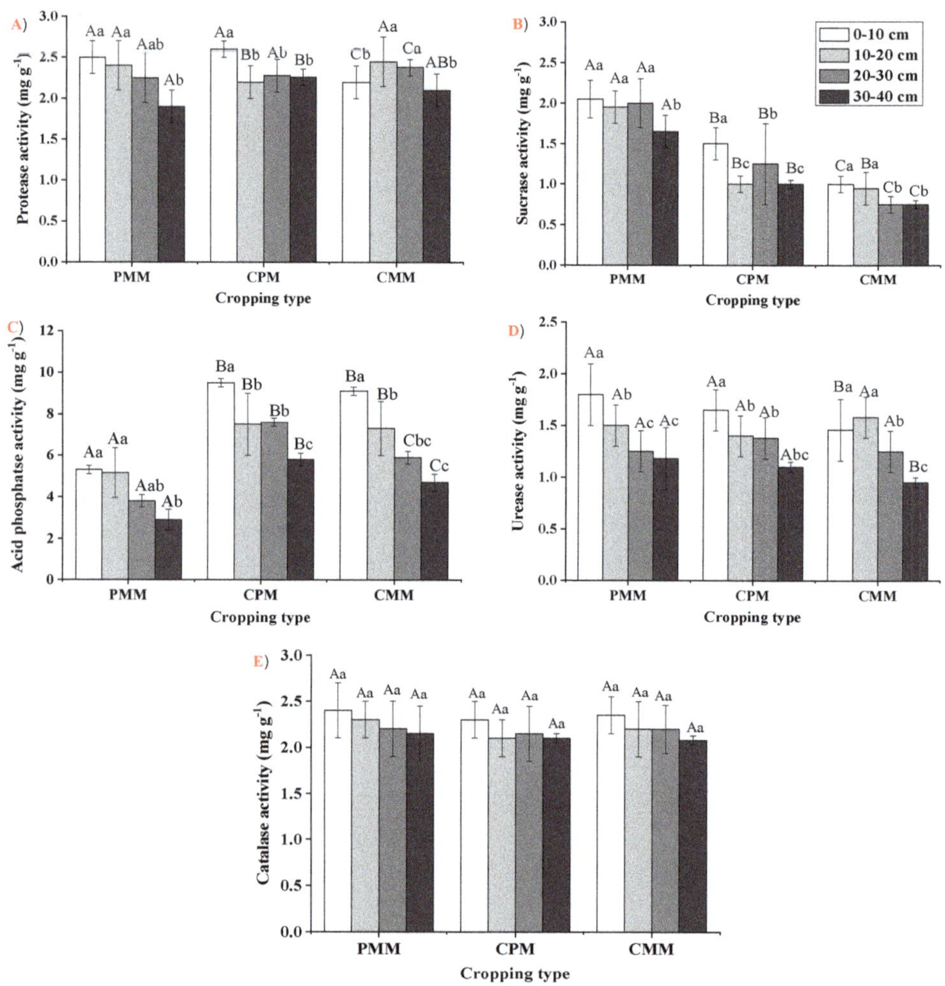

Figure 3. The activity of soil enzymes (**A**) protease; (**B**) sucrase; (**C**) acid phosphatase; (**D**) urease; and (**E**) catalase (mg.g^{-1}) at different soil depths in the three cropping models in the study site. A two-way analysis of variance (ANOVA) test was conducted and the values in the columns were the mean ± SD. Different uppercase letters indicated significant differences between the cropping systems at $p < 0.05$, while different lowercase letters indicated significant differences among soil depths $p < 0.05$.

Table 2. Correlation analysis of soil nutrients and enzyme activities in *C. oleifera* mono-cropping model (CMM), peanut mono-cropping model (PMM), and the Camellia–peanut inter-cropping model (CPM).

Cropping Model	Soil Enzyme	SOC	TN	TP	TK	AN	AP	AK
PMM	Sucrase	0.21	0.22	0.04	−0.27	0.01	0.08	0.07
	Protease	0.45	0.47	−0.05	−0.14	0.29	0.37	0.17
	Urease	**0.68 ***	**0.71 ***	−0.15	**−0.5 ***	−0.14	0.25	**0.58 ***
	Acid phosphatase	**0.94 ***	**0.92 ****	0.13	−0.29	0.19	0.41	**0.57 ***
	Catalase	0.17	0.05	0.06	−0.02	−0.11	0.2	−0.03
CPM	Sucrase	0.33	**0.51 ***	−0.08	0	0.11	−0.18	0.22
	Protease	**0.56 ***	0.21	0.11	0.2	0.39	0.32	0.38
	Urease	**0.75 ***	0.26	0.32	0.23	**0.43 ***	0.03	**0.46 ***
	Acid phosphatase	**0.42 ****	0.34	0.15	−0.11	0.33	−0.03	0.13
	Catalase	**0.44 ***	0.26	0.34	0.12	0.25	−0.14	**0.41 ***
CMM	Sucrase	0.35	0.41	0.38	−0.05	0.03	**0.41 ***	0.31
	Protease	0.05	−0.01	0.17	−0.07	**−0.55 ***	0.19	−0.31
	Urease	**0.41 ***	**0.52 ***	**0.52 ***	0.23	0.23	**0.48 ***	**0.49 ***
	Acid phosphatase	**0.51 ****	**0.62 ***	**0.58 ***	0.11	0.12	**0.62 ***	**0.44 ****
	Catalase	**0.43 ***	**0.52 ***	**0.54 ***	0.01	−0.08	**0.54 ***	**0.42 ***

Note: Significant values are bold. Values given are the Pearson correlation coefficients. * Correlation is significant at the 0.05 level (2-tailed). ** Correlation is significant at the 0.01 level (2-tailed).

Table 3. Two-way ANOVA results for soil variables. *, ** and *** indicate a significant level at $p < 0.05$, $p < 0.01$ and $p < 0.001$, respectively.

	Cropping Model		Soil Depth		Cropping Model × Soil Depth	
Variable	*F*	*P*	*F*	*P*	*F*	*P*
TN	122.6	**	1233.6	*	365.5	**
TP	870.2	*	1364.4	**	102.1	***
TK	683.3	**	65.3	0.274	89.2	**
AN	42.9	**	78.3	**	51.2	**
AP	632.4	**	896.7	**	3546.5	**
AK	536.2	*	336.3	***	1456.5	**
Protease	256.3	0.066	256.5	**	785.4	0.064
Sucrase	158.9	**	447.3	**	1235.9	*
A. phosphatase	1244.3	*	380.3	0.066	3225.0	**
Urease	1563.2	**	125.6	**	2132.5	**
Catalase	853.2	**	0.072	**	125.2	**

3. Discussion

In the present study, SOC was unanimously higher in the 0–10 cm soil layer among all the cropping systems because of the higher litterfall on the topmost layer. Compared to the mono-cropping of tea, SOC was higher in peanut–tea intercropping, possibly due to the higher litterfall and tea biomass decomposition inputs. Growing different crops on the same land simultaneously helps maintain the SOC and improve nutrient cycling [24,25]. A similar kind of increase in tea soil SOC content was also observed by [26,27]. N is vital for plant growth and development [22,28]. It is available in many soil forms, such as nitrate and ammonia [29,30]. Peanut being a legume plant could fix atmospheric nitrogen [31,32] and as a result, we found out that PMM had the highest amount of TN for the upper two layers, while in the remaining two layers, it was observed to be highest in CPM. This could be due to the tea plant deep root system, which makes way in the soil deeper layers for the atmospheric N that the peanut plant fixes. Both TP and TK contents were higher in the CPM among all the soil layers than the tea and peanut mono-cropping system. The probable reason was that intercropping can provide a much more comprehensive ground cover along with better water use efficiency [31,32].

Leguminous plants can convert the unavailable N form into a useable form [33]. The comparison of CPM and PMM depicted that AN was 51.78%, 5.92%, and 15.32% lower in the 10–20 cm, 20–30 cm, and 30–40 cm layers of the intercropping system than in the peanut mono-cropping system because the of higher urease activity in CPM, which led to an increase in the NH_3 loss from soil [34,35]. AP in the intercropping of tea and peanut was higher in all the soil layers than the tea and peanut monoculture cropping. This is because leguminous plant led to the acidification of the rhizosphere with the help of roots, and these roots release the organic acids [36–39]. Due to acidification, the enzyme acid phosphatase starts the dissolution of P-based minerals, which increases the P availability. Maurya and Lal [40] obtained similar results, where the amount of P increased due to the release of acid phosphatase by the chickpea, which led to the conversion of organic P into inorganic P. AK content was higher in CPM compared to CMM and PMM in all soil layers. AK was 11.1%, 43.06%, 46.79% higher in 0–10 cm, 10–20 cm, and 20–30 cm layers of the intercropping system than the mono-cropping of tea. Similarly, AK was 13.4%, 9.29%, and 22.06% higher in 0–10 cm, 10–20 cm, and 20–30 cm layers of intercropping system than the mono-cropping system of peanut. The increase in AK in the intercropping system can be attributed to the increase in the soil enzymatic activity.

Soil enzymes play a significant role in the soil ecosystem's overall biochemical functioning [41–43]. A better understanding of the soil enzymatic activity in different cropping systems with a depth effect can provide better knowledge about how intercropping systems can improve soil fertility. It is evident from the previous studies that the mono-cropping system could potentially harm the soil enzyme mechanism, which results in a significant decrease in the soil enzymatic activity [44]. In our current study, a very considerable increase in the enzyme activities was observed in the intercropping system of the tea and peanut than in the mono-cropping systems. Protease is the main enzyme involved in the catalysis of N minerals and N cycling [45]. The protease activity was higher in the layers of 0–10 cm and 30–40 cm for the intercropping system than both mono-cropping systems. This change in protease activity might be attributed to a higher SOC content in the topsoil [46]. However, in the 30–40 cm layer, the increase might be due to a higher N content. Soil sucrase enzymes catalyze sucrase to glucose and fructose with hydrolysis, and it is also connected with the biomass of the soil microbes [47,48]. The mono-cropping of peanut had the highest sucrase enzyme activity, possibly due to a higher SOM. These findings are supported by Li et al. [41], who observed that soil enzyme activity was significantly enhanced due to intercropping. Acid phosphatase is the main enzyme involved in ester hydrolysis and phosphoric acid anhydrides [49,50]. It converts the esters and anhydrides into phosphate, and is a key enzyme in P cycling in soil [51].

In our current study, acid phosphatase activity was higher in the intercropping of peanut–tea than tea mono-cropping because the peanut was identified as a species whose roots release an ample amount of acid phosphatase in the soil [52,53]. The soil urease enzyme catalyzes urea into NH_3 and CO_2 with hydrolysis [54]. This is a necessary process that regulates N availability to the plants after the application of urea. To some level, urease indicates the availability of N in different cropping systems. The catalase's primary function is to decompose the organic matter into the plant's useable form [55,56]. In our current study, the mono-cropping of peanut had a higher activity of urease and catalase than intercropping because of the SOM content of the mono-cropping of peanut. In this study, the enzyme activity was more or less associated with the content and distribution of SOC, TN and AK among all the cropping systems. Our results are in line with Tian et al. [57] and Udawatta et al. [58]; they also observed a stronger correlation between SOC and enzymes. An increase in the SOM and litter quantity enhances the soil activity [59,60] and this increment has direct involvement in the improvement of nutrient cycling, along with greenhouse emission [61–63], which, in return, has a positive impact on the ecosystem, plant growth and overall SOC.

4. Materials and Methods

4.1. Study Site

The study was conducted in Hunan Botanical Garden in Changsha city, Hunan Province, China (113°02′–113°03′ E, 28°06′–28°07′ N) (Figure 4). The study area had a typical subtropical humid monsoon climate, with a mean annual precipitation of 1378 mm, mean annual temperature of 17.2 °C, and a mean annual relative humidity of 81%. The mean annual average sunshine was 1814.8 h and the frost-free period was 275 days. The site was the hillside, with a slope of about 5–15°. The soil was classified as typical red earth developed from the quaternary red clay reticulated parent material. The soil texture ranged from clay loam to sandy loam, with a depth of about 1 m. The soil was acid with a pH of 4.5–5.5. Soil bulk density was ranged between 1.16 to 1.22 (g/m^3).

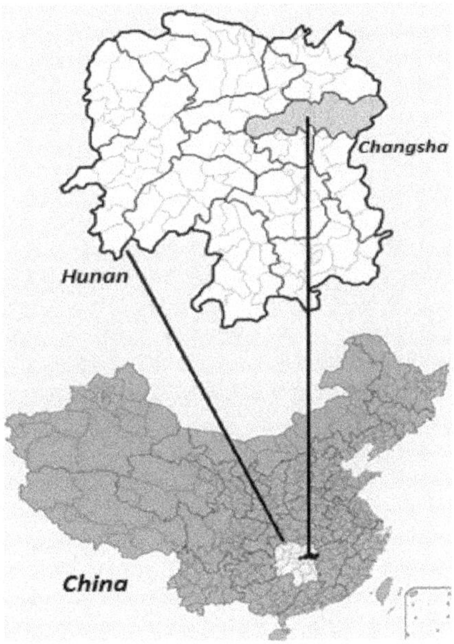

Figure 4. Location of the study area, Hunan botanical garden, Changsha, Hunan.

4.2. Experimental Design

In this study, three cropping models were set up which included: (1) a *C. oleifera* mono-cropping model (CMM); (2) a peanut mono-cropping model (PMM), and (3) a Camellia–peanut inter-cropping model (CPM). The *C. oleifera* monoculture stand was established in 2010, and the line-row spacing of the *C. oleifera* was planted in 4 m × 3 m. In CMM, an area with three lines and three rows of *C. oleifera* trees was selected as a plot (about 110 m^2). Four plots were set up for CMM in this study.

However, in PMM, the peanuts were planted in the open space beside the *C. oleifera* forests according to the 0.25 m × 0.1 m spacing in April 2018. An area with 40 lines and 100 rows of peanut plants was selected as a plot (about 100 m^2). Four plots were set up for PMM in this study. Moreover, in CPM, an area with three lines and three rows of *C. oleifera* trees was selected as a plot (about 110 m^2). Peanuts were intercropped between *C. oleifera* trees in April 2018. The spacing between the peanut and tea plant was 1 m. Four plots of CPM were set up in this study. The three cropping models employed similar field management practices during the study period.

4.3. Soil Sampling

Soil samples were collected in September 2019. Four plots of 20 × 20 m were established in each of CMM, PMM, and CPM plots; four pits were dug diagonally. A soil corer was used to obtain soil samples from 0–10, 10–20, 20–30, and 30–40 cm depths from each pit. Four replicates were taken from each plot for each depth. The rocks and plant residues were removed from the soil samples. Soil samples were placed into self-sealing plastic bags, labelled, and delivered to the laboratory for further analysis. For enzymatic activity analysis, the soil samples were placed in −80 °C until further use. For nutrients analysis, after air drying, the soil samples were passed through a 0.15 mm and 0.20 mm sieve for the determination of soil organic carbon (SOC), total nitrogen (TN), total phosphorus (TP), total potassium (TK), available nitrogen (AN), available phosphorus (AP), and available potassium (AK).

4.4. Soil Chemical Analysis

Urease activity was measured following the method described in [64]. Five grams of soil was incubated with 10 mL of citrate phosphate buffer (pH 6.7) and 5 mL of 10 % urea solution at 38 °C for 3 h. Activity was determined by measuring the released NH_4^+ with a spectrophotometer at 578 nm. Acid phosphatase activity was analyzed with nitrophenyl phosphate disodium (PhOH mg g^1, 37 °C, 24 h), and catalase with $KMnO_4$ (0.1 mol L^{-1} $KMnO_4$ ug g^{-1}, 30 °C, 20 h) [65]. Sucrase activity was determined by the method of [66]. For sucrose, the air-dried soil (5 g) was incubated with 15 mL sucrose. Five microliters (5 mL) of phosphate buffer (pH 5.5) and five drops of toluene at 37 °C for 24 h, and the reaction solution was filtered through the quantitative filter paper as rapidly as possible after incubation. Filtrate (1 mL) was mixed with 3 mL salicylic acid at 100 °C for 5 min in the water bath, and the mixture was adjusted to 50 mL and cooled with deionized water. Sucrase activity was determined spectrophotometrically at 508 nm. The protease activity was determined by ninhydrin colorimetry, expressed in milligrams of amino nitrogen in 1 g of soil cultured for 24 h in a 37 °C incubator [67].

SOC was determined by the hydrated potassium thermo-dichromate oxidation method. TN was determined using the CN elemental analyzer, while the colorimetric method of the molybdenum–antimony solution with royal acid was used to determine TP. The flame photometer method was used to determine TK. AN was analyzed by the Kjeldahl method [68]; AP was determined by the diacid extraction spectrophotometric colorimetry method [69], and AK was determined using a flame photometer method by ammonium acetate extraction [70]. Detailed information about the total and available nutrients was also mentioned in our published paper [6].

4.5. Statistical Analysis

Two-way analysis of variance (ANOVA) was performed to analyze the difference in soil enzymatic activity, soil total nutrient status and soil nutrient availability between different cropping systems at four different soil depths. Correlation analysis was performed using the Pearson statistical method to analyze the association among the studied soil parameters. Moreover, the path analysis using multiple linear analysis was also performed to study the relationships betweesn soil enzyme and soil nutrients. Results were declared statistically significant at $p < 0.05$ and the means that exhibited significant differences were compared using Tukey's significance test. All statistical analyses were performed using the SPSS Statistical Package (SPSS 17.0, Chicago, IL, USA).

The multiple linear regression equations of soil sucrase activity (Y1), protease activity (Y2), urease activity (Y3), acid phosphatase activity (Y4), catalase activity (Y5) and soil nutrient factors are as follows:

The multiple linear regression equations in PMM are:

$$Y1 = -0.534X1 + 0.821X2 + 0.173X3 - 0.258X4 - 0.012X5 - 0.065X6 - 0.257X7 \quad (1)$$

$$Y2 = 0.689X1 + 0.354X2 - 0.392X3 + 0.020X4 + 0.225X5 + 0.524X6 - 0.195X7 \quad (2)$$

$$Y3 = -0.021X1 + 0.443X2 - 0.394X3 - 0.157X4 - 0.115X5 + 0.233X6 + 0.275X7 \quad (3)$$

$$Y4 = 1.045X1 - 0.128X2 - 0.042X3 + 0.080X4 + 0.134X5 + 0.069X6 + 0.048X7 \quad (4)$$

$$Y5 = 2.023X1 - 2.069X2 - 0.181X3 + 0.065X4 + 0.065X5 + 0.244X6 + 0.169X7 \quad (5)$$

The multiple linear regression equations in CPM are:

$$Y1 = 0.104X1 + 0.517X2 - 0.297X3 - 0.046X4 + 0.018X5 - 0.219X6 + 0.329X7 \quad (6)$$

$$Y2 = 0.469X1 - 0.018X2 - 0.283X3 + 0.208X4 + 0.179X5 - 0.037X6 + 0.25X7 \quad (7)$$

$$Y3 = 0.707X1 - 0.038X2 - 0.028X3 + 0.161X4 + 0.0009X5 - 0.017X6 + 0.196X7 \quad (8)$$

$$Y4 = 0.219X1 + 0.239X2 + 0.053X3 - 0.139X4 + 0.154X5 - 0.131X6 + 0.031X7 \quad (9)$$

$$Y5 = 0.243X1 + 0.143X2 + 0.197X3 - 0.037X4 + 0.015X5 - 0.274X6 + 0.264X7 \quad (10)$$

The multiple linear regression equations in CMM are:

$$Y1 = 0.009X1 + 0.213X2 - 0.368X3 - 0.089X4 + 0.001X5 + 0.532X6 + 0.248X7 \quad (11)$$

$$Y2 = 0.070X1 - 0.041X2 + 0.905X3 + 0.110X4 - 0.557X5 - 0.483X6 - 0.506X7 \quad (12)$$

$$Y3 = 0.111X1 + 0.278X2 - 0.142X3 + 0.309X4 + 0.216X5 + 0.368X6 + 0.072X7 \quad (13)$$

$$Y4 = 0.043X1 + 0.445X2 - 0.598X3 + 0.249X4 + 0.148X5 + 0.885X6 - 0.015X7 \quad (14)$$

$$Y5 = -0.200X1 + 0.436X2 + 0.088X3 - 0.132X4 - 0.199X5 + 0.174X6 + 0.274X7 \quad (15)$$

X1, X2, X3, X4, X5, X6 and X7 represent SOC, TN, TP, TK, AN, AP and AK, respectively. These equations present the direct path coefficients. The indirect path coefficients were obtained using the direct path coefficients multiplied by each soil nutrient's correlation coefficient. The direct path coefficient was the direct effect of soil nutrient factors on enzyme activity.

5. Conclusions

In our study, the activity of enzymes, SOC, and TN was higher in the topsoil than the subsoil because of the higher accumulation of SOM on the topsoil and the much favorable moisture and temperature conditions of topsoil. SOC and most of the soil total nutrient content and nutrient availability were significantly higher in the intercropping system. Moreover, apart from soil catalase and urease, other enzymes' soil enzymatic activities were higher in the intercropping than in the mono-cropping system. Path analysis provides the weighted effect of soil nutrients on soil enzymes in each cropping type. Intercropping may also lead to an increase in the other soil quality indicators, such as SOM. Intercropping of peanut with tea could significantly increase soil fertility. The long-term sustainability of the soil ecosystem in tea farming can be achieved with the help of peanut–tea intercropping.

Author Contributions: Conceptualization, T.H.F., J.M., X.C.; Data curation, A.S. and M.H.U.R.; Formal analysis, T.H.F. and J.M.; Funding acquisition, W.Y.; Investigation, J.W. and W.Y.; Methodology, T.H.F., U.K. and J.M.; Project administration, J.W. and W.Y.; Resources, X.C.; Software, A.S., M.H.U.R. and M.A.T.; Writing—original draft, T.H.F. and U.K.; Writing—review & editing, M.A.T., X.C. and W.Y. All authors have read and agreed to the published version of the manuscript.

Funding: The funding sources include the National Science and Technology Support Grant of China (2015BAD07B0503), Forestry Science and Technology Promotion Project of China (No. 12 2017) and Postdoctoral research funding of Central South University of Forestry and Technology (70702-45200003).

Institutional Review Board Statement: Not applicable.

Informed Consent Statement: Not applicable.

Data Availability Statement: No additional data available.

Conflicts of Interest: The authors declare there is no conflict of interest exists.

Appendix A

Table A1. Path coefficients between soil nutrients factors (X) and soil enzyme activity (Y) in PMM.

Factors Code		X1	X2	X3	X4	X5	X6	X7
Y1	X1	−0.534 *	−0.518	−0.048	0.219	−0.021	−0.198	−0.315
	X2	0.796	0.821 *	0.066	−0.320	0.099	0.304	0.550
	X3	0.016	0.014	0.173 *	0.014	0.022	0.112	0.054
	X4	0.106	0.101	−0.021	−0.258 *	−0.080	0.041	0.065
	X5	0.000	−0.001	−0.002	−0.004	−0.012 *	−0.001	0.000
	X6	−0.024	−0.024	−0.042	0.010	−0.007	−0.065 *	−0.027
	X7	−0.152	−0.172	−0.080	0.064	0.005	−0.105	−0.257 *
Y2	X1	0.689 *	0.668	0.062	−0.282	0.028	0.255	0.407
	X2	0.343	0.354 *	0.028	−0.138	0.042	0.131	0.237
	X3	−0.035	−0.031	−0.392 *	−0.031	−0.051	−0.255	−0.122
	X4	−0.008	−0.008	0.002	0.020*	0.006	−0.003	−0.005
	X5	0.009	0.027	0.029	0.070	0.225 *	0.025	−0.005
	X6	0.194	0.194	0.341	−0.084	0.058	0.524 *	0.215
	X7	−0.115	−0.131	−0.060	0.049	0.004	−0.080	−0.195 *
Y3	X1	−0.021 *	−0.020	−0.002	0.009	−0.001	−0.008	−0.012
	X2	0.430	0.443 *	0.035	−0.173	0.053	0.164	0.297
	X3	−0.035	−0.032	−0.394 *	−0.032	−0.051	−0.256	−0.122
	X4	0.064	0.061	−0.013	−0.157 *	−0.049	0.025	0.039
	X5	−0.005	−0.014	−0.015	−0.036	−0.115 *	−0.013	0.002
	X6	0.086	0.086	0.151	−0.037	0.026	0.233 *	0.096
	X7	0.162	0.184	0.085	−0.069	−0.006	0.113	0.275 *
Y4	X1	1.045 *	1.014	0.094	−0.428	0.042	0.387	0.617
	X2	−0.124	−0.128 *	−0.010	0.050	−0.015	−0.047	−0.086
	X3	−0.004	−0.003	−0.042 *	−0.003	−0.005	−0.027	−0.013
	X4	−0.033	−0.031	0.006	0.080 *	0.025	−0.013	−0.020
	X5	0.005	0.016	0.017	0.042	0.134 *	0.015	−0.003
	X6	0.026	0.026	0.045	−0.011	0.008	0.069 *	0.028
	X7	0.028	0.032	0.015	−0.012	−0.001	0.020	0.048 *
Y5	X1	2.023 *	1.962	0.182	−0.829	0.081	0.749	1.194
	X2	−2.007	−0.069 *	−0.166	0.807	−0.248	−0.766	−1.386
	X3	−0.016	−0.014	−0.181 *	−0.014	−0.024	−0.118	−0.056
	X4	−0.027	−0.025	0.005	0.065 *	0.020	−0.010	−0.016
	X5	0.003	0.008	0.008	0.020	0.065 *	0.007	−0.001
	X6	0.090	0.090	0.159	−0.039	0.027	0.244 *	0.100
	X7	0.100	0.113	0.052	−0.042	−0.003	0.069	0.169 *

* indicate a significant level at $p < 0.05$.

Table A2. Path coefficients between soil nutrients factors (X) and soil enzyme activity (Y) in in CPM.

Factors Code		X1	X2	X3	X4	X5	X6	X7
Y1	X1	0.104 *	0.202	−0.086	0.000	0.010	−0.007	0.105
	X2	0.041	0.517 *	−0.050	−0.003	0.004	−0.033	0.039
	X3	0.030	0.088	−0.297 *	−0.015	0.007	−0.068	0.174
	X4	0.000	0.036	−0.098	−0.046 *	−0.001	−0.018	0.132
	X5	0.059	0.114	−0.113	0.002	0.018 *	−0.072	0.102
	X6	0.003	0.078	−0.092	−0.004	0.006	−0.219 *	0.049
	X7	0.033	0.062	−0.157	−0.018	0.006	−0.033	0.329 *

Table A2. *Cont.*

Factors	Code	X1	X2	X3	X4	X5	X6	X7
Y2	X1	0.469 *	−0.007	−0.082	0.000	0.102	−0.001	0.080
	X2	0.183	−0.018 *	−0.048	0.015	0.039	−0.006	0.030
	X3	0.136	−0.003	−0.283 *	0.069	0.068	−0.011	0.133
	X4	0.000	−0.001	−0.093	0.208 *	−0.009	−0.003	0.100
	X5	0.267	−0.004	−0.108	−0.010	0.179 *	−0.012	0.078
	X6	0.014	−0.003	−0.088	0.017	0.059	−0.037 *	0.038
	X7	0.150	−0.002	−0.150	0.083	0.055	−0.006	0.250 *
Y3	X1	0.707 *	−0.015	−0.008	0.000	0.001	−0.001	0.063
	X2	0.276	−0.038 *	−0.005	0.011	0.000	−0.003	0.024
	X3	0.205	−0.006	−0.028 *	0.053	0.000	−0.005	0.104
	X4	0.000	−0.003	−0.009	0.161 *	0.000	−0.001	0.078
	X5	0.403	−0.008	−0.011	−0.008	0.001 *	−0.006	0.061
	X6	0.021	−0.006	−0.009	0.013	0.000	−0.017 *	0.029
	X7	0.226	−0.005	−0.015	0.064	0.000	−0.003	0.196 *
Y4	X1	0.219 *	0.093	0.015	0.000	0.088	−0.004	0.010
	X2	0.085	0.239 *	0.009	−0.010	0.034	−0.020	0.004
	X3	0.064	0.041	0.053 *	−0.046	0.059	−0.041	0.016
	X4	0.000	0.017	0.017	−0.139 *	−0.008	−0.010	0.012
	X5	0.125	0.053	0.020	0.007	0.154 *	−0.043	0.010
	X6	0.007	0.036	0.016	−0.011	0.051	−0.131 *	0.005
	X7	0.070	0.029	0.028	−0.056	0.048	−0.020	0.031 *
Y5	X1	0.243 *	0.056	0.057	0.000	0.009	−0.008	0.084
	X2	0.095	0.143 *	0.033	−0.003	0.003	−0.041	0.032
	X3	0.070	0.024	0.197 *	−0.012	0.006	−0.085	0.140
	X4	0.000	0.010	0.065	−0.037 *	−0.001	−0.022	0.106
	X5	0.139	0.031	0.075	0.002	0.015 *	−0.090	0.082
	X6	0.007	0.021	0.061	−0.003	0.005	−0.274 *	0.040
	X7	0.078	0.017	0.104	−0.015	0.005	−0.041	0.264 *

* indicate a significant level at $p < 0.05$.

Table A3. Path coefficients between soil nutrients factors (X) and soil enzyme activity (Y) in in CMM.

Factors	Code	X1	X2	X3	X4	X5	X6	X7
Y1	X1	0.009 *	0.083	−0.107	0.000	0.001	0.016	0.079
	X2	0.004	0.213 *	−0.063	−0.006	0.000	0.080	0.037
	X3	0.003	0.036	−0.368 *	−0.029	0.000	0.165	0.131
	X4	0.000	0.015	−0.121	−0.089 *	0.000	0.043	0.099
	X5	0.005	0.047	−0.140	0.004	0.001 *	0.176	0.077
	X6	0.000	0.032	−0.114	−0.007	0.000	0.532 *	0.037
	X7	0.003	0.026	−0.195	−0.036	0.000	0.080	0.248 *
Y2	X1	0.070 *	−0.016	0.262	0.000	−0.317	−0.014	−0.162
	X2	0.027	−0.041 *	0.154	0.008	−0.123	−0.072	−0.076
	X3	0.020	−0.007	0.905 *	0.036	−0.212	−0.150	−0.268
	X4	0.000	−0.003	0.299	0.110 *	0.028	−0.039	−0.202
	X5	0.040	−0.009	0.344	−0.006	−0.557 *	−0.159	−0.157
	X6	0.002	−0.006	0.281	0.009	−0.184	−0.483 *	−0.076
	X7	0.022	−0.005	0.480	0.044	−0.173	−0.072	−0.506 *
Y3	X1	0.111 *	0.108	−0.041	0.000	0.123	0.011	0.023
	X2	0.043	0.278 *	−0.024	0.022	0.048	0.055	0.011
	X3	0.032	0.047	−0.142 *	0.102	0.082	0.114	0.038
	X4	0.000	0.019	−0.047	0.309 *	−0.011	0.029	0.029
	X5	0.063	0.061	−0.054	−0.015	0.216 *	0.121	0.022
	X6	0.003	0.042	−0.044	0.025	0.071	0.368 *	0.011
	X7	0.036	0.033	−0.075	0.124	0.067	0.055	0.072 *

Table A3. *Cont.*

Factors	Code	X1	X2	X3	X4	X5	X6	X7
Y4	X1	0.043 *	0.174	−0.173	0.000	0.084	0.027	−0.005
	X2	0.017	0.445 *	−0.102	0.017	0.033	0.133	−0.002
	X3	0.012	0.076	−0.598 *	0.082	0.056	0.274	−0.008
	X4	0.000	0.031	−0.197	0.249 *	−0.007	0.071	−0.006
	X5	0.025	0.098	−0.227	−0.012	0.148 *	0.292	−0.005
	X6	0.001	0.067	−0.185	0.020	0.049	0.885 *	−0.002
	X7	0.014	0.053	−0.317	0.100	0.046	0.133	−0.015 *
Y5	X1	−0.200 *	0.170	0.026	0.000	−0.113	0.005	0.088
	X2	−0.078	0.436 *	0.015	−0.009	−0.044	0.026	0.041
	X3	−0.058	0.074	0.088 *	−0.044	−0.076	0.054	0.145
	X4	0.000	0.031	0.029	−0.132 *	0.010	0.014	0.110
	X5	−0.114	0.096	0.033	0.007	−0.199 *	0.057	0.085
	X6	−0.006	0.065	0.027	−0.011	−0.066	0.174 *	0.041
	X7	−0.064	0.052	0.047	−0.053	−0.062	0.026	0.274 *

* indicate a significant level at $p < 0.05$.

References

1. Brooker, R.W.; Bennett, A.E.; Cong, W.; Daniell, T.J.; George, T.S.; Hallett, P.D.; Hawes, C.; Iannetta, P.P.M.; Jones, H.G.; Karley, A.J. Improving intercropping: A synthesis of research in agronomy, plant physiology and ecology. *New Phytol.* **2015**, *206*, 107–117. [CrossRef]
2. Ananthi, T.; Amanullah, M.M.; Al-Tawaha, A.R.M.S. A review on maize-legume intercropping for enhancing the productivity and soil fertility for sustainable agriculture in India. *Adv. Environ. Biol.* **2017**, *11*, 49–64.
3. Farooq, T.H.; Shakoor, A.; Wu, X.; Li, Y.; Rashid, M.H.U.; Zhang, X.; Gilani, M.M.; Kumar, U.; Chen, X.; Yan, W. Perspectives of plantation forests in the sustainable forest development of China. *iForest* **2021**, *14*, 166. [CrossRef]
4. Farooq, T.H.; Yan, W.; Rashid, M.H.U.; Tigabu, M.; Gilani, M.M.; Zou, X.H.; Wu, P.F. Chinese fir (*Cunninghamia Lanceolata*) a green gold of China with continues decline in its productivity over the successive rotations: A review. *Appl. Ecol. Environ. Res.* **2019**, *17*, 11055–11067. [CrossRef]
5. Farooq, T.H.; Wu, W.; Tigabu, M.; Ma, X.; He, Z.; Rashid, M.H.; Gilani, M.M.; Wu, P.F. Growth, biomass production and root development of Chinese fir in relation to initial planting density. *Appl. Ecol. Environ. Res.* **2019**, *17*, 3553–3566. [CrossRef]
6. Farooq, T.H.; Ma, X.; Rashid, M.H.U.; Wu, W.; Xu, J.; Tarin, M.W.K.; He, Z.; Wu, P. Impact of stand density on soil quality in Chinese Fir (*Cunninghamia Lanceolata*) monoculture. *Appl. Ecol. Environ. Res.* **2019**, *17*, 3553–3566. [CrossRef]
7. Powlson, D.S.; Gregory, P.J.; Whalley, W.R.; Quinton, J.N.; Hopkins, D.W.; Whitmore, A.P.; Hirsch, P.R.; Goulding, K.W.T. Soil management in relation to sustainable agriculture and ecosystem services. *Food Policy* **2011**, *36*, S72–S87. [CrossRef]
8. Ghosh, P.K.; Bandyopadhyay, K.K.; Wanjari, R.H.; Manna, M.C.; Misra, A.K.; Mohanty, M.; Rao, A.S. Legume effect for enhancing productivity and nutrient use-efficiency in major cropping systems–an Indian perspective: A review. *J. Sustain. Agric.* **2007**, *30*, 59–86. [CrossRef]
9. Rowe, H.; Withers, P.J.A.; Baas, P.; Chan, N.I.; Doody, D.; Holiman, J.; Jacobs, B.; Li, H.; MacDonald, G.K.; McDowell, R. Integrating legacy soil phosphorus into sustainable nutrient management strategies for future food, bioenergy and water security. *Nutr. Cycl. Agro Ecosyst.* **2016**, *104*, 393–412. [CrossRef]
10. Gebru, H. A review on the comparative advantages of intercropping to mono-cropping system. *J. Biol. Agric. Healthc.* **2015**, *5*, 1–13.
11. Powell, J.M.; Williams, T.O. *Livestock, Nutrient Cycling and Sustainable Agriculture in the West African Sahel*; International Institute for Environment and Development: London, UK, 1993.
12. Nyawade, S.O.; Gachene, C.K.K.; Karanja, N.N.; Gitari, H.I.; Schulte-Geldermann, E.; Parker, M.L. Controlling soil erosion in smallholder potato farming systems using legume intercrops. *Geoderma Reg.* **2019**, *17*, 00225. [CrossRef]
13. Seran, T.H.; Brintha, I. Review on maize based intercropping. *J. Agron.* **2010**, *9*, 135–145. [CrossRef]
14. Dong, B.; Wu, B.; Hong, W.; Li, X.; Li, Z.; Xue, L.; Huang, Y. Transcriptome analysis of the tea oil camellia (*Camellia oleifera*) reveals candidate drought stress genes. *PLoS ONE* **2017**, *12*, 0181835. [CrossRef] [PubMed]
15. Akram, N.A.; Shafiq, F.; Ashraf, M. Peanut (*Arachis hypogaea* L.): A prospective legume crop to offer multiple health benefits under changing climate. *Compr. Rev. Food Sci. Food Saf.* **2018**, *17*, 1325–1338. [CrossRef] [PubMed]
16. Li, Q.; Wu, L.; Jun, C.; Khan, M.A.; Luo, X.; Lin, W. Biochemical and microbial properties of rhizospheres under maize/peanut intercropping. *J. Integr. Agric.* **2016**, *15*, 101–110. [CrossRef]
17. He, Y.; Ding, N.; Shi, J.; Wu, M.; Liao, H.; Xu, J. Profiling of microbial PLFAs: Implications for interspecific interactions due to intercropping which increase phosphorus uptake in phosphorus limited acidic soils. *Soil Biol. Biochem.* **2013**, *57*, 625–634. [CrossRef]

18. Burns, R.G.; DeForest, J.L.; Marxsen, J.; Sinsabaugh, R.L.; Stromberger, M.E.; Wallenstein, M.D.; Weintraub, M.N.; Zoppini, A. Soil enzymes in a changing environment: Current knowledge and future directions. *Soil Biol. Biochem.* **2013**, *58*, 216–234. [CrossRef]
19. Das, S.K.; Varma, A. Role of enzymes in maintaining soil health. In *Soil Enzymology*; Springer: Berlin/Heidelberg, Germany, 2010; pp. 25–42.
20. Neher, D.A. Role of nematodes in soil health and their use as indicators. *J. Nematol.* **2001**, *33*, 161.
21. Farooq, T.H.; Tigabu, M.; Ma, X.; Zou, X.; Liu, A.; Odén, P.C.; Wu, P. Nutrient uptake, allocation and biochemical changes in two Chinese fir cuttings under heterogeneous phosphorus supply. *IForest* **2018**, *11*, 411–417. [CrossRef]
22. Farooq, T.H.; Yan, W.; Chen, X.; Shakoor, A.; Rashid, M.H.U.; Gilani, M.M.; He, Z.; Wu, P. Dynamics of canopy development of *Cunninghamia lanceolata* mid-age plantation in relation to foliar nitrogen and soil quality influenced by stand density. *Glob. Ecol. Conserv.* **2020**, *24*, 01209. [CrossRef]
23. Wu, P.; Wang, G.; Farooq, T.H.; Li, Q.; Zou, X.; Ma, X. Low phosphorus and competition affect Chinese fir cutting growth and root organic acid content: Does neighboring root activity aggravate P nutrient deficiency? *J. Soils Sediments* **2017**, *17*, 2775–2785. [CrossRef]
24. Loveland, P.; Webb, J. Is there a critical level of organic matter in the agricultural soils of temperate regions: A review. *Soil Tillage Res.* **2003**, *70*, 1–18. [CrossRef]
25. Jiang, F.; Drohan, P.J.; Cibin, R.; Preisendanz, H.E.; White, C.M.; Veith, T.L. Reallocating crop rotation patterns improves water quality and maintains crop yield. *Agric. Syst.* **2021**, *187*, 103015. [CrossRef]
26. Li, W.; Zheng, Z.; Li, T.; Zhang, X.; Wang, Y.; Yu, H.; He, S.; Liu, T. Effect of tea plantation age on the distribution of soil organic carbon fractions within water-stable aggregates in the hilly region of Western Sichuan, China. *Catena* **2015**, *133*, 198–205. [CrossRef]
27. Lu, W.; Shen, X.; Chen, Y. Effects of intercropping peanut on soil nutrient status and microbial activity within young Camellia oleifera plantation. *Commun. Soil Sci. Plant Anal.* **2019**, *50*, 1232–1238. [CrossRef]
28. Hu, C.H.; Wang, P.Q.; Zhang, P.P.; Nie, X.M.; Li, B.B.; Tai, L.; Liu, W.T.; Li, W.Q.; Chen, K.M. NADPH oxidases: The vital performers and center hubs during plant growth and signaling. *Cells* **2020**, *9*, 437. [CrossRef]
29. Cameron, K.C.; Di, H.J.; Moir, J.L. Nitrogen losses from the soil/plant system: A review. *Ann. Appl. Biol.* **2013**, *162*, 145–173. [CrossRef]
30. Dubey, R.S.; Pessarakli, M. Physiological mechanisms of nitrogen absorption and assimilation in plants under stressful conditions. *Handb. Plant Crop Physiol.* **1995**, 605–625.
31. Jensen, E.S. Grain yield, symbiotic N_2 fixation and interspecific competition for inorganic N in pea-barley intercrops. *Plant Soil* **1996**, *182*, 25–38. [CrossRef]
32. Jensen, E.S.; Carlsson, G.; Hauggaard-Nielsen, H. Intercropping of grain legumes and cereals improves the use of soil N resources and reduces the requirement for synthetic fertilizer N: A global-scale analysis. *Agron. Sustain. Dev.* **2020**, *40*, 1–9. [CrossRef]
33. Xu, B.C.; Li, F.M.; Shan, L. Switchgrass and milkvetch intercropping under 2: 1 row-replacement in semiarid region, northwest China: Aboveground biomass and water use efficiency. *Eur. J. Agron.* **2008**, *28*, 485–492. [CrossRef]
34. Ikramul Haq, M.; Maqbool, M.M.; Ali, A.; Farooq, S.; Khan, S.; Saddiq, M.S.; Khan, K.A.; Ali, S.; Ifnan Khan, M.; Hussain, A.; et al. Optimizing planting geometry for Barley-Egyptian clover intercropping system in semi-arid sub-tropical climate. *PLoS ONE* **2020**, *15*, 0233171. [CrossRef] [PubMed]
35. Giller, K.E. *Nitrogen Fixation in Tropical Cropping Systems*; Cabi: Oxfordshire, UK, 2001.
36. Singh, D.K.; Kumar, S. Nitrate reductase, arginine deaminase, urease and dehydrogenase activities in natural soil (ridges with forest) and in cotton soil after acetamiprid treatments. *Chemosphere* **2008**, *71*, 412–418. [CrossRef] [PubMed]
37. Yu, S.; Xue, L.; Feng, Y.; Liu, Y.; Song, Z.; Mandal, S.; Yang, L.; Sun, Q.; Xing, B. Hydrochar reduced NH_3 volatilization from rice paddy soil: Microbial-aging rather than water-washing is recommended before application. *J. Clean. Prod.* **2020**, *268*, 122233. [CrossRef]
38. Li, L.; Li, S.-M.; Sun, J.H.; Zhou, L.L.; Bao, X.G.; Zhang, H.-G.; Zhang, F.-S. Diversity enhances agricultural productivity via rhizosphere phosphorus facilitation on phosphorus-deficient soils. *Proc. Natl. Acad. Sci. USA* **2007**, *104*, 11192–11196. [CrossRef]
39. Shen, J.; Li, C.; Mi, G.; Li, L.; Yuan, L.; Jiang, R.; Zhang, F. Maximizing root/rhizosphere efficiency to improve crop productivity and nutrient use efficiency in intensive agriculture of China. *J. Exp. Bot.* **2013**, *64*, 1181–1192. [CrossRef]
40. Wang, X.; Gao, Y.; Zhang, H.; Shao, Z.; Sun, B.; Gao, Q. Enhancement of rhizosphere citric acid and decrease of NO_3-/NH_4+ ratio by root interactions facilitate N fixation and transfer. *Plant Soil* **2020**, *447*, 169–182. [CrossRef]
41. Rasool, B.; Ramzani, P.M.A.; Zubair, M.; Khan, M.A.; Lewińska, K.; Turan, V.; Karczewska, A.; Khan, S.A.; Farhad, M.; Tauqeer, H.M.; et al. Impacts of oxalic acid-activated phosphate rock and root-induced changes on lead bioavailability in rhizosphere and distribution in mung bean plant. *Environ. Pollut.* **2021**, 116903. [CrossRef]
42. Maurya, P.R.; Lal, R. Effects of different mulch materials on soil properties and on the root growth and yield of maize (*Zea mays*) and cowpea (*Vigna unguiculata*). *F. Crop. Res.* **1981**, *4*, 33–45. [CrossRef]
43. Dick, R.P. Soil enzyme activities as indicators of soil quality. *Defin. Soil Qual. Sustain. Environ.* **1994**, *35*, 107–124.
44. Sherene, T. Role of soil enzymes in nutrient transformation: A review. *Bio. Bull.* **2017**, *3*, 109–131.
45. Han, S.; Delgado-Baquerizo, M.; Luo, X.; Liu, Y.; Van Nostrand, J.D.; Chen, W.; Zhou, J.; Huang, Q. Soil aggregate size-dependent relationships between microbial functional diversity and multifunctionality. *Soil Biol. Biochem.* **2021**, *154*, 108143. [CrossRef]

46. Roohi, M.; Arif, M.S.; Yasmeen, T.; Riaz, M.; Rizwan, M.; Shahzad, S.M.; Ali, S.; Bragazza, L. Effects of cropping system and fertilization regime on soil phosphorous are mediated by rhizosphere-microbial processes in a semi-arid agroecosystem. *J. Environ. Manag.* **2020**, *271*, 111033. [CrossRef] [PubMed]
47. Jan, M.T.; Roberts, P.; Tonheim, S.K.; Jones, D.L. Protein breakdown represents a major bottleneck in nitrogen cycling in grassland soils. *Soil Biol. Biochem.* **2009**, *41*, 2272–2282. [CrossRef]
48. Kwiatkowski, C.A.; Harasim, E.; Feledyn-Szewczyk, B.; Antonkiewicz, J. Enzymatic activity of loess soil in organic and conventional farming systems. *Agriculture* **2020**, *10*, 135. [CrossRef]
49. Chen, L.; Rossi, F.; Deng, S.; Liu, Y.; Wang, G.; Adessi, A.; De Philippis, R. Macromolecular and chemical features of the excreted extracellular polysaccharides in induced biological soil crusts of different ages. *Soil Biol. Biochem.* **2014**, *78*, 1–9. [CrossRef]
50. Wang, Y.; Dong, J.; Zheng, X.; Zhang, J.; Zhou, P.; Song, X.; Song, W.; Wang, S. Wheat straw and biochar effect on soil carbon fractions, enzyme activities, and nutrients in a tobacco field. *Can. J. Soil Sci.* **2021**, *101*, 1–12. [CrossRef]
51. Rashid, M.H.U.; Tigabu, M.; Chen, H.; Farooq, T.H.; Ma, X.; Wu, P. Calcium-mediated adaptive responses to low phosphorus stress in Chinese fir. *Trees* **2020**, *34*, 825–834. [CrossRef]
52. Deng, S.P.; Tabatabai, M.A. Effect of tillage and residue management on enzyme activities in soils: III. Phosphatases and arylsulfatase. *Biol. Fertil. Soils* **1997**, *24*, 141–146. [CrossRef]
53. Zhang, W.; Gruszewski, H.A.; Chevone, B.I.; Nessler, C.L. An Arabidopsis purple acid phosphatase with phytase activity increases foliar ascorbate. *Plant Physiol.* **2008**, *146*, 431–440. [CrossRef] [PubMed]
54. Uren, N.C. Types, amounts, and possible functions of compounds released into the rhizosphere by soil-grown plants. *Rhizosph. Biochem. Org. Subst. Soil Plant Interface* **2007**, *2*, 1–21.
55. Wen, F.; VanEtten, H.D.; Tsaprailis, G.; Hawes, M.C. Extracellular proteins in pea root tip and border cell exudates. *Plant Physiol.* **2007**, *143*, 773–783. [CrossRef] [PubMed]
56. Mobley, H.L.T.; Hu, L.-T.; Foxall, P.A. Helicobacter pylori urease: Properties and role in pathogenesis. *Scand. J. Gastroenterol.* **1991**, *26*, 39–46. [CrossRef]
57. Ansari, M.W.; Trivedi, D.K.; Sahoo, R.K.; Gill, S.S.; Tuteja, N. A critical review on fungi mediated plant responses with special emphasis to Piriformospora indica on improved production and protection of crops. *Plant Physiol. Biochem.* **2013**, *70*, 403–410. [CrossRef] [PubMed]
58. Li, Z.H.; Wang, Q.; Ruan, X.; Pan, C.D.; Jiang, D.A. Phenolics and plant allelopathy. *Molecules* **2010**, *15*, 8933–8952. [CrossRef]
59. Tian, Y.; Cao, F.; Wang, G. Soil microbiological properties and enzyme activity in Ginkgo–tea agroforestry compared with monoculture. *Agrofor. Syst.* **2013**, *87*, 1201–1210. [CrossRef]
60. Udawatta, R.P.; Kremer, R.J.; Adamson, B.W.; Anderson, S.H. Variations in soil aggregate stability and enzyme activities in a temperate agroforestry practice. *Appl. Soil Ecol.* **2008**, *39*, 153–160. [CrossRef]
61. Shakoor, A.; Shakoor, S.; Rehman, A.; Ashraf, F.; Abdullah, M.; Shahzad, S.M.; Farooq, T.H.; Ashraf, M.; Manzoor, M.A.; Altaf, M.; et al. Effect of animal manure, crop type, climate zone, and soil attributes on greenhouse gas emissions from agricultural soils—A global meta-analysis. *J. Clean. Prod.* **2020**, 124019. [CrossRef]
62. Shakoor, A.; Shahzad, S.M.; Chatterjee, N.; Arif, M.S.; Farooq, T.H.; Altaf, M.M.; Tufail, M.A.; Dar, A.A.; Mehmood, T. Nitrous oxide emission from agricultural soils: Application of animal manure or biochar? A global meta-analysis. *J. Environ. Manag.* **2021**, *285*, 112170. [CrossRef]
63. Shakoor, A.; Shahbaz, M.; Farooq, T.H.; Sahar, N.E.; Shahzad, S.M.; Altaf, M.M.; Ashraf, M. A global meta-analysis of greenhouse gases emission and crop yield under no-tillage as compared to conventional tillage. *Sci. Total Environ.* **2021**, *750*, 142299. [CrossRef]
64. Hu, W.; Jiao, Z.; Wu, F.; Liu, Y.; Dong, M.; Ma, X.; Fan, T.; An, L.; Feng, H. Long-term effects of fertilizer on soil enzymatic activity of wheat field soil in Loess Plateau, China. *Ecotoxicology* **2014**, *23*, 2069–2080. [CrossRef] [PubMed]
65. Yang, L.; Li, T.; Li, F.; Lemcoff, J.H.; Cohen, S. Fertilization regulates soil enzymatic activity and fertility dynamics in a cucumber field. *Sci. Hortic.* **2008**, *116*, 21–26. [CrossRef]
66. Zhao, S.; Chen, X.; Deng, S.; Dong, X.; Song, A.; Yao, J.; Fang, W.; Chen, F. The effects of fungicide, soil fumigant, bio-organic fertilizer and their combined application on *chrysanthemum Fusarium* wilt controlling, soil enzyme activities and microbial properties. *Molecules* **2016**, *21*, 526. [CrossRef] [PubMed]
67. Lu, N.; Yu, M.; Cui, M.; Luo, Z.; Feng, Y.; Cao, S.; Sun, Y.; Li, Y. Effects of different ectomycorrhizal fungal inoculates on the growth of *Pinus tabulaeformis* seedlings under greenhouse conditions. *Forests* **2016**, *7*, 316. [CrossRef]
68. Ryan, J.; Estefan, G.; Rashid, A. *Soil and Plant Analysis Laboratory Manual*; ICARDA: Beirut, Lebanon, 2001; ISBN 9291271187.
69. Bray, R.H.; Kurtz, L.T. Determination of total, organic, and available forms of phosphorus in soils. *Soil Sci.* **1945**, *59*, 39–46. [CrossRef]
70. Schollenberger, C.J.; Simon, R.H. Determination of exchange capacity and exchangeable bases in soil—ammonium acetate method. *Soil Sci.* **1945**, *59*, 13–24. [CrossRef]

plants

MDPI

Article

Evaluation of Physiological and Morphological Traits for Improving Spring Wheat Adaptation to Terminal Heat Stress

Hafeez ur Rehman [1], Absaar Tariq [1], Imran Ashraf [1], Mukhtar Ahmed [2,3,*], Adele Muscolo [4], Shahzad M. A. Basra [1] and Matthew Reynolds [5]

[1] Department of Agronomy, University of Agriculture, Faisalabad 38000, Pakistan; hafeezcp@gmail.com (H.u.R.); arid192@gmail.com (A.T.); imran.ashraf@post.com (I.A.); shehzadbasra@gmail.com (S.M.A.B.)
[2] Department of Agricultural Research for Northern Sweden, Swedish University of Agricultural Sciences, 90183 Umeå, Sweden
[3] Department of Agronomy, Pir Mehr Ali Shah, Arid Agriculture University, Rawalpindi 46300, Pakistan
[4] Department of Agriculture, Mediterranea University, Feo di Vito, 89124 Reggio Calabria, Italy; amuscolo@unirc.it
[5] International Maize and Wheat Improvement Center, El Batán, Texcoco CP 56130, Mexico; m.reynolds@cgiar.org
* Correspondence: mukhtar.ahmed@slu.se or ahmadmukhtar@uaar.edu.pk

Abstract: Wheat crop experiences high temperature stress during flowering and grain-filling stages, which is termed as "terminal heat stress". Characterizing genotypes for adaptive traits could increase their selection for better performance under terminal heat stress. The present study evaluated the morpho-physiological traits of two spring wheat cultivars (Millet-11, Punjab-11) and two advanced lines (V-07096, V-10110) exposed to terminal heat stress under late sowing. Early maturing Millet-11 was used as heat-tolerant control. Late sowing reduced spike length (13%), number of grains per spike (10%), 1000-grain weight (13%) and biological yield (15–20%) compared to timely sowing. Nonetheless, higher number of productive tillers per plant (19–20%) and grain yield (9%) were recorded under late sowing. Advanced lines and genotype Punjab-11 had delayed maturity and better agronomic performance than early maturing heat-tolerant Millet-11. Advanced lines expressed reduced canopy temperature during grain filling and high leaf chlorophyll *a* (20%) and *b* (71–125%) contents during anthesis under late sowing. All wheat genotypes expressed improved stem water-soluble carbohydrates under terminal heat stress that were highest for heat-tolerant Millet-11 genotype during anthesis. Improved grain yield was associated with the highest chlorophyll contents showing stay green characteristics with maintenance of high photosynthetic rates and cooler canopies under late sowing. The results revealed that advanced lines and Punjab-11 with heat adaptive traits could be promising source for further use in the selection of heat-tolerant wheat genotypes.

Keywords: canopy temperature; water soluble carbohydrates; heat stress; stay green; seed yield

Citation: Rehman, H.u.; Tariq, A.; Ashraf, I.; Ahmed, M.; Muscolo, A.; Basra, S.M.A.; Reynolds, M. Evaluation of Physiological and Morphological Traits for Improving Spring Wheat Adaptation to Terminal Heat Stress. *Plants* **2021**, *10*, 455. https://doi.org/10.3390/plants10030455

Academic Editor: Ornella Calderini

Received: 25 December 2020
Accepted: 24 February 2021
Published: 28 February 2021

1. Introduction

Wheat (*Triticum aestivum* L.) is a staple crop and nourishes billions of people daily, while its productivity is significantly decreased under high temperature. Owing to global climate changes, wheat yields are expected to decline by 6% for each 1 °C increase in temperature. Therefore, gain yield of wheat crop must be increased by 60% until 2050 to fulfil the food demands of burgeoning global population [1]. According to the International Maize and Wheat Improvement Center (CIMMYT), world wheat producing regions are grouped into eight different mega environments (MEs) [2]. The ME1 and ME5 are highly productive irrigated environments in South Asia; however, wheat crop is exposed to terminal heat stress in these MEs [3]. South Asia includes India, Nepal, Pakistan and

Bangladesh where wheat is sown late due to the delay in the harvesting of previous crops, i.e., rice and cotton in rice-wheat and cotton-wheat zones owing to delayed harvest of long maturing semi-dwarf rice varieties and delayed picking of cotton respectively [4,5]. The delay in wheat cultivation exposes the crop to high temperatures during flowering and grain-filling stages termed as "terminal heat stress" [6]. Nonetheless, wheat grown on $\geq$13.9 million ha in the rice-wheat cropping system of South Asia [1] including rice-wheat and cotton-wheat zones of Pakistan experience terminal heat stress [4,5].

Spring wheat in South Asia is sown in the months of November and December while harvested during April and May particularly in rice-wheat and cotton-wheat cropping zones. The timely planted wheat crop in these zones has temperature requirement of 12–22 °C during vegetative and reproductive growth periods particularly at anthesis and grain filling stages.

Late sown wheat experiences higher canopy temperature (>31 °C) during reproductive period adversely affecting many physiological processes resulting in shortened crop growth cycle and reduced yields [6]. For example, high temperature during anthesis reduces pollen viability by restricting embryo development and promotes anther sterility resulting in less grain numbers [7,8]. High temperature exposure after anthesis decreases grain filling rate which is associated with declined grain yield [7–10]. These decreases in grain filling rate and duration accelerate senescence with the loss of chlorophyll and limits assimilate supply towards developing grains [11].

Nonetheless, wheat plants evolve several physiological mechanisms to cope with terminal heat stress which include early maturity [3,6], stay green [11,12], reduced canopy temperature [13], accumulation of high stem water soluble carbohydrates [14] and high biomass accumulation [13] to translate assimilates into yield. For instance, early maturity as an adaptation strategy provides explanation for variation to physiological responses and grain yield under high temperature among wheat genotypes during reproductive period [3,6]. Stay green trait maintains high leaf chlorophyll contents that are associated with grain yield and its components under heat stress [11,12,15]. Likely, the loss in green area affects grain size that can be compensated by improving the remobilization of water-soluble carbohydrates stored in stem and leaf sheaths to developing grains under high temperature and drought stress [14]. However, studies showing the direct relationship of water-soluble carbohydrates with grain yield under heat or drought stress are limited. In addition, wheat plants with cooler canopy during grain filling, have capability to access subsoil moisture which helps to maintain evaporation and photosynthesis under hot irrigated conditions [16]. Wheat plants with cooler canopies also have positive association with grain yield [13,17,18]. Therefore, these physiological traits can be the best combination for genetic improvement of wheat genotypes to mitigate heat and drought stress because of their common genetic basis [16,19,20].

Exploring combination of physiological traits needs identification of wheat genotypes which produce high yield and express adaptation traits under high temperature exposure. Therefore, developing heat tolerant genotypes through breeding is a major objective in wheat improvement programs [3,6]. In view of global climate changes, these efforts should be accelerated to reduce high temperature effects [1,21] as rise in the global average temperature is a serious threat to wheat production in different MEs prone to terminal drought or heat [3,22,23].

To cope with situation, at present, efforts for evaluation of exotic germplasm developed by CIMMYT are in progress through National Agriculture Research Programs (NARPs). For this, elite nurseries of wheat germplasm are tested in MEs of South Asia for their adaptability to heat and associated traits which are later assembled into mega varieties for both improved productivity and heat tolerance [3,6,24]. Similarly, an international wheat phenotyping network has been established in South Asia for application of phenotyping techniques to accelerate selection and support breeding program for their success by incorporating physiological traits into new generation of lines for heat or drought tolerance [24]. Thus, physiological characterization of existing genotypes or plant genetic resources through phenotyping tools may provide better understanding of heat adaptive traits and their integration into breeding programs may help to translate these genotypes into desirable plant types [25,26]. The present study, therefore, compared the performance of wheat genotypes including one widely grown and two promising advanced lines with early maturing heat tolerant cultivar to identify morpho-physiological traits for adaptation to terminal heat stress under late sowing. Most specifically the association of these traits with grain yield was determined for their further use in cultivar selection.

2. Results

2.1. Crop Phenological Development

High temperature significantly reduced crop development period, including booting (7–8 d), heading (2 d), anthesis (3 d), grain filling (10 d) and maturity period (10–12 d) under late sowing. Among the genotypes, earlier heading (3, 3, 4 d), booting (9–11, 10–12, 11–15 d), anthesis (1–2, 2–5, 4 d) and maturity (10–13, 10–12, 10–11 d) were observed for Punjab-11, V-07096 and V-10110 respectively compared to heat tolerant Millet-11 under late sowing condition. However, delayed heading (2 d), booting (3 d) and anthesis (7 d) were observed for heat tolerant Millet-11 while grain filling period was reduced for all genotypes under late sowing condition. Interactions were significant for all phenological traits (Table 1).

2.2. Physiological Traits

Wheat crop expressed a 20% and 71–125% increase in Chl *a* and *b* contents respectively, under late sowing compared to timely sown crop during anthesis. Heat tolerant check Millet-11 and advanced line V-07096 expressed the highest Chl *a* and *b* contents during the first growing season (2012–2013), while these were highest in Punjab-11 and V-10110 during the second growing season (2013–2014) under late sowing. Reduced canopy temperature was observed in late sown crop during 2012–2013 and vice versa for 2013–2014. Heat tolerant Millet-11 and advanced lines V-07096 and V-10110 expressed reduced canopy temperature during both seasons. Maximum increase (50%) in stem water-soluble carbohydrates was found under late sowing. Nonetheless, all genotypes expressed similar stem water-soluble carbohydrates under late sowing (Table 1).

Table 1. Mean comparison for crop phenology and physiological traits in wheat genotypes under late sown condition.

Wheat Genotypes	Days to Heading			Grain Filling Period (Days)		
	2012–2013			2013–2014		
	Planting Time		Means Genotypes	Planting Time		Means Genotypes
	Normal	Late Sown		Normal	Late Sown	
Millet-11	92.00 g	94.50 f	93.25 D	35.67 a	18.50 e	27.08 AB
Punjab-11	103.00 c	100.17 e	101.58 C	32.50 b	23.00 d	27.75 A
V-07096	104.83 b	101.50 d	103.17 B	30.17 c	22.67 d	26.41 B
V-10110	107.00 a	103.33 c	105.17 A	29.50 c	22.67 d	26.08 B
Means SD	101.71 A	99.87 B		31.96 A	21.71 B	
HSD	SD = 0.36; G = 0.76, SD × G = 1.30			SD = 1.08.; G = 1.00, SD × G = 1.72		
	Days to Booting					
Wheat Genotypes	2012–2013			2013–2014		
	Planting Time		Means Genotypes	Planting Time		Means Genotypes
	Normal	Late Sown		Normal	Late Sown	
Millet-11	77.67 f	80.67 e	79.17 C	78.67 f	81.33 e	80.00 C
Punjab-11	94.33 b	83.33 d	88.83 B	95.67 b	86.33 cd	91.00 B
V-07096	95.67 ab	85.33 c	90.50 A	98.33 a	86.67 c	92.50 A
V-10110	96.67 a	85.67 c	91.17 A	99.33 a	84.67 d	92.00 AB
Means SD	91.08 A	83.75 B		93.00 A	84.75 B	
HSD	SD = 0.71; G = 1.07, SD × G = 1.84			SD = 0.62; G = 1.09, SD × G = 1.88		

Table 1. *Cont.*

Wheat Genotypes	Days to Anthesis					
	2012–2013			2013–2014		
	Normal	Late Sown	Means Genotypes	Normal	Late Sown	Means Genotypes
Millet-11	96.67 f	103.67 e	100.17 D	99.33 d	100.67 d	100.00 D
Punjab-11	107.33 cd	106.67 d	107.00 C	111.33 b	108.33 c	109.83 C
V-07096	110.33 b	108.00 bc	109.17 B	114.33 a	109.33 bc	111.83 B
V-10110	112.00 a	108.67 bc	110.33 A	115.33 a	111.33 b	113.33 A
Means SD	106.58	106.75		110.08 A	107.42 B	
HSD	SD = n.s.; G = 0.76, SD × G = 1.30			SD = 0.36; G = 1.31, SD × G = 2.26		
	Days to Maturity					
Wheat Genotypes	**2012–2013**			**2013–2014**		
	Normal	**Late Sown**	**Means Genotypes**	**Normal**	**Late Sown**	**Means Genotypes**
Millet-11	132.67 b	122.33 c	127.50 B	134.67 b	119.00 d	126.83 D
Punjab-11	140.33 a	130.33 b	135.33 A	143.33 a	130.67 c	137.00 C
V-07096	140.67 a	130.33 b	135.50 A	144.33 a	132.33 b	138.33 B
V-10110	141.67 a	130.67 b	136.17 A	144.67 a	134.67 b	139.67 A
Means SD	138.83 A	128.42 B		141.75 A	129.17 B	
HSD	SD = 1.43; G = 0.99, SD × G = 1.71			SD = 0.95; G = 1.10, SD × G = 1.91		
	Chl *a* (mg/g Fwt)					
Wheat Genotypes	**2012–2013**			**2013–2014**		
	Normal	**Late Sown**	**Means Genotypes**	**Normal**	**Late Sown**	**Means Genotypes**
Millet-11	0.10 d	0.12 a	0.11	0.08 e	0.09 d	0.08 C
Punjab-11	0.10 cd	0.11 b	0.10	0.12 ab	0.12 ab	0.12 A
V-07096	0.11 c	0.12 ab	0.11	0.11 c	0.09 d	0.10 B
V-10110	0.11 c	0.11 b	0.11	0.11 bc	0.13 a	0.12 A
Means SD	0.10 B	0.12 A		0.10	0.10	
HSD	SD = 0.006; G = n.s., SD × G = 0.006			SD = n.s.; G = 0.004, SD × G = 0.006		
	Chl *b* (mg/g Fwt)					
Wheat Genotypes	**2012–2013**			**2013–2014**		
	Normal	**Late Sown**	**Means Genotypes**	**Normal**	**Late Sown**	**Means Genotypes**
Millet-11	0.07	0.15	0.11 D	0.03 e	0.03 e	0.03 D
Punjab-11	0.09	0.19	0.14 B	0.08 d	0.19 a	0.13 A
V-07096	0.10	0.19	0.15 A	0.09 c	0.09 c	0.09 C
V-10110	0.08	0.17	0.12 C	0.08 d	0.17 b	0.12 B
Means SD	0.08 B	0.18 A		0.07 B	0.12 A	
HSD	SD = 0.006; G = 0.004, SD × G = n.s.			SD = 0.006; G = 0.004, SD × G = 0.006		
	Canopy Temperature (°C)					
Wheat Genotypes	**2012–2013**			**2013–2014**		
	Normal	**Late Sown**	**Means Genotypes**	**Normal**	**Late Sown**	**Means Genotypes**
Millet-11	22.53	21.32	21.92 B	23.27	25.15	24.21 AB
Punjab-11	22.67	22.36	22.52 A	23.92	25.53	24.73 A
V-07096	22.93	22.17	22.56 A	23.45	24.57	24.01 B
V-10110	22.82	22.1	22.46 AB	23.31	25.18	24.25 AB
Means SD	22.74 A	21.99 B		23.49 B	25.11 A	
HSD	SD = 0.50; G = 0.56, SD × G = n.s.			SD = 0.88; G = 0.55, SD × G = n.s.		

Table 1. *Cont.*

Wheat Genotypes	Water Soluble Carbohydrates (μmol /g Fwt)		
	Average over two seasons		
	Normal	Late Sown	Means Genotypes
Millet-11	1.90 b	2.37 a	2.14
Punjab-11	1.34 c	1.99 ab	1.66
V-07096	1.29 c	2.04 ab	1.67
V-10110	1.05 d	2.04 ab	1.55
Means SD	1.40	2.11	
HSD	SD = n.s.; G = n.s., SD × G = 0.16		

Letters among columns denote significant differences in means for sowing dates while letters within columns denote significant differences between cultivars at $P \leq 0.05$; SD = Sowing dates; G = Genotype; n.s. = non-significant; Fwt = Fresh weight basis.

2.3. Gas Exchange Traits

No significant difference was observed for photosynthetic (A) and transpiration rates (E) under timely and late sown wheat while intercellular CO_2 concentration (Ci) and stomatal conductance (Gs) were reduced under late sowing. However, lowest photosynthetic and transpiration rates were found for advanced line V-07096. Regarding interactions, the highest Gs was found for timely sown advanced line V-07096 and it was drastically reduced under late sowing condition. On the other hand, highest Ci was found in advanced line V-07096 under both sowing conditions while Gs was significantly reduced under late sowing (Figure 1).

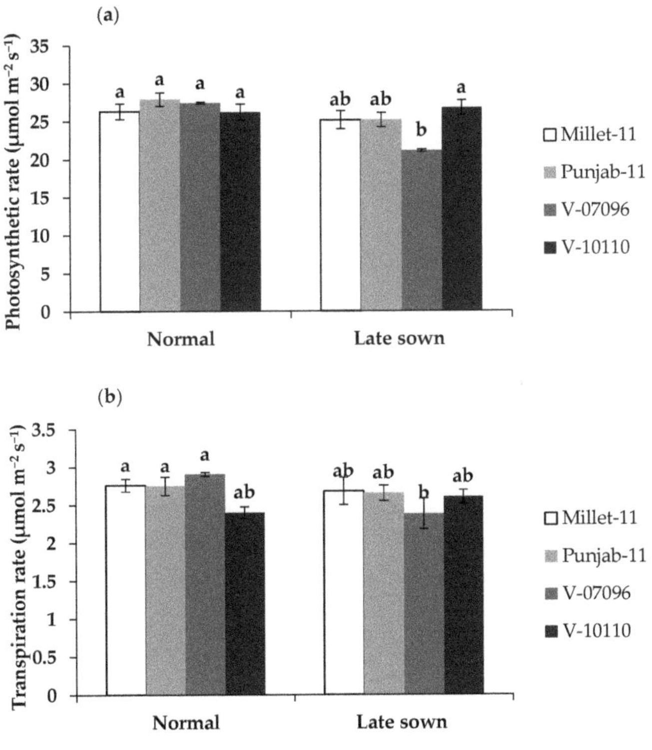

Figure 1. *Cont.*

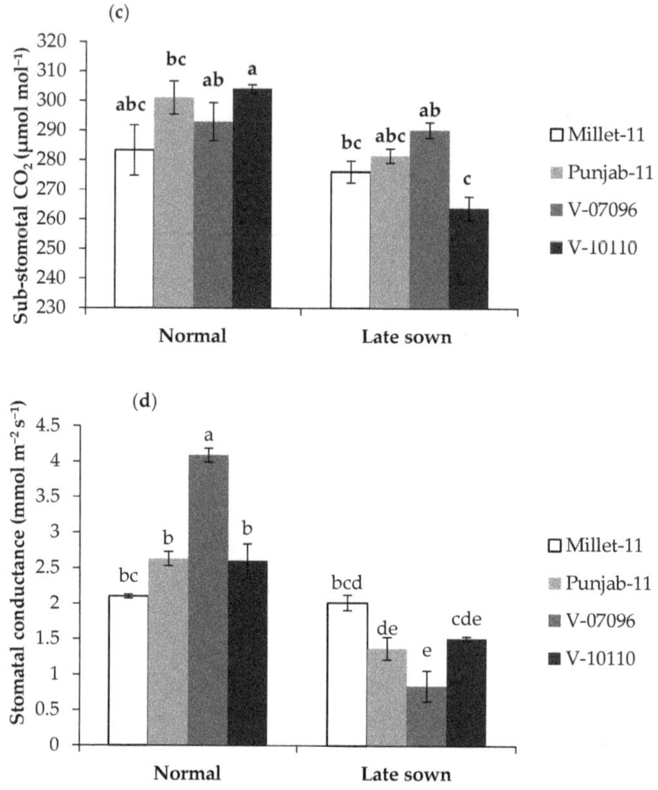

Figure 1. Gas exchange traits (**a**) photosynthetic rate (**b**) transpiration rate (**c**) sub-stomatal CO_2 (**d**) stomatal conductance in wheat genotypes under late sowing. Different lowercase letters indicate significant differences among treatments; Means with same letters show non-significant difference at $p > 0.05$.

2.4. Yield Related Traits

Terminal heat stress significantly reduced yield related traits in all the wheat genotypes under late sowing. Among these traits, plant height, spike length, grains per spike, thousand grain weight and biological yield were reduced more under late sowing than timely sowing (Table 2). Total number of productive tillers were highest during both years, while grain yield was high during first growing season and was similar under both timely and late sowing in 2nd growing season. Advanced lines V-07096 and V-10110 expressed the highest plant height, spike length, number of grains per spike, thousand grain weight including biological and grain yields. Genotype Punjab-11 also expressed highest biological and grain yields significantly similar to advanced line V-07096. For total productive tillers, the differences among genotypes were non-significant. The interactions were significant for spike length, thousand grain weight and biological yield, while these traits decreased during second growing season, grain yield during both years and plant height during first growing season were significant (Table 2).

Table 2. Mean comparison for yield related traits in wheat genotypes under late sowing condition.

Wheat Genotypes	Total Productive Tillers Per Plant					
	2012–2013			2013–2014		
	Planting Time			Planting Time		
	Normal	Late Sown	Means Genotypes	Normal	Late Sown	Means Genotypes
Millet-11	5.03	6.33	5.68	4.50	5.50	5.00
Punjab-11	5.73	6.07	5.90	5.67	6.00	5.78
V-07096	4.87	6.07	5.47	4.67	5.83	5.19
V-10110	5.13	6.20	5.67	5.44	6.67	6.06
Means SD	5.19 B	6.17A		5.01 B	6.00 A	
HSD	SD = 0.63; G = n.s. SD × G = n.s.			NS = 0.91, G = n,s, SD × G = n.s.		
Wheat Genotypes	Spike Length (cm)			Grains Per Spike		
	Average over two seasons			Average over two seasons		
	Normal	Late Sown	Means Genotypes	Normal	Late Sown	Means Genotypes
Millet-11	10.93 b	9.46 c	10.19 B	42.77	37.93	40.35 B
Punjab-11	10.93 b	10.02 bc	10.48 B	41.40	32.97	37.18 B
V-07096	11.91 a	10.49 b	11.20 A	48.87	44.20	46.53 A
V-10110	12.48 a	10.20 b	11.34 A	41.87	41.70	41.78 AB
Means SD	11.57 A	10.04 B		43.72	39.20	
HSD	SD = 0.61; G = 0.42, SD × G = 0.72			SD = n.s.; G = 4.79, SD × G = n.s.		
Wheat Genotypes	Thousand grain weight (g)					
	2012–2013			2013–2014		
	Normal	Late Sown	Means Genotypes	Normal	Late Sown	Means Genotypes
Millet-11	40.67	39.00	39.83 B	43.17 bc	42.00 c	42.58 B
Punjab-11	39.83	41.20	40.51 B	48.83 ab	43.33 bc	46.08 A
V-07096	47.67	47.90	47.78 A	49.50 a	39.83 c	44.67 AB
V-10110	46.00	48.23	47.11 A	51.50 a	43.17 bc	47.33 A
Means SD	43.54	44.08		48.25 A	42.08 B	
HSD	SD = n.s.; G = 3.73, SD × G = n.s.			SD = 4.40; G = 3.43, SD × G = 5.99		
Wheat Genotypes	Biological Yield (g m^{-2})					
	2012–2013			2013–2014		
	Normal	Late Sown	Means Genotypes	Normal	Late Sown	Means Genotypes
Millet-11	1212.20	1049.70	1130.90 B	1110.50 b	988.00 b	1049.30 B
Punjab-11	1448.80	1216.70	1332.80 A	1472.90 a	1053.80 a	1263.40 A
V-07096	1577.20	1247.00	1412.10 A	1462.00 a	1106.30 b	1284.20 A
V-10110	1238.30	1137.80	1188.10 B	1129.00 b	973.80 b	1051.40 B
Means SD	1369.10 A	1162.80 B		1293.60 A	1030.50 B	
HSD	SD = 79.01; G = 122.18, SD × G = n.s.			SD = 107.88; G = 83.03; SD × G = 143.1		
Wheat Genotypes	Grain Yield (g m^{-2})					
	2012–2013			2013–2014		
	Normal	Late Sown	Means Genotypes	Normal	Late Sown	Means Genotypes
Millet-11	199.18 b	228.67 ab	213.93 B	149.37 c	155.93 c	152.65 B
Punjab-11	251.87 a	239.50 ab	245.69 A	198.92 a	183.33 ab	191.13 A
V-07096	245.97 a	253.72 a	249.84 A	194.60 a	190.74 ab	192.67 A
V-10110	195.28 b	255.58 a	225.43 AB	163.24 bc	144.32 c	153.78 B
Means SD	223.08 B	244.37 A		176.53	168.58	
HSD	SD = 13.72.; G = 26.67, SD × G = 45.97			SD = n.s.; G = 12.55, SD × G = 21.63		

Table 2. *Cont.*

Wheat Genotypes	Plant Height (cm)					
	2012–2013			2013–2014		
	Normal	Late sown	Means Genotypes	Normal	Late sown	Means Genotypes
Millet-11	105.88 bc	90.13 e	98.01 B	95.07	88.47	91.77 B
Punjab-11	106.93 b	95.78 de	101.35 B	106.10	95.73	100.92 A
V-07096	115.67 a	104.83 bc	110.25 A	103.60	98.90	101.25 A
V-10110	119.50 a	100.19 cd	109.84 A	101.40	91.50	96.45 AB
Means SD	111.99 A	97.73 B		101.54 A	93.65 B	
HSD	SD = 6.37; G = 3.48, SD × G = 6.69			SD = 0.02.; G = n.s., SD × G = n.s.		

Letters among columns denote significant differences in means for sowing dates while letters within columns denote significant differences between cultivars at $P \leq 0.05$; SD = Sowing dates; G = Genotype; n.s. = non-significant.

Positive correlation was recorded for Chl *a* and *b* contents, and canopy temperature with grain yield while no relationship was noted between stem water-soluble carbohydrates and grain yield under both timely and late sowing (Figure 2). A significantly strong, and positive relationship of 1000-grain weight was found with days to heading during both years while with maturity time during the 2nd year only. Nonetheless, the relationship of grain yield with heading and maturity time was non-significant during both years and negative with maturity time during the 1st growing season (Table 3).

Table 3. Correlation co-efficient of heading and maturity time with thousand seed weight and seed yield.

	Pearson's Correlation (r_p) for Thousand Grain Weight and Grain Yield			
	Thousand Grain Weight		Grain Yield	
	1st Year	2nd Year	1st Year	2nd Year
Days to heading	0.61 (0.0016)	0.58 (0.0034)	0.23 (0.2764)	0.37 (0.0792)
Days to maturity	0.27 (0.1981)	0.81 (>0.0001)	−0.12 (0.5871)	0.39 (0.0604)

3. Discussion

Identification of plant traits and developing wheat cultivars with improved adaptation to terminal heat stress is priority for plant breeders around the globe. The present study compared the performance of wheat genotypes to identify the morpho-physiological traits for terminal heat stress tolerance under late sowing. Results showed that late sowing reduced yield traits, including spike length, number of grains per spike, thousand grain weight and biological yield than timely sown crop, albeit response varied between sowing conditions and years. The reduced number of grains per spike in the present study might be due to low grain fertility and floral spikelets associated with increased temperature even by 1 °C during booting and anthesis stages [27]. Reduction in number of grains depends on developmental stage, at which high temperature occurs and determined by supply of carbohydrates during floral development, which is a sensitive process to high temperature [28]. Higher number of grains per spike in advanced lines of the present study might be attributed to increased availability of stem water-soluble carbohydrates. Advanced lines including genotype Punjab-11 had higher thousand grain weight and grain yield than heat tolerant early maturing check Millet-11 and maintained it under terminal heat stress. Better agronomic and yield performance of the advanced lines and Punjab-11 seemed to be genotypic specific effects [29], while achieving and maintaining the optimal grain weight is considered an index of heat stress tolerance and adaptation [30]. Another possible reason for the decrease in yield traits of the present study might be attributed to reduced duration between different crop stages and less translation of biomass to yield (Tables 1 and 2). Reduced biological yield, thousand grain weight and 6–20% decrease in grain yield is reported in wheat crop exposed to terminal heat stress [3,23,31].

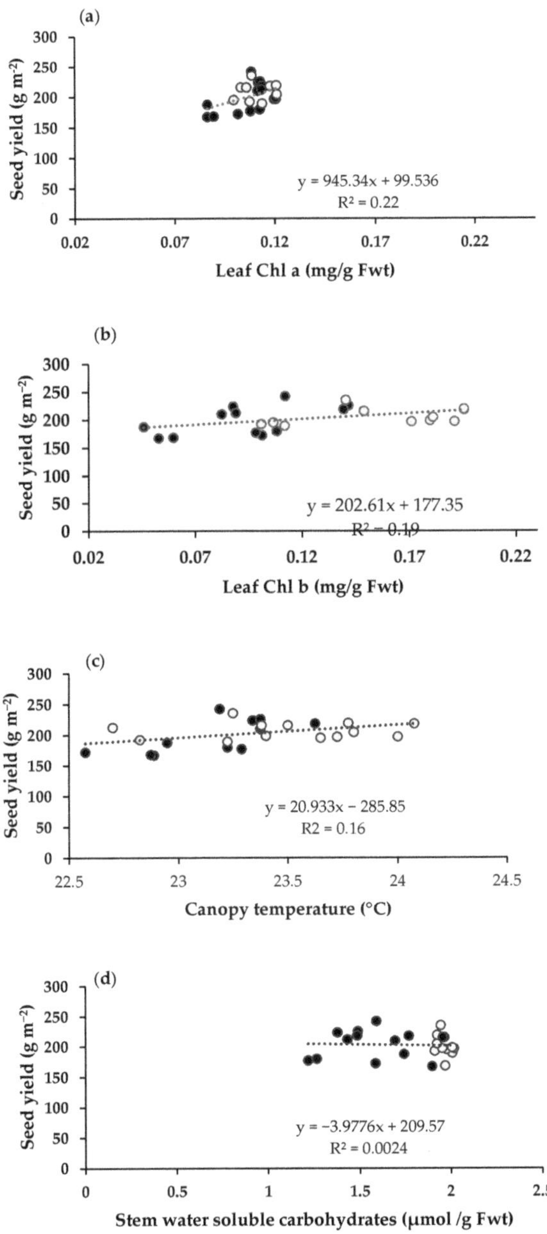

Figure 2. Regression relationship of seed yield with (**a**) leaf Chl a, (**b**) Chl b, (**c**) canopy temper-ature (**d**) stem water soluble carbohydrates. Filled box with black color indicate the data val-ues for timely planted and white filled for late planted wheat crop.

Nonetheless, a higher number of total productive tillers and grain yield during the first growing season under late sowing might be due to superiority in physiological traits of genotypes expressing high leaf chlorophyll contents and cooler canopies during anthesis or grain filling stages. Both traits help the plants to maintain better photosynthetic performance and remobilization of stem reserves to the developing grains during grain filling under high temperature or drought stress [3]. High temperature increased transpiration and keep crop canopies cool and turgid to maintain the photosynthetic performance [16]. Although relatively low transpiration and photosynthetic rates were found especially for advanced line V 07096 under late sowing, these were significantly similar with timely sowing supporting the hypothesis that genotypes maintained photosynthetic performance under terminal heat stress.

Reduced canopy temperature response of wheat lines V-07096 and V-10110 was also reflected with a decrease in gas exchange traits including A, E and Ci under late sowing in the present study [16]. Higher leaf chlorophyll contents at anthesis in wheat genotypes are an indicative of delayed senescence, high photosynthetic rate and remobilization of assimilates under terminal heat stress. Both traits contributed to higher grain yield in wheat genotypes under late sowing as evident from direct association of leaf chlorophyll contents and canopy temperature with grain yield (Figure 2a,b). Stay green trait is highly dependent on the environment and has a strong positive relationship with grain filling rate, duration and grain yield under heat stress [20]. Grain yield and stay grain are controlled by similar quantitative traits loci (QTLs) which are co-localized for productivity enhancement under heat stressed environments [20]. Heat tolerant wheat lines developed with physiological traits having cooler canopies and stay green showed superior yield and higher thousand grain weight with better adaptability under terminal heat stress [3,24].

The stem water-soluble carbohydrate is a potential adaptive trait for developing heat or drought tolerant wheat and 10–50% variation in stem water-soluble carbohydrates to total stem dry weight has been reported in wheat [32,33]. Higher accumulation of stem water-soluble carbohydrates in advanced lines along with early maturing heat tolerant check Millet-11 of the present study indicated increased buffering capacity of these genotypes to remobilize carbon reserves towards the developing grains accumulated during stem elongation period under terminal heat stress [14].

Nonetheless, no relationship of stem water-soluble carbohydrates with grain yield in the present study (Figure 2d) validates that weak association of water-soluble carbohydrates with yield as reported earlier and response is dependent on environment [16,26]. Higher stem water-soluble carbohydrates expressed under late sowing in the present study validate the potential of trait under stress condition and should be further investigated. In addition to superior physiological traits, wheat genotypes showed delayed maturity, however, response was compensated with higher yields ranging 5.38–23.82% compared to early maturing heat tolerant check Millet-11. Early maturity is as considered a breeding criterion to escape the effects of terminal heat stress, however, short duration may be accompanied with grain yield losses [3]. Interestingly, association of thousand grain weight with days to heading was significant and strong during both years and with days to maturity in the 1st year (Table 3).

Nonetheless, superior performance of advanced lines and Punjab-11 for grain yield and physiological traits demonstrated their potential to be used for the physiological breeding programs. Thus genotypes with heat adaptive traits should be considered in parent selection for a targeted environment or for the identification of one or more adaptive traits [24–46].

4. Materials and Methods

4.1. Study Site

This study was conducted during the winter (rabi) growing season of 2013–2014 and 2104–2015 at University of Agriculture Faisalabad (latitude, 31° 26′ N; longitude, 73° 06′ E; altitude 184.4 m). The soil had sandy loam texture with Lyallpur series, an Aridisol, a fine silty, mixed, hyperthermic Ustalfic Haplargid according to USDA classification. Seeds of wheat genotypes were obtained from Wheat Research Institute, Ayub Agriculture Research Institute (AARI), Faisalabad.

4.2. Plant Material

The four wheat genotypes used in the present study were two widely cultivated (Millet-11, Punjab-11) and two promising lines (V-07096, V-10110).

4.3. Creation of Heat Stress Environment and Experimental Design

Genotypes were cultivated at two sowing times viz. timely and late sowing. The crop planted on the 10th and 13th of November was considered as normal sowing while on the 10th and 13th of December was recorded as late sowing during 2013 and 2014, respectively. The delayed sowing was done with the objective to create a heat stress environment at anthesis and during reproductive stages. The early maturing genotype Millet-11 was used as heat tolerant check while Punjab-11 was selected for its high yield potential [47]. The experimental treatments were randomized in complete block design (RCBD) with split-plot arrangement with sowing dates into main plots and wheat genotypes into sub-plots and each replicated thrice. Wheat genotypes were planted into net plot size of 5.7 m × 3.5 m at inter-row spacing of 22.5 cm.

4.4. Climate and Weather Conditions

The Faisalabad features semi-arid sub-tropical climate and is located in Punjab-Pakistan. The rice-wheat rotation is generally practiced as a cropping pattern in this ecological zone. Wheat is grown as a spring crop starting from November and harvested in April. The average maximum and minimum temperatures during winter were 21 °C and 6 °C, respectively. May, June and July are the hottest months during summer and December, January and February are the coldest ones during winter [48]. Average annual rainfall is about 300 mm and that is highly seasonal and 50% of it is received during monsoon in July and August. However, in the present study, the temperature from sowing to booting ranged between 16 to 28 °C during 2012–2013 and 9.6 to 28.1 °C during 2013–2014. The maximum temperature of wheat crop sown under normal and late sowing from anthesis to maturity including grain filling period ranged from 16.8 to 31.5 °C in 2012–2013 and 17.3 to 33.6 °C in 2013–2014 of present study and similar trend was observed for low temperature (Figure 3).

4.5. Crop Husbandry

The seed rate of 125 kg ha^{-1} was used for wheat cultivation. Recommended fertilizers doses of nitrogen (N), phosphorus (P) and potassium (K) were applied at rate of 100–85–60 kg ha^{-1} using urea, single super phosphate and sulfate of potash, respectively. Whole of the P and K were applied during land preparation and while N was applied in three splits with the 1st half during land preparation as basal, 2nd at the 1st irrigation and 3rd at the 2nd irrigation. Crop was irrigated four times including the 1st as pre-saturated, 2nd at crown initiation, 3rd at tillering and 4th during anthesis stages. All plant protection measures were performed as recommended and weeds were controlled manually.

4.6. Observations

4.6.1. Crop Phenology

Phenological measurements at different crop developmental stages such as days to booting, heading, anthesis and maturity including grain filling period were recorded following Zadoks scale [49].

4.6.2. Physiological Traits

For canopy temperature, measurements were taken during mid grain-filling stage using Infrared Thermometer between 10:00 h to 14:00 h under clear bright sky with no wind.

Two readings per plot were taken and averaged. Leaf chlorophyll contents were measured according to Arnon et al. [50]. Five flag leaves were harvested from each plot at 10th days after anthesis and 0.5 g leaf sample was extracted in acetone (80% v/v) and kept overnight in sealed falcon tubes at 5 °C. The absorbance of the supernatant was determined at 645 nm and 663 nm using a spectrophotometer (Hitachi-220 Japan). Water soluble carbohydrates i.e., stem reserves were measured according to Yemm and Willis [51] and stem from main primary tillers was collected at the 7th day after anthesis. Fresh material (0.5 g) was boiled in 5 mL distilled water for 1 h. The extract was filtered and distilled water was added up to 50 mL. The 5 mL anthrone reagent was added into 1 mL of the extract along the sidewall of the test tube and vortexed. The mixture was heated in a water bath at 90–95 °C for 20 min, cooled down and absorbance was taken at 620 nm using a spectrophotometer (Hitachi-220 Japan).

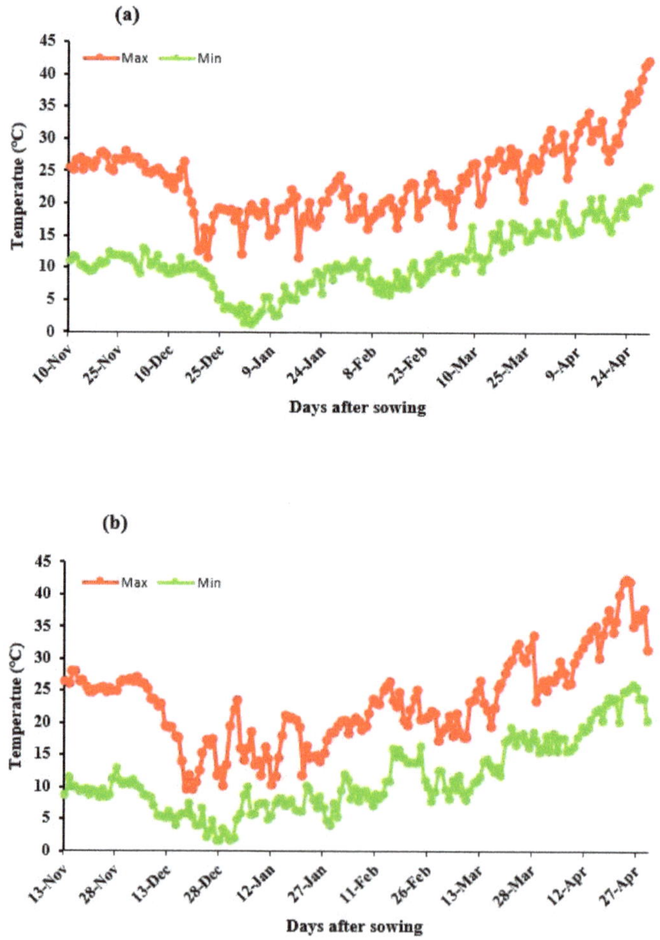

Figure 3. Maximum and minimum temperatures during crop growing period (**a**) 2013–2014 and (**b**) 2014–2015.

4.6.3. Gas Exchange Traits

Stomatal conductance (gs), sub-stomatal CO_2 concentration (Ci), photosynthetic (A), and transpiration rates (E) were recorded at anthesis stage from top the 3rd leaf of every plant using infra-red gas analyzer (IRGA) (model, LCA-4; Analytical Development Company, Hoddesdon, England) [52]. All these measurements were recorded at 13.00–14.00 h. During measurements, leaf chamber molar gas flow rate was 248 μmol s^{-1}, ambient CO_2 conc. (Cref) 352 μmol mol^{-1}, ambient pressure (P) of 98.01 k Pa, molar flow of air/leaf area 221.06 mol $m^{-2}s^{-1}$ and leaf chamber volume with gas flow rate (v) of 380 mL/min.

4.6.4. Yield Related Traits

Average plant height, spike length and number of grains per spike were recorded at maturity by threshing 10 primary tillers manually. Numbers of productive tillers of each genotype from selected plants were counted at maturity and averaged. The grains of threshed plants for each replication were bulked separately for 1000-grains and weighed. For straw yield, an area of 2 m^2 was randomly harvested twice per plot, tagged and weighed separately with the help of an electric balance. The harvested samples were threshed manually and grain yield for each genotype was measured. Both straw and grain yields were expressed as g m^{-2}.

4.6.5. Statistical Analysis

Analysis of variance (ANOVA) was performed for data comparison using Statistix 8.1 software. The year was taken as factor to find the homogeneity or heterogeneity and where year was significant, the data of both years were analyzed and presented separately, otherwise pooled. MS-Excel was used to present data graphically. To overcome confounding effects of maturity among cultivars for physiological traits, co-variance analysis was performed. The significant difference among treatment means was compared following Tukey's range test at 5% probability level. Pearson's correlation was also performed for heading and maturity time with thousand grain wight and grain yield to determine relationship between them.

5. Conclusions

The present study revealed necessary variation among genotypes for various morpho-physiological traits associated with adaptation to terminal heat stress. Advanced lines and cv. Punjab-11 had better agronomic performance for most of the traits than heat tolerant check Millet-11 under late sowing. A positive relationship of leaf chlorophyll contents and canopy temperature was found with grain yield. Therefore, advanced lines and Punjab-11 with high biomass and grain yield, reduced canopy temperature and high leaf chlorophyll contents can be promising sources to be utilized in physiological breeding for developing climate resilience in wheat.

Author Contributions: Conceptualization, H.u.R. and M.R.; Data curation, A.T.; Formal analysis, H.u.R. and A.T.; Investigation, H.u.R.; Methodology, H.u.R. and S.M.A.B.; Project administration, H.u.R. and S.M.A.B.; Supervision, H.u.R., S.M.A.B. and M.R.; Visualization, M.A.; Writing—original draft, H.u.R., A.T. and M.A.; Writing—review & editing, I.A., M.A. and A.M. All authors have read and agreed to the published version of the manuscript.

Funding: Higher Education Commission, Pakistan.

Data Availability Statement: Data is contained within the article.

Acknowledgments: The authors acknowledge the Higher Education Commission, Pakistan for providing financial support in completion of these studies and Shahid Farooq, Academic Editor, Plos One for critically editing the manuscript.

Conflicts of Interest: The authors declare no conflict of interest.

Abbreviations

CIMMYT: International Maize and Wheat Improvement Center, MEs: mega environments, NARPs: National Agriculture Research Programs, SD: Sowing dates, G: Genotype, n.s: non-significant, Fwt: Fresh weight, A: Photosynthetic rate, E: Transpiration rate, Ci: Intercellular CO_2 concentration, Gs: Stomatal conductance, AARI: Ayub Agriculture Research Institute, RCBD: Randomized complete block design, N: Nitrogen, P: Phosphorus, K:Potassium, IRGA: Infra-red gas analyzer and ANOVA: Analysis of variance

References

1. Asseng, S.; Ewert, F.; Martre, P.; Rotter, R.P.; Lobell, D.B.; Cammarano, D.; Kimball, B.A.; Ottman, M.J.; Wall, G.W.; White, J.W.; et al. Rising temperatures reduce global wheat production. *Nat. Clim. Chang.* **2015**, *5*, 143–147. [CrossRef]
2. Braun, H.J.; Pfeiffer, W.H.; Pollmer, W.G. Environments for selecting widely adapted spring wheat. *Crop Sci.* **1992**, *32*, 1420–1427. [CrossRef]
3. Mondal, S.; Singh, R.P.; Crossa, J.; Huerta-Espinoab, J.; Sharmac, I.; Chatrathc, R.; Singhd, G.P.; Sohue, V.S.; Mavie, G.S.; Sukuru, V.S.P.; et al. Earliness in wheat: A key to adaptation under terminal and continual high temperature stress in South Asia. *Field Crops Res.* **2013**, *151*, 19–26. [CrossRef]
4. Nawaz, A.; Farooq, M.; Nadeem, F.; Siddique, K.H.M.; Lal, R. Rice–wheat cropping systems in South Asia: Issues, options and opportunities. *Crop Past. Sci.* **2019**, *70*, 395–427. [CrossRef]
5. Shah, M.A.; Farooq, M.; Hussain, M. Productivity and profitability of cotton–wheat system as influenced by relay intercropping of insect resistant transgenic cotton in bed planted wheat. *Eur. J. Agron.* **2016**, *75*, 33–41. [CrossRef]
6. Joshi, A.K.; Mishra, B.; Chatrath, R.; Ferrara, G.O.; Singh, R.P. Wheat improvement in India: Present status, emerging challenges and future prospects. *Euphytica* **2007**, *157*, 431–446. [CrossRef]
7. Al-Khatib, K.; Paulsen, G.M. Model of high temperature injury to wheat during grain development. *Physiol. Plant.* **1984**, *61*, 363–368. [CrossRef]
8. Tashiro, T.; Wardlaw, I.F. The response to high temperature shock and humid-ity changes prior to and during the early stages of grain development in wheat. *Aust. J. Plant Physiol.* **1990**, *17*, 551–561.
9. Wardlaw, I.; Wrigley, C. Heat tolerance in temperate cereals: An overview. *Funct. Plant Biol.* **1994**, *21*, 695–703. [CrossRef]
10. Weigand, C.L.; Cueller, J.A. Duration of grain filling and kernel weight of wheat as affected by temperature. *Crop Sci.* **1981**, *21*, 95–101. [CrossRef]
11. Lopes, M.S.; Reynolds, M.P. Stay-green in spring wheat can be determined by spectral reflectance measurements (normalized difference vegetation index) independently from phenology. *J. Exp. Bot.* **2012**, *63*, 3789–3798. [CrossRef] [PubMed]
12. Talukder, S.K.; Babar, M.A.; Vijayalakshmi, K.; Poland, J.; Prasad, P.V.; Bowden, R.; Fritz, A. Mapping QTL for the traits associated with heat tolerance in wheat (*Triticum aestivum* L.). *BMC Genet.* **2014**, *15*, 97. [CrossRef]
13. Pinto, R.S.; Reynolds, M.P. Common genetic basis for canopy temperature depression under heat and drought stress associated with optimized root distribution in bread wheat. *Theor. Appl. Genet.* **2015**, *128*, 575–585. [CrossRef] [PubMed]
14. Blum, A.; Sinmena, B.; Mayer, J.; Golan, G.; Shpiler, L. Stem reserve mobilisation supports wheat-grain filling under heat stress. *Funct. Plant Biol.* **1994**, *21*, 771–781. [CrossRef]
15. Kumari, M.; Singh, V.P.; Tripathi, R.; Joshi, A.K. Variation for staygreen trait and its association with canopy temperature depression and yield traits under terminal heat stress in wheat. In *Wheat Production in Stressed Environments*; Springer: Dordrecht, The Netherlands, 2007; pp. 357–363.
16. Pinto, R.S.; Reynolds, M.P.; Mathews, K.L.; McIntyre, C.L.; Olivares-Villegas, J.-J.; Chapman, S.C. Heat and drought adaptive QTL in a wheat population designed to minimize confounding agronomic effects. *Theor. Appl. Genet.* **2010**, *121*, 1001–1021. [CrossRef] [PubMed]
17. Reynolds, M.P.; Balota, M.; Delgado, M.I.B.; Amani, J.; Fischer, R.A. Physiological and morphological traits associated with spring wheat yield under hot irrigated conditions. *Aust. J. Plant Physiol.* **1994**, *2*, 717–730. [CrossRef]
18. Kumari, M.; Pudake, R.N.; Singh, V.P.; Joshi, A.K. Association of stay green trait with canopy temperature depression and yield traits under terminal heat stress in wheat (*Triticum aestivum* L.). *Euphytica* **2013**, *190*, 87–97. [CrossRef]
19. Reynolds, M.P.; Tuberosa, R. Translational research impacting on crop productivity in drought-prone environments. *Curr. Opin. Plant Biol.* **2008**, *11*, 171–179. [CrossRef]
20. Pinto, R.S.; Lopes, M.S.; Collins, N.C.; Reynolds, M.P. Modelling and genetic dissection of stay green under heat stress. *Theor. Appl. Genet.* **2016**, *129*, 2055–2074. [CrossRef]
21. Kumar, S.; Kumari, P.; Kumar, U.; Grover, M.; Singh, A.K.; Singh, R.; Sengar, R.S. Molecular approaches for designing heat tolerant wheat. *J. Plant Biochem. Biotechnol.* **2013**, *22*, 359–371. [CrossRef]
22. Ortiz, R.; Sayre, K.D.; Govaerts, B.; Gupta, R.; Subbarao, G.; Ban, T.; Hodson, D.; Dixon, J.M.; Iva ́n, O.M. Climate change: Can wheat beat the heat? *Agric. Ecosys. Environ.* **2008**, *126*, 46–58. [CrossRef]
23. Lobell, D.B.; Burke, M.B.; Tebaldi, C.; Mastrandrea, M.D.; Falcon, W.P.; Naylor, R.L. Prioritizing climate change adaptation needs for food security in 2030. *Science* **2008**, *319*, 607–610. [CrossRef] [PubMed]
24. Pask, A.; Joshi, A.K.; Manès, Y.; Sharma, I.; Chatrath, R.; Singh, G.P.; Sohu, V.S.; Mavi, G.S.; Sakuru, V.S.P.; Kalappanavar, I.K.; et al. A wheat phenotyping network to incorporate physiological traits for climate change in South Asia. *Field Crops Res.* **2014**, *168*, 156–167. [CrossRef]
25. Reynolds, M.P.; Foulkes, J.; Furbank, R.; Griffiths, S.; King, J.; Murchie, E.; Parry, M.; Slafer, G. Achieving yield gains in wheat. *Plant Cell Environ.* **2012**, *35*, 1799–1823. [CrossRef] [PubMed]
26. Reynolds, M.P.; Langridge, P. Physiological breeding. *Curr. Opin. Plant Biol.* **2016**, *31*, 162–171. [CrossRef] [PubMed]

27. Ahmed, K.; Shabbir, G.; Ahmed, M.; Shah, K.N. Phenotyping for drought resistance in bread wheat using physiological and biochemical traits. *Sci. Total Environ.* **2020**, *729*, 139082. [CrossRef] [PubMed]
28. Ahmed, M. Introduction to Modern Climate Change. Andrew E. Dessler: Cambridge University Press, 2011, 252 pp, ISBN-10: 0521173159. *Sci. Total Environ.* **2020**, *734*, 139397. [CrossRef]
29. Ahmed, M.; Aslam, M.A.; Hassan, F.; Hayat, R.; Ahmad, S. Biochemical, Physiological and Agronomic Response of Wheat to Changing Climate of Rainfed Pakistan. *Pak. J. Bot* **2014**, *51*, 2.
30. Ahmed, M.; Hassan, F.; Asif, M. Physiological response of bread wheat ('*Triticum aestivum*'L.) to high temperature and moisture stresses. *Aust. J. Crop Sci.* **2012**, *6*, 749.
31. Ahmed, M.; Hassan, F.; Aslam, M.A.; Akram, M.N.; Akmal, M. Regression model for the study of sole and cumulative effect of temperature and solar radiation on wheat yield. *Afr. J. Biotechnol.* **2011**, *10*, 9114–9121.
32. Tariq, M.; Ahmed, M.; Iqbal, P.; Fatima, Z.; Ahmad, S. Crop Phenotyping. In *Systems Modeling*; Ahmed, M., Ed.; Springer Singapore: Singapore, 2020; pp. 45–60. [CrossRef]
33. Liu, B.; Martre, P.; Ewert, F.; Porter, J.R.; Challinor, A.J.; Müller, C.; Ruane, A.C.; Waha, K.; Thorburn, P.J.; Aggarwal, P.K.; et al. Global wheat production with 1.5 and 2.0 °C above pre-industrial warming. *Glob. Chang. Biol.* **2019**, *25*, 1428–1444. [CrossRef]
34. Asseng, S.; Martre, P.; Maiorano, A.; Rötter, R.P.; O'Leary, G.J.; Fitzgerald, G.J.; Girousse, C.; Motzo, R.; Giunta, F.; Babar, M.A.; et al. Climate change impact and adaptation for wheat protein. *Glob. Chang. Biol.* **2019**, *25*, 155–173. [CrossRef] [PubMed]
35. Khan, A.; Ahmad, M.; Ahmed, M.; Hussain, M.I. Rising Atmospheric Temperature Impact on Wheat and Thermotolerance Strategies. *Plants* **2021**, *10*, 43. [CrossRef]
36. Khan, A.; Ahmad, M.; Shah, M.K.N.; Ahmed, M. Genetic manifestation of physio-morphic and yield related traits conferring thermotolerance in wheat. *Pak. J. Bot.* **2020**, *52*, 1545–1552. [CrossRef]
37. Ahmad, S.; Abbas, G.; Ahmed, M.; Fatima, Z.; Anjum, M.A.; Rasul, G.; Khan, M.A.; Hoogenboom, G. Climate warming and management impact on the change of phenology of the rice-wheat cropping system in Punjab, Pakistan. *Field Crops Res.* **2019**, *230*, 46–61. [CrossRef]
38. Ahmed, M.; Fayyazul, H. Response of Spring Wheat (*Triticum aestivum* L.) Quality Traits and Yield to Sowing Date. *PLoS ONE* **2015**, *10*, e0126097. [CrossRef] [PubMed]
39. Ahmed, M.; Farooq, S. Growth and physiological responses of wheat cultivars under various planting windows. *JAPS J. Anim. Plant Sci.* **2013**, *23*, 1407–1414.
40. Wheeler, T.R.; Hong, T.D.; Ellis, R.H.; Batts, G.R.; Morrison, J.I.L.; Hadley, P. The duration and rate of grain growth, and harvest index, of wheat (*Triticum aestivum* L.) in response to temperature and carbon dioxide. *J. Exp. Bot.* **1996**, *47*, 623–630. [CrossRef]
41. Demotes-Mainard, S.; Jeuffroy, M.H. Effects of nitrogen and radiation on dry matter and nitrogen accumulation in the spike of winter wheat. *Field Crops Res.* **2004**, *87*, 221–233. [CrossRef]
42. Anjum, F.; Wahid, A.; Javed, F.; Arshad, M. Influence of foliar applied thiourea on flag leaf gas exchange and yield parameters of bread wheat (*Triticum aestivum* L.) cultivars under salinity and heat stresses. *Int. J. Agric. Biol.* **2008**, *10*, 619–626.
43. Singha, P.; Bhowmick, J.; Chaudhury, B.K. Effect of temperature on yield and yield components of fourteen wheat (*Triticum aestivum* L.) genotypes. *Environ. Ecol.* **2006**, *24*, 550–554.
44. Shirdelmoghanloo, H.; Taylor, J.D.; Lohraseb, I.; Rabie, H.; Brien, C.; Timmins, A.; Martin, P.; Mather, D.E.; Emebiri, L.; Collins, N.C. A QTL on the short arm of wheat (*Triticum aestivum* L.) chromosome 3B affects the stability of grain weight in plants exposed to a brief heat shock early in grain filling. *BMC Plant Biol.* **2016**, *16*, 100. [CrossRef] [PubMed]
45. Reynolds, M.P.; Pierre, C.S.; Saad, A.S.I.; Vargas, M.; Condon, A.G. Evaluating potential genetic gains in wheat associated with stress-adaptive trait expression in elite genetic resources under drought and heat stress. *Crop Sci.* **2007**, *47*, S-172–S-189. [CrossRef]
46. Rebetzke, G.J.; van Herwaarden, A.F.; Jenkins, C.; Weiss, M.; Lewis, D.; Ruuska, S.; Tabe, L.; Fettell, N.A.; Richards, R.A. Quantitative trait loci for water-soluble carbohydrates and associations with agronomic traits in wheat. *Aust. J. Agric. Res.* **2008**, *59*, 891–905. [CrossRef]
47. AARI. *Wheat: An Overview*; Ayub Agriculture Research Institute: Faisalabad, Pakistan, 2018; pp. 1–13.
48. Nawaz, A.; Farooq, M.; Lal, R.; Rehman, A.; Rehman, H. Comparison of conventional and conservation rice-wheat systems in Punjab, Pakistan. *Soil Till. Res.* **2017**, *169*, 35–43. [CrossRef]
49. Zadoks, J.C.; Chang, T.T.; Konzak, C.F. A decimal code for growth stages of cereals. *Weed Res.* **1974**, *14*, 415–421. [CrossRef]
50. Arnon, D.I. Copper, enzyme in isolated chloroplasts polyphenol oxidase in *Beta vulgaris*. *Plant Physiol.* **1949**, *24*, 1–15. [CrossRef] [PubMed]
51. Yemm, E.W.; Willis, A.J. The estimation of carbohydrates in plant extracts by anthrone. *Biochem. J.* **1954**, *57*, 508–514. [CrossRef]
52. Long, S.P.; Bernacchi, C.J. Gas exchange measurements, what they can tell us about the underlying limitations to photosynthesis? Procedures and sources of error. *J. Exp. Bot.* **2003**, *54*, 2393–2401. [CrossRef]

plants

MDPI

Article

Elevated Nitrogen Priming Induced Oxinitro-Responses and Water Deficit Tolerance in Rice

Kamolchanok Umnajkitikorn [1,*], Mitsutaka Fukudome [2,3], Toshiki Uchiumi [2] and Neung Teaumroong [4]

1 School of Crop Production Technology, Institute of Agricultural Technology, Suranaree University of Technology, Nakhon Ratchasima 30000, Thailand
2 Graduate School of Science and Engineering, Kagoshima University, 1-21-35 Korimoto, Kagoshima 890-0065, Japan; fukudome@nibb.ac.jp (M.F.); uttan@sci.kagoshima-u.ac.jp (T.U.)
3 Division of Symbiotic Systems, National Institute for Basic Biology, Myodaiji, Okazaki, Aichi 444-8585, Japan
4 School of Biotechnology, Institute of Agricultural Technology, Suranaree University of Technology, Nakhon Ratchasima 30000, Thailand; neung@sut.ac.th
* Correspondence: k.umnajkitikorn@sut.ac.th; Tel.: +66-44223706

Abstract: Under water deficit conditions, the essential macronutrient nitrogen becomes limited as a result of reduced dissolved nitrogen and root nitrogen uptake. An elevated nitrogen level might be able to mitigate these effects, integrated with the idea of using nitric oxide as abiotic stress tolerant inducers. In this study, we evaluated the potential of using elevated nitrogen priming prior to water shortage to mitigate plant stress through nitric oxide accumulation. We grew rice plants in 300 mg L^{-1} nitrogen for 10 weeks, then we primed plants with four different nitrogen concentrations: 100, 300 (control), 500 and 1000 mg L^{-1} nitrogen prior to inducing water deficit conditions. Plants primed with 500 mg L^{-1} nitrogen possessed a higher photosynthetic rate, relative water content, electrolyte leakage and lipid peroxidation under water deficit conditions, compared to control plants. The induction of water deficit tolerance was supported with the activation of antioxidant defense system, induced by the accumulation of nitric oxide in leaves and roots of rice plants. We originally demonstrated the accumulation of nitric oxide in leaves of rice plants. The elevated nitrogen priming can be used to enhance water deficit tolerance in irrigated paddy fields, instead of nitric oxide donors.

Keywords: rice; nitrogen; water stress; drought; antioxidant; reactive oxygen species; reactive nitrogen species

Citation: Umnajkitikorn, K.; Fukudome, M.; Uchiumi, T.; Teaumroong, N. Elevated Nitrogen Priming Induced Oxinitro-Responses and Water Deficit Tolerance in Rice. *Plants* **2021**, *10*, 381. https://doi.org/10.3390/plants10020381

Academic Editor: M. Iftikhar Hussain
Received: 27 January 2021
Accepted: 11 February 2021
Published: 17 February 2021

1. Introduction

Water deficit conditions cause widespread crop loss globally due to abiotic stress. By the end of the 21st century, drought and water deficit stress may account for more than 70 percent of crop loss below optimal productivity [1]. There are many methods to cope with stresses from water deficit conditions: breeding strategies, cultural practices, and irrigation approaches [2,3]. Drought stress induces mature and young leaf senescence. The primary site of damage during stress is the chloroplasts, where more than 70% of the leaf nitrogen (N) is sequestrated. In many annual crop species, chloroplast degradation requires a main strategy to cope with drought stress—known as an escape strategy [4]. This strategy aims to utilize the accumulated nutrients, particularly nitrogen and energy for acclimation and survival during stress episodes [5].

Under water deficit conditions, the loss of transpiration and cell turgor pressure lead to a reduction in uptake and transport of nitrogen, silicon, magnesium and calcium and other important nutrients [6]. Water deficit does not only limit nitrogen uptake but also restricts nitrogen assimilation through the inhibition of enzymes involved in nitrogen metabolism, such as nitrate reductase and glutamine synthetase [7]. Leaf yellowing commonly appears as the consequence of water deficit due to low chlorophyll biosynthesis and high chloroplast and chlorophyll degradation [5].

Nitric oxide (NO) has been proposed as a regulator of the antioxidant defense system during stress conditions [8–11]. Cai et al. 2015 found that the activity of nitric oxide synthase (NOS) and the accumulation of NO in rice plants were activated under salinity and drought stress [9]. Transgenic rice plants with overexpression of rat neuronal NO synthase (nNOS) showed an improvement in drought and salt tolerance [9]. External application of nitric oxide (NO) donor S-nitroso-N-acetylpenicillamine (SNAP) also alleviated the damage from salinity stress on chickpea (Cicer arietinum L.) plants [11]. The application of sodium nitroprussiate (SNP) (NO donor) also improved the chilling tolerance of winter wheat via the accumulation of fructan [12]. Spraying rice seedlings with SNP also enhanced drought tolerance, maintained leaf water potential, enhanced the antioxidant defense system and improved the stability of membranes [13]. However, NO donors or the transgenic approach are costly and inapplicable in many parts of the world. We were eager to search for a simplified method to utilize NO-triggered tolerant mechanisms. NO can be generated as a byproduct from a reaction catalyzed with nitrate reductase (NR) [14,15]. NR-generated NO had a pivotal role in lateral root formation and nitrogen uptake under partial nitrate nutrition in rice [16]. A high level of nitrogen application therefore becomes an interesting option due to the fact that farmers can access and afford nitrogen fertilizer.

Long-term application of high amounts of chemical N fertilizer lessened the water and nutrient use efficiency as well as inducing environmental problems and sensitivity to insects and diseases [17,18]. However, with moderate drought stress an appropriate level of N application can benefit the growth and yield of rice plants. Applying nitrogen can induce leaf production and thus accelerate transpiration and drought stress severity, and elevated nitrogen application levels may further stimulate rice shoot/root growth, promote photosynthesis, transpiration and yield [19,20]. High nitrogen levels also enhanced plasticity of roots, allowing plants to extract more of the water available from the soil [21]. Moreover, high N application induced a reduction in mesophyll conductance (g_m), which limits photosynthetic capacity and N-use efficiency [17]. Some recent studies have also shown that a high level of N in growing media enhanced the antioxidant defense system and some nitrogen assimilation enzymes, such as glutamine synthetase (GS) and glutamate synthase (GOGAT) but decreased N-use efficiency [22]. Although the effect of a large amount of N in growing media or in soil has long been seen to promote photosynthesis and N assimilation in various crop plants, regulation of high N levels triggering drought tolerance through NO signaling has never been investigated.

We therefore hypothesized that if we increased the nitrogen concentration in the fertilizer for only a short-period of time (priming), NO accumulation would be activated. Thus, the antioxidant defense system would also be activated. An addition of nitrogen could delay protein and chlorophyll degradation during a short period of water scarcity. In this study we aim to identify the appropriate level of N for priming treatment prior to stress episodes and to assess the mechanisms of nitrogen induced water deficit tolerance.

2. Results

2.1. Elevated Nitrogen Priming Enhanced Photosynthesis and Leaf Growth under Water Deficit Conditions

An elevated nitrogen level is well-known for promoting plant growth. However, the effects of an elevated nitrogen level for short term application (priming) on photosynthesis and growth have not been investigated. Rice plants primed with an elevated N level (500 mg L^{-1} and 1000 mg L^{-1} N) maintained higher photosynthetic activity (Figure 1a), but only the 500 mg L^{-1} N promoted rice plant growth as indicated by increasing leaf area under water deficit conditions (Figure 1b,c) when compared to the control plants (at 300 mg L^{-1} N). Elevated nitrogen priming failed to alter the net photosynthetic rate or leaf area of rice plants under well-watered conditions. Low N level priming had lower photosynthetic activity and growth under well-watered conditions (Figure 1).

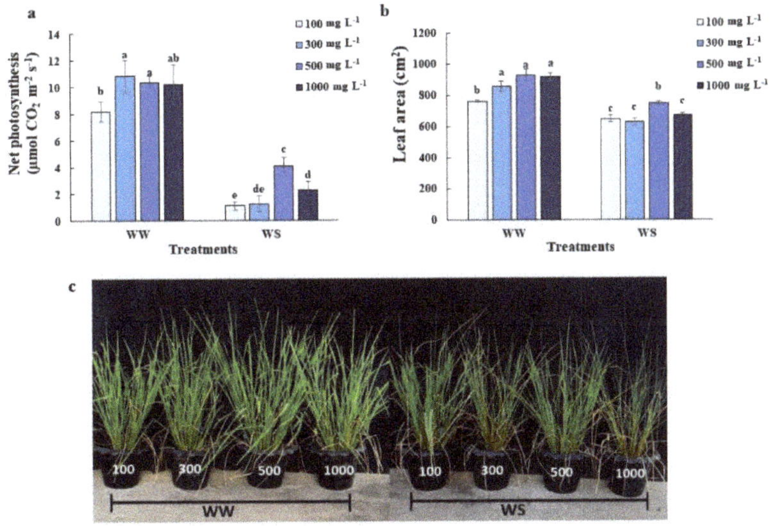

Figure 1. Growth and photosynthesis responses of rice plants to different nitrogen concentration priming. (**a**) Net photosynthesis of the youngest fully expanded leaves of the greenhouse-grown 11-week old rice plants after maintaining relative soil moisture content at 10–15% for 7 days (WS), compared with those under well-watered conditions (WW). (**b**) Leaf area of plants subjected to different level of nitrogen priming. (**c**) Plant growth under different level of nitrogen priming. The values shown are the Mean ± SE (n = 6 and 10, respectively). The different letters above the bars indicate significant differences by one-way ANOVA and Duncan's test ($p \leq 0.05$).

2.2. *Elevated Nitrogen Priming Promoted Relatively Higher Leaf Relative Water Content (RWC) under Water Deficit Conditions*

Elevated N-primed rice had higher RWC under water deficit conditions than normal N and lower N-primed rice (Figure 2a). Stomata conductance and transpiration rate from different priming nitrogen concentrations under water deficit conditions showed no significant differences. However, under well-watered conditions N-primed plants showed significantly lower stomata conductance (Figure 2b,c). A dramatically increased instantaneous water use efficiency occurred only in rice plants subjected to 500 mg L^{-1} N-priming treatment (Figure 2d).

2.3. *Elevated Nitrogen Priming Alleviated Chlorophyll and Protein Degradation as Well as Cell Damage under Water Deficit Conditions*

The elevated N-primed leaf maintained significantly higher chlorophyll and protein content under water deficit conditions (Figure 3a,b) but had only significantly higher protein content under well-watered conditions (Figure 3b). Electrolyte leakage (EL) was lower in the 500 mg L^{-1} N-primed plants and low N-primed plants (Figure 3c). However, excessive N (1000 mg L^{-1} N) had significantly higher electrolyte leakage (Figure 3c).

2.4. *Elevated Nitrogen Priming Enhanced Antioxidant Defense Mechanisms under Water Deficit Conditions*

To elucidate the mechanisms of elevated-N priming in delaying leaf senescence, we investigated the antioxidant defense system. Under well-watered conditions, SOD and CAT activity were the same between different N priming treatments (Figure 4a,b), while APX activity was slightly lower 100 mg L^{-1} and 500 mg L^{-1} N-primed plants and significantly lower in 1000 mg L^{-1} N-primed plants, compared to nonprimed (300 mg L^{-1} N) plants (Figure 4c). In contrast, under water deficit conditions, SOD and APX activities significantly increased in plants primed with 500 mg L^{-1} N, compared to control (Figure 4a,b). However, CAT activity in elevated N-primed plants significantly increased under water deficit conditions, compared to control (300 mg L^{-1} N) under well-watered conditions (Figure 4b).

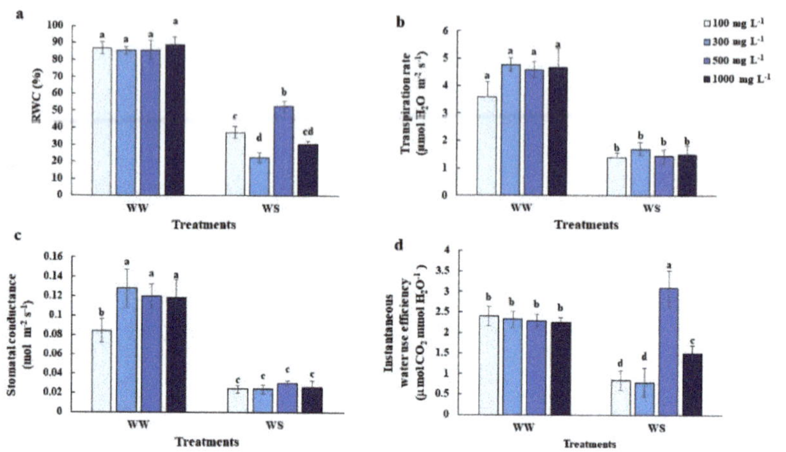

Figure 2. Physiological responses associated with plant water status. (**a**) relative leaf water content (RWC), (**b**) stomatal conductance, (**c**) transpiration rate and (**d**) instantaneous water use efficiency of the youngest fully expanded leaves of the greenhouse-grown 11-week old rice plants after maintaining relative soil moisture content at 10–15% for 7 days, compared with those under well-watered conditions (WW). Values are the Mean ± SE (n = 6). The different letters above the bars indicate significant differences by one-way ANOVA and Duncan's test ($p \leq 0.05$).

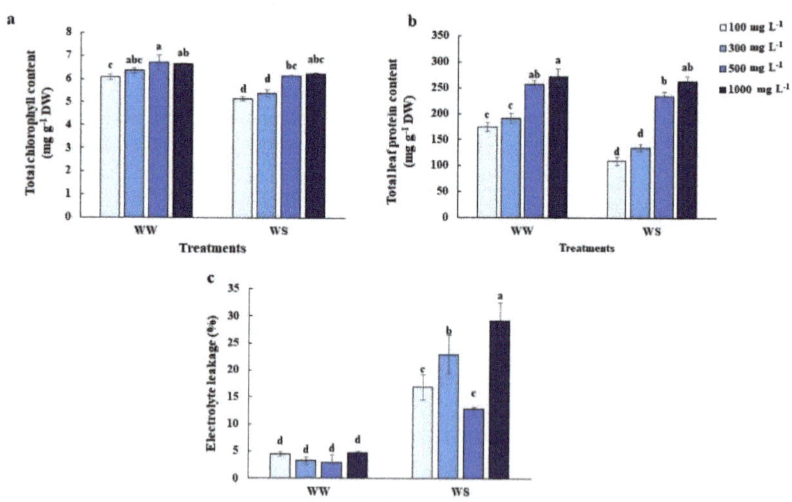

Figure 3. Biochemical responses associated with leaf senescence: (**a**) Total chlorophyll content, (**b**) total leaf protein content and (**c**) electrolyte leakage of the youngest fully expanded leaves of the greenhouse-grown 11-week old rice plants after maintaining relative soil moisture content at 10–15% for 7 days (WS), compared with those under well-watered conditions (WW). Values shown are the Mean ± SE (n = 4, 6 and 4, respectively). The different letters above the bars indicate significant differences by one-way ANOVA and Duncan's test ($p \leq 0.05$).

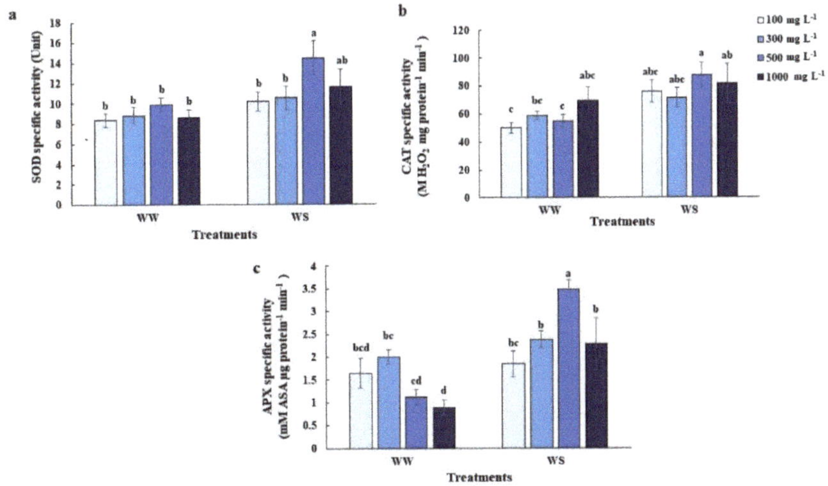

Figure 4. Enzymatic reactive oxygen species (ROS)-scavenging antioxidant responses: (**a**) Superoxide dismutase (SOD), (**b**) catalase (CAT) and (**c**) ascorbate peroxidase (APX) specific activities of the youngest fully expanded leaves of the 5 greenhouse-grown 11-week old rice plants after maintaining relative soil moisture content at 10–15% for 7 days (WS), compared with those under well-watered conditions (WW). Values are the Mean ± SE (n = 4). The different letters above the bars indicate significant differences by one-way ANOVA and Duncan's test ($p \leq 0.05$).

2.5. *Nitric Oxide Accumulation in Elevated Nitrogen Primed-Plants Alleviated Reactive Oxygen Species (ROS) Accumulation under Water Deficit Conditions*

To further investigate the mechanisms underlining the elevated nitrogen priming in alleviation of oxidative damage, we therefore focused on the nitric oxide signaling which has been reported to trigger antioxidant defense mechanisms and set up an in-lab experiment to investigate the process with the application of nitric oxide donor and scavenger. Nitric oxide (NO) accumulation occurred under the PEG-induced water deficit conditions for elevated nitrogen primed plant roots and leaves (500D), and plants without elevated nitrogen priming but supplied with a nitric oxide donor, sodium nitroprusside (SNP) (300D + SNP) (Figure 5). In contrast, the released NO and histological NO accumulation remained low under PEG-induced water deficit in the plant without elevated N priming (300D) and plants-primed with 500 mg L^{-1} N but supplied with a NO scavenger, 2-(4-carboxyphenyl)-4,4,5,5-tetramethylimidazoline-1-oxyl-3-oxide (cPTIO) (500D + cPTIO) (Figure 5).

Additionally, H_2O_2 content and staining showed low ROS levels in roots and leaves in plants without PEG treatment (control condition) (Figures 6 and 7a,b). The accumulation of ROS and lipid peroxidation were induced by PEG as shown in 300D. However, ROS and MDA content remained low in 500D plants and 300D +SNP (Figures 6 and 7), while the endogenous NO and released NO were more pronounced in these treatments (Figures 6 and 7). The cPTIO also inverted elevated nitrogen priming effects, leading to ROS accumulation and lipid peroxidation (Figures 6 and 7).

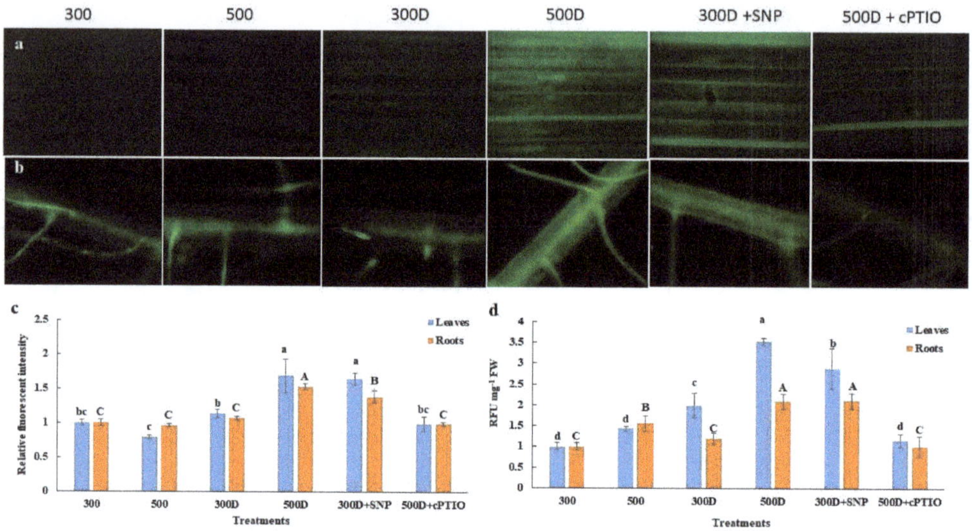

Figure 5. Reactive nitrogen species accumulation responses. The epifluorescent images of rice (**a**) leaves and (**b**) roots after 1-h incubation in 20 μM DAF-FM DA. The plants were subjected to the different NO-associated treatments for 3. days: 300 mg L^{-1} N-primed (300C); 500 mg L^{-1} N-primed (500C); 300 mg L^{-1} N-primed with PEG water deficit induction (300D); 500 mg L^{-1} N-primed with PEG water deficit induction (500D); 300 mg L^{-1} N-primed with PEG water deficit induction + Nitric oxide donor (300D + sodium nitroprussiate—SNP); 500 mg L^{-1} N-primed with PEG water deficit induction + Nitric oxide scavenger (500D + cPTIO) (**c**) Relative fluorescent intensity of the histochemical nitric oxide (NO) accumulation of the leaves and roots (**d**) Relative fluorescent units (RFUs) of the released NO after 2-h incubation in 7 μM DAF-FM. Values shown are the Mean ± SE (n ≥ 30 and n = 4, respectively). The different letters above the bars indicate significant differences by one-way ANOVA and Duncan's test ($p \leq 0.05$), and the small letters represent the statistics of the leaves and the capital letters represent the statistics of the roots.

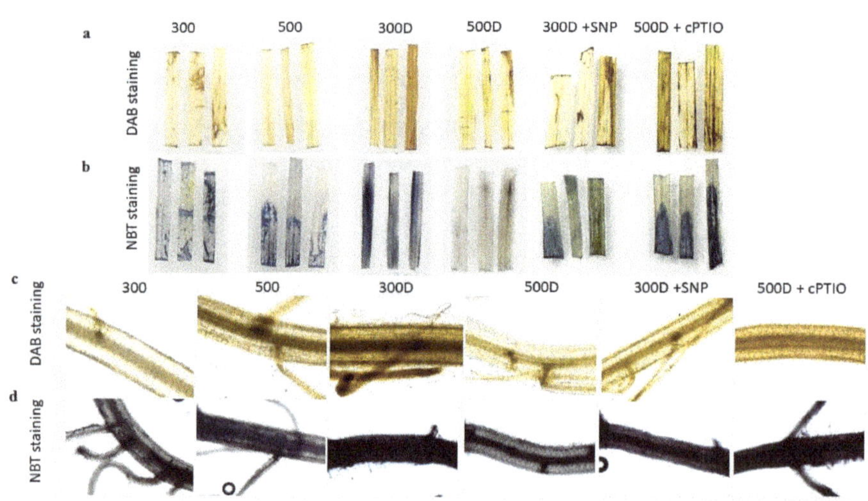

Figure 6. Reactive oxygen species accumulation responses. Hydrogen peroxide accumulation detected by 3,3′-diaminobenzidine (DAB) staining of (**a**) leaves and (**c**) roots. Superoxide radical accumulation detected by nitro blue tetrazolium (NBT) staining of (**b**) leaves and (**d**) roots of rice plants after 3-day growth in the different NO-associated treatments.

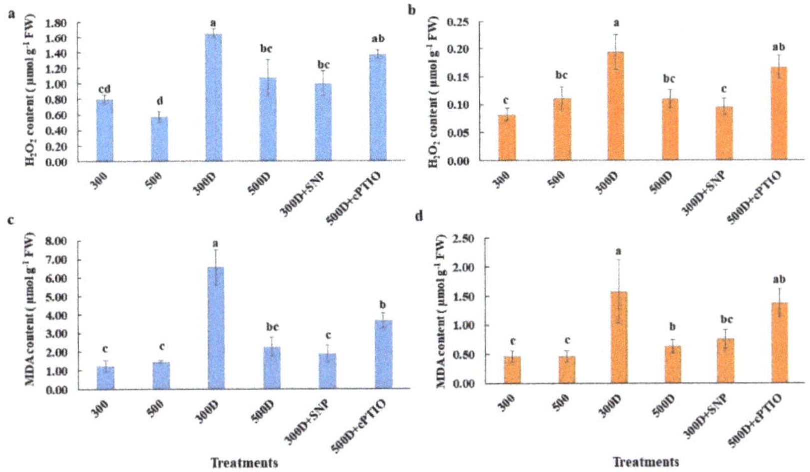

Figure 7. Quantitative ROS accumulation and lipid peroxidation responses: quantitative hydrogen peroxide accumulation in (**a**) leaves and (**b**) roots, and malondialdehyde (MDA) content of (**c**) leaves and (**d**) roots of rice plants after 3-day growth in the different NO-associated treatments. Values shown are the Mean ± SE (n = 4). The different letters above the bars indicate significant differences by one-way ANOVA and Duncan's test ($p \leq 0.05$).

3. Discussion

Nitrogen is an essential nutrient that plays the main role in leaf development. In many cases, high nitrogen application induced leaf area increment and total plant transpiration increment while N also increases root growth, leading to the added water and nutrient uptakes to support the increased transpiration and maintain high photosynthetic activity and chloroplast functions during water deficit conditions [23]. However, high nitrogen priming during a short period of nitrogen application effects were first investigated here (prior to drought episodes). We found rice plants primed with 500 mg L^{-1} N prior to water deficit conditions maintained higher photosynthetic rates (Figure 1a) but did not alter the transpiration rate (Figure 2c), resulting in a dramatic increase in instantaneous water use efficiency (Figure 2d). Elevated nitrogen priming in our studies also induced plant responses similar to the high soil-incorporated nitrogen application [20,21]. Haefele et al. [24] and Zhong et al. [25] both suggested the use of high nitrogen fertilizer increases water-use efficiency, drought tolerance and survival rates for C3 and C4 crops. The improvement of photosynthetic capacity mainly depended on the activity of ribulose-1,5-bisphosphate carboxylase (Rubisco). Around 15 to 30% of leaf nitrogen are invested in Rubisco [26]. The high nitrogen levels induced higher Rubisco content, which was generally degraded during stress episodes [27]. The elevated N-primed rice might delay the degradation of Rubisco and maintain a high photosynthetic rate under water deficit conditions, as well as less sensitive stomatal behavior [24,25]. The maintenance of relative water content in elevated N-primed plants may be explained by the accumulation of NO as shown in Li et al. 2013 where the NO donor could have induced fructans accumulation, which resulted in plant cell water content regulation [12].

On the other hand, Gao et al. 2019 indicated that low nitrogen priming levels reduced the transpiration rate, stomatal conductance and maintained a higher leaf relative water content to mitigate drought-induced damage [28]. We also found slightly lower transpiration rates (Figure 2c) and significantly higher leaf relative water content (Figure 2a) in 100 mg L^{-1}-N primed plants. Low nitrogen priming may also have induced lower chlorophyll and protein content, slowing down the process of stress recovery [29].

Although elevated N (500 mg L^{-1}-N) priming benefits the tolerance of rice plants, excessive N (1000 mg L^{-1}-N) priming leads to stress symptoms, like decreasing photosynthetic activity (Figure 1a), lower RWC (Figure 2a) and higher electrolyte leakage (Figure 3c), compared to plants primed with 500 mg L^{-1} N under control conditions. Excessive nitrogen application can also generate an imbalance in the carbon/nitrogen ratio which causes mature leaf senescence, leading to yield reduction after drought episodes [30]. Excessive nitrogen has impacts on the antioxidant defense systems in wheat [31], and in our study (Figure 4). Many nonenzymatic antioxidant molecules, such as GABA, 4-hydroxybenzoylcholine and several phenolic compounds were downregulated in the leaves of rice plants grown under excessive N, which led to the overaccumulation of ROS [31].

Redox imbalance under water deficit conditions is accountable for significant plant cell damage. The antioxidant induction system is a viable approach to alleviate cellular damage and a higher tolerability of plants to water deficit conditions [10,32,33]. We observed enhanced antioxidant enzyme activity in 500 mg L^{-1} N primed rice (Figures 1a and 2a). Perhaps the one-day elevated N priming was sufficient to activate antioxidant systems. Five-hundred mg L^{-1} N priming reduced membrane damage as demonstrated by the lower electrolyte leakage (Figure 3c) and MDA content (Figure 7d), which were supported by increasing SOD and APX activities (Figure 4a,c). Antioxidant defense systems act as drought tolerance mechanisms in many crop species, including rice, creeping bentgrass (*Agrostis stolonifera* L.), wheat (*Triticum aestivum* L.) seedlings and peanuts (*Arachis hypogaea* L.) [34–38].

The antioxidant defense system that controls ROS status in plant cells has also been reported to be regulated by the level of reactive nitrogen species (RNS) from both endogenous production and exogenous application [8,9,35,36,39]. Cai et al. 2015 [9] found that the overexpression of rat neuronal NO synthase (nNOS) in rice enhanced drought and salt tolerance of rice with a higher NOS activity and the accumulation of NO, together with a reduction in H_2O_2 accumulation, electrolyte leakage and MDA content. Using NO donors, SNP as a seed priming solution and foliar spray has previously induced rice drought tolerance in rice [13]. Because NO acts as a signaling molecule it can enhance antioxidant capacity thus reducing oxidative damage [13]. The 500 mg L^{-1} N leaves and roots accumulated higher endogenous NO and released NO, which was inhibited by the NO scavenger, cPTIO (Figure 5b–d). The level of NO accumulation was close to that of the normal N-primed (300 mg L^{-1} N) plants supplied with 1 mM SNP (Figure 5b,d). In contrast, 500 mg L^{-1} N primed plants possessed less $O_2^{\bullet-}$ and H_2O_2 accumulation evidenced by the NBT, DAB staining and H_2O_2 quantification (Figures 6 and 7a,b). We measured only $O_2^{\bullet-}$ and H_2O_2 because these two species have been reported to directly interact with NO [40,41]. Adding cPTIO may mitigate such effects (Figures 6 and 7a,b). Control plants (300 mg L^{-1} N) had less $O_2^{\bullet-}$ and H_2O_2 further suggesting a relationship between elevated N priming and increased NO (Figures 6 and 7a,b). This phenomenon might be explained by the fact that NO in plants can be generated by NO synthase (NOS), nitrate reductase (NR) and xanthine oxidoreductase [10]. When priming rice with elevated N, it induced NR activity, promoting greater nitrite levels (precursors for both NO production and N assimilation [42]. Higher NO levels in elevated N-primed rice promoted the antioxidant defense system, causing less accumulated ROS and less oxidative damage while maintaining photosynthetic functions. NO acted as a potent inhibitor of lipid peroxidation by scavenging lipid alcoxyl (LO$^{\cdot}$) and peroxyl (LOO$^{\cdot}$) radicals [43]. NO also directly quenches the ROS, such as superoxide radical ($O_2^{\bullet-}$) [44], limits oxidative damage and prevents the onset of cell death [45]. Moreover, NO maintained membrane fluidness, cell wall relaxation, cell enlargement and plant growth under stressful conditions [11,42]. Therefore, elevated nitrogen priming demonstrated a water deficit tolerance through NO-mediated antioxidant defense mechanisms.

4. Materials and Methods

4.1. Plant Material and Greenhouse Growth Conditions

Seeds of Thai rice (*Oryza sativa* L. subspecies indica cv. Pathumthani 1) were germinated on moist germination paper for 10 d at 28 °C in the dark. Seedlings were transplanted into 2-L pots filled with 1:1 vermiculite: perlite. The seedlings were grown in the greenhouse with day lengths of 11–12 h and daily average temperatures of 35 ± 4/24 ± 7 °C day/night and 700–1500 μmol m^{-2} s^{-1} mid-day photosynthetically active radiation (PAR). Plants were fertilized every other day with (N 300 mg L^{-1} equal in nitrate and ammonium forms), P 20 mg L^{-1}, K 75 mg L^{-1}, Ca 25 mg L^{-1}, Mg 17 mg L^{-1}, S 55 mg L^{-1}, Fe 3.30 mg L^{-1}, Mn 0.50 mg L^{-1}, Zn 0.05 mg L^{-1}, Mo 0.01 mg L^{-1}, Cu 0.02 mg L^{-1}) for 10 weeks until the plants were one week before panicle initiation. Then, different concentrations of nitrogen (N) fertilizer were applied by using N 100 mg L^{-1} (equal in nitrate and ammonium forms), P 20 mg L^{-1}, K 75 mg L^{-1}, Ca 25 mg L^{-1}, Mg 17 mg L^{-1}, S 55 mg L^{-1}, Fe 3.30 mg L^{-1}, Mn 0.50 mg L^{-1}, Zn 0.05 mg L^{-1}, Mo 0.01mg L^{-1}, Cu 0.02 mg L^{-1} as base fertilizer and added ammonium nitrate (NH_4NO_3) to reach target concentrations of 300, 500 and 1000 mg L^{-1} N and used double volume of the field capacity of the pots to wash the previous nitrogen fertilizer. Water deficit stress was applied by withholding water for ~7 d when visual stress symptoms (i.e., leaf rolling) appeared (10–15% relative soil water content) [7], when the gas exchange parameters were measured and the leaf samples were taken for biochemical assays. The leaves were collected between 9.00 am and 11.00 am and immediately frozen in liquid nitrogen and kept at −80 °C until use.

4.2. Gas-Exchange Measurements

Net photosynthetic rate and stomatal conductance were measured with the photosynthesis system ADC LCi-SD (BioScientific, UK). The measurements were conducted under 900 ± 50 μmol m^{-2} s^{-1} light intensity and 380 ± 10 μmol mol^{-1} CO_2 surrounding the leaf (Ca) at 32 ± 2 °C. To calculate instantaneous transpiration, the transpiration rate was divided by the net photosynthetic rate.

4.3. Leaf Area

After one-day post recovery, all remaining leaves were removed and all surface dust was wiped away prior to measuring the area with an LI-3100C area meter (LI-COR, Lincoln, NE, USA) in square centimeters.

4.4. Relative Water Content (RWC) of Leaves

RWC was determined according to Gao et al. 2019 [28]. Leaves were weighed immediately to obtain fresh weight (FW), soaked in water overnight in the dark and weighed again to obtain turgid fresh weight (TW), and then dried at 75 °C until constant weight (DW). RWC was then calculated as follows:

$$RWC = (FW - DW)/(TW - DW) \quad (1)$$

4.5. Electrolyte Leakage (EL)

The electrolyte leakage measurement was adjusted based on the method described previously by Cai et al. 2015 [9]. After placing six 1 cm^2 leaves from each treatment into a 50-mL-tube with 20 mL distilled deionized water, the tubes were shaken and electro-conductivity (EC_i) was immediately measured with a pH/cond meter (WTW, inoLab, Germany). After soaking the leaves for 12 h, the second conductivity was measured (EC_f). Then, the leaves were boiled for 1 h and the total electro-conductivity was measured (EC_t). The electrolyte leakage was calculated as follows:

$$\%\,EL = \left(\frac{EC_f - EC_i}{EC_t - EC_i}\right) \times 100 \quad (2)$$

4.6. Determining SOD, CAT and APX Activities

Enzymes were extracted by the modified method of Umnajkitikorn et al. 2013 [46]. The frozen leaves were ground in liquid nitrogen with a mortar and pestle. Two hundred fifty mg of the leaf powder were homogenized in 1 mL of the extraction buffer containing 50 mM potassium phosphate buffer (pH 7.8), 0.1 mM disodium EDTA, 1 mM ascorbic acid and, 2% PVPP (*w*/*v*). The homogenate was centrifuged at 15,000× *g* for 20 min at 4 °C. The supernatant was collected for SOD, CAT and APX activity assays.

SOD activity was assayed based on the method described by Beauchamp and Fridovich 1971 [47] and Vaidyanathan et al. 2003 [48]. One point two milliliters of the reaction mixture, containing 50 mM sodium phosphate buffer (pH 7.8), 10 mM EDTA, 1 mM NBT, 5 mM L-methionine, 0.2 mM riboflavin and mixed with 80 μL of extracts. Each reaction was carried out at 25 ± 2 °C under a light intensity, which is sufficient to increase absorbance of 0.110/10 min (in the absence of the enzyme) for 30 min. Two hundred microliters of the reactions were taken to Nunc microwell 96-well plate (Thermo ScientificTM, Shanghai, China) for the absorbance at 560 nm (A560), using spectrophotometer (Biotex, Epoch, Winooski, VT, USA). The nonirradiated reaction mixture served as a blank and was deducted from A560. One unit of SOD activity was defined as the amount of enzyme required to inhibit the reduction of NBT by 50%.

CAT activity was assayed based on the method of Sunohara and Matsumoto 2004 [49]. One hundred and ninety microliters of the assay mixture contained 20 mM H_2O_2 in 50 mM potassium phosphate buffer (pH 7.0), 0.1 mM disodium-EDTA and 10 μL of extracts. The reactions were carried in Nunc 96 well UV transparent plate (Thermo ScientificTM, Vantaa, Finland). The absorbance was measured at 240 nm with a UV/VIS spectrophotometer (Biotex, Epoch, Winooski, VT, USA). The enzyme activity was defined as the amount of H_2O_2 decomposed min^{-1} mg $protein^{-1}$. The molar coefficient of H_2O_2 (E) is 39.4 M^{-1} cm^{-1}) at 240 nm.

APX activity was assayed based on the method of Sunohara and Matsumoto 2004 [49]. One hundred and eighty microliters of the assay mixture contained120 μL of 50 mM potassium phosphate buffer (pH 7.0), 20 μL of 1 mM EDTA, 20 μL of 5 mM L-ascorbic acid, 20 μL of 1 mM H_2O_2 and 20 μL of extracts. The subsequent decrease in ascorbic acid was observed at 290 nm (E = 2.8 mM^{-1} cm^{-1}) with a UV/VIS spectrophotometer (Biotex, Epoch, Winooski, VT, USA). The enzyme activity was defined as the amount of ascorbic acid (ASA) which decomposed min^{-1} mg $protein^{-1}$.

4.7. Protein Quantification

The Bradford assay [50] was used for protein quantification using bovine serum albumin as the standard.

4.8. The RNS and ROS Experiment

Rice seeds (*Oryza sativa* subspecies indica cv. Pathumthani 1) were germinated and grown for 5 d at 25 °C in darkness. Seedlings were transplanted into 9-cm pots filled with vermiculite 4 plants/pot and grown in the growth room with 12 h/12 h light/dark cycles and 25/20 °C day/night cycles for 3 weeks with fertilizer containing 300 mg L^{-1} N. Then, the plants were randomly divided into 6 groups and transferred to glass tubes with the following treatments: (1) 300 mg L^{-1} N (300 C); (2) 500 mg L^{-1} a day and switched back to the 300 mg L^{-1} (500 C); (3) 300 mg L^{-1} N + gradually increased polyethylene glycol (PEG) 6000 (300 D); (4) 500 mg L^{-1} a day and switched back to the 300 mg L^{-1} + gradually increased PEG (500 D); (5) 300 mg L^{-1} N + PEG 6000 + nitric oxide donor, 1 mM SNP (300 D + SNP); (6) 500 mg L^{-1} N + PEG 6000 + 3 mM of Nitric oxide scavenger, 2-(4-carboxyphenyl)-4,4,5,5-tetramethylimidazoline-1-oxyl-3-oxide (cPTIO) (500 D + cPTIO). After 3 days under each treatment, the seedlings were harvested for further ROS and RNS analyses.

4.9. NO Determination

To histologically detect NO, the leaf and root segments were soaked for 1 h with 20 μM 4-amino-5-methylamino-2′,7′-difluorescein diacetate (DAF-FM DA; Sekisui Medical, Tokyo, Japan). The epifluorescence images were captured with an Eclipse 90i microscope (Nikon, Tokyo, Japan). The intensity of the fluorescent signal from at least 30 pictures of each treatment was quantified by using Image J software (by Wayne Rasband, NIH, MD, USA).

The concentration of NO released from the roots was assessed in leaf and root segments of the seedlings in each treatment after incubation in 7 μM DAF-FM for 2 h. The fluorescence intensity of the DAF-FM solution was measured as described in Fukudome et al. 2016 [51] but adapted by adding 100 μL of the solution into the 96 well black plate (SPL life sciences, Gyeonggi-do, Korea), then measured with a fluorescence spectrophotometer (Varioskan Lux, Finland) at 495 nm and 519 nm as excitation and emission wavelengths, respectively. The released NO was expressed as relative fluorescence units (RFUs) per fresh weight of leaves or roots.

4.10. Histochemical Detection of H_2O_2 and $O_2^{\bullet -}$

H_2O_2 was detected in situ according to Fukudome et al. 2019 [39] and Signorelli et al. 2013 [52]. Detached leaves and roots were vacuum-infiltrated in the dark with 10 mM potassium phosphate buffer, and 0.1% (*w*/*v*) 3,3′-diaminobenzidine (DAB), at pH 7.8. Samples were incubated overnight in the dark. The leaf segments were boiled in 95% ethanol at 90oC to remove chlorophyll. The clear leaf segments were then photographed. The staining areas were calculated by Image J software (by Wayne Rasband, NIH, USA), in the red channel according to procedure described in [53]. Data were shown in Supplementary Figure S2. The detailed method of staining area identification is given in Supplementary file 2.

Superoxide radical ($O_2^{\bullet -}$) was detected in situ essentially as described by to Fukudome et al. 2019 [39] and Signorelli et al. 2013 [52]. Detached leaves and roots were vacuum-infiltrated with 10mM potassium phosphate buffer, 0.1% (*w*/*v*) nitro blue tetrazolium (NBT), and 0.05% (*v*/*v*) Tween 20, pH 7.8. Treated samples were incubated overnight in the dark, cleared and photographed as described above. The staining areas were calculated by Image J software (by Wayne Rasband, NIH, MD, USA), in the blue channel according to the procedure described in [53]. Data are shown in Supplementary Figure S2.

4.11. H_2O_2 Quantification

The frozen leaves were ground and 5–10 mg of samples were extracted with 1 mL of 20 mM K_2HPO_4 (pH 6.5), homogenized at 1500 rpm for 5 min at 4 °C and then centrifuged for 5 min at 16,200× *g* at 4 °C. The H_2O_2 was colorimetrically quantified by Amplex Red detection assay (Thermo Fisher Scientific). The measurement was performed on the supernatant as described in Brumbarova et al. 2016 [54] with some modification of the incubation time from 30 min to 15 min. The assay solutions were analyzed at 560 nm with a UV/VIS spectrophotometer (Biotex, Epoch, Winooski, VT, USA). The standard curve was also generated with H_2O_2 concentration from 0–5 μM.

4.12. Malondialdehyde (MDA) Measurements

The youngest fully expanded leaves were homogenized with 5 mL of 50 mM potassium phosphate buffer pH 7.5 and centrifuged at 20,000× *g* for 25 min. For measurements of MDA concentration, 4 mL of 20% trichloroacetic acid containing 0.5% thiobarbituric acid were added to a 1 mL aliquot of the supernatant. The mixture was heated at 95 °C for 30 min, quickly cooled in ice and then centrifuged at 10,000× *g* for 10 min. The absorbance of the supernatant was measured at 532 nm. The results are shown as MDA content mg^{-1} FW, using malondialdehyde tetrabutylammonium salt as a standard (Sigma-Aldrich, Cat 63287, Singapore).

4.13. Statistical Analysis

The SPSS 25 statistical package was used for statistical analyses. The experiments were based on a randomized complete block design.

5. Conclusions

Elevated N priming at approximately 60% more than normal N level enhanced water deficit tolerance with NO-triggered antioxidant defense systems. The elevated N priming also supported the sustainability of photosynthetic activity and relative water contentof the leaves, together with the reduction of membrane damage. This approach has potential for in-situ investigation in aerobic rice fields with fertigation systems as a potential mitigating factor for enhancing drought tolerance in rice.

Supplementary Materials: The following are available online at https://www.mdpi.com/2223-7747/10/2/381/s1, Supplementary file 1: Fertilizer Information and Formulation sheet. Supplementary file 2: Figure S1: Examples of the picture calibrated for staining area evaluation. Figure S2: The ROS staining area of tissue segments of rice plants after 3–day growth in the different NO-associated treatments; Procedure for staining area quantification by image J.

Author Contributions: Experiment planning, K.U. and T.U.; Performing experiments K.U. and M.F.; Statistical analyses and data interpretation, K.U.; Manuscript preparation, K.U., M.F., T.U and N.T. All authors have read and agreed to the published version of the manuscript.

Funding: This work was supported by Suranaree University of Technology (SUT), Thailand.

Institutional Review Board Statement: Not applicable.

Informed Consent Statement: Not applicable.

Data Availability Statement: Data available in a publicly accessible repository.

Acknowledgments: The authors would like to thank Colin T. Strine, School of Biology, Institute of Science, Suranaree University of Technology, for English corrections in this manuscript.

Conflicts of Interest: The authors declare that there is no conflict of interest.

References

1. Leng, G.; Hall, J. Crop yield sensitivity of global major agricultural countries to droughts and the projected changes in the future. *Sci. Total Environ.* **2019**, *654*, 811–821. [CrossRef]
2. Bodner, G.; Nakhforoosh, A.; Kaul, H.P. Management of crop water under drought: A review. *Agron. Sustain. Dev.* **2015**, *35*, 401–442. [CrossRef]
3. Oladosu, Y.; Rafii, M.Y.; Samuel, C.; Fatai, A.; Magaji, U. Drought Resistance in Rice from Conventional to Molecular Breeding: A Review. *Int. J. Mol. Sci.* **2019**, *20*, 3519. [CrossRef]
4. Shavrukov, Y.; Kurishbayev, A.; Jatayev, S.; Shvidchenko, V.; Zotova, L.; Koekemoer, F.; Groot, S.; De Soole, K.; Lan-gridge, P. Early Flowering as a Drought Escape Mechanism in Plants: How Can It Aid Wheat Production? *Front. Plant Sci.* **2017**, *8*, 1–8. [CrossRef] [PubMed]
5. Sade, N.; del Mar Rubio-Wilhelmi, M.; Umnajkitikorn, K.; Blumwald, E. Stress-induced senescence and plant tolerance to abiotic stress. *J. Exp. Bot.* **2017**, *69*, 1–9. [CrossRef] [PubMed]
6. Fahad, S.; Bajwa, A.A.; Nazir, U.; Anjum, S.A.; Farooq, A.; Zohaib, A.; Sadia, S.; Nasim, W.; Adkins, S.; Saud, S.; et al. Crop production under drought and heat stress: Plant responses and management options. *Front. Plant Sci.* **2017**, *8*, 1–16. [CrossRef]
7. Sade, N.; Umnajkitikorn, K.; Wilhelmi, M.d.M.R.; Wright, M.; Wang, S.; Blumwald, E. Delaying chloroplast turn-over increases water-deficit stress tolerance through the enhancement of nitrogen assimilation in rice. *J. Exp. Bot.* **2017**, *69*, 867–878. [CrossRef]
8. Nabi, R.B.S.; Tayade, R.; Hussain, A.; Kulkarni, K.P.; Imran, Q.M.; Mun, B.G.; Yun, B.W. Nitric oxide regulates plant re-sponses to drought, salinity, and heavy metal stress. *Environ. Exp. Bot.* **2019**, *161*, 120–133. [CrossRef]
9. Cai, W.; Liu, W.; Wang, W.; Fu, Z.; Han, T.; Lu, Y. Overexpression of Rat Neurons Nitric Oxide Synthase in Rice En-hances Drought and Salt Tolerance. *PLoS ONE* **2015**, *10*, e0131599. [CrossRef]
10. Groß, F.; Durner, J.; Gaupels, F. Nitric oxide, antioxidants and prooxidants in plant defence responses. *Front. Plant Sci.* **2013**, *4*, 1–15. [CrossRef]
11. Ahmad, P.; Latef, A.A.A.; Hashem, A.; Abd Allah, E.F.; Gucel, S.; Tran, L.S.P. Nitric oxide mitigates salt stress by regu-lating levels of osmolytes and antioxidant enzymes in chickpea. *Front. Plant Sci.* **2016**, *7*, 1–11. [CrossRef]
12. Li, C.; Li, T.; Zhang, D.; Jiang, L.; Shao, Y. Exogenous nitric oxide effect on fructan accumulation and FBEs expression in chilling-sensitive and chilling-resistant wheat. *Environ. Exp. Bot.* **2013**, *86*, 2–8. [CrossRef]

13. Farooq, M.; Basra, S.M.A.; Wahid, A.; Rehman, H. Exogenously Applied Nitric Oxide Enhances the Drought Tolerance in Fine Grain Aromatic Rice (Oryza sativa L.). *J. Agron. Crop Sci.* **2009**, *195*, 254–261. [CrossRef]
14. Wildt, J.; Kley, D.; Rockel, A.; Rockel, P.; Segschneide, H.J. Emission of NO from several higher plant species. *J. Geophys. Res.* **1997**, *102*, 5919–5927. [CrossRef]
15. Tejada-jimenez, M.; Llamas, A.; Galv, A.; Fern, E. Photosynthetic Eukaryotes. *Plants* **2019**, *8*, 56. [CrossRef]
16. Sun, H.; Li, J.; Song, W.; Tao, J.; Huang, S.; Chen, S.; Hou, M.; Xu, G.; Zhang, Y. Nitric oxide generated by nitrate re-ductase increases nitrogen uptake capacity by inducing lateral root formation and inorganic nitrogen uptake under par-tial nitrate nutrition in rice. *J. Exp. Bot.* **2015**, *66*, 2449–2459. [CrossRef]
17. Cao, X.; Zhong, C.; Sajid, H.; Zhu, L. Effects of watering regime and nitrogen application rate on the photosynthetic parameters, physiological characteristics, and agronomic traits of rice. *Acta Physiol. Plant.* **2017**, *39*, 1–12. [CrossRef]
18. Sun, Y.; Wang, M.; Alejandro, L.; Mur, J.; Shen, Q.; Guo, S. Unravelling the Roles of Nitrogen Nutrition in Plant Disease Defences. *Int. J. Mol. Sci.* **2020**, *21*, 572. [CrossRef] [PubMed]
19. Menge, D.M.; Onyango, J.C.; Yamauchi, A.; Kano-Nakata, M.; Asanuma, S.; Thi, T.T.; Inukai, Y.; Kikuta, M.; Makihara, D. Effect of nitrogen application on the expression of drought-induced root plasticity of upland NERICA rice. *Plant Prod. Sci.* **2019**, *22*, 180–191. [CrossRef]
20. Sedri, M.H.; Amini, A.; Golchin, A. Evaluation of Nitrogen Effects on Yield and Drought Tolerance of Rainfed Wheat using Drought Stress Indices. *J. Crop Sci. Biotechnol.* **2019**, *22*, 235–242. [CrossRef]
21. Tran, T.T.; Kano-Nakata, M.; Takeda, M.; Menge, D.; Mitsuya, S.; Inukai, Y.; Yamauchi, A. Nitrogen application enhanced the expression of developmental plasticity of root systems triggered by mild drought stress in rice. *Plant Soil* **2014**, *378*, 139–152. [CrossRef]
22. Li, S.; Zhou, L.; Daniel, S.D.A.; Guochang, D.; Sun, M.; Wu, S.; Lin, S. Nitrogen supply enhances the physiological resistance of Chinese fir plantlets under polyethylene glycol (PEG)-induced drought stress. *Sci. Rep.* **2020**, *10*, 1–8. [CrossRef] [PubMed]
23. Dinh, T.H.; Watanabe, K.; Takaragawa, H.; Nakabaru, M. Photosynthetic response and nitrogen use efficiency of sugarcane under drought stress conditions with different nitrogen application levels. *Plant Prod. Sci.* **2017**, *1008*, 1–11. [CrossRef]
24. Haefele, S.M.; Kato, Y.; Singh, S. Field Crops Research Climate ready rice: Augmenting drought tolerance with best management practices. *Crop. Res.* **2016**, *190*, 60–69. [CrossRef]
25. Zhong, S.; Xu, Y.; Meng, B.; Loik, M.E.; Ma, J.Y.; Sun, W. Nitrogen addition increases the sensitivity of photosynthesis to drought and re-watering differentially in C3 versus C4 grass species. *Front. Plant Sci.* **2019**, *10*, 1–17. [CrossRef]
26. Evans, J.R. Photosynthesis and nitrogen relationships in leaves of C3 plants. *Oecologia* **1989**, *78*, 9–19. [CrossRef]
27. Abid, M.; Tian, Z.; Ata-ul-karim, S.T.; Cui, Y.; Liu, Y. Nitrogen Nutrition Improves the Potential of Wheat (*Triticum aestivum* L.) to Alleviate the Effects of Drought Stress during Vegetative Growth Periods. *Front Plant Sci.* **2016**, *7*, 1–14. [CrossRef] [PubMed]
28. Gao, J.; Luo, Q.; Sun, C.; Hu, H.; Wang, F.; Tian, Z.; Jiang, D.; Cao, W.; Dai, T. Low nitrogen priming enhances photo-synthesis adaptation to water-deficit stress in winter wheat (*Triticum aestivum* L.) seedlings. *Front. Plant Sci.* **2019**, *10*, 1–11. [CrossRef]
29. Heckathorn, S.A.; DeLucia, E.H.; Zielinski, R.E. The contribution of drought-related decreases in foliar nitrogen concentration to decreases in photosynthetic capacity during and after drought in prairie grasses. *Physiol. Plant.* **1997**, *101*, 173–182. [CrossRef]
30. Chen, D.; Wang, S.; Xiong, B.; Cao, B.; Deng, X. Carbon/nitrogen imbalance associated with drought-induced leaf senescence in sorghum bicolor. *PLoS ONE* **2015**, *10*, e0137026. [CrossRef]
31. Kong, L.; Xie, Y.; Hu, L.; Si, J.; Wang, Z. Excessive nitrogen application dampens antioxidant capacity and grain filling in wheat as revealed by metabolic and physiological analyses. *Sci. Rep.* **2017**, *7*, 1–14. [CrossRef]
32. Iqbal, A.; Dong, Q.; Wang, X.; Gui, H.; Zhang, H.; Zhang, X.; Song, M. High Nitrogen Enhances Drought Tolerance in Cotton through Antioxidant Enzymatic Activities, Nitrogen Metabolism and Osmotic Adjustment Asif. *Plants* **2020**, *9*, 178. [CrossRef] [PubMed]
33. Laxa, M.; Liebthal, M.; Telman, W.; Chibani, K.; Dietz, K.J. The role of the plant antioxidant system in drought tolerance. *Antioxidants* **2019**, *8*, 94. [CrossRef] [PubMed]
34. Chang, Z.; Liu, Y.; Dong, H.; Teng, K.; Han, L.; Zhang, X. Effects of cytokinin and nitrogen on drought tolerance of creeping bentgrass. *PLoS ONE* **2016**, *11*, e0154005. [CrossRef]
35. Hasanuzzaman, M.; Fujita, M. Exogenous sodium nitroprusside alleviates arsenic-induced oxidative stress in wheat (*Triticum aestivum* L.) seedlings by enhancing antioxidant defense and glyoxalase system. *Ecotoxicology* **2013**, *22*, 584–596. [CrossRef]
36. He, H.; Oo, T.L.; Huang, W.; He, L.; Gu, M. Nitric oxide acts as an antioxidant and inhibits programmed cell death induced by aluminum in the root tips of peanut (Arachis hypogaea L.). *Sci. Rep.* **2019**, *9*, 1–12. [CrossRef]
37. Guo, Z.; Ou, W.; Lu, S.; Zhong, Q. Differential responses of antioxidative system to chilling and drought in four rice cultivars differing in sensitivity. *Plant Physiol. Biochem.* **2006**, *44*, 828–836. [CrossRef]
38. Shehab, G.G.; Ahmed, O.K.; El-beltagi, H.S. Effects of Various Chemical Agents for Alleviation of Drought Stress in Rice Plants (*Oryza sativa* L.). *Not. Bot. Horti Agrobot. Cluj-Napoca* **2010**, *38*, 139–148.
39. Fukudome, M.; Watanabe, E.; Osuki, K.I.; Uchi, N.; Uchiumi, T. Ectopic or over-expression of class 1 phytoglobin genes confers flooding tolerance to the root nodules of lotus japonicus by scavenging nitric oxide. *Antioxidants* **2019**, *8*, 206. [CrossRef]
40. Del Río, L.A. ROS and RNS in plant physiology: An overview. *J. Exp. Bot.* **2015**, *66*, 2827–2837. [CrossRef]
41. Radi, R. Oxygen radicals, nitric oxide, and peroxynitrite: Redox pathways in molecular medicine. *Proc. Natl. Acad. Sci. USA* **2018**, *115*, 5839–5848. [CrossRef]

42. Chamizo-ampudia, A.; Sanz-luque, E.; Llamas, A.; Galvan, A.; Fernandez, E. Nitrate Reductase Regulates Plant Nitric Oxide Homeostasis. *Trends Plant Sci.* **2017**, *22*, 163–174. [CrossRef] [PubMed]
43. Rubbo, H.; Radi, R.; Anselmi, D.K.; Marion, B.; Stephen, B.; John, E.; Jason, P. Nitric oxide reaction with lipid peroxyl radicals spares alpha-tocopherol during lipid peroxidation. *J. Biol. Chem.* **2000**, *275*, 10812–10818. [CrossRef]
44. Jain, P.; Singh, N.; Kaur, H. Mechanisms of nitric oxide crosstalk with reactive oxygen species scavenging enzymes during abiotic stress tolerance in plants. *Free Radic Res.* **2015**, *50*, 291–303. [CrossRef]
45. Hasanuzzaman, M.; Hossain, M.A.; Fujita, M. Nitric oxide modulates antioxidant defense and the methylglyoxal detoxi-fication system and reduces salinity-induced damage of wheat seedlings. *Plant Biotechnol. Rep.* **2011**, *5*, 353–365. [CrossRef]
46. Umnajkitikorn, K.; Faiyue, B.; Saengnil, K. Enhancing Antioxidant Properties of Germinated Thai rice (Oryza sativa L.) cv. Kum Doi Saket with Salinity. *Rice Res. Open Access* **2013**, *4*, 1–8. [CrossRef]
47. Beauchamp, C.; Fridovich, I. Superoxide dismutase: Improved assays and an assay applicable to acrylamide gels. *Anal. Biochem.* **1971**, *44*, 276–287. [CrossRef]
48. Vaidyanathan, H.; Sivakumar, P.; Chakrabarty, R.; Thomas, G. Scavenging of reactive oxygen species in NaCl-stressed rice (Oryza sativa L.)—Differential response in salt-tolerant and sensitive varieties. *Plant Sci.* **2003**, *165*, 1411–1418. [CrossRef]
49. Sunohara, Y.; Matsumoto, H. Oxidative injury induced by the herbicide quinclorac on Echinochloa oryzicola Vasing and the involvement of antioxidative ability in its highly selective action in grass species. *Plant Sci.* **2004**, *167*, 597–606. [CrossRef]
50. Bradford, M.M. A rapid and sensitive method for the quantitation of microgram quantities of protein utilizing the prin-ciple of protein-dye binding. *Anal. Biochem.* **1976**, *72*, 248–254. [CrossRef]
51. Fukudome, M.; Calvo-Begueria, L.; Kado, T.; Osuki, K.I.; Rubio, M.C.; Murakami, E.I.; Nagata, M.; Kucho, K.I.; Sandal, N.; Stougaard, J.; et al. Hemoglobin LjGlb1-1 is involved in nodulation and regulates the level of nitric oxide in the Lotus japonicus-Mesorhizobium loti symbiosis. *J. Exp. Bot.* **2016**, *67*, 5275–5283. [CrossRef] [PubMed]
52. Signorelli, S.; Corpas, F.J.; Borsani, O.; Barroso, J.B.; Monza, J. Water stress induces a differential and spatially distributed nitro-oxidative stress response in roots and leaves of Lotus japonicus. *Plant Sci.* **2013**, *201–202*, 137–146. [CrossRef] [PubMed]
53. Available online: https://imagej.nih.gov/ij/docs/examples/stained-sections/index.html (accessed on 16 February 2021).
54. Brumbarova, T.; Le, C.T.T.; Bauer, P. Hydrogen Peroxide Measurement in Arabidopsis Root Tissue Using Amplex Red. *Bio-protocol* **2016**, *6*, 1–11. [CrossRef]

plants

Review

Rising Atmospheric Temperature Impact on Wheat and Thermotolerance Strategies

Adeel Khan [1], Munir Ahmad [1], Mukhtar Ahmed [2,3,*] and M. Iftikhar Hussain [4,5]

[1] Department of Plant Breeding and Genetics, PMAS-Arid Agriculture University Rawalpindi, Rawalpindi 46300, Pakistan; adeelkhanbreeder@gmail.com (A.K.); muneer.ahmad@uaar.edu.pk (M.A.)
[2] Department of Agricultural Research for Northern Sweden, Swedish University of Agricultural Sciences, 90183 Umeå, Sweden
[3] Department of Agronomy, PMAS Arid Agriculture University, Rawalpindi 46300, Pakistan
[4] Department of Plant Biology & Soil Science, Faculty of Biology, University of Vigo, Campus As Lagoas Marcosende, 36310 Vigo, Spain; mih786@gmail.com
[5] CITACA, Agri-Food Research and Transfer Cluster, Campus da Auga, University of Vigo, 32004 Ourense, Spain
* Correspondence: mukhtar.ahmed@slu.se

Abstract: Temperature across the globe is increasing continuously at the rate of 0.15–0.17 °C per decade since the industrial revolution. It is influencing agricultural crop productivity. Therefore, thermotolerance strategies are needed to have sustainability in crop yield under higher temperature. However, improving thermotolerance in the crop is a challenging task for crop scientists. Therefore, this review work was conducted with the aim of providing information on the wheat response in three research areas, i.e., physiology, breeding, and advances in genetics, which could assist the researchers in improving thermotolerance. The optimum temperature for wheat growth at the heading, anthesis, and grain filling duration is 16 ± 2.3 °C, 23 ± 1.75 °C, and 26 ± 1.53 °C, respectively. The high temperature adversely influences the crop phenology, growth, and development. The pre-anthesis high temperature retards the pollen viability, seed formation, and embryo development. The post-anthesis high temperature declines the starch granules accumulation, stem reserve carbohydrates, and translocation of photosynthates into grains. A high temperature above 40 °C inhibits the photosynthesis by damaging the photosystem-II, electron transport chain, and photosystem-I. Our review work highlighted that genotypes which can maintain a higher accumulation of proline, glycine betaine, expression of heat shock proteins, stay green and antioxidant enzymes activity viz., catalase, peroxidase, super oxide dismutase, and glutathione reductase can tolerate high temperature efficiently through sustaining cellular physiology. Similarly, the pre-anthesis acclimation with heat treatment, inorganic fertilizer such as nitrogen, potassium nitrate and potassium chloride, mulches with rice husk, early sowing, presoaking of a 6.6 mM solution of thiourea, foliar application of 50 ppm dithiothreitol, 10 mg per kg of silicon at heading and zinc ameliorate the crop against the high temperature. Finally, it has been suggested that modern genomics and omics techniques should be used to develop thermotolerance in wheat.

Keywords: heat stress; photosynthesis; antioxidant enzymes; HSPs; QTLs; omics

Citation: Khan, A.; Ahmad, M.; Ahmed, M.; Iftikhar Hussain, M. Rising Atmospheric Temperature Impact on Wheat and Thermotolerance Strategies. *Plants* **2021**, *10*, 43. https://doi.org/10.3390/plants10010043

Received: 21 November 2020
Accepted: 22 December 2020
Published: 27 December 2020

1. Introduction

Climate change is the result of a higher level of greenhouse gases such carbon dioxide (CO_2), nitrous oxide, and methane (CH_4). These gases can entrap the sun rays leading towards the severity of extreme events for crops development [1,2]. It has been observed that CO_2 was increased 0.6 ± 0.1 ppm/year in the early 1960s and 2.3 ± 0.6 ppm/year during the last decade. Meanwhile, the CH_4 gas was doubled after the industrial revolution until the 1980s and it is increasing at the rate of 12 parts per billion per year. However, during the last three decades it was increasing 2–5 parts per billion per year. The nitrous oxide concentration was enhanced 18% more than the 1970s and increased 0.8 parts per billion per year [3,4].

The escalating global warming evokes an extreme weather pattern, increases disease incidences, insect pest survival, and ultimately influences crop productivity [5,6]. Global warming potential (GWP) is the contribution of one molecule of compound over 100 years to global warming as compared to CO_2. It allows the comparison of different gas contributions to global warming and how much energy emissions of 1 ton of gas absorbs more than 1 ton of CO_2 over a given time period. The larger global warming potential represents more potential of the given gas to persist and the ability to warm the Earth temperature over a given time period. The GWP of carbon dioxide is 1, CH_4 28–36, and nitrous oxide 265–298 over 100 years. However, these gases possess more potential and persistency to entrap the sun rays than CO_2 but a major contributor in global warming is CO_2 [7].

Agricultural crop productivity depends on biotic (diseases and insect pest) and abiotic (heat, drought, and salinity) factors [8]. Among the abiotic stresses, the higher temperature is a major concern influencing crop growth and development. The global temperature roughly increased by 1.5 °C with the same accelerating trend in all regions from the 1970s, as reported by the intergovernmental panel of climatic change (IPCC) and was predicted to increase 2.5–5.8 °C until the 2100s [3]. The global average temperature annually increased by 0.04–0.07 °C and 0.15–0.17 °C per decade since the 1880s and 1970s, respectively according to the National Oceanic and Atmospheric Administration (NOAA, 2018). Therefore, global warming characterized by an extreme temperature possesses the challenge to improve the yield potential of crops.

Terminal and continual heat stresses are two major constraints influencing crops growth and development. The temperature threshold levels were reported at different stages for crops viz., cotton [9–13], rice [14–19], sorghum [20,21], barley [12], maize [9,22,23], and soybean [23]. Wheat is an imperative staple food, the cheapest energy source, provides 8–20% of protein, and 70–75% of calories in our average diet [24], but a high temperature restricts the wheat crop to express its full genetic potential. Therefore, there is a dire need to understand the wheat response against the high temperature and a suitable strategy to improve its productivity.

2. Impact of High Temperature on Wheat

High temperature influences the wheat productivity in tropical, subtropical, arid and semi-arid regions of the world. The optimum temperature for wheat growth and development are given in Table 1. The high temperature in the tropical region is an inevitable constraint for wheat during germination and early growth stages, whereas in the Mediterranean region, the reproductive stage is highly sensitive [25]. A high temperature of 3–4 °C above the optimum temperature at grain filling reduces 10–50% of the wheat yield in Asia with the current production technology and varieties [26]. High temperature declines 0.07% per °C grain yield depending on the wheat variety [27]. Each degree increase in the temperature at the grain filling duration reduces 6% of wheat yield globally [28,29] and 3–17% in South Asia including India and Pakistan [30]. It accredited directly or indirectly the disturbance in different cellular, physiological functions and metabolic pathways associated with the grain yield in wheat (Figure 1).

Table 1. Optimal temperature requirements of wheat at different growth stages.

Stages	Optimum Temperature (°C)	Minimum Temperature (°C)	Maximum Temperature (°C)
Root growth	17.2 ± 0.87	3.5 ± 0.73	24 ± 1.21
Shoot growth	18.5 ± 1.90	4.5 ± 0.76	20.1 ± 0.64
Leaf initiation	20.5 ± 1.25	1.5 ± 0.52	23.5 ± 0.95
Terminal Spikelet	16 ± 2.3	2.5 ± 0.49	20 ± 1.6
Anthesis	23 ± 1.75	10 ± 1.12	26 ± 1.01
Grain Filling Duration	26 ± 1.53	13 ± 1.45	30 ± 2.13

[28,30–47].

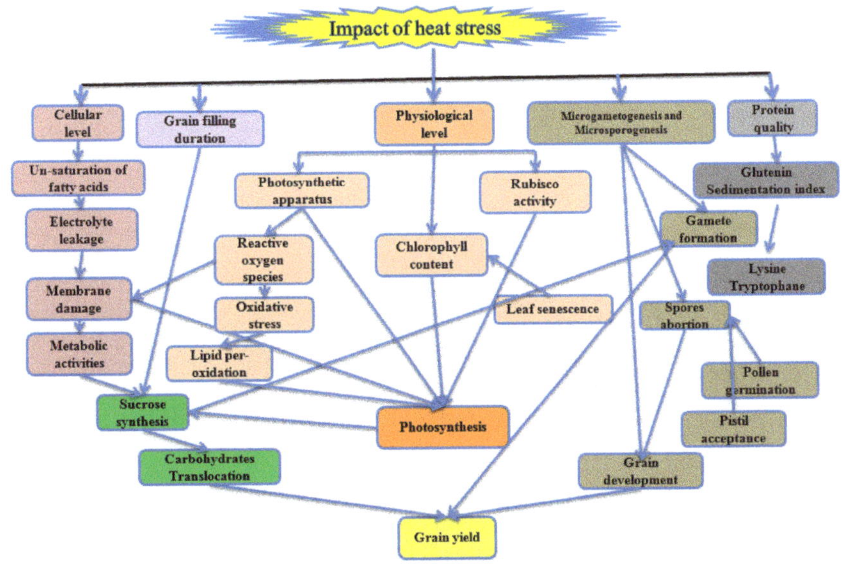

Figure 1. Schematic illustration of the high temperature impact on wheat associated with the grain yield.

2.1. Cellular Metabolism

The plasma membrane is a highly organized structure composed of lipids and proteins. It regulates the enzymatic activity and transport of ions. High temperature alters the microtubules organization, expansion, elongation, and cell differentiation [48]. It increases the kinetic energy of hydrogen bonds between adjacent fatty acids, weakens the bonds, and leads to membrane fluidity. This fluidity, unsaturation of fatty acids, and disruption of different proteins trigger the electrolyte leakage [49,50]. High temperature causes 25–55% electrolyte leakage at 45 °C for 1 h [51], while 21–40% leakage at 40 °C for 30 min [52]. Therefore, the cell damages its internal composition and sustainable physiological processes (e.g., photosynthesis, respiration, and transpiration) associated with the synthesis and translocation of carbohydrates into the grains.

2.2. Grain Filling Duration

High temperature enforces the plant to complete the growing degree days earlier, which results in early maturity and shorter life cycle of plant, lesser biosynthetic products accumulation, and ultimately poor grain development [32,53,54]. Vernalization (*VRN1, VRN2*) and the photoperiodic (*PPD-A1, PPD-D1*) sensitive gene determines the developmental phases at volatile temperature events and triggers earliness in wheat by limiting various growth phases [55,56]. The longer grain filling duration determines the appropriate grains development associated with the grain yield [57]. However, high temperature reduces the duration to uptake the available nutrients and translocation of photosynthates.

2.3. Grain Formation and Development

Vital events at the reproductive stage such as flowering initiation, pollen germination, pistil receptiveness, and embryo development determine the florets fertility [58,59]. The embryo sac and embryo formation are sensitive to high temperature [60]. Microgametogenesis and microsporogenesis are sensitive to high temperature, which hinder the gametes development and cause spores abortion [61,62].

Wheat grain contains 60–70% starch content and gradually drops under high temperature [63,64]. High temperature inhibits the starch accumulation into grains ascribable to the enzymes inactivity viz., granule bound starch, soluble starch, and sucrose synthase during the grain filling phase [65,66]. It also declines the starch content synthesis [67,68], stem reserves carbohydrates translocation [69,70], alters the structure of aleurone layer, and endosperm of seed [71,72], which ultimately influences grain development.

2.4. Leaf Senescence

Leaf senescence is the reduction in green leaf area during the reproductive phase due to the retardation in the chlorophyll content and carotenoids [73,74]. The chlorophyll content and carotenoid have an indispensable role in harvesting sunlight for photosynthesis [75]. High temperature disturbs the chloroplast integrity, leaf senescence, and ultimately photosynthesis in wheat [76].

Leaf senescence during the grain filling duration degrades the leaf chlorophyll content. Initially, chlorophyll-b is converted into chlorophyll-a during the chlorophyll cycle (Figure 2). The chlorophyllase enzyme catalyzes chlorophyll-a into chlorophyllide-a or pheophytin and subsequently into pheophorbide-a. Pheophorbide-a monooxygenase catalyzes the pheophorbide-a and is converted to red chlorophyll catabolites ensuing fluorescent and non-fluorescent chlorophyll catabolites [77,78]. A high temperature of 42 °C declines the enzymes efficiency viz., 5-aminolevulinate dehydrogenase (45%), mg-protoporphyrin IX methyltransferase (65%), protochlorophyllide oxidoreductase (70%), and increases chlorophyllase (46%) in wheat [79].

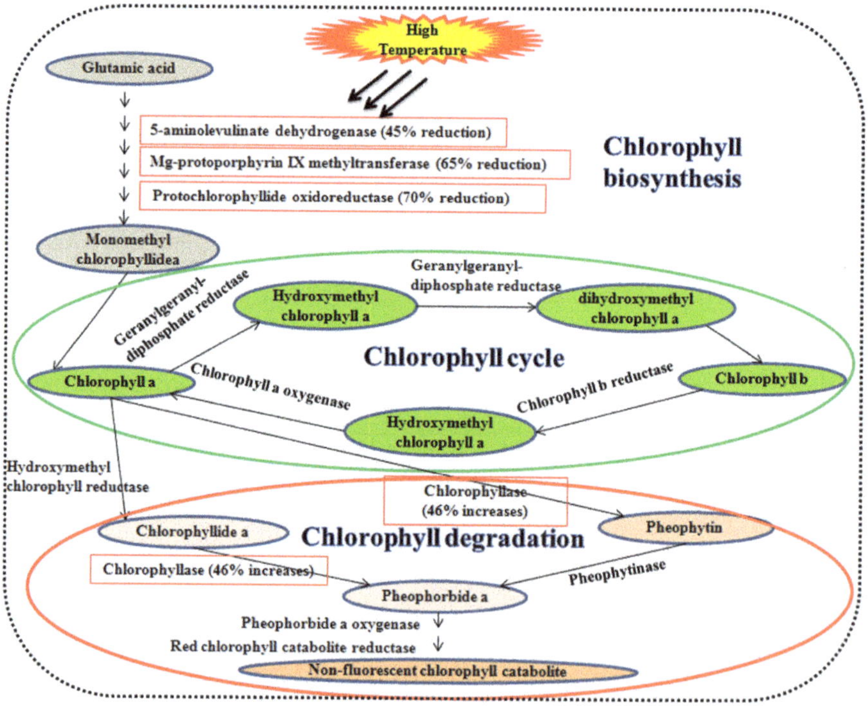

Figure 2. Impact on the high temperature of leaf senescence. Enzymes associated with chlorophyll synthesis viz., 5- aminolevulinate dehydrogenase, mg-protoporphyrin IX methyltransferase, and protochlorophyllide oxidoreductase, whereas chlorophyllase is responsible for chlorophyll degradation.

Chlorophyll deficiency reduces the absorbance of light energy and transfer to the reaction centers (RCs) of PS-II and PS-I at high temperature in wheat [80,81]. Carotenoids dissipate the excess light and protect the reaction centers against stress conditions [82]. Carotenoids viz., xanthophylls, and isoprene maintain the thylakoid membrane from leakage [83]. However, thylakoid components are sensitive at a temperature above 40 °C and inhibit the carotenoids biosynthesis pathways in the chloroplast [46,84], which interrupt the photosynthesis stability and ultimately reduce the grain yield in wheat [25].

2.5. Protein Quality

The protein content, protein quality, and glutenin/gliadin determine the backing quality of bakery products [85,86]. High temperature enhances the total protein content but reduces the end use of protein quality [87,88], which is more or less dependent on the grain protein concentration [89]. Protein fractions (albumin, globulin, gliadin, and glutenin) are important components for the end use quality of wheat grain [90]. High temperature at the grain filling duration decreases the albumin and globulin content [91], whereas it increases the gliadin content at the expense of glutenin content in wheat [92]. Furthermore, high temperature increases the protein content but reduces the production of glutenin, sedimentation index [71], and essential amino acids such as lysine, methionine, and tryptophan content, which determines the viscoelastic properties of wheat loaf [45].

2.6. Physiological Process

Heat stress inhibits the photosynthesis, damaging photosynthetic apparatus, and synthesis of ROS (reactive oxygen species) as discussed below.

2.6.1. Photosynthesis Response to High Temperature

A high temperature of 35/25 °C (day/night) at the grain filling duration inhibits the leaf photosynthesis up to 50% in wheat (Figures 3 and 4). The net photosynthesis during the wheat crop cycle is essential in controlling the crop biomass and grain yield under a high temperature. The optimum temperature for net photosynthesis is 20–30 °C, but a high temperature above 32 °C declines the photosynthetic rate rapidly in wheat [46]. The photosynthesis in wheat leaves is more sensitive than those, which are associated with the synthesis and mobilization of stem reserves into developing grains during grain filling. Photosynthesis is associated with the activity of photosynthetic apparatus, Rubisco (Ribulose bisphosphate carboxylase/oxygenase) enzyme, and various green organs of the plants such as chlorophyll content and carotenoids [76,93].

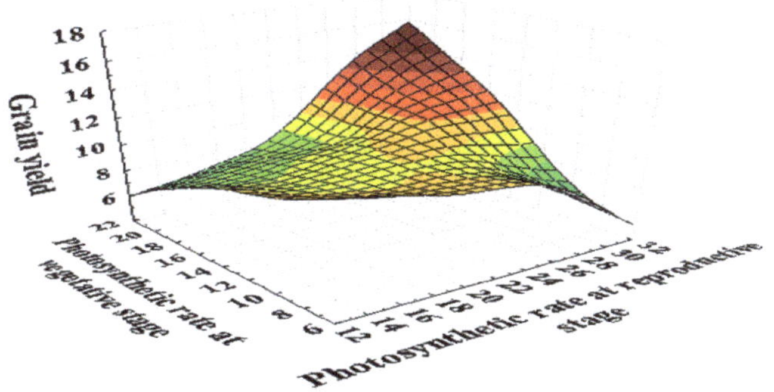

Figure 3. Photosynthetic (μmol m^{-2} s^{-1}) response at the seedling and reproductive stage of 180 wheat genotypes with the grain yield per plant (g). Photosynthetic rate was recorded on a clear day between 10:00 a.m. to 12:00 p.m. with the help of infrared gas analyzer (IRGA ADC, LCA-4, Hoddesdon, UK). Data collected under normal and heat stress conditions at the vegetative (Zadoks scale 39) and reproductive stages (Zadoks scale 69) [94]. It represented that the photosynthesis is directly associated with the grain yield at both stages. As the photosynthetic rate decreases, it reduces the grain yield in wheat.

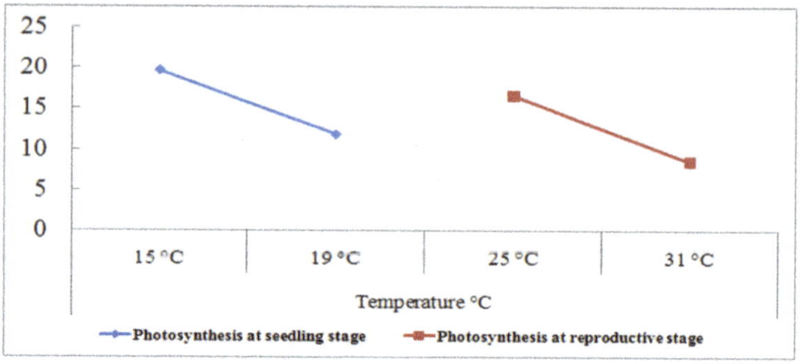

Figure 4. Photosynthetic (μmol m^{-2} s^{-1}) response of 180 wheat genotypes with the grain yield per plant (g) at the seedling and reproductive stages. Data collected under normal and heat stress (4–5 °C above normal) conditions [94].

2.6.2. Photosynthetic Apparatus

High temperature disturbs the photosystem-II (PS-II) and photosystem-I (PS-I) mediated electron transport chain (ETC). A high temperature of 35–40 °C at the grain filling phase directly damages the photosynthetic apparatus including the PS-II and PS-I mediated electron transport chain [46]. PS-II is a complex subunit of chlorophylls and proteins and is more sensitive than PS-I [73,95]. It harvests the light energy to oxidize the water molecule and transfer electrons to plastoquinone (PQ) ensuing the cytochrome b6f complex, but a high temperature declines the efficiency of PQ and Cytochrome b6f [96].

The light harvesting complex-II (LHC-II) is an assortment of proteins associated with the PS-II core complex. It harvests the sunlight energy and transfers it to the PS-II core complex to form multi-complex proteins [97]. High basal florescence separates the LHC-II from the PS-II core complex and alters the energy distribution to PS-I [98]. A high temperature of 32–38 °C also synthesizes the zeaxanthin, which destabilizes the thylakoid membrane composition and photosynthetic apparatus [48].

2.6.3. Rubisco Activity

Rubisco is an essential light activated enzyme, which possesses the binding sites for CO_2 and Rubisco activase for the regulation of the Calvin cycle, but its efficiency gradually declines under a high temperature of 25–40 °C in wheat [99]. Sugar phosphate inhibitors viz., XuBP (D-xylulose-1, 5-bisphosphate), RuBP (Ribulose-1, 5-bisphosphate), CA1P (2-Carboxy-D-arabinitol 1-phosphate), and CTBP (2-Carboxytetritol-1, 4-bisphosphate) impaired with the active site, which modulate the Rubisco activity for photosynthesis [100,101]. Rubisco activase removes these inhibitors from the active site and facilitates the carboxylation reaction modulated by the Rubisco enzyme [102]. It also protects the nascent proteins from aggregation but it is heat labile. Therefore, a high temperature of >32 °C alters the composition for the accessibility of carbamylation [103,104].

High temperature declines the solubility of CO_2 and enhances the O_2 level from the compensation point due to the reduction in evapotranspiration [105–107] and specificity of the Rubisco enzyme activity, which is poor in discriminating O_2 and CO_2 [108,109] (Figure 5). These factors stimulate the photorespiration and consume ATPs, release the fixed CO_2, and produce the photorespiratory metabolite (glyoxylate), which consume $NADH_2$ [110,111] and ultimately reduce the yield up to 20% in wheat [112].

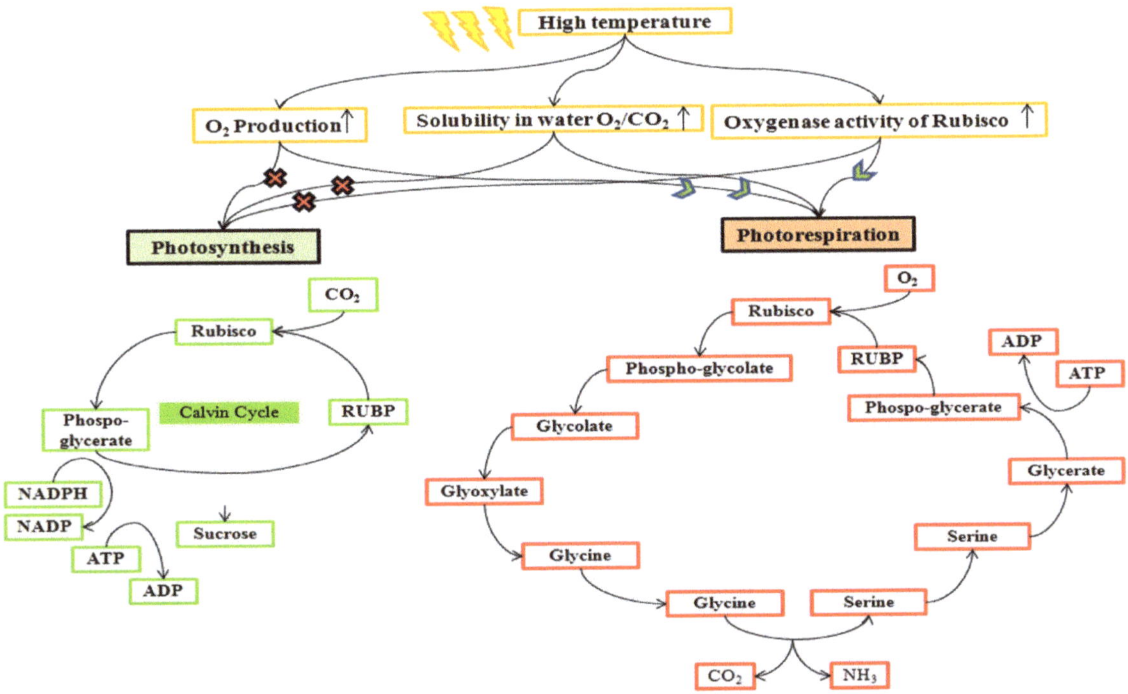

Figure 5. Rubisco enzyme activity pathway alteration at a high temperature. Rubisco has a characteristic of both oxygenation and carboxylation activities. High temperature increases the synthesis of oxygen through photosynthesis, which enhanced the solubility of oxygen than carbon dioxide. Therefore, it promotes the oxygenase activity of Rubisco and stimulates photorespiration, which compartmentalized in chloroplast, peroxisomes, and mitochondria.

2.6.4. Reactive Oxygen Species

Reactive oxygen species (ROS) are synthesized during the malfunction of PS-II and the Calvin cycle of photosynthesis [113] causing lipid per-oxidation and cell membrane damage in wheat [114,115]. ROS such as super oxides (O^{-2}), hydroxyl radical (OH^{-}), and hydrogen peroxide (H_2O_2) commonly synthesize at high temperatures. The manganese superoxide dismutase (Mn-SOD) catalysis in mitochondria produces hydrogen peroxides, whereas the auto-oxidation of ubisemiquinone complex-I and complex-III generates super oxides radicals ensuing the oxidative stress in the cell, as well as DNA damage, protein modification, and membrane instability [48,116].

Super oxides synthesize by the reduction of one electron, whereas further electrons reduction generates peroxide, which is neutralized by two protons of hydrogen atom and form H_2O_2, as shown in Figure 6. Hydrogen peroxide is produced by incomplete water molecules oxidation, which is reduced by the manganese to form the hydroxyl radical [117]. The hydrogen peroxide concentration gradually increases from vegetative to milky dough stage at a high temperature and negatively influences the photosynthesis [118].

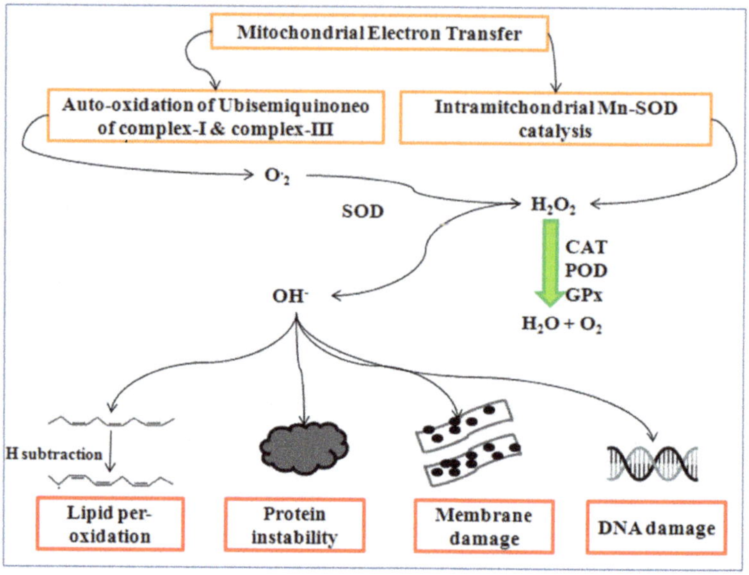

Figure 6. Synthesis of the reactive oxygen species and their consequences.

3. Tolerance Mechanism against High Temperature

The plant's tolerance to high temperature facilitates adaptation in adverse conditions through maintaining their physiology and ameliorate grain yield.

3.1. Phytohormones and Bioregulators

Phytohormones inevitably associated with the antioxidant enzymes activity and growth regulation under heat stress conditions [119]. Phytohormones viz., proline, glycine betaine, salicylic acid, abscisic acid, and ethylene maintain the physiology at a high temperature through soluble salts accumulation in the cell and reducing H_2O_2 production in wheat, as displayed in Figure 7.

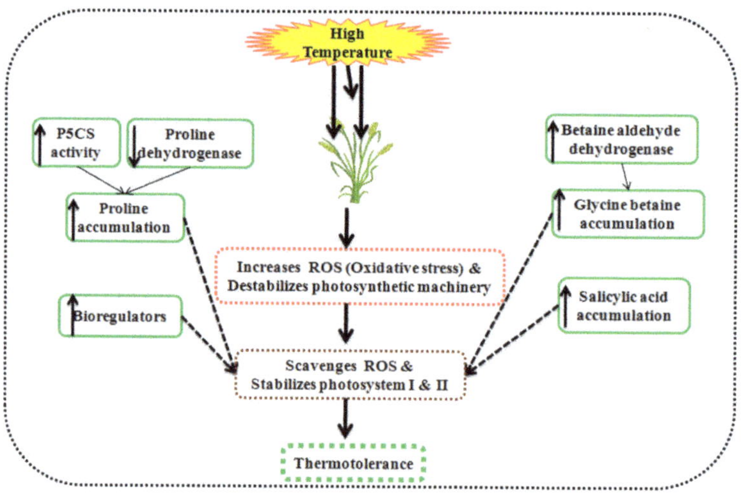

Figure 7. Schematic illustration of osmolytes associated with thermotolerance in wheat.

A high temperature of 30–35 °C discolorizes the chlorophyll, beta-carotene, and damages the photochemical activity. Glycine betaine accumulates in the chloroplast of leaves and stabilizes PS-II, reaction centers in the thylakoid membrane [120,121], Rubisco enzyme, and inhibits the ROS production [122]. It adjusts the osmotic pressure, ameliorate antioxidant enzymes activity, and photosynthesis under high temperature in wheat [123]. Salicylic acid acts as a phenolic hormone in plants and is responsible for osmoregulation, scavenges ROS, and maintains the photosynthesis in wheat [124]. It also triggers the osmolytes synthesis viz., glycine betaine, proline, and sugars under heat stress conditions [125–127].

Proline accumulation is determined by the proline dehydrogenase activity and Δ1-pyrroline-5-carboxylate synthetase/reductase (P5CS) [128]. High temperature increases the *P5CS* and decreases proline dehydrogenase in tolerant wheat seedlings. Proline dehydrogenase catalyzes the proline degeneration in mitochondria. However, glutamate acts as a precursor in the presence of *P5CS1* for the proline synthesis and accumulates in plant under heat stress conditions [129,130]. The proline content is directly linked to a high temperature of 35–40 °C and ameliorates the defense mechanism in wheat seedlings [131]. A high temperature of 35 °C than 25 °C accumulates a higher proline content (up to 200%) and improves the photosynthetic efficiency and yield [132].

Bioregulators upregulate the antioxidant defense mechanism and maintains the PS-II under high temperature. Foliar application during the grain filling period and seed priming with a 6.6 mM solution of thiourea intensifies the antioxidant enzyme activity, chlorophyll content, total soluble protein, amino acid, and grain weight in wheat [133]. Foliar application of 50 ppm dithiothreitol also ameliorates the adverse effect of high temperature in wheat [134].

3.2. *Stay Green*

Stay green represents the chlorophyll retention and longevity of photosynthetic apparatus for the adaptation of wheat under high temperature [135–137]. Stay green associated with the stabilized photosynthetic apparatus of chloroplast viz., scavenges of ROS, and maintaining the photosynthetic apparatus indicates the slow degeneration of tissues in wheat.

The stay green trait has the potential to protect photosystem-II in the chloroplast and inhibits the ROS synthesis in wheat [138,139]. It maintains the green pigment at a high temperature of >30 °C during the grain filling phase. The short grain filling duration and high canopy temperature are associated with non-stay green genotypes in wheat [140]. Stay green is positively associated with the normal grain filling phase, membrane stability, photosynthesis, stem reserve carbohydrates, and grain development [141,142].

Chlorophyll biosynthesis enzymes determine the senescence in wheat, which influences the assimilation and translocation of photosynthates into grains during grain filling [37,143]. For example, the *SGR* mutant of Arabidopsis and rice exhibit the stay green phenotype due to the suppression of Mg dechelatase enzyme, which is responsible for chlorophyll degradation [144]. *SGR* mutants have also been reported in other species viz., pea, tomato, and pepper [142]. The *NYC* gene suppression also delays the senescence of crops that catalyzed the chlorophyll breakdown for the conversion of chlorophyll-b into chlorophyll-a [145]. The *PPH* genes mutant removes the phytol from phaeophytin in Arabidopsis and results in stay green [146]. Genes *NYC*, *PPH*, and *SGR* have a potential role for stay green in arbidopsis and rice that must be explored in wheat to improve thermotolerance.

3.3. *Antioxidant Enzymes*

Antioxidant enzymes protect the plant from ROS, convert the free radicals of oxygen and hydroxyl into H_2O_2 followed by the water molecule. These enzymes scavenge the ROS, balance the production/elimination of ROS from oxidative stress, maintain the growth, development, metabolism, and overall productivity [147]. Antioxidant enzymes viz., POD (peroxidase), SOD (superoxide dismutase), CAT (catalase), and GR (glutathione reductase) usually generate under a high temperature of 35/28 °C day/night in wheat [147–149].

The SOD enzyme converts the O^{-2} to H_2O_2, which is a less toxic form than the free radicals [150,151]. CAT and POD convert H_2O_2 into H_2O, but the CAT activity is higher than other antioxidant enzymes in wheat [152,153]. CAT reduces several millions of H_2O_2 molecules into H_2O and oxygen per minute [154,155]. GR protects the plant from oxidative stress by reducing oxidized glutathione [156,157]. Glutathione peroxidase (GPx) efficiency depends on high γ-glutamyl cysteine synthetase and glutathione synthetase activity for the reduction of H_2O_2 into H_2O [158].

3.4. Heat Shock Proteins

Wheat plant produces heat shock proteins (HSPs) at 32–34 °C and provides protection against high temperature [159,160]. High temperature disturbs the membrane proteins in plants but upregulates the translation of heat shock genes, which encodes for HSPs [132,161,162]. These HSPs protect the cell from adverse effects of heat stress by maintaining photosynthesis, upregulation of other proteins, and cell metabolism [163]. There are different families of ATP dependent HSPs viz., HSP60, HSP70, HSP90, and HSP100 except small HSPs based on molecular weight.

The small HSP (smHSP) in wheat genome assembles with other homo-oligomers and facilitates binding in ATP independent manners. It assembles with HSP90 to prevent unfolding and refolding of proteins under high temperature [159,160]. HSP60 expresses constitutively in chloroplast and mitochondria [164,165]. The Rubisco large subunit binding protein (chHSP60) is a cofactor of HSP60, which regulates the Rubisco enzyme folding at high temperature [166].

HSP70 is a highly conserved protein, which recognizes only a short sequence of the polypeptide chain, temporal and inhibits aggregation of non-native protein at high temperature [167]. HSP110 is a sub family of HSP70 and inhibits the aggregation with a greater capability than HSP70 [168]. HSP90 regulates transcription, cellular signaling, and managing protein folding through assembling molecular proteins including HSP40 and HSP70 [118,168,169], whereas HSP100 interacts with different smHSPs and HSP70 to prevent the aggregation of protein [170].

4. Tolerance Strategies against High Temperature

Strategies against heat stress viz., crop management, conventional, non-conventional, and molecular approaches ameliorate the thermotolerance in wheat. These strategies are further elaborated below.

4.1. Crop Management

Agronomic practices including seed priming, organic mulches, inorganic fertilizers, and timely sowing with recommended management practices mitigate the heat stress in wheat. Wheat seed priming in the aerated solution of $CaCl_2$ (1.2%) for 12 h improves the germination, growth, leaf area index, chlorophyll content, assimilation rate, and grain yield [171–173]. Mulching with rice husk conserves water, improves water use efficiency, maintains the water status in soil, and slows down the release of nitrogen for plant uptake [174,175].

The application of inorganic fertilizers viz., nitrogen, and potassium maintain the chlorophyll content, osmoregulation, cytokinin biosynthesis, protein stability, redox homeostasis, and photosynthesis at high temperature [25,176]. Zinc improves the superoxide dismutase activity, membrane integrity, chlorophyll content, chlorophyll florescence, and kernel growth at high temperature [27,177]. The silicon application at 10 mg/kg of soil at heading improves the osmotic potential (26%), photosynthetic rate (21%), catalase activity (38%), superoxide dismutase activity (35%), stomatal conductance (20%), and transpiration rate (32%) in wheat under high temperature [178,179].

Sowing time is a counteract strategy against high temperature. Delayed planting compels the plants to complete their growing degree days earlier, but they have to face high temperature during the anthesis and grain filling phase [53,180]. Wheat planted in normal sowing dates utilizes a longer duration to capture the available reserves/carbohydrates and improve the grain development [70,181,182].

4.2. Conventional Approaches for Thermotolerance

Thermotolerance is an inherited component stabilizing economic yield against heat stress. Tolerance to high temperature is characterized as the least effect on growth, development, and productivity. Screening of wheat genotypes is difficult in a spatial environment under natural heat stress conditions. This is due to the consistent selection criteria that have not been developed to screen diverse germplasm. The selection criteria based on traits directly associated with the grain yield facilitates better improvement in the genetic material for thermotolerance (Table 2).

Breeding has made considerable advances in the genetic basis, diversity, and development of thermotolerant varieties. However, utilization and explorations of novel genetic diversity facilitates the genetic improvement for thermotolerance in the breeding program. However, the genetic gain is limited due to the narrow genetic basis [183,184]. Therefore, utilizing landraces and wild relatives increases the genetic variation in wheat for developing thermotolerance. Breeding for thermotolerance utilizing land races and wild relatives viz., *Aegilops speltoides, Aegilops tauschii, Triticum turgidum,* and *Triticum durum* have the ability to maintain chlorophyll content, canopy temperature depression, membrane stability, and photosynthesis under stress conditions [74,185–187].

Table 2. Major desirable selection criteria for the screening heat tolerance in wheat.

Traits	References
Cell membrane stability	[50,51,188,189]
Proline content	[131,190–192]
Heat susceptibility index for grain yield	[25,193–196]
Chlorophyll content	[76,131,188,189,197,198]
Photosynthesis	[48,106,107,117,199]
Stay green	[70,137,140,142,143,200]
Grain filling duration	[70,181,201]
Grains formation	[59,67,202–204]
Grain development	[67,71,203–205]
Early heading	[46,64,204,206,207]
Canopy temperature depression	[30,140,201,208–212]

4.3. Non-Conventional Approaches

Plants development utilizing genetic engineering or the indirect selection of traits through molecular markers or omics technology facilitates the improvement against heat stress in wheat.

4.3.1. Biotechnological Approach and Heat Shock Factors

Genetic engineering is the development of cultivar through incorporating the individual gene [213]. Advances in biotechnology enable the faster genetic gain than conventional breeding methods. Several genes encoding heat shock factors have been identified in wheat, but novel genes identification for thermotolerance remains a challenge (Table 3). The identification of novel genes and their altered expression under high temperature in wheat crop provides the molecular basis for improving thermotolerance.

Table 3. List of genes encoding transcription factors related to thermotolerance.

TFs/Genes	Source	Function	Reference
Hsf6A/wheat	HVA1s promoter of barley	Regulation of heat shock protein genes and improve thermotolerance	[214]
EF-Tu	Ubiquitin 1 promoter of maize	Overexpression reduces the thermal aggregation of leaf proteins, photosynthetic membrane, and increases CO_2 fixation	[215]
HvSUT1	Hordein B1 promoter of barley	Increase sucrose transport into grains	[216]
TaFER-5B	Ubiquitin 1 of maize	Reduces oxidative stress by scavenging ROS and improves leaf iron content	[217]
TaGASR1	Wheat variety TM107	Reduces ROS and hormonal signal transduction pathway	[218]
TaHsfC2a	Monocot-specific HsfC2 subclass	Thermotolerance development via the ABA-mediated regulatory pathway	[219]
TaHSP23.9	Chinese wheat based on proteomic analysis	Upregulation under heat stress facilitates in seed development during the grain filling phase	[220]
TaFBA1	F-box gene from wheat	Upregulation improves photosynthesis and the antioxidant enzyme activity	[221]
TaHsfA2-1	Wheat	Overexpression of heat shock proteins and chlorophyll content	[222]
SGR	Arbidopsis and rice	Binding of light harvesting complex during photosystem-II	[142,144]
NYC	Arbidopsis and rice	Responsible for the activity of chlorophyll reductase to convert chl-b into chl-a	[142,145]
PPH	Arbidopsis and rice	Responsible for the activity pheophytinase for dephytylation to phaeophytin	[142,146]

Quantitative Trait Loci (QTL)

Heat tolerance is under polygenic control and the QTL analysis enlightens the genetic basis of thermotolerance in wheat. It facilitates the indirect selection of traits rather than the selection based on phenotypic traits. Many QTLs have been identified for physio-morphic traits in wheat, but few were identified against heat stress (Table 4), which facilitates in gene pyramiding and marker assisted selection in wheat breeding programs for developing thermotolerance.

Table 4. Major quantitative trait loci (QTL) identified for traits against heat stress.

Traits	Chromosome	References
Chlorophyll content	*2A, 3A, 6A, 7A, 2B, 5B, 2D*	[223,224]
Chlorophyll florescence	*1A, 2A, 3A, 3B, 2D, 1D*	[224,225]
Plasma membrane damage	*7A, 2B*	[223]
Thylakoid membrane damage	*6A, 7A, 1D*	[223]
Canopy temperature	*7A, 1B, 2B, 3B*	[226]
Grain weight	*1A, 2A, 4A, 1B, 2B, 3B, 4B, 6B, 6D*	[226–228]
Grains formation	*1A, 4A, 2B, 3B, 5B*	[228,229]
Chlorophyll florescence	*1A, 4A, 1B, 2B, 7D*	[230]
Senescence	*2A, 3A, 6A, 7A, 3B, 6B*	[231,232]
Stay green	*1A, 3B, 7D*	[233,234]

4.3.2. Omics Technology

Omics techniques facilitate the development of thermotolerance in wheat through the identification of transcriptional, translational, and post translational mechanisms (Figure 8). Transcriptomics represent the alteration in transcriptome factors under different environmental conditions through the DNA microarray technology [235,236]. It has already been used to study the glumes [237], grain development [238], and quality traits [239] for the identification of candidate gene expression under heat stress conditions. MicroRNAs (miRNAs) are non-coding small RNA that serve as the regulation of post-transcriptional gene expression in plants. Micromics assist in the candidate miRNA identification and their role in transcriptome homeostasis, developmental, and cellular tolerance of plants under high temperature [197].

Proteomics is the analysis of candidate proteins, the expression when they translated from mRNA to functional proteins, and a further characterization of their role in the heat tolerance mechanism [240]. Proteomic analysis revealed heat shock proteins, protein synthesis, detoxification, photosynthesis, and protein quality under heat stress conditions [115,241–245]. Hence, the omics technology provides us with a novel opportunity for the identification of genes, their expression, and pathways linked to these genes. However, the further genetic network and their component identification will be a challenge to adapt plants in a high temperature environment.

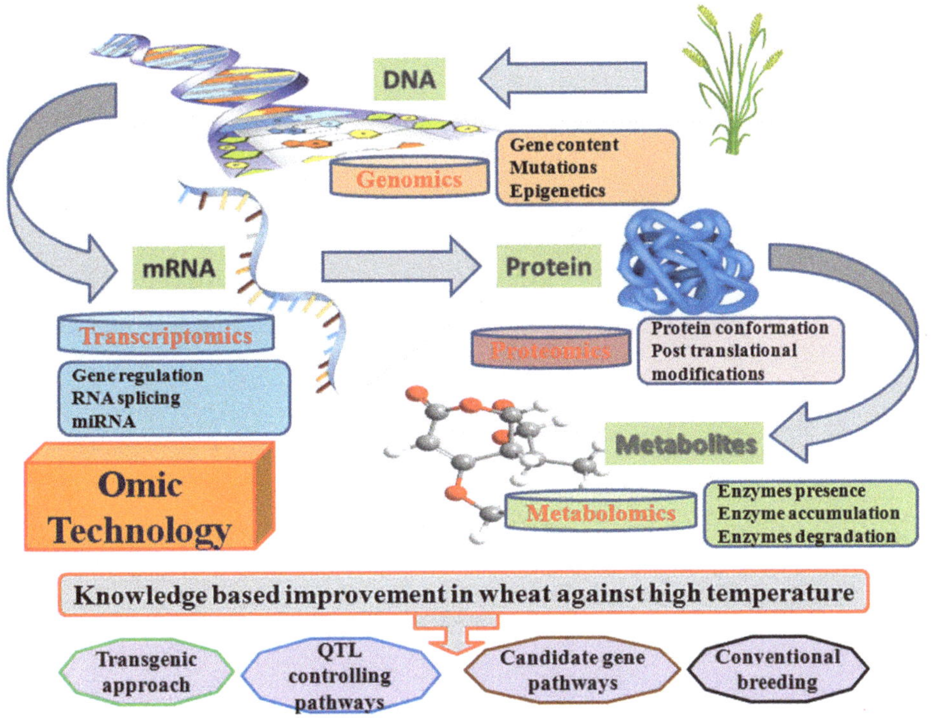

Figure 8. Schematic diagram representing the omics techniques associated with thermotolerance development in wheat at the molecular genetics level.

5. Conclusions and Future Prospects

Temperature is gradually increasing and affecting crop productivity. The impact of high temperature on wheat crop has been extensively studied, but understanding the mechanism of thermotolerance remains elusive. High temperature disrupts membrane stability, declines grain filling duration, grain formation, and starch accumulation into grains. Inhibition in the physiological process has been observed due to the high temperature stress. It disturbs the photosynthetic apparatus and generates the reactive oxygen species leading towards oxidative stress. The strategy against high temperature requires systematically understanding the physiological, metabolic, and development process associated with thermotolerance. The tolerance mechanism including more accumulation of proline, glycine betaine, antioxidant enzyme activity, heat shock protein, and stay green could be a useful indicator for thermotolerance.

Crop management stabilizes the physiological process and metabolic pathways through mulches, extra irrigation, inorganic fertilizers, early sowing, exogenous application of micronutrient, osmoprotectants, and bioregulators. Integrating crop management practices with molecular genetics tools can ameliorate the adverse effects of high temperature, but need to further explore the strategies associated with high yield under heat stress [246–258]. The tolerance development can be achieved through a selection based on thermotolerant traits from existing germplasm and breeding utilizing land races and their wild relatives. The suitable selection criteria based on thermotolerant traits requires developing germplasm against heat stress. Recently, the canopy temperature depression at the reproductive stage, grain filling duration, heat susceptibility index for grain yield, and stay green have been established for screening germplasm against heat stress conditions. Stay green with other useful traits provide the solution of the burning problem due to the high temperature in wheat. Therefore, the contribution of the synthesis of chlorophyll turnover equation in photosynthesizing leaves for the stay green trait expression has a good future against high temperature stress.

The marker assisted breeding programs must be pooled with the transgenic approach for thermotolerance QTLs and genes. Understanding the QTLs and omics techniques pave the way to develop thermotolerance in wheat, but a further understanding of the genes network and their regulation of expression related to high temperature would be a challenge. There is a need to understand the molecular and biochemical basis of thermotolerance from the upcoming changing climate for crop improvement. Functional genomics also proved to be supportive against high temperature, but the alteration in transcriptomes and proteomes needs to be further investigated against high temperature. Noteworthy, molecular and genetic approaches facilitate crop adaptability coupled with the economic yield under high temperature, but the expression of yield potential requires the estimation of yield at the crop level. Therefore, the application of incorporating a future scenario into crop models provides model-based recommendations to improve thermotolerance in wheat.

Author Contributions: The research was designed by M.A. (Munir Ahmad) and M.A. (Mukhtar Ahmed); M.A. (Munir Ahmad), M.A. (Mukhtar Ahmed), A.K., and M.I.H. performed the review; A.K. analyzed the data with input from M.A. (Munir Ahmad) and M.A. (Mukhtar Ahmed); A.K., M.A. (Mukhtar Ahmed), and M.A. (Munir Ahmad) wrote the review. All authors have read and agreed to the published version of the manuscript.

Funding: This research was funded by Higher Education Commission (HEC) Pakistan grant number NRPU and the APC was funded through Discount Voucher: 1a55b19f0b99ca1b (1600.00 CHF) given to Mukhtar Ahmed for his review services to MDPI journals.

Data Availability Statement: The data presented in this study are available in this study as well as in the PhD thesis of Adeel Khan i.e., first author.

Acknowledgments: All authors are thankful to the Higher Education Commission (HEC) Pakistan, for the financial support of this study. The authors are thankful to all the valuable reviewers who putted lot of efforts to improve this review work. Furthermore, the authors dedicate this work to the COVID-19 patients, front line doctors, paramedic staff, and all others who are directly or indirectly fighting against this disease.

Conflicts of Interest: The authors declare no conflict of interest.

References

1. Knox, J.; Daccache, A.; Hess, T.; Haro, D. Meta-analysis of climate impacts and uncertainty on crop yields in Europe. *Environ. Res. Lett.* **2016**, *11*, 113004. [CrossRef]
2. Xiao, D.; Bai, H.; Liu, D.L. Impact of future climate change on wheat production: A simulated case for China's wheat system. *Sustainabilty* **2018**, *10*, 1277. [CrossRef]

3. IPCC. Climate Change. Mitigation of Climate Change. Contribution of Working Group III to the Fifth Assessment Report of the Intergovernmental Panel on Climate Change. 2014. Available online: https://www.ipcc.ch/site/assets/uploads/2018/02/ipcc_wg3_ar5_full.pdf (accessed on 25 February 2019).
4. Portner, H.O.; Roberts, D.C.; Masson-Delmotte, V.; Zhai, P.; Tignor, M.; Poloczanska, E.; Mintenbeck, K.; Alegría, A.; Nicolai, M.; Okem, A.; et al. *IPCC Special Report on the Ocean and Cryosphere in a Changing Climate*; Intergovernmental Panel on Climate Change (IPCC): Geneva, Switzerland, 2019.
5. Grace, M.A.; Achick, T.-F.E.; Bonghan, B.E.; Bih, M.E.; Ngo, N.V.; Ajeck, M.J.; Prudence, G.T.B.; Ntungwen, F.C. An overview of the impact of climate change on pathogens, pest of crops on sustainable food biosecurity. *Int. J. Ecotoxicol. Ecobiol.* **2019**, *4*, 114–119. [CrossRef]
6. Porfirio, L.L.; Newth, D.; Finnigan, J.J.; Cai, Y. Economic shifts in agricultural production and trade due to climate change. *Palgrave Commun.* **2018**, *4*, 1–9. [CrossRef]
7. Skytt, T.; Nielsen, S.N.; Jonsson, B.G. Global warming potential and absolute global temperature change potential from carbon dioxide and methane fluxes as indicators of regional sustainability-A case study of Jamtland, Sweden. *Ecol. Ind.* **2020**, *110*, 105831. [CrossRef]
8. Pareek, A.; Dhankher, O.P.; Foyer, C.H. Mitigating the impact of climate change on plant productivity and ecosystem sustainability. *J. Exp. Bot.* **2020**, *71*, 451–456. [CrossRef]
9. Schlenker, W.; Roberts, M.J. Nonlinear temperature effects indicate severe damages to US crop yields under climate change. *Proc. Natl. Acad. Sci. USA* **2009**, *106*, 15594–15598. [CrossRef]
10. Conaty, W.; Burke, J.; Mahan, J.; Neilsen, J.; Sutton, B. Determining the optimum plant temperature of cotton physiology and yield to improve plant-based irrigation scheduling. *Crop Sci.* **2012**, *52*, 1828–1836. [CrossRef]
11. Sawan, Z.M. Cotton production and climatic factors: Studying the nature of its relationship by different statistical methods. *Cogent Biol.* **2017**, *3*, 1292882. [CrossRef]
12. Fromme, D.D. Effect of an Upper Temperature Threshold on Heat Unit Calculations, Defoliation Timing, Lint Yield, and Fiber Quality in Cotton. *J. Cotton Sci.* **2014**, *18*, 108–121.
13. Jans, Y.; von Bloh, W.; Schaphoff, S.; Muller, C. Global cotton production under climate change-implications for yield and water consumption. *Hydrol. Earth Syst. Sci. Discuss.* **2020**, 1–27. [CrossRef]
14. Wu, Y.C.; Chang, S.J.; Lur, H.S. Effects of field high temperature on grain yield and quality of a subtropical type japonica rice-Pon-Lai rice. *Plant Prod. Sci.* **2016**, *19*, 145–153. [CrossRef]
15. Rathnayake, W.; De Silva, R.; Dayawansa, N. Assessment of the suitability of temperature and relative humidity for rice cultivation in rainfed lowland paddy fields in Kurunegala district. *Trop. Agric. Res.* **2016**, *27*, 370–388. [CrossRef]
16. Hatfield, J.L.; Prueger, J.H. Temperature extremes: Effect on plant growth and development. *Weather Clim. Extrem.* **2015**, *10*, 4–10. [CrossRef]
17. Yang, Z.; Zhang, Z.; Zhang, T.; Fahad, S.; Cui, K.; Nie, L.; Peng, S.; Huang, J. The effect of season-long temperature increases on rice cultivars grown in the central and southern regions of China. *Front. Plant Sci.* **2017**, *8*, e01908. [CrossRef]
18. Mukamuhirwa, A.; Persson Hovmalm, H.; Bolinsson, H.; Ortiz, R.; Nyamangyoku, O.; Johansson, E. Concurrent drought and temperature stress in rice-a possible result of the predicted climate change: Effects on yield attributes, eating characteristics, and health promoting compounds. *Int. J. Environ. Res. Public Health* **2019**, *16*, 1043. [CrossRef]
19. Kumar, R.R.; Pathak, H.; Sharma, S.K.; Kala, Y.K.; Nirjal, M.K.; Singh, G.P.; Goswami, S.; Rai, R.D. Novel and conserved heat-responsive microRNAs in wheat (*Triticum aestivum* L.). *Funct. Integr. Genom.* **2015**, *15*, 323–348. [CrossRef]
20. Wiebbecke, C.E.; Graham, M.A.; Cianzio, S.R.; Palmer, R.G. Day temperature influences the male-sterile locus ms9 in soybean. *Crop Sci.* **2012**, *52*, 1503–1510. [CrossRef]
21. Prasad, P.V.; Boote, K.J.; Allen Jr, L.H. Adverse high temperature effects on pollen viability, seed-set, seed yield and harvest index of grain-sorghum [*Sorghum bicolor* (L.) Moench] are more severe at elevated carbon dioxide due to higher tissue temperatures. *Agric. Forest Meteorol.* **2006**, *139*, 237–251. [CrossRef]
22. Crafts-Brandner, S.J.; Salvucci, M.E. Sensitivity of photosynthesis in a C4 plant, maize, to heat stress. *Plant Physiol.* **2002**, *129*, 1773–1780. [CrossRef]
23. Zhao, C.; Liu, B.; Piao, S.; Wang, X.; Lobell, D.B.; Huang, Y.; Huang, M.; Yao, Y.; Bassu, S.; Ciais, P. Temperature increase reduces global yields of major crops in four independent estimates. *Proc. Natl. Acad. Sci. USA* **2017**, *114*, 9326–9331. [CrossRef] [PubMed]
24. Day, L. Proteins from land plants–potential resources for human nutrition and food security. *Trends Food Sci. Tech.* **2013**, *32*, 25–42. [CrossRef]
25. Akter, N.; Islam, M.R. Heat stress effects and management in wheat. A review. *Agron. Sustain. Dev.* **2017**, *37*, 1–19. [CrossRef]
26. Hussain, J.; Khaliq, T.; Ahmad, A.; Akhter, J.; Asseng, S. Wheat responses to climate change and its adaptations: A focus on arid and semi-arid environment. *Int. J. Environ. Res.* **2018**, *12*, 117–126. [CrossRef]
27. Nuttall, J.G.; Barlow, K.M.; Delahunty, A.J.; Christy, B.P.; O'Leary, G.J. Acute high temperature response in wheat. *Agron. J.* **2018**, *110*, 1296–1308. [CrossRef]
28. Asseng, S.; Ewert, F.; Martre, P.; Rotter, R.P.; Lobell, D.; Cammarano, D.; Kimball, B.; Ottman, M.; Wall, G.; White, J.W. Rising temperatures reduce global wheat production. *Nat. Clim. Chang.* **2015**, *5*, 143–147. [CrossRef]
29. Lobell, D.B.; Burke, M.B.; Tebaldi, C.; Mastrandrea, M.D.; Falcon, W.P.; Naylor, R.L. Prioritizing climate change adaptation needs for food security in 2030. *Science* **2008**, *319*, 607–610. [CrossRef]

30. Pask, A.; Joshi, A.; Manes, Y.; Sharma, I.; Chatrath, R.; Singh, G.; Sohu, V.; Mavi, G.; Sakuru, V.; Kalappanavar, I. A wheat phenotyping network to incorporate physiological traits for climate change in South Asia. *Field Crops Res.* **2014**, *168*, 156–167. [CrossRef]
31. Tack, J.; Barkley, A.; Nalley, L.L. Effect of warming temperatures on US wheat yields. *Proc. Natl. Acad. Sci. USA* **2015**, *112*, 6931–6936. [CrossRef]
32. Mondal, S.; Singh, R.P.; Mason, E.R.; Huerta-Espino, J.; Autrique, E.; Joshi, A.K. Grain yield adaptation and progress in breeding for early-maturing and heat-tolerant wheat lines in South Asia. *Field Crops Res.* **2016**, *192*, 78–85. [CrossRef]
33. Farooq, M.; Bramley, H.; Palta, J.A.; Siddique, K.H. Heat stress in wheat during reproductive and grain-filling phases. *Crit. Rev. Plant Sci.* **2011**, *30*, 491–507. [CrossRef]
34. Ortiz, R.; Braun, H.J.; Crossa, J.; Crouch, J.H.; Davenport, G.; Dixon, J.; Joshi, A.K. Wheat genetic resources enhancement by the International Maize and Wheat Improvement Center (CIMMYT). *Genet. Resour. Crop Evol.* **2008**, *55*, 1095–1140. [CrossRef]
35. Tahir, M.; Ali, A.; Nadeem, M.A.; Hussain, A.; Khalid, F. Effect of different sowing dates on growth and yield of wheat (*Triticum aestivum* L.) varieties in District Jhang, Pakistan. *Pak. J. Life Soc. Sci.* **2009**, *7*, 66–69.
36. Zhao, H.; Dai, T.; Jiang, D.; Cao, W. Effects of high temperature on key enzymes involved in starch and protein formation in grains of two wheat cultivars. *J. Agron. Crop Sci.* **2008**, *194*, 47–54. [CrossRef]
37. Zhao, H.; Dai, T.B.; Jing, Q.; Jiang, D.; Cao, W.X. Leaf senescence and grain filling affected by post-anthesis high temperatures in two different wheat cultivars. *Plant Growth Regul.* **2007**, *51*, 149–158. [CrossRef]
38. Spiertz, J.H.J.; Hamer, R.J.; Xu, H.; Martin, C.P.; Don, C.; Putten, P.E.L.V. Heat stress in wheat (*Triticum aestivum* L.) effects on grain growth and quality traits. *Eur. J. Agron.* **2006**, *25*, 89–95. [CrossRef]
39. Acevedo, E.; Silva, P.; Silva, H. Wheat growth and physiology. In *Bread Wheat*; Curtis, B.C., Ed.; FAO Plant Production and Protection Series: Rome, Italy, 2002.
40. Stone, P.J.; Nicolas, M.E. Wheat cultivars vary widely in their responses of grain-yield and quality to short periods of postanthesis heat-stress. *Aust. J. Plant Physiol.* **1994**, *21*, 887–900.
41. Asseng, S.; Foster, I.; Turner, N.C. The impact of temperature variability on wheat yields. *Glob. Chang. Biol.* **2011**, *17*, 997–1012. [CrossRef]
42. Buriro, M.; Oad, F.C.; Keerio, M.I.; Tunio, S.; Gandahi, A.W.; Hassan, S.W.U.; Oad, S.M. Wheat seed germination under the influence of temperature regimes. *Sarhad J. Agric.* **2011**, *27*, 539–543.
43. Tahir, I.; Nakata, N.; Yamaguchi, T.; Nakano, J.; Ali, A. Influence of high shoot and root-zone temperatures on growth of three wheat genotypes during early vegetative stages. *J. Agron. Crop Sci.* **2008**, *194*, 141–151. [CrossRef]
44. Porter, J.R.; Gawith, M. Temperatures and the growth and development of wheat: A review. *Eur. J. Agron.* **1999**, *10*, 23–36. [CrossRef]
45. Nuttall, J.; O'Leary, G.; Panozzo, J.; Walker, C.; Barlow, K.; Fitzgerald, G. Models of grain quality in wheat-A review. *Field Crops Res.* **2017**, *202*, 136–145. [CrossRef]
46. Narayanan, S. Effects of high temperature stress and traits associated with tolerance in wheat. *Open Access J. Sci.* **2018**, *2*, 177–186. [CrossRef]
47. Mukherjee, A.; Wang, S.-Y.S.; Promchote, P. Examination of the climate factors that reduced wheat yield in northwest India during the 2000s. *Water* **2019**, *11*, 343. [CrossRef]
48. Djanaguiraman, M.; Boyle, D.; Welti, R.; Jagadish, S.; Prasad, P. Decreased photosynthetic rate under high temperature in wheat is due to lipid desaturation, oxidation, acylation, and damage of organelles. *BMC Plant Biol.* **2018**, *18*, 55–63. [CrossRef] [PubMed]
49. Bita, C.E.; Gerats, T. Plant tolerance to high temperature in a changing environment scientific fundamentals and production of heat stress-tolerant crops. *Front. Plant Sci.* **2013**, *4*, 273–286. [CrossRef]
50. Niu, Y.; Xiang, Y. An overview of biomembrane functions in plant responses to high-temperature stress. *Front. Plant Sci.* **2018**, *9*, 915–923. [CrossRef]
51. ElBasyoni, I.; Saadalla, M.; Baenziger, S.; Bockelman, H.; Morsy, S. Cell membrane stability and association mapping for drought and heat tolerance in a worldwide wheat collection. *Sustainaibilty* **2017**, *9*, 1606. [CrossRef]
52. Khan, S.U.; Gurmani, A.R.; Qayyum, A.; Khan, H. Heat tolerance evaluation of wheat (*Triticum aestivum* L.) genotypes based on some potential heat tolerance indicators. *J. Chem. Soc. Pak.* **2013**, *35*, 647–653.
53. Aslam, M.A.; Ahmed, M.; Stöckle, C.O.; Higgins, S.S.; Hayat, R. Can growing degree days and photoperiod predict spring wheat phenology. *Front. Environ. Sci.* **2017**, *5*, 57–68. [CrossRef]
54. Prajapat, A.L.; Saxena, R. Thermal requirements of wheat (*Triticum aestivum* L.) cultivars under different growing environments. *Int. J. Chem. Stud.* **2018**, *6*, 17–22.
55. Dubcovsky, J.; Loukoianov, A.; Fu, D.; Valarik, M.; Sanchez, A.; Yan, L. Effect of photoperiod on the regulation of wheat vernalization genes VRN1 and VRN2. *Plant Mol. Biol.* **2006**, *60*, 469–480. [CrossRef] [PubMed]
56. Whittal, A.; Kaviani, M.; Graf, R.; Humphreys, G.; Navabi, A. Allelic variation of vernalization and photoperiod response genes in a diverse set of North American high latitude winter wheat genotypes. *PLoS ONE* **2018**, *13*, e0203068. [CrossRef] [PubMed]
57. Moshatati, A.; Siadat, S.; Alami-Saeid, K.; Bakhshandeh, A.; Jalal-Kamali, M. The impact of terminal heat stress on yield and heat tolerance of bread wheat. *Int. J. Plant Prod.* **2017**, *11*, 549–560.
58. Cleland, E.E.; Chuine, I.; Menzel, A.; Mooney, H.A.; Schwartz, M.D. Shifting plant phenology in response to global change. *Trends Ecol. Evol.* **2007**, *22*, 357–365. [CrossRef]

59. Raja, M.M.; Vijayalakshmi, G.; Naik, M.L.; Basha, P.O.; Sergeant, K.; Hausman, J.F.; Khan, P.S.S.V. Pollen development and function under heat stress: From effects to responses. *Acta Physiol. Planta.* **2019**, *41*, 1–20. [CrossRef]
60. García, C.C.; Nepi, M.; Pacini, E. It is a matter of timing: Asynchrony during pollen development and its consequences on pollen performance in angiosperms-a review. *Protoplasma* **2017**, *254*, 57–73. [CrossRef]
61. Thakur, P.; Kumar, S.; Malik, J.A.; Berger, J.D.; Nayyar, H. Cold stress effects on reproductive development in grain crops. *Environ. Exp. Bot.* **2010**, *67*, 429–443. [CrossRef]
62. Hyun, Y.; Richter, R.; Coupland, G. Competence to flower: Age-controlled sensitivity to environmental cues. *Plant Physiol.* **2017**, *173*, 36–46. [CrossRef]
63. Sterbova, L.; Bradova, J.; Sedlacek, T.; Holasova, M.; Fiedlerova, V.; Dvoracek, V.; Smrckova, P. Influence of technological processing of wheat grain on starch digestibility and resistant starch content. *Starch Starke* **2016**, *68*, 593–602. [CrossRef]
64. Balla, K.; Karsai, I.; Bónis, P.; Kiss, T.; Berki, Z.; Horváth, Á.; Mayer, M.; Bencze, S.; Veisz, O. Heat stress responses in a large set of winter wheat cultivars (*Triticum aestivum* L.) depend on the timing and duration of stress. *PLoS ONE* **2019**, *14*, e0222639. [CrossRef] [PubMed]
65. Lu, H.; Hu, Y.; Wang, C.; Liu, W.; Ma, G.; Han, Q.; Ma, D. Effects of high temperature and drought stress on the expression of gene encoding enzymes and the activity of key enzymes involved in starch biosynthesis in wheat grains. *Front. Plant Sci.* **2019**, *10*, 1414. [CrossRef] [PubMed]
66. Zi, Y.; Ding, J.; Song, J.; Humphreys, G.; Peng, Y.; Li, C.; Zhu, X.; Guo, W. Grain yield, starch content and activities of key enzymes of waxy and non-waxy wheat (*Triticum aestivum* L.). *Sci. Rep.* **2018**, *8*, 1–12.
67. Khatun, S.; Ahmed, J.U.; Hossain, T.; Islam, M.R.; Mohi-Ud-Din, M. Variation of wheat cultivars in their response to elevated temperature on starch and dry matter accumulation in grain. *Int. J. Agron.* **2016**, *2016*, 1–6. [CrossRef]
68. Thalmann, M.; Santelia, D. Starch as a determinant of plant fitness under abiotic stress. *New Phytol.* **2017**, *214*, 943–951. [CrossRef]
69. MacNeill, G.J.; Mehrpouyan, S.; Minow, M.A.; Patterson, J.A.; Tetlow, I.J.; Emes, M.J.; Raines, C. Starch as a source, starch as a sink: The bifunctional role of starch in carbon allocation. *J. Exp. Bot.* **2017**, *68*, 4433–4453. [CrossRef]
70. Schittenhelm, S.; Langkamp-Wedde, T.; Kraft, M.; Kottmann, L.; Matschiner, K. Effect of two-week heat stress during grain filling on stem reserves, senescence, and grain yield of European winter wheat cultivars. *J. Agron. Crop Sci.* **2020**, *206*, 722–733. [CrossRef]
71. Dias, A.S.; Bagulho, A.S.; Lidon, F.C. Ultrastructue and biochemical traits of bread and durum wheat grains under heat stress. *Brazz. J. Plant Physiol.* **2008**, *20*, 323–333. [CrossRef]
72. Tetlow, I.J.; Emes, M.J. Starch biosynthesis in the developing endosperms of grasses and cereals. *Agronomy* **2017**, *7*, 81. [CrossRef]
73. Wang, X.; Cai, J.; Jiang, D.; Liu, F.; Dai, T.; Cao, W. Pre-anthesis high-temperature acclimation alleviates damage to the flag leaf caused by post-anthesis heat stress in wheat. *J. Plant Physiol.* **2011**, *168*, 585–593. [CrossRef]
74. Pradhan, G.P.; Prasad, P.V.; Fritz, A.K.; Kirkham, M.B.; Gill, B.S. Effects of drought and high temperature stress on synthetic hexaploid wheat. *Funct. Plant Biol.* **2012**, *39*, 190–198. [CrossRef] [PubMed]
75. Zhang, X.; Wollenweber, B.; Jiang, D.; Liu, F.; Zhao, J. Water deficits and heat shock effects on photosynthesis of a transgenic Arabidopsis thaliana constitutively expressing ABP9, a bZIP transcription factor. *J. Exp. Bot.* **2008**, *59*, 839–848. [CrossRef] [PubMed]
76. Haque, M.S.; Kjaer, K.H.; Rosenqvist, E.; Sharma, D.K.; Ottosen, C.O. Heat stress and recovery of photosystem II efficiency in wheat (*Triticum aestivum* L.) cultivars acclimated to different growth temperatures. *Environ. Exp. Bot.* **2014**, *99*, 1–8. [CrossRef]
77. Ohmiya, A.; Hirashima, M.; Yagi, M.; Tanase, K.; Yamamizo, C. Identification of genes associated with chlorophyll accumulation in flower petals. *PLoS ONE* **2014**, *9*, e113738. [CrossRef]
78. Yang, L.Y.; Yang, S.L.; Li, J.Y.; Ma, J.H.; Pang, T.; Zou, C.M.; He, B.; Gong, M. Effects of different growth temperatures on growth, development, and plastid pigments metabolism of tobacco (*Nicotiana tabacum* L.) plants. *Bot. Stud.* **2018**, *59*, 1–13. [CrossRef]
79. Tewari, A.K.; Tripathy, B.C. Temperature-stress-induced impairment of chlorophyll biosynthetic reactions in cucumber and wheat. *Plant Physiol.* **1998**, *117*, 851–858. [CrossRef]
80. Almeselmani, M.; Abdullah, F.; Hareri, F.; Naaesan, M.; Ammar, M.A.; ZuherKanbar, O.; Saud, A.A. Effect of drought on different physiological characters and yield component in different varieties of Syrian durum wheat. *J. Agric. Sci.* **2011**, *3*, 127–134. [CrossRef]
81. Balouchi, H. Screening wheat parents of mapping population for heat and drought tolerance detection of wheat genetic variation. *Int. J. Biol. Life Sci.* **2010**, *6*, 56–66.
82. Nagy, L.; Kiss, V.; Brumfeld, V.; Osvay, K.; Börzsönyi, Á.; Magyar, M.; Szabó, T.; Dorogi, M.; Malkin, S. Thermal effects and structural changes of photosynthetic reaction centers characterized by wide frequency band hydrophone: Effects of carotenoids and terbutryn. *Photochem. Photobiol.* **2015**, *91*, 1368–1375. [CrossRef]
83. Shah, S.H.; Houborg, R.; McCabe, M.F. Response of chlorophyll, carotenoid and SPAD-502 measurement to salinity and nutrient stress in wheat (*Triticum aestivum* L.). *Agronomy* **2017**, *7*, 61. [CrossRef]
84. Ristic, Z.; Bukovnik, U.; Momcilovic, I.; Fu, J.; Prasad, P.V.V. Heat-induced accumulation of chloroplast protein synthesis elongation factor, EF-Tu, in winter wheat. *J. Plant Physiol.* **2008**, *165*, 192–202. [CrossRef] [PubMed]
85. Labuschagne, M.T.; Moloi, J.; Biljon, A.V. Abiotic stress induced changes in protein quality and quantity of two bread wheat cultivars. *J. Cereal Sci.* **2016**, *69*, 259–263. [CrossRef]

86. Balla, K.; Rakszegi, M.; Li, Z.; Bekes, F.; Bencze, S.; Veisz, O. Quality of winter wheat in relation to heat and drought shock after anthesis. *Czech J. Food Sci.* **2011**, *29*, 117–128. [CrossRef]
87. Li, Y.; Wu, Y.; Espinosa, H.; Pena, R.J. The influence of drought and heat stress on the expression of end-use quality parameters of common wheat. *J. Cereal Sci.* **2013**, *57*, 73–78. [CrossRef]
88. Castro, M.; Peterson, C.; Dalla Rizza, M.; Dellavalle, P.D.; Vázquez, D.; Ibanez, V.; Ross, A. *Wheat Production in Stressed Environments*; Springer: Berlin/Heidelberg, Germany, 2007; pp. 365–371.
89. Xue, C.; Matros, A.; Mock, H.P.; Mühling, K.H. Protein composition and baking quality of wheat flour as affected by split nitrogen application. *Front. Plant Sci.* **2019**, *10*, 642–650. [CrossRef]
90. Zhang, Y.; Hu, X.; Juhasz, A.; Islam, S.; Yu, Z.; Zhao, Y.; Li, G.; Ding, W.; Ma, W. Characterising avenin-like proteins (ALPs) from albumin/globulin fraction of wheat grains by RP-HPLC, SDS-PAGE, and MS/MS peptides sequencing. *BMC Plant Biol.* **2020**, *20*, 45–55. [CrossRef]
91. Hurkman, W.J.; Vensel, W.H.; Tanaka, C.K.; Whitehand, L.; Altenbach, S.B. Effect of high temperature on albumin and globulin accumulation in the endosperm proteome of the developing wheat grain. *J. Cereal Sci.* **2009**, *49*, 12–23. [CrossRef]
92. Branlard, G.; Lesage, V.S.; Bancel, E.; Martre, P.; Méleard, B.; Rhazi, L. *Advances in Wheat Genetics: From Genome to Field*; Springer: Tokyo, Japan, 2015; pp. 255–264.
93. Xu, X.L.; Zhang, Y.H.; Wang, Z.M. Effect of heat stress during grain filling on phosphoenolpyruvate carboxylase and ribulose-1,5-bisphosphate carboxylase/oxygenase activities of various green organs in winter wheat. *Photosynthetica* **2003**, *42*, 317–320. [CrossRef]
94. Khan, A. Plant Breeding and Genetics. Ph.D. Thesis, PMAS-Arid Agriculture University Rawalpindi, Rawalpindi, Pakistan, 2020.
95. Mathur, S.; Allakhverdiev, S.I.; Jajoo, A. Analysis of high temperature stress on the dynamics of antenna size and reducing side heterogeneity of Photosystem II in wheat leaves (*Triticum aestivum* L.). *Biochim. Biophys. Acta* **2011**, *1807*, 22–29. [CrossRef]
96. Mathur, S.; Agrawal, D.; Jajoo, A. Photosynthesis: Response to high temperature stress. *J. Photochem. Photobio.* **2014**, *137*, 116–126. [CrossRef]
97. Drop, B.; Webber-Birungi, M.; Yadav, S.K.; Filipowicz-Szymanska, A.; Fusetti, F.; Boekema, E.J.; Croce, R. Light-harvesting complex II (LHCII) and its supramolecular organization in Chlamydomonas reinhardtii. *Biochim. Biophys. Acta BBA Bioenergetics* **2014**, *1837*, 63–72. [CrossRef] [PubMed]
98. Zhang, S.; Scheller, H.V. Light-harvesting complex II binds to several small subunits of photosystem I. *J. Biol. Chem.* **2004**, *279*, 3180–3187. [CrossRef] [PubMed]
99. Sage, R.F.; Kubien, D.S. The temperature response of C3 and C4 photosynthesis. *Plant Cell Environ.* **2007**, *30*, 1086–1106. [CrossRef] [PubMed]
100. Hobson, P.D.; Keys, A.J.; Parry, M.A.; Lea, P.J. Regulation of Rubisco during drought and heat stress in wheat. *Comp. Biochem. Physiol. A Mol. Integr. Physiol.* **2002**, *132*, 173–180.
101. Yin, Z.; Zhang, Z.; Deng, D.; Chao, M.; Gao, Q.; Wang, Y.; Yang, Z.; Bian, Y.; Hao, D.; Xu, C. Characterization of RuBisCo activase genes in maize: An α-isoform gene functions alongside a β-isoform gene. *Plant Physiol.* **2014**, *164*, 2096–2106. [CrossRef]
102. Portis, A.R.; Li, C.; Wang, D.; Salvucci, M.E. Regulation of Rubisco activase and its interaction with Rubisco. *J. Exp. Bot.* **2007**, *59*, 1597–1604. [CrossRef]
103. Kumar, R.R.; Goswami, S.; Dubey, K.; Singh, K.; Singh, J.P.; Kumar, A.; Rai, G.K.; Singh, S.D.; Bakshi, S.; Singh, B. RuBisCo activase-a catalytic chaperone involved in modulating the RuBisCo activity and heat stress-tolerance in wheat. *J. Plant Biochem. Biotech.* **2019**, *28*, 63–75. [CrossRef]
104. Scafaro, A.P.; Bautsoens, N.; den Boer, B.; Van Rie, J.; Gallé, A. A conserved sequence from heat-adapted species improves Rubisco activase thermostability in wheat. *Plant Physiol.* **2019**, *181*, 43–54. [CrossRef]
105. Aliyev, J.A. Photosynthesis, photorespiration and productivity of wheat and soybean genotypes. *Physiol. Plant.* **2012**, *145*, 369–383. [CrossRef]
106. Posch, B.C.; Kariyawasam, B.C.; Bramley, H.; Coast, O.; Richards, R.A.; Reynolds, M.P.; Trethowan, R.; Atkin, O.K. Exploring high temperature responses of photosynthesis and respiration to improve heat tolerance in wheat. *J. Exp. Bot.* **2019**, *70*, 5051–5069. [CrossRef]
107. Morales, F.; Ancín, M.; Fakhet, D.; González-Torralba, J.; Gámez, A.L.; Seminario, A.; Soba, D.; Ben Mariem, S.; Garriga, M.; Aranjuelo, I. Photosynthetic metabolism under stressful growth conditions as a bases for crop breeding and yield improvement. *Plants* **2020**, *9*, 88. [CrossRef] [PubMed]
108. Tabita, F.R.; Hanson, T.E.; Li, H.; Satagopan, S.; Singh, J.; Chan, S. Function, structure, and evolution of the RubisCO-like proteins and their RubisCO homologs. *Microbiol. Mol. Biol. Rev.* **2007**, *71*, 576–599. [CrossRef]
109. Badger, M.R.; Bek, E.J. Multiple Rubisco forms in proteobacteria: Their functional significance in relation to CO_2 acquisition by the CBB cycle. *J. Exp. Bot.* **2008**, *59*, 1525–1541. [CrossRef] [PubMed]
110. Long, S.P.; Ainsworth, E.A.; Rogers, A.; Ortiz, D.R. Rising atmospheric carbon dioxide: Plants face the future. *Annu. Rev. Plant Biol.* **2004**, *55*, 591–628. [CrossRef] [PubMed]
111. Dusenge, M.E.; Duarte, A.G.; Way, D.A. Plant carbon metabolism and climate change: Elevated CO_2 and temperature impacts on photosynthesis, photorespiration and respiration. *New Phytol.* **2019**, *221*, 32–49. [CrossRef]
112. Walker, B.J.; VanLoocke, A.; Bernacchi, C.J.; Ort, D.R. The costs of photorespiration to food production now and in the future. *Annu. Rev. Plant Biol.* **2016**, *67*, 107–129. [CrossRef]

113. Asada, K. Production and scavenging of reactive oxygen species in chloroplasts and their functions. *Plant Physiol.* **2006**, *141*, 391–396. [CrossRef]
114. Miller, G.; Suzuki, N.; Yilmaz, S.C.; Mittler, R. Reactive oxygen species homeostasis and signalling during drought and salinity stresses. *Plant Cell Environ.* **2010**, *33*, 453–467. [CrossRef]
115. Wang, X.; Dinler, B.S.; Vignjevic, M.; Jacobsen, S.; Wollenweber, B. Physiological and proteome studies of responses to heat stress during grain filling in contrasting wheat cultivars. *Plant Sci.* **2015**, *230*, 33–50. [CrossRef]
116. Pospíšil, P. Production of reactive oxygen species by photosystem II as a response to light and temperature stress. *Front. Plant Sci.* **2016**, *7*, e01950. [CrossRef]
117. Foyer, C.H. Reactive oxygen species, oxidative signaling and the regulation of photosynthesis. *Environ. Exp. Bot.* **2018**, *154*, 134–142. [CrossRef] [PubMed]
118. Kumar, R.R.; Goswami, S.; Sharma, S.K.; Singh, K.; Gadpayle, K.A.; Singh, S.D.; Pathak, H.; Rai, R.D. Differential expression of heat shock protein and alteration in osmolyte accumulation under heat stress in wheat. *J. Plant Biochem. Biotechnol.* **2013**, *22*, 16–26. [CrossRef]
119. Hasanuzzaman, M.; Nahar, K.; Alam, M.M.; Roychowdhury, R.; Fujita, M. Physiological, biochemical and molecular mechanisms of heat stress tolerance in plants. *Int. J. Mol. Sci.* **2013**, *14*, 9643–9684. [CrossRef] [PubMed]
120. Allakhverdiev, S.I.; Kreslavski, V.D.; Klimov, V.V.; Los, D.A.; Carpentier, R.; Mohanty, P. Heat stress: An overview of molecular responses in photosynthesis. *Photosyn. Res.* **2008**, *98*, 541–550. [CrossRef]
121. Park, E.-J.; Jeknic, Z.; Pino, M.T.; Murata, N.; Chen, T.H.-H. Glycinebetaine accumulation is more effective in chloroplasts than in the cytosol for protecting transgenic tomato plants against abiotic stress. *Plant Cell Environ.* **2007**, *30*, 994–1005. [CrossRef]
122. Annunziata, M.G.; Ciarmiello, L.F.; Woodrow, P.; Dell'Aversana, E.; Carillo, P. Spatial and temporal profile of glycine betaine accumulation in plants under abiotic stresses. *Front. Plant Sci.* **2019**, *10*, 230. [CrossRef]
123. Wang, Y.; Liu, S.; Zhang, H.; Zhao, Y.; Zhao, H.; Liu, H. Glycine betaine application in grain filling wheat plants alleviates heat and high light-induced photoinhibition by enhancing the psbA transcription and stomatal conductance. *Acta Physiol. Planta.* **2014**, *36*, 2195–2202. [CrossRef]
124. Sukumar, T.; Neeraj, J.; Samal, S.; Mishra, B. Salicylic acid and high temperature stress. *Ann. Biol.* **2015**, *31*, 18–23.
125. Chen, T.H.; Murata, N. Glycinebetaine protects plants against abiotic stress: Mechanisms and biotechnological applications. *Plant Cell Environ.* **2011**, *34*, 1–20. [CrossRef]
126. Aldesuquy, H.S.; Abbas, M.A.; Abo-Hamed, S.A.; Elhakem, A.H.; Alsokari, S.S. Glycine betaine and salicylic acid induced modification in productivity of two different cultivars of wheat grown under water stress. *J. Stress Physiol. Biochem.* **2012**, *8*, 72–89.
127. Khan, M.I.R.; Iqbal, N.; Masood, A.; Per, T.S.; Khan, N.A. Salicylic acid alleviates adverse effects of heat stress on photosynthesis through changes in proline production and ethylene formation. *Plant Signal. Behav.* **2013**, *8*, e26374. [CrossRef] [PubMed]
128. Sharma, A.; Shahzad, B.; Kumar, V.; Kohli, S.K.; Sidhu, G.P.S.; Bali, A.S.; Handa, N.; Kapoor, D.; Bhardwaj, R.; Zheng, B. Phytohormones regulate accumulation of osmolytes under abiotic stress. *Biomolecules* **2019**, *9*, 285. [CrossRef] [PubMed]
129. Hare, P.; Cress, W.; Van Staden, J. Proline synthesis and degradation: A model system for elucidating stress-related signal transduction. *J. Exp. Bot.* **1999**, *50*, 413–434. [CrossRef]
130. Fichman, Y.; Gerdes, S.Y.; Kovács, H.; Szabados, L.; Zilberstein, A.; Csonka, L.N. Evolution of proline biosynthesis: Enzymology, bioinformatics, genetics, and transcriptional regulation. *Biol. Rev.* **2015**, *90*, 1065–1099. [CrossRef] [PubMed]
131. Sattar, A.; Sher, A.; Ijaz, M.; Ul-Allah, S.; Rizwan, M.S.; Hussain, M.; Jabran, K.; Cheema, M.A. Terminal drought and heat stress alter physiological and biochemical attributes in flag leaf of bread wheat. *PLoS ONE* **2020**, *15*, e0232974. [CrossRef]
132. Gupta, N.K.; Agarwal, S.; Agarwal, V.P.; Nathawat, N.S.; Gupta, S.; Singh, G. Effect of short-term heat stress on growth, physiology and antioxidative defence system in wheat seedlings. *Acta Physiol. Plant.* **2013**, *35*, 1837–1842. [CrossRef]
133. Asthir, B.; Thapar, R.; Farooq, M.; Bains, N.S. Exogenous application of thiourea improves the performance of late sown wheat by inducing terminal heat resistance. *Int. J. Agric. Biol.* **2013**, *15*, 1337–1342.
134. Agarwal, V.; Gupta, N.; Gupta, P.; Rizwan, M.; Singh, G. Sulfhydral compounds mitigate the adverse effect of high temperature stress in contrasting wheat genotypes. *Vegetos Int. J. Plant Res.* **2017**, *30*, 87–91. [CrossRef]
135. Lopes, M.S.; Reynolds, M.P. Stay-green in spring wheat can be determined by spectral reflectance measurements (normalized difference vegetation index) independently from phenology. *J. Exp. Bot.* **2012**, *63*, 3789–3798. [CrossRef]
136. Adu, M.O.; Sparkes, D.L.; Parmar, A.; Yawson, D.O. Stay green in wheat: Comparative study of modern bread wheat and ancient wheat cultivars. *ARPN J. Agric. Biol. Sci.* **2011**, *6*, 16–24.
137. Kamal, N.M.; Gorafi, Y.S.A.; Abdelrahman, M.; Abdellatef, E.; Tsujimoto, H. Stay-green trait: A prospective approach for yield potential, and drought and heat stress adaptation in globally important cereals. *Int. J. Mol. Sci.* **2019**, *20*, 5837. [CrossRef] [PubMed]
138. Guo, M.; Liu, J.-H.; Ma, X.; Luo, D.-X.; Gong, Z.-H.; Lu, M.-H. The plant heat stress transcription factors (HSFs): Structure, regulation, and function in response to abiotic stresses. *Front. Plant Sci.* **2016**, *7*, e00114. [CrossRef] [PubMed]
139. Gururani, M.A.; Venkatesh, J.; Tran, L.S.P. Regulation of photosynthesis during abiotic stress-induced photoinhibition. *Mol. Plant* **2015**, *8*, 1304–1320. [CrossRef] [PubMed]
140. Kumari, M.; Pudake, R.N.; Singh, V.P.; Joshi, A.K. Association of staygreen trait with canopy temperature depression and yield traits under terminal heat stress in wheat (*Triticum aestivum* L.). *Euphytica* **2013**, *190*, 87–97. [CrossRef]

141. Jagadish, K.S.; Kavi Kishor, P.B.; Bahuguna, R.N.; von Wirén, N.; Sreenivasulu, N. Staying alive or going to die during terminal senescence-an enigma surrounding yield stability. *Front. Plant Sci.* **2015**, *6*, e01070. [CrossRef]
142. Thomas, H.; Ougham, H. The stay-green trait. *J. Exp. Bot.* **2014**, *65*, 3889–3900. [CrossRef]
143. Shantanu, D.; Nabarun, R.; Indrani, C.; Monoj, S.; Debojit, S. Significance of stay-green to foster crop production under stress environment-a mini-review. *J. Exp. Biol. Agric. Sci.* **2018**, *6*, 639–651.
144. Sakuraba, Y.; Schelbert, S.; Park, S.-Y.; Han, S.-H.; Lee, B.-D.; Andrès, C.B.; Kessler, F.; Hörtensteiner, S.; Paek, N.-C. STAY-GREEN and chlorophyll catabolic enzymes interact at light-harvesting complex II for chlorophyll detoxification during leaf senescence in Arabidopsis. *Plant Cell* **2012**, *24*, 507–518. [CrossRef]
145. Horie, Y.; Ito, H.; Kusaba, M.; Tanaka, R.; Tanaka, A. Participation of chlorophyll b reductase in the initial step of the degradation of light-harvesting chlorophyll a/b-protein complexes in Arabidopsis. *J. Biol. Chem.* **2009**, *284*, 17449–17456. [CrossRef]
146. Morita, R.; Sato, Y.; Masuda, Y.; Nishimura, M.; Kusaba, M. Defect in non-yellow coloring 3, an α/β hydrolase-fold family protein, causes a stay-green phenotype during leaf senescence in rice. *Plant J.* **2009**, *59*, 940–952. [CrossRef]
147. Caverzan, A.; Casassola, A.; Brammer, S.P. Antioxidant responses of wheat plants under stress. *Genet. Mol. Biol.* **2016**, *39*, 1–6. [CrossRef] [PubMed]
148. Almeselmani, M.; Deshmukh, P.; Sairam, R. High temperature stress tolerance in wheat genotypes: Role of antioxidant defence enzymes. *Acta Agron. Hungarica* **2009**, *57*, 1–14. [CrossRef]
149. Wang, C.; Wen, D.; Sun, A.; Han, X.; Zhang, J.; Wang, Z.; Yin, Y. Differential activity and expression of antioxidant enzymes and alteration in osmolyte accumulation under high temperature stress in wheat seedlings. *J. Cereal Sci.* **2014**, *60*, 653–659. [CrossRef]
150. Xu, L.; Han, L.; Huang, B. Antioxidant enzyme activities and gene expression patterns in leaves of Kentucky bluegrass in response to drought and postdrought recovery. *J. Am. Soc. Hortic. Sci.* **2011**, *136*, 247–255. [CrossRef]
151. Awasthi, R.; Bhandari, K.; Nayyar, H. Temperature stress and redox homeostasis in agricultural crops. *Front. Environ. Sci.* **2015**, *3*, e00011. [CrossRef]
152. Tripathy, B.C.; Oelmuller, R. Reactive oxygen species generation and signaling in plants. *Plant Signal Behav.* **2012**, *7*, 1621–1633. [CrossRef]
153. Jing, J.; Guo, S.; Li, Y.; Li, W. The alleviating effect of exogenous polyamines on heat stress susceptibility of different heat resistant wheat (*Triticum Aestivum* L.) varieties. *Sci. Rep.* **2020**, *10*, 1–12. [CrossRef]
154. Wu, K.; Xiao, S.; Chen, Q.; Wang, Q.; Zhang, Y.; Li, K.; Yu, Y.; Chen, L. Changes in the activity and transcription of antioxidant enzymes in response to Al stress in black soybeans. *Plant Mol. Biol. Rep.* **2013**, *31*, 141–150. [CrossRef]
155. Sarfraz-Ardakani, M.R.; Khavari-Nejad, R.A.; Moradi, F.; Najafi, F. Abscisic acid and cytokinin-induced osmotic and antioxidant regulation in two drought-tolerant and drought-sensitive cultivars of wheat during grain filling under water deficit in field conditions. *Not. Sci. Biol.* **2014**, *6*, 354–362. [CrossRef]
156. Gill, S.S.; Anjum, N.A.; Hasanuzzaman, M.; Gill, R.; Trivedi, D.K.; Ahmad, I.; Tuteja, N. Glutathione and glutathione reductase: A boon in disguise for plant abiotic stress defense operations. *Plant Physiol. Biochem.* **2013**, *70*, 204–212. [CrossRef]
157. Sairam, R.K.; Srivastava, G.C.; Saxena, D.C. Increased antioxidant activity under elevated temperatures: A mechanism of heat stress tolerance in wheat genotypes. *Biol. Plant* **2000**, *43*, 245–251. [CrossRef]
158. Weydert, C.J.; Cullen, J.J. Measurement of superoxide dismutase, catalase and glutathione peroxidase in cultured cells and tissue. *Nat. Protoc.* **2010**, *5*, 51–66. [CrossRef] [PubMed]
159. Grigorova, B.; Vaseva, I.; Demirevska, K.; Feller, U. Combined drought and heat stress in wheat changes in some heat shock proteins. *Biol. Planta.* **2011**, *55*, 105–111. [CrossRef]
160. Wang, W.; Vinocur, B.; Shoseyov, O.; Altman, A. Role of plant heat-shock proteins and molecular chaperones in the abiotic stress response. *Trends Plant Sci.* **2004**, *9*, 244–252. [CrossRef]
161. Lin, Q.; Wang, Y.M.; Nose, A.; Hong, H.T.K.; Agarie, S. Effect of high night temperature on lipid and protein compositions in tonoplasts isolated from Ananas comosus and Kalanchoe pinnata leaves. *Biol. Plant* **2008**, *52*, 59–65. [CrossRef]
162. Caeiro, A.S.; Ramos, P.C.; Teixeira, A.R.; Ferreira, R.B. The ubiquitin/proteasome pathway from Lemna minor subjected to heat shock. *Biol. Plant* **2008**, *52*, 695–702. [CrossRef]
163. Nadeem, M.; Li, J.; Wang, M.; Shah, L.; Lu, S.; Wang, X.; Ma, C. Unraveling field crops sensitivity to heat stress: Mechanisms, approaches, and future prospects. *Agronomy* **2018**, *8*, 128. [CrossRef]
164. Kotak, S.; Larkindale, J.; Lee, U.; von Koskull-Döring, P.; Vierling, E.; Scharf, K.-D. Complexity of the heat stress response in plants. *Plant Biol.* **2007**, *10*, 310–316. [CrossRef]
165. Hemantaranjan, A.; Nishant Bhanu, A.; Singh, M.; Yadav, D.; Patel, P. Heat stress responses and thermotolerance. *Adv. Plants Agric. Res.* **2014**, *1*, e00012. [CrossRef]
166. Xu, Y.; Zhan, C.; Huang, B. Heat shock proteins in association with heat tolerance in grasses. *Int. J. Proteom.* **2011**. [CrossRef]
167. Ye, J.; Yang, X.; Hu, G.; Liu, Q.; Li, W.; Zhang, L.; Song, X. Genome-wide investigation of heat shock transcription factor family in wheat (*Triticum aestivum* L.) and possible roles in anther development. *Int. J. Mol. Sci.* **2020**, *21*, 608. [CrossRef] [PubMed]
168. Sumesh, K.V.; Sharma, N.P.; Ghildiyal, M.C. Starch synthase activity and heat shock protein in relation to thermal tolerance of developing wheat grains. *Biol. Plant* **2008**, *52*, 749–753. [CrossRef]
169. Lu, Y.; Zhao, P.; Zhang, A.; Ma, L.; Xu, S.; Wang, X. Alternative splicing diversifies the heat response and evolutionary strategy of conserved Heat Shock Protein 90 in bread wheat (*Triticum aestivum* L.). *Res. Square* **2019**. [CrossRef]

170. Kumar, A.; Sharma, S.; Chunduri, V.; Kaur, A.; Kaur, S.; Malhotra, N.; Kumar, A.; Kapoor, P.; Kumari, A.; Kaur, J. Genome-wide identification and cCharacterization of heat shock protein family reveals role in development and stress conditions in *Triticum aestivum* L. *Sci. Rep.* **2020**, *10*, 1–12.
171. Wajid, M.; Khan, M.; Shirazi, M.; Summiya, F.; Saba, M. Seed priming induced high temperature tolerance in wheat by regulating germination metabolism and physio-biochemical properties. *Int. J. Agric. Biol.* **2018**, *20*, 2140–2148.
172. Banerjee, A.; Roychoudhury, A. Seed priming as a method to generate heat-stress tolerance in plants: A minireview. *Heat Stress Toler. Plants Physiol. Mol. Genet. Perspect.* **2020**, 23–32. [CrossRef]
173. Hussain, I.; Ahmad, R.; Farooq, M.; Wahid, A. Seed priming improves the performance of poor quality wheat seed. *Int. J. Agric. Biol.* **2013**, *15*, 1343–1348.
174. Rummana, S.; Amin, A.; Islam, M.; Faruk, G. Effect of irrigation and mulch materials on growth and yield of wheat. *Bangladesh Agron. J.* **2018**, *21*, 71–76. [CrossRef]
175. Humphreys, E.; Gaydon, D.; Eberbach, P. Evaluation of the effects of mulch on optimum sowing date and irrigation management of zero till wheat in central Punjab, India using APSIM. *Field Crops Res.* **2016**, *197*, 83–96.
176. Kajla, M.; Yadav, V.K.; Chhokar, R.S.; Sharma, R.K. Management practices to mitigate the impact of high temperature on wheat. *J. Wheat Res.* **2015**, *7*, 1–12.
177. Tao, Z.; Wang, D.M.; Chang, X.H.; Wang, Y.J.; Yang, Y.S.; Zhao, G.C. Effects of zinc fertilizer and short-term high temperature stress on wheat grain production and wheat flour proteins. *J. Integrat. Agric.* **2018**, *17*, 1979–1990. [CrossRef]
178. Fahad, S.; Bajwa, A.A.; Nazir, U.; Anjum, S.A.; Farooq, A.; Zohaib, A.; Sadia, S.; Nasim, W.; Adkins, S.; Saud, S. Crop production under drought and heat stress: Plant responses and management options. *Front. Plant Sci.* **2017**, *8*, e01147. [CrossRef] [PubMed]
179. Cooke, J.; Leishman, M.R. Consistent alleviation of abiotic stress with silicon addition: A meta-analysis. *Funct. Ecol.* **2016**, *30*, 1340–1357. [CrossRef]
180. Baloch, M.; Baloch, A.; Shaikh, M.; Baloch, M.; Asad, M.; Depar, M.; Baloch, G.; Gandahi, N.; Mallano, I.; Baloch, A. Evaluation of wheat genotypes for earliness and growth traits exposed to high temperature through delayed planting. *Sindh Univ. Res. J.* **2015**, *47*, 81–89.
181. Wu, X.; Tang, Y.; Li, C.; Wu, C. Characterization of the rate and duration of grain filling in wheat in southwestern China. *Plant Prod. Sci.* **2018**, *21*, 358–369. [CrossRef]
182. Khan, A.; Ahmad, M.; Shah, M.K.N.; Ahmed, M. Genetic manifestation of physio-morphic and yield related traits conferring thermotolerance in wheat. *Pak. J. Bot.* **2020**, *52*, 1545–1552. [CrossRef]
183. Driedonks, N.; Rieu, I.; Vriezen, W.H. Breeding for plant heat tolerance at vegetative and reproductive stages. *Plant Reprod.* **2016**, *29*, 67–79. [CrossRef]
184. Janni, M.; Gullì, M.; Maestri, E.; Marmiroli, M.; Valliyodan, B.; Nguyen, H.T.; Marmiroli, N. Molecular and genetic bases of heat stress responses in crop plants and breeding for increased resilience and productivity. *J. Exp. Bot.* **2020**, *71*, 3780–3802. [CrossRef]
185. Gupta, S.; Kaur, S.; Sehgal, S.; Sharma, A.; Chhuneja, P.; Bains, N.S. Genotypic variation for cellular thermotolerance in Aegilops tauschii Coss., the D genome progenitor of wheat. *Euphytica* **2010**, *175*, 373–381. [CrossRef]
186. Rampino, P.; Gullì, M.; De Pascali, M.; De Caroli, M.; Marmiroli, N.; Perrotta, C. Wild and cultivated Triticum species differ in thermotolerant habit and HSP gene expression. *Plant Biosyst. Int. J. Deal. Asp. Plant Biol.* **2019**, *153*, 337–343. [CrossRef]
187. Kishii, M. An update of recent use of Aegilops species in wheat breeding. *Front. Plant Sci.* **2019**, *10*, e00585. [CrossRef] [PubMed]
188. Bala, P.; Sikder, S. Evaluation of heat tolerance of wheat genotypes through membrane thermostability test. *MAYFEB J. Agric. Sci.* **2017**, *2*, 1–6.
189. Rehman, S.U.; Bilal, M.; Rana, R.M.; Tahir, M.N.; Shah, M.K.N.; Ayalew, H.; Yan, G. Cell membrane stability and chlorophyll content variation in wheat (*Triticum aestivum*) genotypes under conditions of heat and drought. *Crop Past. Sci.* **2016**, *67*, 712–718. [CrossRef]
190. Ahmed, M.; Qadir, G.; Shaheen, F.A.; Aslam, M.A. Response of proline accumulation in bread wheat (*Triticum aestivum* L.) under rainfed conditions. *J. Agric. Meteorol.* **2017**, *73*, 147–155. [CrossRef]
191. Saeedipour, S. Relationship of grain yield, ABA and proline accumulation in tolerant and sensitive wheat cultivars as affected by water stress. *Proc. Natl. Acad. Sci. USA* **2013**, *83*, 311–315. [CrossRef]
192. Qaseem, M.F.; Qureshi, R.; Shaheen, H. Effects of pre-anthesis drought, heat and their combination on the growth, yield and physiology of diverse wheat (*Triticum aestivum* L.) genotypes varying in sensitivity to heat and drought stress. *Sci. Rep.* **2019**, *9*, 1–12. [CrossRef]
193. Suresh, S.; Bishnoi, O.P.; Behl, R.K. Use of heat susceptibility index and heat response index as a measure of heat tolerance in wheat and triticale. *Ekin J. Crop Breed. Genet.* **2018**, *4*, 39–44.
194. Elbashir, A.A.E.; Gorafi, Y.S.A.; Tahir, I.S.A.; Kim, J.S.; Tsujimoto, H. Wheat multiple synthetic derivatives: A new source for heat stress tolerance adaptive traits. *Breed. Sci.* **2017**, *67*, 248–256. [CrossRef]
195. Tricker, P.J.; ElHabti, A.; Schmidt, J.; Fleury, D. The physiological and genetic basis of combined drought and heat tolerance in wheat. *J. Exp. Bot.* **2018**, *69*, 3195–3210. [CrossRef]
196. Poudel, A.; Thapa, D.B.; Sapkota, M. Cluster analysis of wheat (*Triticum aestivum* L.) genotypes based upon response to terminal heat stress. *Int. J. App. Sci. Biotech.* **2017**, *5*, 188–193. [CrossRef]
197. Ni, Z.; Li, H.; Zhao, Y.; Peng, H.; Hu, Z.; Xin, M.; Sun, Q. Genetic improvement of heat tolerance in wheat: Recent progress in understanding the underlying molecular mechanisms. *Crop J.* **2018**, *6*, 32–41. [CrossRef]

198. Pandey, G.C.; Mamrutha, H.; Tiwari, R.; Sareen, S.; Bhatia, S.; Siwach, P.; Tiwari, V.; Sharma, I. Physiological traits associated with heat tolerance in bread wheat (*Triticum aestivum* L.). *Physiol. Mol. Biol. Plants* **2015**, *21*, 93–99. [CrossRef] [PubMed]
199. Chovancek, E.; Zivcak, M.; Botyanszka, L.; Hauptvogel, P.; Yang, X.; Misheva, S.; Hussain, S.; Brestic, M. Transient heat waves may affect the photosynthetic capacity of susceptible wheat genotypes due to insufficient photosystem I photoprotection. *Plants* **2019**, *8*, 282. [CrossRef] [PubMed]
200. Abdelrahman, M.; El-Sayed, M.; Jogaiah, S.; Burritt, D.J.; Tran, L.S.P. The "STAY-GREEN" trait and phytohormone signaling networks in plants under heat stress. *Plant Cell Rep.* **2017**, *36*, 1009–1025. [CrossRef]
201. Thapa, S.; Jessup, K.E.; Pradhan, G.P.; Rudd, J.C.; Liu, S.; Mahan, J.R.; Devkota, R.N.; Baker, J.A.; Xue, Q. Canopy temperature depression at grain filling correlates to winter wheat yield in the US Southern High Plains. *Field Crops Res.* **2018**, *217*, 11–19. [CrossRef]
202. Gong, W.; Han, R.; Li, H.; Song, J.; Yan, H.; Li, G.; Liu, A.; Cao, X.; Guo, J.; Zhai, S. Agronomic traits and molecular marker identification of Wheat-Aegilops caudata addition lines. *Front. Plant Sci.* **2017**, *8*, e01743. [CrossRef]
203. Sehgal, A.; Sita, K.; Siddique, K.H.; Kumar, R.; Bhogireddy, S.; Varshney, R.K.; HanumanthaRao, B.; Nair, R.M.; Prasad, P.; Nayyar, H. Drought or/and heat-stress effects on seed filling in food crops: Impacts on functional biochemistry, seed yields, and nutritional quality. *Front. Plant Sci.* **2018**, *9*, e01705. [CrossRef]
204. Khan, A.; Khaliq, I.; Ahmad, M.; Ahmed, H.G.M.D.; Khan, A.G.; Farooq, M.S. Comparative performance of spring wheat (*Triticum aestivum* L.) through heat stress indices. *Pak. J. Bot.* **2018**, *50*, 481–488.
205. Baloch, M.S.; Nadim, M.A.; Zubair, M.; Awan, I.U.; Khan, E.A.; Ali, S. Evaluation of wheat under normal and late sowing conditions. *Pak. J. Bot.* **2012**, *44*, 1727–1732.
206. Mondal, S.; Singh, R.P.; Crossa, J.; Huerta-Espino, J.; Sharma, I.; Chatrath, R.; Singh, G.P.; Sohu, V.S.; Mavi, G.S.; Sukuru, V.S.P.; et al. Earliness in wheat: A key to adaptation under terminal and continual high temperature stress in South Asia. *Field Crops Res.* **2013**, *151*, 19–26. [CrossRef]
207. Hossain, A.; Silva, J.A.T.D. Phenology, growth and yield of three wheat (*Triticum aestivum* L.) varieties as affected by high temperature stress. *Notulae Sci. Biol.* **2012**, *4*, 97–109. [CrossRef]
208. Deery, D.M.; Rebetzke, G.; Jimenez-Berni, J.A.; Bovill, B.; James, R.A.; Condon, A.G.; Furbank, R.; Chapman, S.; Fischer, R. Evaluation of the phenotypic repeatability of canopy temperature in wheat using continuous-terrestrial and airborne measurements. *Front. Plant Sci.* **2019**, *10*, 875. [CrossRef] [PubMed]
209. Neukam, D.; Ahrends, H.; Luig, A.; Manderscheid, R.; Kage, H. Integrating wheat canopy temperatures in crop system models. *Agronomy* **2016**, *6*, 7. [CrossRef]
210. Gautam, A.; Prasad, S.S.; Jajoo, A.; Ambati, D. Canopy temperature as a selection parameter for grain yield and its components in durum wheat under terminal heat stress in late sown conditions. *Agric. Res.* **2015**, *4*, 238–244. [CrossRef]
211. Mason, R.E.; Singh, R.P. Considerations when deploying canopy temperature to select high yielding wheat breeding lines under drought and heat stress. *Agronomy* **2014**, *4*, 191–201. [CrossRef]
212. Khan, A.; Ahmad, M.; Shah, M.K.N.; Ahmed, M. Performance of wheat genotypes for Morpho-Physiological traits using multivariate analysis under terminal heat stress. *Pak. J. Bot.* **2020**, *52*, 1981–1988. [CrossRef]
213. Asthir, B. Protective mechanisms of heat tolerance in crop plants. *J. Plant Int.* **2015**, *10*, 202–210. [CrossRef]
214. Xue, G.-P.; Drenth, J.; McIntyre, C.L. TaHsfA6f is a transcriptional activator that regulates a suite of heat stress protection genes in wheat (*Triticum aestivum* L.) including previously unknown Hsf targets. *J. Exp. Bot.* **2015**, *66*, 1025–1039. [CrossRef]
215. Fu, J.; Momčilović, I.; Clemente, T.E.; Nersesian, N.; Trick, H.N.; Ristic, Z. Heterologous expression of a plastid EF-Tu reduces protein thermal aggregation and enhances CO_2 fixation in wheat (*Triticum aestivum*) following heat stress. *Plant Mol. Biol.* **2008**, *68*, 277–288. [CrossRef]
216. Weichert, H.; Högy, P.; Mora-Ramirez, I.; Fuchs, J.; Eggert, K.; Koehler, P.; Weschke, W.; Fangmeier, A.; Weber, H. Grain yield and quality responses of wheat expressing a barley sucrose transporter to combined climate change factors. *J. Exp. Bot.* **2017**, *68*, 5511–5525. [CrossRef]
217. Zang, X.; Geng, X.; Wang, F.; Liu, Z.; Zhang, L.; Zhao, Y.; Tian, X.; Ni, Z.; Yao, Y.; Xin, M. Overexpression of wheat ferritin gene TaFER-5B enhances tolerance to heat stress and other abiotic stresses associated with the ROS scavenging. *BMC Plant Biol.* **2017**, *17*, 14. [CrossRef] [PubMed]
218. Zhang, L.; Geng, X.; Zhang, H.; Zhou, C.; Zhao, A.; Wang, F.; Zhao, Y.; Tian, X.; Hu, Z.; Xin, M. Isolation and characterization of heat-responsive gene TaGASR1 from wheat (*Triticum aestivum* L.). *J. Plant Biol.* **2017**, *60*, 57–65. [CrossRef]
219. Hu, X.J.; Chen, D.; Lynne McIntyre, C.; Fernanda Dreccer, M.; Zhang, Z.B.; Drenth, J.; Kalaipandian, S.; Chang, H.; Xue, G.P. Heat shock factor C2a serves as a proactive mechanism for heat protection in developing grains in wheat via an ABA-mediated regulatory pathway. *Plant Cell Environ.* **2018**, *41*, 79–98. [CrossRef] [PubMed]
220. Wang, J.; Gao, X.; Dong, J.; Tian, X.; Wang, J.; Palta, J.A.; Xu, S.; Fang, Y.; Wang, Z. Over-expression of the heat-responsive wheat gene TaHSP23.9 in transgenic Arabidopsis conferred tolerance to heat and salt stress. *Front. Plant Sci.* **2020**, *11*, e00243. [CrossRef] [PubMed]
221. Li, Q.; Wang, W.; Wang, W.; Zhang, G.; Liu, Y.; Wang, Y.; Wang, W. Wheat F-box protein gene TaFBA1 is involved in plant tolerance to heat stress. *Front. Plant Sci.* **2018**, *9*, e00521. [CrossRef]
222. Liu, Z.; Li, G.; Zhang, H.; Zhang, Y.; Zhang, Y.; Duan, S.; Sheteiwy, M.S.A.; Zhang, H.; Shao, H.; Guo, X. TaHsfA2-1, a new gene for thermotolerance in wheat seedlings: Characterization and functional roles. *J. Plant Physiol.* **2020**, 153135. [CrossRef]

223. Talukder, S.K.; Babar, M.A.; Vijayalakshmi, K.; Poland, J.; Prasad, P.V.V.; Bowden, R.; Fritz, A. Mapping QTL for the traits associated with heat tolerance in wheat (*Triticum aestivum* L.). *BMC Genet.* **2014**, *15*, 1–13. [CrossRef]
224. Bhusal, N.; Sharma, P.; Sareen, S.; Sarial, A. Mapping QTLs for chlorophyll content and chlorophyll fluorescence in wheat under heat stress. *Biol. Plant.* **2018**, *62*, 721–731. [CrossRef]
225. Sharma, D.K.; Torp, A.M.; Rosenqvist, E.; Ottosen, C.O.; Andersen, S.B. QTLs and potential candidate genes for heat stress tolerance identified from the mapping populations specifically segregating for Fv/Fm in wheat. *Front. Plant Sci.* **2017**, *8*, e01668. [CrossRef]
226. Pinto, R.S.; Reynolds, M.P.; Mathews, K.L.; McIntyre, C.L.; Villegas, J.O.; Chapman, S.C. Heat and drought adaptive QTL in a wheat population designed to minimize confounding agronomic effects. *Theor. Appl. Genet.* **2010**, *121*, 1001–1021. [CrossRef]
227. Shirdelmoghanloo, H.; Taylor, J.D.; Lohraseb, I.; Rabie, H.; Brien, C.; Timmins, A.; Martin, P.; Mather, D.E.; Emebiri, L.; Collins, N.C. A QTL on the short arm of wheat (*Triticum aestivum* L.) chromosome 3B affects the stability of grain weight in plants exposed to a brief heat shock early in grain filling. *BMC Plant Biol.* **2016**, *16*, 1–15. [CrossRef] [PubMed]
228. Mason, R.E.; Mondal, S.; Beecher, F.W.; Pacheco, A.; Jampala, B.; Ibrahim, A.M.; Hays, D.B. QTL associated with heat susceptibility index in wheat (*Triticum aestivum* L.) under short-term reproductive stage heat stress. *Euphytica* **2010**, *174*, 423–436. [CrossRef]
229. Shi, W.; Hao, C.; Zhang, Y.; Cheng, J.; Zhang, Z.; Liu, J.; Yi, X.; Cheng, X.; Sun, D.; Xu, Y. A combined association mapping and linkage analysis of kernel number per spike in common wheat (*Triticum aestivum* L.). *Front. Plant Sci.* **2017**, *8*, e01412. [CrossRef]
230. Azam, F.; Chang, X.; Jing, R. Mapping QTL for chlorophyll fluorescence kinetics parameters at seedling stage as indicators of heat tolerance in wheat. *Euphytica* **2015**, *202*, 245–258. [CrossRef]
231. Vijayalakshmi, K.; Fritz, A.K.; Paulsen, G.M.; Bai, G.; Pandravada, S.; Gill, B.S. Modeling and mapping QTL for senescence-related traits in winter wheat under high temperature. *Mol. Breed.* **2010**, *26*, 163–175. [CrossRef]
232. Shi, S.; Azam, F.I.; Li, H.; Chang, X.; Li, B.; Jing, R. Mapping QTL for stay-green and agronomic traits in wheat under diverse water regimes. *Euphytica* **2017**, *213*, 246–254. [CrossRef]
233. Kumar, U.; Joshi, A.K.; Kumari, M.; Paliwal, R.; Kumar, S.; Roder, M.S. Identification of QTLs for stay green trait in wheat (*Triticum aestivum* L.) in the 'Chirya 3'×'Sonalika'population. *Euphytica* **2010**, *174*, 437–445. [CrossRef]
234. Yang, D.; Li, M.; Liu, Y.; Chang, L.; Cheng, H.; Chen, J.; Chai, S. Identification of quantitative trait loci and water environmental interactions for developmental behaviors of leaf greenness in wheat. *Front. Plant Sci.* **2016**, *7*, e00273. [CrossRef]
235. Sita, K.; Sehgal, A.; HanumanthaRao, B.; Nair, R.M.; Vara Prasad, P.; Kumar, S.; Gaur, P.M.; Farooq, M.; Siddique, K.H.; Varshney, R.K. Food legumes and rising temperatures: Effects, adaptive functional mechanisms specific to reproductive growth stage and strategies to improve heat tolerance. *Front. Plant Sci.* **2017**, *8*, e01658. [CrossRef]
236. Gonzalez-Schain, N.; Dreni, L.; Lawas, L.M.; Galbiati, M.; Colombo, L.; Heuer, S.; Jagadish, K.S.; Kater, M.M. Genome-wide transcriptome analysis during anthesis reveals new insights into the molecular basis of heat stress responses in tolerant and sensitive rice varieties. *Plant Cell Physiol.* **2016**, *57*, 57–68. [CrossRef]
237. Zou, H.; Tzarfati, R.; Hubner, S.; Krugman, T.; Fahima, T.; Abbo, S.; Saranga, Y.; Korol, A.B. Transcriptome profiling of wheat glumes in wild emmer, hulled landraces and modern cultivars. *BMC Genom.* **2015**, *16*, 1–21. [CrossRef] [PubMed]
238. Wan, Y.; Poole, R.L.; Huttly, A.K.; Toscano-Underwood, C.; Feeney, K.; Welham, S.; Gooding, M.J.; Mills, C.; Edwards, K.J.; Shewry, P.R. Transcriptome analysis of grain development in hexaploid wheat. *BMC Genom.* **2008**, *9*, 1–16. [CrossRef] [PubMed]
239. Singh, A.; Mantri, S.; Sharma, M.; Chaudhury, A.; Tuli, R.; Roy, J. Genome-wide transcriptome study in wheat identified candidate genes related to processing quality, majority of them showing interaction (quality x development) and having temporal and spatial distributions. *BMC Genom.* **2014**, *15*, 1–26. [CrossRef]
240. Keller, M.; Simm, S.; Consortium, S.I. The coupling of transcriptome and proteome adaptation during development and heat stress response of tomato pollen. *BMC Genom.* **2018**, *19*, 447. [CrossRef] [PubMed]
241. Rollins, J.; Habte, E.; Templer, S.; Colby, T.; Schmidt, J.; Von Korff, M. Leaf proteome alterations in the context of physiological and morphological responses to drought and heat stress in barley (*Hordeum vulgare* L.). *J. Exp. Bot.* **2013**, *64*, 3201–3212. [CrossRef] [PubMed]
242. Liu, G.-T.; Ma, L.; Duan, W.; Wang, B.-C.; Li, J.-H.; Xu, H.-G.; Yan, X.-Q.; Yan, B.-F.; Li, S.-H.; Wang, L.-J. Differential proteomic analysis of grapevine leaves by iTRAQ reveals responses to heat stress and subsequent recovery. *BMC Plant Biol.* **2014**, *14*, 110. [CrossRef] [PubMed]
243. Lu, Y.; Li, R.; Wang, R.; Wang, X.; Zheng, W.; Sun, Q.; Tong, S.; Dai, S.; Xu, S. Comparative proteomic analysis of flag leaves reveals new insight into wheat heat adaptation. *Front. Plant Sci.* **2017**, *8*, e01086. [CrossRef]
244. Kumar, R.R.; Singh, K.; Ahuja, S.; Tasleem, M.; Singh, I.; Kumar, S.; Grover, M.; Mishra, D.; Rai, G.K.; Goswami, S. Quantitative proteomic analysis reveals novel stress-associated active proteins (SAAPs) and pathways involved in modulating tolerance of wheat under terminal heat. *Funct. Int. Genom.* **2019**, *19*, 329–348. [CrossRef]
245. Wang, X.; Xu, Y.; Hu, Z.; Xu, C. Genomic selection methods for crop improvement: Current status and prospects. *Crop J.* **2018**, *6*, 330–340. [CrossRef]
246. Ahmed, M.; Stockle, C.O. *Quantification of Climate Variability, Adaptation and Mitigation for Agricultural Sustainability*; Springer: Berlin/Heidelberg, Germany, 2016.
247. Zahid, A.; Ali, S.; Ahmed, M.; Iqbal, N. Improvement of Soil Health through Residue Management and Conservation Tillage in Rice-Wheat Cropping System of Punjab, Pakistan. *Agronomy* **2020**, *10*, 1844. [CrossRef]

248. Sheirdil, R.A.; Hayat, R.; Zhang, X.X.; Abbasi, N.A.; Ali, S.; Ahmed, M.; Khattak, J.Z.K.; Ahmad, S. Exploring potential soil bacteria for sustainable wheat (*Triticum aestivum* L.) Production. *Sustainability* **2019**, *11*, 3361. [CrossRef]
249. Qadeer, U.; Ahmed, M.; Hassan, F.U.; Akmal, M. Impact of nitrogen addition on physiological, crop total nitrogen, efficiencies and agronomic traits of the wheat crop under rainfed conditions. *Sustainability* **2019**, *11*, 6486. [CrossRef]
250. Asseng, S.; Martre, P.; Maiorano, A.; Rotter, R.P.; O'Leary, G.J.; Fitzgerald, G.J.; Girousse, C.; Motzo, R.; Giunta, F.; Babar, M.A.; et al. Climate change impact and adaptation for wheat protein. *Glob. Chang. Biol.* **2019**, *25*, 155–173. [CrossRef] [PubMed]
251. Ahmed, M. Introduction to Modern Climate Change. *Sci. Total Environ.* **2020**, *734*, 139397. [CrossRef]
252. Ahmad, S.; Abbas, G.; Ahmed, M.; Fatima, Z.; Anjum, M.A.; Rasul, G.; Khan, M.A.; Hoogenboom, G. Climate warming and management impact on the change of phenology of the rice-wheat cropping system in Punjab, Pakistan. *Field Crop. Res.* **2019**, *230*, 46–61. [CrossRef]
253. Fatima, Z.; Ahmed, M.; Hussain, M.; Abbas, G.; Ul-Allah, S.; Ahmad, S.; Ahmed, N.; Ali, M.A.; Sarwar, G.; Haque, E.U.; et al. The fingerprints of climate warming on cereal crops phenology and adaptation options. *Sci. Rep.* **2020**, *10*, 18013. [CrossRef]
254. Ahmed, M.; Qadeer, U.; Fayayz-ul-Hassan; Fahad, S.; Naseem, W.; Duangpan, S.; Ahmad, S. Abiotic stress tolerance in wheat and the role of silicon: An experimental evidence. In *Agronomic Crops: Volume 3: Stress Responses and Tolerance*; Hasanuzzaman, M., Ed.; Springer: Singapore, 2020; pp. 443–479. [CrossRef]
255. Ahmed, M.; Hassan, F.U.; Aslam, M.; Aslam, M.A. Physiological attributes based resilience of wheat to climate change. *Int. J. Agric. Biol.* **2012**, *14*, 407–412.
256. Ahmed, M.; Aslam, M.A.; Hassan, F.U.; Hayat, R.; Ahmad, S. Biochemical, physiological and agronomic response of wheat to changing climate of rainfed Pakistan. *Pak. J. Bot.* **2019**, *51*, 535–551. [CrossRef]
257. Ahmed, M.; Akram, M.N.; Asim, M.; Aslam, M.; Hassan, F.-U.; Higgins, S.; Stöckle, C.O.; Hoogenboom, G. Calibration and validation of APSIM-Wheat and CERES-Wheat for spring wheat under rainfed conditions: Models evaluation and application. *Comput. Electron. Agric.* **2016**, *123*, 384–401. [CrossRef]
258. Ahmed, K.; Shabbir, G.; Ahmed, M.; Shah, K.N. Phenotyping for drought resistance in bread wheat using physiological and biochemical traits. *Sci. Total Environ.* **2020**, *729*, 139082. [CrossRef]

plants

MDPI

Article

Agro-Morphological, Yield and Quality Traits and Interrelationship with Yield Stability in Quinoa (*Chenopodium quinoa* Willd.) Genotypes under Saline Marginal Environment

M. Iftikhar Hussain [1,2,3,*], Adele Muscolo [4], Mukhtar Ahmed [5,6], Muhammad Ahsan Asghar [7] and Abdullah J. Al-Dakheel [8]

1 International Center for Biosaline Agriculture (ICBA), Crop Diversification and Genetic Improvement Section, Dubai P.O. Box 14660, UAE
2 Department of Plant Biology & Soil Science, Faculty of Biology, Campus As Lagoas Marcosende, University of Vigo, 36310 Vigo, Spain
3 CITACA, Agri-Food Research and Transfer Cluster, Campus da Auga, University of Vigo, 32004 Ourense, Spain
4 Department of Agriculture, Mediterranea University, Feo di Vito, 89122 Reggio Calabria, Italy; amuscolo@unirc.it
5 Department of Agricultural Research for Northern Sweden, Swedish University of Agricultural Sciences, SE-901 83, Umea, Sweden; mukhtar.ahmed@slu.se
6 Department of Agronomy, PMAS Arid Agriculture University, Rawalpindi 46000, Pakistan
7 CAS Key Laboratory of Mountain Ecological Restoration and Bioresource Utilization & Ecological Restoration Biodiversity Conservation Key Laboratory of Sichuan Province, Chengdu Institute of Biology, Chinese Academy of Sciences, University of Chinese Academy of Sciences, Chengdu 610000, China; ahsanasghar2017@mails.ucas.ac.cn
8 Research Farms, United Arab Emirates University, Al-Ain P.O. Box 15551, UAE; a.dakheel@uaeu.ac.ae
* Correspondence: iftikhar@uvigo.es

Received: 10 November 2020; Accepted: 11 December 2020; Published: 13 December 2020

Abstract: Quinoa (*Chenopodium quinoa* Willd.) is a halophytic crop that shows resistance to multiple abiotic stresses, including salinity. In this study we investigated the salinity tolerance mechanisms of six contrasting quinoa cultivars belonging to the coastal region of Chile using agro-physiological parameters (plant height (PH), number of branches/plant (BN), number of panicles/plant (PN), panicle length (PL), biochemical traits (leaf C%, leaf N%, grain protein contents); harvest index and yield (seed yield and plant dry biomass (PDM) under three salinity levels (0, 10, and 20 d Sm^{-1} NaCl). The yield stability was evaluated through comparision of seed yield characteristics [(static environmental variance (S^2) and dynamic Wricke's ecovalence (W^2)]. Results showed that significant variations existed in agro-morphological and yield attributes. With increasing salinity levels, yield contributing parameters (number of panicles and panicle length) decreased. Salt stress reduced the leaf carbon and nitrogen contents. Genotypes Q21, and AMES13761 showed higher seed yield (2.30 t ha^{-1}), more productivity and stability at various salinities as compared to the other genotypes. Salinity reduced seed yield to 44.48% and 60% at lower (10 dS m^{-1}) and higher salinity (20 dS m^{-1}), respectively. Grain protein content was highest in NSL106398 and lowest in Q29 when treated with saline water. Seed yield was positively correlated with PH, TB, HI, and C%. Significant and negative correlations were observed between N%, protein contents and seed yield. PH showed significant positive correlation with APL, HI, C% and C:N ratio. HI displayed positive correlations with C%, N% and protein content., All measured plant traits, except for C:N ratio, responded to salt in a genotype-specific way. Our results indicate that the genotypes (Q21 and AMES13761) proved their suitability under sandy desert soils of Dubai, UAE as they exhibited higher seed yield while NSL106398 showed an higher seed protein content. The present research highlights the need to preserve quinoa biodiversity for a better seedling establishment, survival and stable yield in the sandy desertic UAE environment.

Keywords: salinity; *Chenopodium quinoa*; biomass; functional plant traits; biochemical traits; genotypes; yield; salt stress

1. Introduction

Soil degradation due to salinity is a big issue in agriculture and forestry, especially in marginal environments. Several factors, such as scarce water resources, loss of topsoil due to wind erosion, sandy soils and high temperature in desert ecosystems are major constraints for crop production [1–3] affecting 250 million people [4]. Many arid and semi-arid regions of the world exhibit a significant portion of salt and degraded marginal lands. According to Wang et al. [4], approximately 20% of the land area is degraded. Global Assessment of Soil degradation (GLASOD) reported that 12 million heactares of land are degraded each year in arid ecosystems, at a cost to the global economy of up to US $42 billion per annum [5]. Furthermore, the continuous increase in global demand for food, fuel and feed has shifted the focus towards degraded lands, because land suitable for food production is shrinking worldwide [6]. This situation is further aggreviated by strong winds in arid and semi arid areas that cause erosion and land degradation, removing the top productive soil layers that can support plant growth [7].

Salinity is a complex phenomenon that critically impacts the morphological and physiological traits of crops through modifications in the osmotic balance, ion homeostasis, and reactive oxygen species regulation, each having a complex and less understandable genetic basis. High accumulation of salts in saline soils results into reduced water potential of soil solutions which causes difficulty for plants to extract water from soil experiencing "osmotic stress" [8]. Specific ion toxicity, the result of excessive uptake of certain ions (Na^+ and Cl^-) is the primary cause of growth reduction [2,9–11]. Toxic ions in salt-affected soils are usually sodium, chloride and sulphate [2,12]. The excessive Na^+ accumulation causes ion toxicity and interferes with plant metabolism, while accumulation of K^+ can alleviate Na^+ toxicity by adjusting osmotic potential and through ion balance. Crop performance may also be adversely affected by salinity-induced nutritional imbalances [13,14]. These imbalances may result from the effect of salinity on nutrient availability, competitive uptake, transport and/or partitioning within the plant caused by physiological inactivation of a given nutrient resulting in increased plant's internal requirement for that essential element [2,15,16]. One or more of these processes may occur at the same time, but whether they ultimately affect crop yield or quality depends on the level of salt stress, composition of salts, crop species, the nutrient in question and a number of environmental factors [15,16]. Therefore, it is imperative to advance our knowledge regarding the identification and evaluation of genotypes and landraces that can be cultivated in nutrient poor and sandy marginal soils. Moreover, these germplasms might have the potential for the restoration of these regions. Several studies concerning salinity tolerance in cultivated crops have been reported, while considerable advances have been made in the development of crop genotypes resistant to drought or salinity [17,18]. Therefore, salt-affected soils can be utilized by growing salt tolerant plants, whether halophytes or non-halophyte crops. The use of halophytic crop species can be considered as a valuable option to sustain agricultural production under saline and dry conditions and potentially under irrigation with saline waters [19,20]. However, it is imperative to explore intra-specific (inter-cultivar) variation for salt tolerance of a crop by screening its available germplasm.

Halophytes have shown potential to be useful resources for global food production, and contribute to the rehabilitation of salt-degraded lands. Quinoa (*Chenopodium quinoa* Willd.) is an important facultative halophyte, and its demand has increased recently in all continents [21]. Quinoa is a highly nutritious pseudo-cereal and has a wide potential to enhance food security through tolerating saline and marginal lands and to alleviate pressure on fertile agricultural soils [22]. Quinoa seeds are enriched in proteins, amino acids (lysine, methionine, threonine) [23]. Different fatty acids (linoleic and linole) and oleic acids have been reported in the seed oil [23]. From a human health perspective these fatty acids are of high quality as compared to those reported from other cereals [24]. Moreover, quinoa seeds are free from any allergic effects that may be caused by harmful chemicals or gluten, which is present in other cereals [25]. Quinoa exhibit various minerals like calcium, phosphorus, potassium, magnesium, phosphorus and zinc in sufficient quantity and protein contents (12–17%) [26]. The quinoa crop has been internationally recognized as a contributor to global food security because it is tolerant to many environmental constraints, such as frost [27,28], drought [29], and salinity [30]. In many developing African countries, quinoa has been significantly introduced in agro-ecosystem and has contributed to the regional food security [31,32]. However, evaluation of salinity tolerance potential of different quinoa genotypes and the possible evaluation of their growth, development and yield behaviour in salt-degraded marginal Arabian Peninsula Sandy desert soils are rarely studied.

The phenotypic plasticity, genotype variability and agronomic adaptation of quinoa are extremely wide and varied significantly from hot arid to subtropical humid climates [33]. Under these circumstances, it is viable to select, introduce, adapt and breed different quinoa genotypes in a wide range of environments. However, to increase food security and agriculture productivity in resource poor and degraded marginal lands like those of United Arab Emirates, it is imperative to study the salt tolerance potential of different quinoa genotypes and assess their yield stability without losing the grain quality. Jacobsen et al. [33], the quinoa plant photoperiod is a critical functional trait and should be evaluated before introducing it in a particular environment. Most of the previously referred studies have been carried out in quinoa germination and seedling growth responses under the control growth chambers. It is important to understand that germination and early seedling establishment stage is critical in the life cycle of plants, and many colleagues have documented these phenomena [34,35]. However, yield response factors of different quinoa genotypes to saline and marginal soil environment and their salinity tolerance mechanisms that may regulate seedling growth, yield and grain quality attributes need to be evaluated under marginal and nutrient poor sandy soils.

Studying the salinity tolerance potential of different quinoa genotypes and the elucidation of seed yield and yield stability is of paramount importance. Therefore, the evaluation of salinity tolerance potential of different quinoa genotypes was conducted under the hyper arid climate of Dubai, UAE with the following main objectives: (i) evaluation of the adaptation of accessions of different origins to the UAE sandy desert marginal environment with spring sowing, (ii) assessment of the salinity tolerance potential of six quinoa genotypes, (iii) determination of the variation and heritability of quinoa morphological and quality traits and their interrelationship with yield attributes and (iv) identification of the impact of saline water in growth, yield stability and grain protein content. The study will help to discriminate between resistant and sensitive quinoa genotypes and to understand the mechanisms of adaptation, for selecting genotypes tolerant to saline water irrigation in order to adapt to salt-degraded marginal sandy desert in the hyper arid UAE environment.

2. Material and Methods

2.1. Experimental Site

The study was carried out at the International Center for Biosaline Agriculture (ICBA) research station in Dubai (United Arab Emirates). The site has latitude of 25°13 N, a longitude of 55°17 E and is 16 m above mean sea level. The local climate is dominated by hot dry weather (April–October), with no rainfall at all during this transitional period and higher air temperature (exceeding 50 °C) during peak season (June, July, August) and high humidity. The winter season (December–February) is cool and dry. The field plot soil is a well drained sandy coarse (97% Sand, 2.2% Silt, 0.2% clay). The soil, classified as Carbonatic, Hyperthermic typic Torripsamment has an EC of 0.2 dS m^{-1}. The soil physical and chemical properties are presented in Table 1 as the mean of two consecutive years (2013 and 2014).

Table 1. Soil classification, chemical and physical properties from 0–60 cm depth os soils at pre-sowing 2013 and post-harvest 2014 from each sub-plot, irrigated with 10 and 20 dS m^{-1}.

Parameters	Values	
Pre-Sowing 2013	Post-Harvest 2014	Post-Harvest 2014
-	10 dS m^{-1}	20 dS m^{-1}
Sand (%)	97.60	97.60
Silt (%)	2.20	2.20
Clay (%)	0.20	0.20
Soil textural class	Sand	Sand
Ece dS m^{-1}	2.04	4.10
pHs	7.04	7.31
Total N mg kg^{-1}	52.00	51.59
P mg kg^{-1}	41.51	46.74
K mg kg^{-1}	45.95	41.61
%Organic matter	1.46	1.32

Ece: Electrical Conductivity.

2.2. Field Soil Physiological and Chemical Analysis

For the analysis of soil salinity, soil samples were taken from each of the sub-plot from 0–60 cm depth prior to the experiment and after harvesting the crop. The air-dried samples were analyzed for physiochemical analysis at the Central Analytical Laboratory, International Center for Biosaline Agriculture, Dubai (UAE). Standard soil analysis methods [36] were used and the results were presented on oven dried soil weight basis (Table 1). The electrical conductivity (E_C) was measured in the soil extract collected from the saturated soil paste (dS m^{-1}). Soil pH of the saturated paste was measured with a standard pH meter calibrated using buffer solutions (pH 4, 7 and 10). Soil texture was determined by using standard Pipette method and wet sieving and sand, silt and clay contents were used on USDA triangle to determine soil textural class. The total amount of nitrogen was determined using standard Kjeldahl method and phosphorous through colorimetri procedure. Available K was detected in 1 N ammonium acetate extract using a flame photometer. Organic carbon (OC) was measured through dichromate oxidation and converted to organic matter by multiplying by a factor of 1.72 [37].

2.3. Land Preparation, Sowing, Plant Growth and Agronomic Practices

The experiemental soil was prepared by a disc plough, harrowed in order to obtain a good seed bed. The organic fertilizer (40 t ha^{-1}) was applied and incorporated into the soil to improve the soil fertility. The chemical fertilizer {NPK (20:20:20)} at the rate of 50 Kg ha^{-1} was applied in two split doses by banding alongside the rows. Experimental design was a randomized complete block with split plot arrangement replicated three times. Quinoa genotypes (Table 2) were manually planted on 26 November 2013. Three salinity treatments (0, 10 and 20 dS m^{-1}) were assigned to each main plot (the salinity treatments were applied one month after the seedling establishment), and six genotypes were assigned to the sub plots. Seeding was carried out by burying 3–4 seeds into the soil to a depth of 1–2 cm close to the dripper. Each genotype was planted in five rows (3 m long) per plot, 25-cm interplant spacing and 50 cm between the rows and one meter between two accessions. The plots were drip irrigated through out the study period with water of different salinities through drop laterals. The field was covered with acryl sheet, following sowing, to protect the plants from bird attacks (Figure 1). The seedlings were thinned to one plant per point, later, following plant establishment.

Table 2. A list of six quinoa (*Chenopodium quinoa* Willd.) genotypes native to the Chilian coastal region that were used to identify ecophysiological indicators suitable for quantifying seedling salinity responses.

S. No.	Code	Germination Line	Source	Origin	Status	Seed Color
2	**Q 18**	*C. quinoa PI614886*	USDA	Maule, Chile	Cultivated	Yellow
4	**Q 21**	*C. quinoa PI614889*	USDA	Bio-Bio, Chile	Cultivated	Orche
5	**Q 22**	*C. quinoa PI634918*	USDA	Chile	//	Yellow
8	**Q 29**	*C. quinoa PI634925*	USDA	Chile	//	Yellow
11	**AMES 13761**	*C. quinoa*	USDA	USA	-	-
12	**NSL 106398**	*C. quinoa*	USDA	USA	-	-

Figure 1. Photos to illustrate the sandy desert marginal soil of ICBA (**a**); irrigation system installation (**b**); quinoa seed sowing and covering with acryl sheet to protect from bird attacks (**c**); quinoa crop at panicle stage (**d**).

2.4. Irrigation System and Treatment Application

During each growing season, saline water treatments (0, 10 and 20 dS m^{-1}) were established and applied after one month from sowing (to ensure plant establishment) using drip irrigation system. Irrigation was supplied via a drip system (drip laterals of 16 mm in diameter), 0.25 cm emitters and were delivering 4.0 L h^{-1}. The experimental plots were equipped with three irrigation valves from RainBird Company (Azusa, CA, USA). One valve was handling fresh, other saline and third valve (a solenoid valve of 2″ size) controlling both saline and fresh water after the main control valve. The 3rd valve was controlling the irrigation water according to the main controller instructions. Saline irrigation water was provided from seedlings (one month old) to the grain filling stage. The climate data in terms of temperature, rainfall, wind speed and humidity was collected from weather station established at the ICBA (Figure 2).

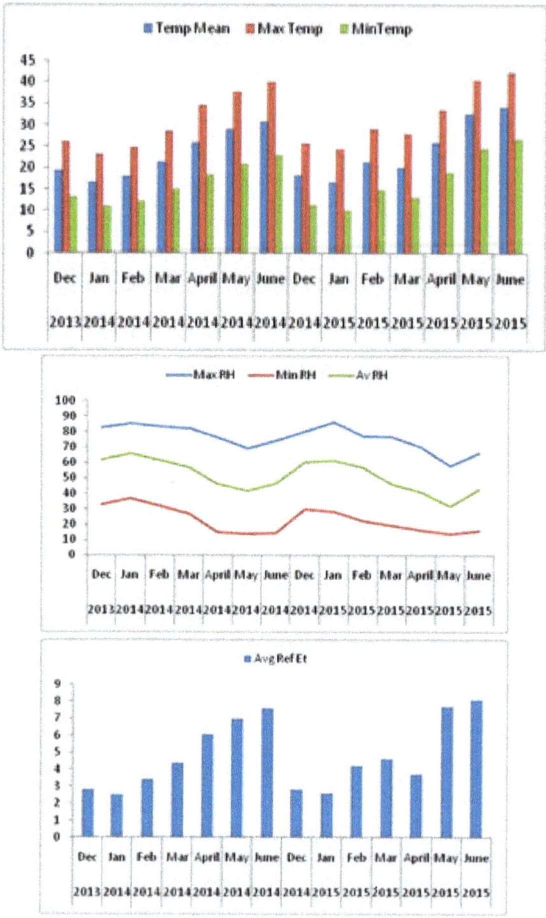

Figure 2. Monthly values of mean (T mean), maximum (T max), and minimum (T min) air temperature, relative humidity and reference evapotranspiration (ETo) in the ICBA weather station, Dubai, UAE.

2.5. Crop Phenology and Morphological Measurements

During the whole crop season, hand weeding was carried out when needed, without applying any herbicide. The data was collected from the middle 1 m of the two central rows. Following the completion of the physiological attributes, other morphological measurements were collected from five plants from each subplot. The average plant height (cm) from the ground level to the tip of the panicle on the main stem was measured at physiological maturity. The total number of branches from the main stem at different node positions, including the basal branches were recorded.

2.6. Harvesting, Biomass and Yield Traits

The yield data was collected at physiological maturity (when the seeds from panicle became hard [38]. From each plot, the number of panicles per plant was counted from five plants. The mean length (cm) of three panicles was taken randomly and averaged. The plant fresh biomass was recorded and then plant material was oven dried at 80 °C for three days and weighted. From each sub-plot, a sample line of 1 m length from the central rows was harvested and seeds were removed from the panicles of the plants, threshed, and weighed (g m^{-2}) and then converted into t ha^{-1}.

2.7. Leaf Carbon (%) and Nitrogen Contents (%) Analysis

The leaf samples from each treatment/plot and control were collected, oven dried and ground into a fine powder. Total N and C contents (% dry matter) were measured by elemental analysis, Flash EA-1112 (Thermo Fisher Scientific, Schwerte, Germany).

2.8. Harvest Index (%)

Harvest index was calculated by using the following formula:

$$\text{Harvest index (\%)} = \text{Grain yield/dry biomass} \times 100 \quad (1)$$

2.9. Grain Protein Contents Measurements

The grains (FW = 200 mg) from each genotype (three replicates/treatment) were crushed into liquid nitrogen and protein contents were measured using Commercial bovine seroalbumin (BSA) through Bradford assays [39] as reported by Hussain and Reigosa [40].

2.10. Statistical Analysis

The agro-physiological parameters were divided into two categories; yield and biochemical traits (carbon (C%), nitrogen (N%), seed yield (SY), harvest index (HI), grain protein contents); agro-morphological traits (plant dry matter (PDM), number of branches (BN), number of panicles (PN)). For each trait, the genotype-treatment combinations (i.e., six genotypes crossed with three salinity treatments) were subjected to analysis in order to summarize the relative merit of genotypic effects and growing conditions as causes of changes in the plant ecophysiological attributes and Dunnett test was employed as post-hoc test for multiple comparison. All analysis were conducted through General Linear Modeling (GLM) procedure and analysis of variance (ANOVA) using the SPSS for Windows version 17.0 software (SPSS Inc., Chicago, IL, USA). Difference between treatments means were compared using Tukey's HSD test. A Pearson's correlation matrix was conducted to assess the relative contribution of ecophysiological trait associations towards the seed yield at overall salinity.

A static yield stability index was calculated according to environmental variance (S^2) [41]. Meanwhile, a dynamic yield stability index was presented following Wricke's ecovalence (W^2) [42].

3. Results

3.1. The Effect of Treatments and Genotypes on Growth Parameters

Quinoa plants were significantly shorter (69.44 cm) after treatment with salinity level of 20 dS m^{-1} as compared to the control (102.55 cm) (Table 3). Genotypes also showed a significant variation in PH that was higher in quinoa genotypes AMES 13,761 (120.44 cm), followed by Q21 (87 cm), Q29 (79.88 cm) and NSL106398 (77 cm) (Table 4). The Q22 showed PH of 73.33 cm (Table 4). The Q18 had the maximum number of branches/plant (19.0), followed by Q29 (15.66), Q21 (15.33) and AMES13761 (15.0) (Table 4). The lowest number of branches/plant was recorded in Q22 (14.11). Water salinity decreased the number of branches/plant that was significantly reduced at each salinity level as compared to control. The highest numbers of panicles/plant were observed in Q18 (17.44), followed by Q29 (14.22) and NLS106398 (13.56). Q21 and Q22 demonstrated in average a similar number of panicles (12), while the lowest number of panicles was produced by AMES13761 (10.56). Quinoa genotypes, AMES 13761, Q21, and Q29 produced longest panicles i.e., 20.78, 19.89, 18.78 cm respectively. Genotype Q18, Q22 and NSL106398 produced the smaller size panicle (17.0–17.74 cm). The application of S3 (20 dS m^{-1}) caused retardation in number of panicles/plant. However, average panicle length was unaffected by salinity treatment. Quinoa genotype Q21 and Q22 produced the highest plant dry biomass (9.0 t/ha) followed by Q18 and Q29 that on average produced 6.28 and 6.74 t/ha. AMES 13,761 (7.94 t/ha) and NSL106398 produced on average 5.91 and 5.45 t/ha dry biomass, respectively (Table 4). Furthermore, plant dry biomass was unaffected by saline water treatment at all salinity level compared to control.

Table 3. Genotype and treatment effects on biomass and agro-physiological traits, and yield components of six quinoa genotypes grown under different water salinity levels.

Variables	Plant Height (cm)	Number of Branches/Plant	No. of Panicles/Plant	Average Panicles Length (cm)	Plant Dry Biomass (t/ha)	Leaf N%	Leaf C%	C:N Ratio	Grain Protein Content	SY (t/ha)	HI
S1–0 (Control)	102.55a	16.88a	14.56a	18.23a	6.86b	1.41a	28.46a	13.73a	8.87c	2.63a	37.56a
S2–10 dS m^{-1}	84.22b	14.72b	12.39c	18.57a	6.64b	1.61a	26.54b	7.94b	10.06b	1.46b	24.59b
S3–20 dS m^{-1}	69.44c	15.33b	13.5b	18.85a	6.71a	1.75a	25.97b	8.4b	10.98a	1.06c	14.88c
Genotype (G)	0.00	0.29	0.05	0.22	0.00	0.00	0.265	0.105	0.000	0.04	0.012
Treatment (T)	0.00	0.35	0.35	0.88	0.36	0.00	0.00	0.00	0.00116	0.00	0.00
G × T interaction	0.79	0.296	0.35	0.48	0.37	0.00	0.91	0.85	0	0.02	0.016

Genotype values are the mean of nine measurements (three treatments and three replications per treatment), while treatment values are the mean of the 54 measurements (six genotypes and three replications per genotype). Means followed by different letters are significantly different ($p \leq 0.05$) according to Tukey's honestly significant difference (HSD) test. Treatments: S1—0, (control); medium salinity—S2, 10 dS m^{-1}; high salinity—S3, 20 dS m^{-1}; ns, not significant; G, Genotypes; T, Treatment.

Table 4. Genotype and treatment effects on biomass and agro-physiological traits, and yield components of six quinoa genotypes grown under different water salinity levels.

Variables	Plant Height (cm)	Number of Branches/Plant	No. of Panicles/Plant	Average Panicles Length (cm)	Plant Dry Biomass (t/ha)	Leaf N%	Leaf C%	C:N Ratio	Grain Protein Content	SY (t/ha)	HI
Genotypes (G)	-	-	-	-	-	-	-	-	-	-	-
Q 18	74.88d	19.0a	17.44a	17.0b	6.74b	1.58b	27.05a	9.88c	9.89b	1.27b	20.19c
Q 21	87.0b	15.33b	12.88c	19.89a	9.0a	1.43c	27.82a	8.04cd	8.96c	2.3a	26.98b
Q 22	73.33d	14.11c	12.22c	17.12b	9.01a	1.45c	26.45b	11.97a	9.12b	1.53b	19.05c
Q 29	79.88c	15.66b	14.22b	18.78a	6.28b	1.4c	26.76b	10.33b	8.76c	1.72b	26.04b
AMES 13761	120.44a	15b	10.56d	20.78a	5.91c	1.47c	27.16a	10.56b	9.19b	1.77b	31.8a
NSL 106398	77.0c	14.78c	13.56b	17.74b	5.45c	2.23a	26.74b	9.35c	13.9a	1.69b	30a

Genotype values are the mean of nine measurements (three treatments and three replications per treatment), while treatment values are the mean of the 54 measurements (six genotypes and three replications per genotype). Means followed by different letters were significantly different ($p \leq 0.05$) according to Tukey's honestly significant difference (HSD) test. Treatments: S1—0, (control); medium salinity—S2, 10 dS m^{-1}; high salinity—S3, 20 dS m^{-1}; ns, not significant; G, Genotypes; T, Treatment.

3.2. Saline Water Irrigation Impact on Carbon (C%) and Nitrogen (N%) and C/N Ratios

Genotypes differed for all traits, whereas a significant interaction existed between genotype and treatments (GT) for leaf N%, grain yield and harvest index (Table 3). The highest salinity treatment was lethal and affected all growth studied traits. There was significant difference in the leaf N% among the genotypes and NSL106398 exhibited the highest N% (2.23%), followed by Q18 (1.58%). Genotype Q29 showed the lowest N% (1.4%) in the dry leaf samples. The C% value was the highest in genotype Q21 (27.82%), followed by AMES13761 (27.16%) and Q18 (27.05%) (Table 4). The genotype Q22 exhibited the lowest values of C% (26.45%). Both salinity levels significantly decreased the C% as compared to the control. There was significant variation in C:N ratio values that were highest in Q22, Q29, AMES13761 and lowest in Q21. The results showed that N concentration was unaffected at all salinity treatments compared to control (Table 4).

3.3. Seed Yield and Harvest Index

The yield components (seed yield, harvest index) were all affected by salinity (Table 3). Genotypes Q21 and AMES13761 exhibited higher seed yield (2.30 t ha^{-1} and 1.77 t ha^{-1}) than the other genotypes (Table 4). The lowest yield was produced by Q18 (1.27 t ha^{-1}) that was 44.7% less than that of Q21 (Table 4). Salinity significantly decrease the seed yield that was 44.48% and 60% lower following exposure to 10 dS m^{-1} and 20 dS m^{-1}, respectively (Table 3). Harvest index (HI) was significantly decreased following saline water treatment. Our results indicated that there was 34.53% and 60.38% reduction after 10 dS m^{-1} and 20 dS m^{-1} salinity treatment as compared to control, respectively (Table 3). Harvest index greatly varied among the genotypes and ranged between 31.8–19.05%. The highest HI values were observed in genotype AMES13761 followed by NSL106398 and the lowest in Q22. The genotype × treatment interaction (G × T) was also significant for seed yield and HI (Table 4).

3.4. Grain Protein Contents

There was significant impact of salt stress on grain protein contents and it was genotype dependent. The GP contents were the highest in NSL106398 (13.9 mg/g DW), followed by Q18 (9.89 mg/g DW), AMES13761 (9.19 mg/g DW) and Q22 (9.12 mg/g DW), respectively. The lowest GP was recorded in Q29 (8.76 mg/g DW) following salt stress treatment. There was a tendency of stimulation in GP following salinity treatment as compared to control (Table 3).

3.5. Yield Stability Trend in Quinoa Genotypes

The grain yield stability among different quinoa genotypes is shown in Table 5. The static environmental variance (S^2) and dynamic Wricke's ecovalence (W^2) for different quinoa genotypes was different. S^2 was in the range of 1.019–3.461.

Table 5. Environmental variance (S^2i) and Wricke's ecovalence (W^2i) over the ambient treatments and three climate treatments for the six quinoa genotypes with highest averaged mean yield across treatments (mi).

S. No.	Genotypes Name	Mi	S^2i	W^2i
1	Q 18	1.274	1.019	1.677
2	Q 21	1.082	3.461	2.080
3	Q 22	1.919	2.310	2.470
4	Q 29	1.494	2.591	2.349
5	AMES 13761	1.398	2.659	5.362
6	NSL 106398	2.577	2.540	9.586

However, a different trend was observed among six quinoa genotypes for W^2 that varied from 1.67–9.58. In stability analysis, the lowest values demonstrate the stability in yield over saline environments. The variety 'Q18' was static stable and high yielder, ranking 1st for S^2i grain yield index across all saline environments. The quinoa 'Q22', 'Q29' and 'NSL106398' were found to produce the the 2nd, 3rd and 4th highest static yield index among these tested genotypes and across all salinity treatments. The genotype 'Q18' showed stable mean yield (W^2i) and ranked 1st among all the genotypes across all environments. Moreover, genotype 'Q22' was S^2i and high yielder, ranking the 2nd for W^2i grain yield index (Table 5).

3.6. Correlations between Seed Yield, Agro-physiological and Yield Attributes

Pearson's correlations analysis showed significantly positive relationships between PH, TB, HI, C% with SY. However, significant and negative correlations were observed between N% and protein contents and seed yield (Table 6). The PH showed significant +ve correlation with AIL, HI, C% and C:N ratio. The NOB and NOI exhibited significant +ve correlation with NOI and AIL. The NOI showed + ve relation with AIL, N%, and protein contents. Harvest index displayed positive correlations with the C%, N% and protein content. N% exhibited +ve relation with C%, C: N ratio and protein contents.

Table 6. Pearson's correlations among physiological and seed yield traits of quinoa as a result of genotypic collective response for all three salinity levels.

Traits	SY	PH	NOB	NOP	APL	TB	HI	N%	C%	CN Ratio	Protein
SY	1	-	-	-	-	-	-	-	-	-	-
PH	0.46 **	1	-	-	-	-	-	-	-	-	-
NOB	−0.001	0.212	1	-	-	-	-	-	-	-	-
NOI	−0.039	0.034	0.832 **	1	-	-	-	-	-	-	-
AIL	−0.118	0.414 **	0.267 **	0.223 **	1	-	-	-	-	-	-
TB	0.209	−0.071	−0.026	0.08	0.058	1	-	-	-	-	-
HI	0.837 **	0.503 **	−0.012	−0.107	−0.167	−0.268 **	1	-	-	-	-
N%	−0.311 **	−0.201	0.136	0.244 **	0.043	−0.174	−0.249 **	1	-	-	-
C%	0.637 **	0.422 **	0.176	0.141	−0.031	0.025	0.637 **	−0.219 **	1	-	-
CN Ratio	0.119	0.264 **	0.051	0.019	0.01	−0.153	0.186	−0.277 **	0.182	1	-
Protein	−0.311 **	−0.201	0.136	0.244 **	0.043	−0.174	−0.249 **	0.000 **	−0.219 **	−0.277	1

** Correlation significant at $p \leq 0.05$ according to Tukey's HSD test; SY: Seed Yield, PH: Plant Height, NOB, Branch number; NOP: Panicles number, APL: Average Panicles Length, TB: Total Biomass, HI: Harvest Index, N%, Nitrogen concentration; C%, Carbon concentration; Protein: Grain Protein contents.

4. Discussion

The rising global demand for nutritious and healthy food has challenged scientists to look for alternate crops, especially for the marginal areas where agricultural production is inefficient due to unfavorable climatic conditions, low soil fertility and lack of good quality irrigation water. In the Arabian Peninsula, scientists are experimenting with quinoa production because it is rich in nutrients, tolerant to salinity and uses much less water than other crops. Against this backdrop, this study identified the agro-morphological traits (PDM, BN, PN) that showed a decreasing trend with increasing salinity. Gómez-Pando et al. [13] showed that germination and plant height were highly decreased in quinoa plants that were subject to different salinity levels. Additionally, some genotypes (total 15 genotypes) were tolerant and less affected while others were susceptible and their agro-physiological attributes were decreased significantly following salinity treatment.

Under the saline and marginal UAE environment, some agro-physiological traits (PH, NBP, and NPP) were decreased after salt stress while PL was unaffected at various salinity levels. A decrease in dry matter yield with increasing soil salinity levels might be due to the inhibition of water availability and hydrolysis of reserved foods and their translocation to the growing shoots [43,44]. Other factors responsible for lower dry biomass yield may include panicle length, chlorophyll concentrations, number of productive tillers, number of primary branches per panicle, and fertility percentage [45]. The reduction in number and size of the panicles per plant is directly related to lower seed yield [46]. Plant biomass, height and seed yield, number of branches, number of panicles, panicle weight and harvest index were reduced in response to saline water irrigation and were subject to genotypic variation [8]. Genotypic variability in seed yield and biomass has been reported before for quinoa plants growing under comparable agroecological conditions [8,43,46]. Moreover, the significant interaction between genotype and irrigation conditions for seed yield, biomass and different agronomical traits highlights not only the genotypic plasticity available in the species but also the need to assess genotypic performance within each growing condition [47]. In the present study, PH was decreased after salinity treatment and AMES 13,761 was the tallest variety while Q22 was dwarf genotype. This was due to the stunted growth of plants caused by the high salt concentration in the nutrient medium. Adolf et al. [48] evaluated 14 quinoa genotypes at different salinity levels and recorded growth and biomass attributes. Pandela rosada and Utusaya were least affected and had capacity to be adapted to the harsh climate of southern altiplano of Bolivia (3600 m above sea level). Another quinoa variety, Amarilla de Maranganí was more tolerant and was not affected with respect to height and biomass production. The results of this study also suggest that plant morphological and agro-physiological characteristics are interrelated factors that highly impact the plant establishment, seedling growth, and yield. Meanwhile, the responses of different quinoa genotypes against salinity were different indicating their genetic diversity. According to different researchers, several mechanisms might contribute towards genotypic differences in salinity tolerance in quinoa. These mechanisms may include Na^+ exclusion from leaf mesophyll cells, better H^+ pumping to restore membrane potential, and Na^+ exclusion from leaf cells demonstrating salinity tolerance of quinoa [32,33,49]. The information about these functional attributes might facilitate the restoration programs of degraded marginal salt affected soils and will help to convert waste to wealth assets.

The evaluation and selection of salt tolerant quinoa genotypes is an important step to pursue their adaptation uder marginal and sandy soils and to check the effect of salinity on grain yield and grain protein contents. The present results demonstrated that significant genetic diversity exists between different quinoa genotypes. Yield components like NPP and PL were different among genotypes. Salinity significantly affected the number of panicles per plant and average panicle length also varies from one to other genotype. According to the report of Long Nguyen [50]; panicle length in quinoa is interconnected with grain yield and variation in this trait lead to significant variation in the final yield. Several researchers have noticed that long panicle bearing genotypes demonstrated higher yield than genotypes with shorter panicles [45,46]. From the results of present study, it was concluded that differences in panicle length were connected with genotypic difference rather than salinity impact.

The seed yield was significantly decreased (60%) at high salinity. It is a typical phenomenon of plants affected by environmental stresses that showed a restricted supply of CO_2 as well as reduced activity of RuBisCO. These processes leads to reduced photosynthesis, carbon assimilation, growth and yield of the plants [16,51–54]. Previously published data showed that the small panicle length, chlorophyll concentrations, number of productive tillers, and lower number of primary branches per panicle were responsible factors in the low yield of quinoa [45].

The salinity treatment did not affect very much the leaf C and N. However, the allocation of N and C was different among the different genotypes and changed in respect to the genetic variability. Grain protein contents (GP) differed significantly after salinity treatment that was stimulated to a cetain extent. Genotype, NSL106398 showed higher grain protein contents while the lowest was observed in Q29. The reduction in quinoa plant dry biomass was 23.7% and 36% after treatment at 10 and 20 dS m^{-1} respectively. The growth of quinoa (cv. Hualhuas) was slightly increased following salinity treatment [55]. The present results demonstrated the competitive advantage of salinity tolerant quinoa genotypes in terms of morphological and ecophysiological attributes [6,13,18,21,43,45,56].

Our results indicate a difference in grain and biological yield (harvest index) among quinoa genotypes, showing that Q21 had a higher seed yield, followed by AMES13761, and both genotypes showed a typical genetic variation. Adolf et al. [48] reported similar results. They found that quinoa genotype, Utusaya (Bolivian origin) had high stomatal conductance compared to control plants and showed a reduction (25%) in CO_2 while "Titicaca" (from Denmark) demonstrated higher reduction (67%) in CO_2 assimilation. To counteract the salinity, Utusaya variety possessed some genetically improved salt tolerance mechanism (osmoprotectants) and water loss through transpiration was much less than other genotypes. Recent studies highlighted that some adaptation mechanisms exist in certain quinoa genotypes that control transpiration and thus, WUE, under saline conditions. Lately, it was correlated with morphological features (stomatal size reduction, density or both) [22,23,47–49]. A significant variation was also oberved in HI among genotypes while AMES13761 showed higher HI, followed by NSL106398 and it was lowest in Q22. Morover, salinity also reduced HI after treatment at 10 dS m^{-1} and 20 dS m^{-1}. These results demonstrated a greater adaptability of quinoa genotypes to the agro-climatic conditions of UAE. Other researchers also documented that Chilean varieties were less sensitive to photoperiod and hence more adaptable to saline and marginal environments [27–29].

The correlation between diggerent physiological, yield and quality characteristics of quinoa are presented in Table 6. Several parameters were positively correlated with other contributing attributes while some also have negative correlation. More prominent +ve correlation was found betweeen NOB, NOI and AIL, TB, C:N ratio). The correlation among grain yield, biological yield and protein contents of grains is often misleading. This can lead to wrong conclusions and policy guidlines for future breeding strategies for marginal environments [57]. Confusion was largely provoked by the fact that the relations between biomass, number of panicle and yield were positive or negative, and sometimes there was no correlation depending on the crop and growing conditions.

Lessons from Different Stress Levels: Trade-off between Survival and Growth

In the present study, the Chilean-based quinoa genotypes displayed remarkable variation in plant establishment, seedling growth, biomass yield, and panicle attributes and salinity responses yield potential. Different salinity treatments (10, 20 dSm^{-1}) caused a significant reduction in biological and grain yield. Even though seed yield was reduced, quinoa was still able to perform relatively well under these sandy, nutrient poor and marginal soil conditions as compared to other high productive soil environments. Under low salt stress (10 dSm^{-1}), average panicle length of quinoa varieties did not differ in their responses to salinity. However, at high salinity (20 dSm^{-1}), seed yield was not highly reduced in all the tested varieties. In this regard, Q21 was highest yielder while Q18 was the most affected with lowest seed yield. We speculate that Q18 employed a "survival" strategy with a more reduced growth rate, biological yield (HI) but higher number of branches and number of panicles and medium level of leaf C% and N%. The grain protein contents were comparative to AMES13761 and Q21 but its seed yield was highly reduced at overall salinity, compared to other varieties. These adaptations allowed Q18 to survive longer, but at the trade-off of the very high reduction in growth rate and seed production. Several authors documented that there is great variation in salinity tolerance among quinoa genotypes [22,23,43,47–49,58–60]. Previously, it was assumed that only Bolivian Salares originated genotypes are high salt tolerance while, now, it is well known that salinity tolerance does not related with geographic distribution and genotypes from Chile coastal areas and highlands are even more salt tolerant [58–60]. Furthermore, it was recently reported that a wild relative of quinoa (*Chenopodium hircinum*) was found to have a much higher salinity tolerance level than quinoa cultivars [59]. Photoperiod and temperature attributes also played a significant role in the growth, development and final grain yield of quinoa. Some researchers demonstrated that quinoa genotypes originated from dry and cold environments were sensitive to the temperature as compared to varieties originated from hot and humid climates [38]. It was evidenced previously that solar rays affects phyllochron in the quinoa varieties. Thus varieties from these regions (Peru, Southern Chile and Bolivia) were more sensitive to solar radiation than Ecudorian varieties. While, it was demonstrated that Ecuadorian quinoas were more sensitives to photoperiod becuase they posess longest phyllochron. However, in the present studies, we did not observe any growth and yield loss due to photoperiod and most of the yield reduction was due to nutrient poor sandy soils and salt stress.

Salinity tolerance evaluation of quinoa indicated a significant diversity in growth strategies, agro-physiological attributes, yield stability and salinity coping phenomena's that will assist in the restoration of sat-degraded marginal sandy soils. Quinoa as a highly nutritious crop and can be a potential candidate in nutrient poor sandy and salt-degraded marginal soils in arid and semi arid regions and will help to adapt the harsh environment, as well as also offer food security. The present studies also suggest the supply of essential plant nutrients through chemical fertilizers and organic manures as soil amendments may be a possible strategy to increase nutrient availability and soil water holding capacity in salt-degraded marginal sandy soils. Other management strategies could include strong seeding establishment at the start of the sowing using fresh water encouraging seedling survival for later irrigation with saline water under nutrient poor, sandy and vulnerable environment.

5. Conclusions

This study concluded that agro-physiological and yield stability attributes can be used to choose quinoa genotypes of contrasting performance across different saline environment and provide a platform to select certain quinoa genotypes for a broad or a specific adaptation. Overall, seedling performance throughout the growth and yield period were genotype dependent. Results highlighted that Chilian based quinoa genotypes showed sufficient adaptation under nutrient poor marginal soils and harsh environment of UAE and were not sensitive to photoperiod conversely to other quinoa genotypes. These genotypes might be useful for the rehabilitation of Dubai marginal soils. Therefore, we suggest that regional farmers might use identified and optimally adapted genotypes of quinoa, crop management strategies and adequate agronomic practices that can help to recover and use these marginal lands for sustainable crop production and food security. This study can be useful for selection, breeding, and up scaling quinoa genotypes in the Arabian Peninsula.

Author Contributions: M.I.H. and A.J.A.-D. conceived and wrote the manuscript. A.M., M.A. and M.A.A. edited the manuscript and provided significant additional information and helps to revise and improve the initial and final draft. All authors have read and agreed to the published version of the manuscript.

Funding: This research was funded by International Fund for Agricultural Development (IFAD), Arab Fund for Economic and Social Development through several regional projects.

Acknowledgments: We thank ICBA for providing tools, guidance, and research facilities for this work. Additionally, we thank Karam for his assistance in the field work and Eng. Qaser Mahmood Khan for maintain irrigation facilities. Authors are very grateful to FAO, Regional Office, Cairo, Egypt for the provision of seeds of different quinoa genotypes to ICBA, Dubai, UAE.

Conflicts of Interest: The authors declare no conflict of interest. The funders had no role in the design of the study; in the collection, analyses, or interpretation of data; in the writing of the manuscript, or in the decision to publish the results.

References

1. Al-Dakheel, A.J.; Hussain, M.I. Genotypic variation for salinity tolerance in *Cenchrus ciliaris* L. *Front. Plant Sci.* **2016**, *7*, 1090. [CrossRef]
2. Bateman, A.; Lewandrowski, W.; Stevens, J.; Muñoz-Rojas, M. Ecophysiological indicators to assess drought responses of arid zone native seedlings in reconstructed soils. *Land Degrad. Dev.* **2016**, *29*, 984–993. [CrossRef]
3. Hussain, M.I.; Farooq, M.; Muscolo, A.; Rehman, A. Crop diversification and saline water irrigation as potential strategies to save freshwater resources and reclamation of marginal soils—A review. *Environ. Sci. Poll. Res.* **2020**, *27*, 28695–28729. [CrossRef] [PubMed]
4. Wang, T.; Xue, X.; Zhou, L.; Guo, J. Combating aeolian desertification in Northern China. *Land Degrad. Dev.* **2015**, *26*, 118–132. [CrossRef]
5. Ambalam, K. United Nations Convention to Combat Desertification: Issues and Challenges. *E-Int. Rel.* **2014**, *30*.
6. Gelfand, I.; Sahajpal, R.; Zhang, X.; Izaurralde, R.C.; Gross, K.L.; Robertson, G.P. Sustainable bioenergy production from marginal lands in the US Midwest. *Nature* **2013**, *493*, 514–517. [CrossRef]
7. El-Keblawy, A.; Abdelfattah, M.A.; Khedr, A.H.A. Relationships between landforms, soil characteristics and dominant xerophytes in the hyper-arid northern United Arab Emirates. *J. Arid Environ.* **2015**, *117*, 28–36. [CrossRef]
8. Hussain, M.I.; Al-Dakheel, A.J.; Reigosa, M.J. Genotypic differences in agro-physiological, biochemical and isotopic responses to salinity stress in quinoa (*Chenopodium quinoa* Willd.) plants: Prospects for salinity tolerance and yield stability. *Plant Physiol. Biochem.* **2018**, *129*, 411–420. [CrossRef]
9. Golos, P.J.; Dixon, K.W.; Erickson, T.E. Plant recruitment from the soil seed bank depends on topsoil stockpile age, height, and storage history in an arid environment. *Restor. Ecol.* **2016**, *24*, S53–S61. [CrossRef]

10. Hussain, M.I.; Al-Dakheel, A.J. *Using Alternate Water Resources for Cultivation of Salt Tolerant Perennial Grasses under Marginal Environment*; TROPENTAG, Management of Land Use Systems for Enhanced Food Security-Conflicts, Controversies and Resolutions: Berlin, Germany, 2015; pp. 16–18.
11. Hussain, M.I.; Lyra, D.A.; Farooq, M.; Nikoloudakis, N.; Khalid, N. Salt and drought stresses in safflower: A Review. *Agron. Sustain. Dev.* **2016**, *36*, 4. [CrossRef]
12. Munns, R.; Tester, M. Mechanisms of salinity tolerance. *Annu. Rev. Plant Biol.* **2008**, *59*, 651–681. [CrossRef] [PubMed]
13. Gómez-Pando, L.R.; Álvarez-Castro, R.; Eguiluz-De La Barra, A. Effect of salt stress on Peruvian germplasm of *Chenopodium quinoa* Willd.: A promising crop. *J. Agron. Crop Sci.* **2010**, *196*, 391–396. [CrossRef]
14. Gonzalez, J.A.; Gallardo, M.; Hilal, M.; Rosa, M.; Prado, F.E. Physiological responses of quinoa (*Chenopodium quinoa* Willd.) to drought and waterlogging stresses: Dry matter partitioning. *Bot. Studies.* **2009**, *50*, 35–42.
15. Lewandrowski, W.; Erickson, T.E.; Dixon, K.W.; Stevens, J.C. Increasing the germination envelope under water stress improves seedling emergence in two dominant grass species across different pulse rainfall events. *J. Appl. Ecol.* **2016**, *54*, 997–1007. [CrossRef]
16. Hussain, M.I.; Al-Dakheel, A.J. Effect of salinity stress on phenotypic plasticity, yield stability and signature of stable isotopes of carbon and nitrogen in Safflower. *Environ. Sci. Pollut. Res.* **2018**, *25*, 23685–23694. [CrossRef] [PubMed]
17. Naeem, M.; Iqbal, M.; Shakeel, A.; Ul-Allah, S.; Hussain, M.; Rehman, A.; Zafar, Z.U.; Ashraf, M. Genetic basis of ion exclusion in salinity stressed wheat: Implications in improving crop yield. *Plant Growth Regul.* **2020**, *92*, 479–496. [CrossRef]
18. Al-Dakheel, A.J.; Hussain, M.I.; Al-Gailani, A.Q.M. Impact of irrigation water salinity on agronomical and quality attributes of *Cenchrus ciliaris* L. accessions. *Agric. Water Manag.* **2015**, *159*, 148–154. [CrossRef]
19. Koyro, H.W.; Khan, M.A.; Lieth, H. Halophytic crops: A resource for the future to reduce the water crisis? *Emir. J. Food Agric.* **2011**, *23*, 1–16. [CrossRef]
20. Koyro, H.W. Effect of salinity on growth, photosynthesis, water relations and solute composition of the potential cash crop halophyte *Plantago coronopus* (L.). *Environ. Exp. Bot.* **2006**, *56*, 136–146. [CrossRef]
21. Hinojosa, L.; González, J.A.; Barrios-Masias, F.H.; Fuentes, F.; Murphy, K.M. Quinoa abiotic stress responses: A review. *Plants* **2018**, *7*, 106. [CrossRef]
22. Ruiz, K.B.; Aloisi, I.; Del Duca, S.; Canelo, V.; Torrigiani, P.; Silva, H.; Biondi, S.S.V. coastal ecotypes of quinoa: Salinity responses in Chilean landraces from contrasting habitats. *Plant Physiol. Biochem.* **2016**, *101*, 1–13. [CrossRef] [PubMed]
23. Wu, G.; Peterson, A.J.; Morris, C.F.; Murphy, K.M. Quinoa seed quality response to sodium chloride and sodium sulfate salinity. *Front. Plant Sci.* **2016**, *7*, 790. [CrossRef] [PubMed]
24. Alvarez-Jubete, L.; Arendt, E.K.; Gallagher, E. Nutritive value and chemical composition of pseudocereals as gluten-free ingredients. *Int. J. Food Sci. Nutr.* **2009**, *60*, 240–257. [CrossRef] [PubMed]
25. Zuidmeer, L.; Goldhahn, K.; Rona, R.J.; Gislason, D.; Madsen, C.; Summers, C.; Sodergren, E.; Dahlstrom, J.; Lindner, T.; Sigurdardottir, S.T. The prevalence of plant food allergies: A systematic review. *J. Allergy Clin. Immunol.* **2008**, *121*, 1210–1218. [CrossRef]
26. Rojas, W.; Soto, J.L.; Pinto, M.; Jäger, M.; Padulosi, S.; Granos, A. *Avances, Logros y Experiencias Desarrolladas en Quinua, Cañahua y Amaranto en Bolivia*; Bioversity International: Roma, Italia, 2010; p. 178.
27. Jacobsen, S.E.; Monteros, C.; Christiansen, J.L.; Bravo, L.A.; Corcuera, L.J.; Mujica, A. Plant responses of quinoa (*Chenopodium quinoa* Willd.) to frost at various phenological stages. *Eur. J. Agron.* **2005**, *22*, 131–139. [CrossRef]
28. Jacobsen, S.E.; Monteros, C.; Corcuera, L.J.; Bravo, L.A.; Christiansen, J.L.; Mujica, A. Frost resistance mechanisms in quinoa (*Chenopodium quinoa* Willd.). *Eur. J. Agron.* **2007**, *26*, 471–475. [CrossRef]
29. Razzaghi, F.; Jacobsen, S.E.; Jensen, C.R.; Andersen, M.N. Ionic and photosynthetic homeostasis in quinoa challenged by salinity and drought–mechanisms of tolerance. *Func. Plant Biol.* **2015**, *42*, 136–148. [CrossRef]
30. Koyro, H.W.; Eisa, S.S. Effect of salinity on composition, viability and germination of seeds of *Chenopodium quinoa* Willd. *Plant Soil* **2008**, *302*, 79–90. [CrossRef]
31. Babatunde, R.O.; Olungunju, F.I.; Fakayode, S.B.; Sola-Ojo, F.E. Prevalence and determinants of malnutrition among under five children of farming households in Kwara State, Nigeria. *J. Agric. Sci.* **2011**, *3*, 173–181. [CrossRef]

32. Bazile, D.; Pulvento, C.; Verniau, A.; Al-Nusairi, M.S.; Ba, D.; Breidy, J.; Hassan, L.; Mohammed, M.I.; Mambetov, O.; Otambekova, M.; et al. Worldwide Evaluations of Quinoa: Preliminary Results from Post International Year of Quinoa FAO Projects in Nine Countries. *Front. Plant Sci.* **2016**, *7*, 850. [CrossRef]
33. Jacobsen, S.E.; Mujica, A.; Jensen, C.R. The resistance of quinoa (*Chenopodium quinoa* Willd.) to adverse abiotic factors. *Food Rev. Int.* **2003**, *19*, 99–109. [CrossRef]
34. Brakez, M.; Harrouni, M.C.; Tachbibi, N.; Daoud, S. Comparative effect of NaCl and seawater on germination of quinoa seed (*Chenopodium quinoa* Willd.). *Emir. J. Food Agric.* **2014**, *26*, 1091–1096. [CrossRef]
35. Panuccio, M.R.; Jacobsen, S.E.; Akhtar, S.S.; Muscolo, A. Effect of saline water on seed germination and early seedling growth of the halophyte quinoa. *AoB Plants* **2014**, *6*, plu047. [CrossRef] [PubMed]
36. Burt, R. *Soil Survey Laboratory Methods Manual*; Soil Survey Investigation Report No. 42, Version 4.0; U.S. Department of Agriculture, Natural Resources Conservation Service, National Soil Survey Center: Lincoln, NE, USA, 2004.
37. Walkley, A.; Black, I. An examination of the Degtjareff method for determining soil organic matter and a proposed modification of the chromic acid titration method. *Soil Sci.* **1934**, *37*, 29–38. [CrossRef]
38. Bertero, H.D.; de la Vega, A.J.; Correa, G.; Jacobsen, S.E.; Mujica, A. Genotype and genotype-by-environment interaction effects for grain yield and grain size of quinoa (*Chenopodium quinoa* Willd.) as revealed by pattern analysis of international multi-environment trials. *Field Crop. Res.* **2004**, *89*, 299–318. [CrossRef]
39. Bradford, M.M. A rapid, sensitive method for the quantitation of microgram quantities of protein utilizing the principle of protein–dye binding. *Anal. Biochem.* **1976**, *72*, 248–254. [CrossRef]
40. Hussain, M.I.; Reigosa, M.J. Allelochemical stress inhibits growth, leaf water relations, PSII photochemistry, non-photochemical fluorescence quenching and heat energy dissipation in three C3 perennial species. *J. Exp. Bot.* **2011**, *62*, 4533–4545. [CrossRef]
41. Roemer, T. Sind die ertragsreichen Sorten ertragssicherer? Mitteilung Deutsche Landwirtschafts-Gesellschaft. *DGL-Mitt* **1917**, *32*, 87–89.
42. Wricke, G. Uber eine Methode zur Erfassung der okologischen Streubreite in Feldversuchen. *Z. Pflanzenzuchtg* **1962**, *47*, 92–96.
43. Cocozza, C.; Pulvento, C.; Lavini, A.; Riccardi, M.; d'Andria, R.; Tognetti, R. Effects of increasing salinity stress and decreasing water availability on ecophysiological traits of Quinoa (*Chenopodium quinoa* Willd.) grown in a Mediterranean-type agroecosystem. *J. Agron. Crop Sci.* **2013**, *199*, 229–240. [CrossRef]
44. Roman, V.J.; den Toom, L.A.; Gamiz, C.C.; van der Pijl, N.; Visser, R.G.; van Loo, E.N.; van der Linden, C.G. Differential responses to salt stress in ion dynamics, growth and seed yield of European quinoa varieties. *Environ. Exp. Bot.* **2020**, *177*, 104146. [CrossRef]
45. Ali, Y.; Aslam, Z.; Ashraf, M.Y.; Tahir, G.R. Effect of salinity on chlorophyll concentration, leaf area, yield and yield components of rice genotypes grown under saline environment. *Int. J. Environ. Sci. Tech.* **2004**, *3*, 221–225. [CrossRef]
46. Hirich, A.; Choukr-Allah, R.; Jacobsen, S.E. Deficit irrigation and organic compost improve growth and yield of quinoa and pea. *J. Agron. Crop Sci.* **2014**, *200*, 390–398. [CrossRef]
47. Shabala, S.; Hariadi, Y.; Jacobsen, S.-E. Genotypic difference in salinity tolerance in quinoa is determined by differential control of xylem Na^+ loading and stomatal density. *J. Plant Physiol.* **2013**, *170*, 906–914. [CrossRef] [PubMed]
48. Adolf, V.I.; Shabala, S.; Andersen, M.N.; Razzaghi, F.; Jacobsen, S.E. Varietal differences of quinoa's tolerance to saline conditions. *Plant Soil* **2012**, *357*, 117–129. [CrossRef]
49. Bonales-Alatorre, E.; Pottosin, I.; Shabala, L.; Chen, Z.-H.; Zeng, F.; Jacobsen, S.-E.; Shabala, S. Differential activity of plasma and vacuolar membrane transporters contributes to genotypic differences in salinity tolerance in a halophyte species, *Chenopodium quinoa*. *Int. J. Mol. Sci.* **2013**, *14*, 9267–9285. [CrossRef]
50. Long Nguyen, V. Effects of salinity stress on growth and yield of quinoa (*Chenopodium quinoa* Willd.) at flower initiation stages. *Viet. J. Agric. Sci.* **2016**, *14*, 321–327.
51. Yousfi, S.; Serret, M.D.; Márquez, A.J.; Voltas, J.; Araus, J.L. Combined use of $\delta^{13}C$, $\delta^{18}O$ and $\delta^{15}N$ tracks nitrogen metabolism and genotypicadaptation of durum wheat to salinity and water deficit. *N. Phytol.* **2012**, *194*, 230–244. [CrossRef]
52. Hussain, M.I.; El-Keblawy, A.; Mitterand Tsombou, F. Leaf age, canopy position, and habitat affect the carbon isotope discrimination and water-use efficiency in three C_3 leguminous *Prosopis* species from a hyper-arid climate. *Plants* **2019**, *8*, 402. [CrossRef]

53. Hussain, M.I.; Al-Shamsi, N.S.; El-Keblawy, A. Physiological responses of the xerohalophyte *Suaeda vermiculata* to salinity in its hyper-arid environment. *Flora* **2020**, *273*, 151705.
54. Hussain, M.I.; Elnaggar, A.; El-Keblawy, A. Eco-physiological adaptations of *Salsola drummondii* to soil salinity: Role of reactive oxygen species, ion homeostasis, carbon isotope signatures and anti-oxidant feedback. *Plant Biosyst. Int. J. Deal. All Asp. Plant Biol.* **2020**, 1–20. [CrossRef]
55. Eisa, S.; Hussein, S.; Geissler, N.; Koyro, H.W. Effect of NaCl salinity on water relations, photosynthesis and chemical composition of Quinoa (*Chenopodium quinoa* Willd.) as a potential cash crop halophyte. *Aust. J. Crop Sci.* **2012**, *6*, 357–368.
56. Jacobsen, S.E. The worldwide potential of quinoa (*Chenopodium quinoa* Willd.). *Food Rev. Int.* **2003**, *19*, 167–177. [CrossRef]
57. Blum, A. Effective use of water (EUW) and not wateruse efficiency (WUE) is the target of crop yield improvement under drought stress. *Field Crop. Res.* **2009**, *112*, 119–123. [CrossRef]
58. Aloisi, I.; Parrotta, L.; Ruiz, K.B.; Landi, C.; Bini, L.; Cai, G.; Biondi, S.; Del Duca, S. New insight into quinoa seed quality under salinity: Changes in proteomic and amino acid profiles, phenolic content, and antioxidant activity of protein extracts. *Front. Plant Sci.* **2016**, *7*, 656. [CrossRef]
59. Schmöckel, S.M.; Lightfoot, D.J.; Razali, R.; Tester, M.; Jarvis, D.E. Identification of putative transmembrane proteins involved in salinity tolerance in *Chenopodium quinoa* by integrating physiological data, RNAseq, and SNP analyses. *Front. Plant Sci.* **2017**, *8*, 1023. [CrossRef]
60. Orsini, F.; Accorsi, M.; Gianquinto, G.; Dinelli, G.; Antognoni, F.; Carrasco, K.B.R.; Martinez, E.A.; Alnayef, M.; Marotti, I.; Bosi, S.; et al. Beyond the ionic and osmotic response to salinity in *Chenopodium quinoa*: Functional elements of successful halophytism. *Funct. Plant Biol.* **2011**, *38*, 818–831. [CrossRef]

plants

MDPI

Review

Agro-Techniques for Lodging Stress Management in Maize-Soybean Intercropping System—A Review

Ali Raza [1,2,†], Muhammad Ahsan Asghar [3,†], Bushra Ahmad [4], Cheng Bin [1,2], M. Iftikhar Hussain [5], Wang Li [1,2], Tauseef Iqbal [1,2], Muhammad Yaseen [6], Iram Shafiq [1,2], Zhang Yi [1,2], Irshan Ahmad [1,2], Wenyu Yang [1,2] and Liu Weiguo [1,2,*]

1. Key Laboratory of Crop Ecophysiology and Farming System in Southwest China, Ministry of Agriculture, Sichuan Agricultural University, Chengdu 611130, China; 2018601001@stu.sicau.edu.cn (A.R.); 2018201013@stu.sicau.edu.cn (C.B.); S20176512@stu.sicau.edu.cn (W.L.); alisicau.edu@gmail.com (T.I.); sicau.edu.433@gmail.com (I.S.); 2019201010@stu.sicau.edu.cn (Z.Y.); 2019605002@stu.sicau.edu.cn (I.A.); mssiyangwy@sicau.edu.cn (W.Y.)
2. Institute of Ecological Agriculture, Sichuan Agricultural University, Chengdu 611130, China
3. CAS Key Laboratory of Mountain Ecological Restoration and Bioresource Utilization & Ecological Restoration Biodiversity Conservation Key Laboratory of Sichuan Province, Chengdu Institute of Biology, Chinese Academy of Sciences, University of Chinese Academy of Sciences, Wuhou 610000, China; ahsanasghar2017@mails.ucas.ac.cn
4. Department of Plant Breeding and Genetics, University of Agriculture, Faisalabad 38000, Punjab, Pakistan; L201702011@stu.sicau.edu.cn
5. Department of Plant Biology & Soil Science, Universidad de Vigo, 36310 Vigo, Spain; mih786@gmail.com
6. State Key Laboratory of Crop Gene Exploration and Utilization in Southwest China, Institute of Rice Research, Sichuan Agricultural University, Wenjiang, Chengdu 625014, China; 2018511004@stu.sicau.edu.cn

* Correspondence: lwgsy@126.com; Tel.: +86-18628958371

† These authors contributed equally to this work.

Received: 3 October 2020; Accepted: 11 November 2020; Published: 17 November 2020

Abstract: Lodging is one of the most chronic restraints of the maize-soybean intercropping system, which causes a serious threat to agriculture development and sustainability. In the maize-soybean intercropping system, shade is a major causative agent that is triggered by the higher stem length of a maize plant. Many morphological and anatomical characteristics are involved in the lodging phenomenon, along with the chemical configuration of the stem. Due to maize shading, soybean stem evolves the shade avoidance response and resulting in the stem elongation that leads to severe lodging stress. However, the major agro-techniques that are required to explore the lodging stress in the maize-soybean intercropping system for sustainable agriculture have not been precisely elucidated yet. Therefore, the present review is tempted to compare the conceptual insights with preceding published researches and proposed the important techniques which could be applied to overcome the devastating effects of lodging. We further explored that, lodging stress management is dependent on multiple approaches such as agronomical, chemical and genetics which could be helpful to reduce the lodging threats in the maize-soybean intercropping system. Nonetheless, many queries needed to explicate the complex phenomenon of lodging. Henceforth, the agronomists, physiologists, molecular actors and breeders require further exploration to fix this challenging problem.

Keywords: intercropping; lodging tolerance; agronomical management; lignin metabolism; resistance genes

1. Introduction

Climatic change and population explosions are the major threats to food security in the future [1,2]. It has been projected that, by 2050, the present world population will be enhanced by up to 30%, which will make the world population nine billion or more people [3]. This problem can only be solved through multiple cropping systems to fulfill the food demand and supply requirements which leads to sustainable agriculture [4]. The maize-soybean intercropping is one of the most important systems that plays a key role in the sustainability of food production systems [5]. The maize-soybean intercropping system has great importance among the legume-cereal intercropping systems because of its maximum yield and efficient use of resources [6,7]. Initially, it was developed in the South-Western region of China and now it is progressing throughout the globe. It is estimated that from the total cropped area of 182.3 million hectares, about 83% area is used for intercropping in Africa [8]. In China, half of the total grain yield is gained through multiple cropping systems [9]. The maize-soybean intercropping system has been adopted in different parts of China due to its maximum production and land use efficiency [7]. The biggest challenge of this century has to be met by China to boost the production of cereals by approximately 600 Mt by 2030 to achieve food security [7]. In another study, it is reported that annually 2.8–3.4 × 10^7 ha area was grown under the intercropping system in China [10]. Over 667 thousand hectares of soybean are being intercropped with maize in the south-west of China [11,12].

Although this system has many advantages, in this intercropping system the spatial light pattern affects the growth of soybean due to the shading of maize plants during the co-growth period [13]. In order to capture more light, the soybean plants increase their heights and this phenomenon is called shade avoidance. Shade avoidance causes several morpho-physiological changes such as low photosynthetic activities, increased intermodal length, decreased stem diameter and higher rate of lodging [14]. Previous studies revealed that continuous and periodic prevailing of shade had decreased the total grain production, approximately 25–45% [15,16]. However, the climatic factors such as continuous storms and heavy rainfall comprise about 8% and 19% lodging to crops, respectively [17,18]. Many studies focused on the monocropping conditions of the soybean, fertilizer regulation and lodging resistant cultivars [19,20]. In addition, higher N application rates could enhance the lodging threat due to excessive canopy growth and stem elongation [21]. It is also noticed that excessive canopy growth decreases the light interception which in turn elongates the stem length [22,23]. Chen et al. [24] briefed that higher application of basal N decreased the lignin content in internodes. Moreover, it has been seen that high planting densities also reduced the lignin content in stem which leads to weaker stem and hence causes the plant lodging [25].

It has been described that the lodging of stem significantly impeded the photosynthetic activities of the plant at the grain filling stage [26]. Some authors reported that Silicon (Si) had enhanced the cell wall thickness of rice stems, shortened the internode length, changed the canopy structure of plant, increased the photosynthesis and hence prompted the lodging resistance [27]. It has been observed that Si improved the stem strength of rice at the reproductive stage [28] and also increased the content of soluble sugars in maize [29]. Another speculation had revealed that the lodging angle at 25–90° from the right angle could decrease the grain production by approximately 20–61% during the grain filling stage in wheat [30]. Shading is an inevitable factor in intercropping systems; many researches have been conducted to mitigate the adverse effects through breeding of shade tolerant cultivars of soybean [31]. However, developing the shade resistance potential of existing cultivars and clarifying the resistance of different genotypes to shade, is one of the most economical and proficient ways to resolve this problem. It has been reported that compounds of Ti can promote the growth of crops, biomass, enzyme activity, chlorophyll content, iron (Fe) uptake and also compensated the nitrogen deficiency [32]. In addition, it was observed that Ti nanoparticles transformed the expression of miRNAs 16 and modified the root architecture [33,34]. The relationship between the heritability of lodging resistance and the height of plant or seed yield has been elucidated previously [35]. In studies of multiple populations, a modest to high lodging resistance heritability has been examined [36]. Although intensive QTLs (Quantitative Trait Loci) techniques have directed to enhance the lodging resistance in soybean [37]. It is extensively reported that shade avoidance had increased the lodging rate in the maize-soybean intercropping system. However, lodging depends on the stem breaking strength, anchorage of roots (root lodging) and structural carbohydrates accumulation in the stem [38,39].

To date, many studies have been conducted about the relationship between lignin biosynthesis and stem strength in intercropping but the approaches that explore how to manipulate the lignin metabolic pathway that enhance the lodging resistance of intercropped soybean have not been elucidated yet. This review endorsed the recent approaches and future prospects to enhance the stem mechanical strength and lodging resistance of soybean under the maize-soybean intercropping system. Therefore, the following modern agronomical, genetics, chemical and genomic approaches are described in this review; (i) to mitigate the shade effect of tall stature plants on the small plants, (ii) how to create a better light environment for intercropped soybean? (iii) to reduce the stem elongation and increased the soybean stem diameter (iv) to enhance the structural carbohydrates contents, which ultimately increased the lodging resistance and yield of soybeans under the maize-soybean intercropping system.

2. Lodging; A Serious Threat to Intercropping System

The maize-soybean intercropping system is widely adopted in the south-west of China. Over 667,000 hectares area is under soybean maize intercropping system [11]. However, there are still some drawbacks in this system likewise; taller plants (maize) shaded the short stature plants (soybean) during the middle and later growth stages of soybean [12]. Shading of maize caused the stem elongation, slender stem and lower amount of lignin content in the stem which resulting in the soybean lodging [19,40]. It is previously documented that, lodging is one of the main factors that result in the reduction of soybean yield up to 50% [41].

2.1. *Stem Development and Lignin Metabolism under Intercropping*

In intercropping system, the plant height, petiole length and internodal length of soybean plant increased and stem diameter decreased due to the more allocation of carbon to the stem and petiole elongation instead of leaves and roots development [42,43]. This mechanism helps the plants to escape and found more light which results in increasing plant height and low stem diameter [44,45]. Furthermore, it was documented that shade conditions or low intensities of light under intercropping, change the sheath of vascular bundles and stem mechanical layers of tissues which play an important role in the stem anatomical structure to increase plant lodging resistance [46].

Another study had shown that shading of maize had increased the length of soybean stems by up to 45.75% that resulted in lodging and ultimately reduced the final yield up to 20–40% [47]. It has been described that the lodging of stem significantly hindered the photosynthetic activities of the plant at the grain filling stage [26]. It is also reported that shade reduces the area of xylem, pith and ratio of xylem tissues [48]. Moreover, it also decreases the stem cross-sectional area, number of vascular bundles and thickness of stem skin [49]. Lignin consists of complex aromatic polymers and constitutes the cell wall of vascular plants [50]. Previous studies observed that cell wall mechanical strength was increased through crosslinking of hemicellulose and cellulose polymers [51]. Under shade conditions the enzymatic activities that is, cinnamyl alcohol dehydrogenase (CAD), phenylalanine ammonialyase (PAL), peroxidase (POD) and 4-coumarate: CoA ligase (4-Cl) were limited which resulted in lower lignin production during lignin biosynthesis [52,53].

A significant and positive correlation of plant height with yield and yield associated parameters has been reported [54]. Previous studies also found that higher lodging percentage with weaker stems caused reduction in the transportation of carbohydrates and dry matter which resulted in the enhancement of the risk of pest and pathogen attacks that resulted in lower yield [55]. However, it was documented that non-structural carbohydrates are moved to seeds during the seed filling stage for seed formation, the mechanical stability of plants depends on the structural carbohydrates (cellulose and lignin) in the lower part of the stems [56]. Stem development relies on the primary constituents of cell wall (lignin and cellulose) and they significantly correlated with stem mechanical strength and lodging resistance [44]. It has been described that the lodging of stem significantly hindered the photosynthetic activities of the plants at the grain filling stage [26]. The consequence of stem lodging caused a reduction in grain production, increased the harvest cost and lower the quality of grains [30]. Another study revealed that 80% of the grain production reduced due to the impact of lodging naturally and artificially on the total crop yield [57].

Stem development and lodging resistance also depend on the various environmental and field management factors as depicted in Table 1. Moreover, the stem mechanical strength is influenced by non- structural carbohydrates and structural carbohydrates mainly cellulose and lignin [58,59]. The accumulation of lignin and cellulose in the stem had significantly enhanced the stem mechanical strength resulting in the increased lodging resistance which is previously reported in rice, wheat and buckwheat stems [25,59]. The lignin metabolic pathway depends on the chemical activities of lignin related enzymes PAL, 4-Cl, CAD and POD [60,61]. The lignin biosynthesis depends on the genotype and environmental changes, likewise, a significant correlation has been found between the lignin biosynthesis and lignin related enzyme activities (PAL, CAD, POD) in the stems of different lodging resistance wheat and buckwheat cultivars [62,63]. Shade impacts the lignin biosynthesis by affecting the activities of lignin related enzymes in the metabolic pathway of lignin [64–66]. However, further studies revealed that shade resistant cultivars have maximum activities of lignin related enzyme (POD, 4-Cl, CAD and PAL) as compared to shade sensitive cultivars [59,67]. Furthermore, it was found that shade had slowed down the activities of lignin related enzyme that caused a lower lignin accumulation in stem of soybean and ultimately leads to weaker stem and lodging [59].

As reported in earlier studies that the accumulation of cellulose and lignin in lodging resistant varieties of rice were higher than that in lodging susceptible varieties of rice varieties [62,68]. It has been found that shading caused the stem elongation and weaker stem which decreased the plant mechanical strength [69]. Shading had also amplified the rate of lodging and conversely increment in the intensity of light had improved the stem strength and lower the rate of lodging in maize [70]. Lodging tolerance can be increased by reducing the plant height and lower the center of gravity point and decreased the aboveground weight of the plant on the basal stem [71–73]. Although all the above posted findings had given some clues to understand the lodging mechanism in the intercropping system, still there is a big gap of information that needs to be explored yet to get better insights on the lodging phenomenon and its management under intercropping systems.

Table 1. Soybean constituents correlated with lodging resistance under maize-soybean intercropping system.

Trait (s)	Cropping System	Behavior	Reference (s)
Morphological Traits			
Plant Height		Positively Correlated with lodging	[48]
Internodal-length		Negatively correlated with lodging resistance	[47,73]
Stem Diameter	**Maize-soybean Intercropping**	Strongly correlated with resistance to lodging	[40,74]
Anatomical Features			
Number of vascular bundles		Positively correlated with lodging resistance	[75]
Width of vascular bundle sheath		Strongly associated with resistance to lodging	[51,76]

Table 1. *Cont.*

Trait (s)	Cropping System	Behavior	Reference (s)
Cell Wall Thickness		Strongly correlated with lodging resistance	[59]
Physical Aspects			
Wind		Significantly associated with lodging	[40]
Rainfall		Strongly correlated with lodging	[74]
Shade		Negatively associated with lodging resistance	[59,77,78]
Biochemical Features			
Lignin and Cellulose Content		Positively associated with cell wall thickness and lodging resistance	[19,79]

2.2. *Role of Carbohydrates and Lignin Biosynthesis*

Lignin and cellulose, the structural carbohydrates, are the main constituents of the cell wall and their components play a vital role in plant vigor, stem strength and the lodging resistance of plants [80]. High lignin concentrations in vascular bundles can increase the strength of the cell wall and boost up the physical resilience of plant stem. The overall lignin content in the basal second internode was significantly correlated with stem strength and elasticity [25,71]. During the development of secondary cell walls, lignin is deposited in the carbohydrate matrix of the cell wall, which makes the whole plant body rigid and allows the plant to develop upwards [81,82]. In addition, by increasing the physical strength of the stem the threat of lodging could be minimized. Lignin was considered to be a macromolecule that played a supporting and fundamental role in enhancing the stem mechanical ability, increasing stem strength, maintaining stem stability and ultimately decreasing the lodging rate by preserving the stem verticality [83–85]. Previous research data revealed that the lignin deposition in the stem was positively correlated with stem breaking strength per section area which indicated that a high concentration of lignin in stem had enhanced the stem physical strength. Furthermore, it was revealed that the activities of the enzymes, that is, PAL, POD, CAD and 4CL played a key role in lignin metabolic pathway [86].

Lignin biosynthesis mainly depends upon the multiple enzymatic activities. Among the lot, the PAL enzyme is one of the most important enzymes that plays a role as rate limiting enzyme and catalyst which convert the L-phenylalanine dehydrogenase into trans-cinnamic acid in the shikimic acid pathway [76,87]. CAD is another important enzyme that takes part in the final reaction of the reduction of lignin biosynthesis [61]. It is revealed that the lignin is biosynthesized through lignin monomers polymerization through POD activity and oxidation reaction of monolignols carried by peroxidase [88,89]. The previously conducted experiment had described the significant positive correlation between the activities of lignin related enzymes (PAL, POD, CAD and 4CL) and lignin content in stem of soybean [55]. Furthermore, some studies had also shown that lodging resistance of plants and their varying degree of lodging tolerant is significantly depended on the level of genes transcript abundance and their contribution rate to lignin related enzymatic activities that enhanced the stem mechanical strength [90]. However, recent research had found that a strong correlation between the expression levels of cinnamoyl-CoA reductase2 *(CCR2)*, ferulate 5- hydroxylase *(F5H2)*, caffeic acid O-methyltransferase2 *(COMT2)*, p-coumarate 3-hydroxylase *(C3H1)* and 4-coumarate: CoA ligase1 *(4CL1)* and lignin contents that enhanced the lodging resistance in wheat [91]. In addition, it is also described that auxin deposition in *Arabidopsis* was prompted by hyper-gravity which led to particular lignin metabolic genes expression at a high level and successively led to the process of lignification in stem inflorescence [92]. Moreover, it has been reported previously that auxin and cytokinins up-regulated the genes expression to lignin related enzymes and secondary cell wall growth and development/lignification [93]. However, in maize–soybean intercropping, more research from biologists and physiologists is still required to understand the shading phenomenon; how shade stress of tall stature crops affects the biochemical and physiological activities of short stature crops that related to lignin biosynthesis and lodging resistance? On the other hand, some biotic and abiotic factors including cellular signaling such as hormones activities that play an essential role in the up-regulation of lignin related enzymes and lignin biosynthesis (i.e., lignin and auxin relationship) under the maize-soybean intercropping system have not been elucidated yet (Figure 1).

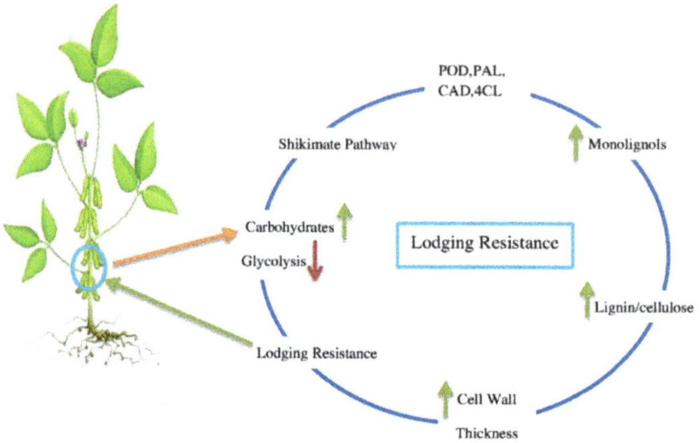

Figure 1. Role of carbohydrates and lignin enzymes in lodging resistance, green arrows represented the increasing trend while red arrows represent the decreasing trend.

2.3. *Plant Hormonal Activities and Shade Avoidance under Intercropping SYSTEM*

Plant hormones are the major elements that take part in the regulation of many plant characters which have a vital role in lodging resistance [94–96]. It has been reported that shade avoidance is a complicated phenomenon that regulates the metabolic and transcriptional factors and ultimately prompted the elongation of stem and apical dominance which in turn supports the young tissues escaping from shade [44,97,98]. Some plants evolve and adapt the shade tolerance strategies to counter the shading effect and enable them to survive under low light conditions [42]. The shade avoidance had enhanced the lodging rate in maize-soybean intercropping system, however, lodging depends on the stem breaking strength, anchorage of roots (root lodging) and structural carbohydrates accumulation in the stem [38,39]. WU Yu-shan et al. investigated the relationship between shade avoidance responses and yield analysis of various soybean cultivars under the relay intercropping system [78]. They concluded that under shady conditions the stem length and specific leaf area were increased by 0.78 and 0.65% as compared to full light conditions. On the other hand, the diameter of stem, leaf area, total biomass, number of nodes and branches were reduced than that of normal light conditions [99]. However, the scientists have no comprehensive knowledge yet about the relationship mechanism between the hormones and shade avoidance response under maize-soybean intercropping. In addition, shade avoidance response is also a crucial factor that affects the normal growth of crops in dense canopies [100,101]. In a recent study, Cui et al. [102] investigated the effect of Gibberellin (GA) application on the maize hormonal and antioxidant activities and grain filling under high planting densities. They illustrated the significant correlation between GA application and the level of endogenous hormones and antioxidant activities of maize under high planting densities. They further found that the exogenous application of GA had enhanced the antioxidant contents (SOD, CAT, MDA and POD) and hormones level (IAA, ABA, ZR and GA3) and finally grain yield of high density maize. Furthermore, it is also investigated that the exogenous application of indole-3-acetic acid (IAA) repressed the bud growth of tillers while the zeatin (Z) hormone significantly promoted the growth of buds under low nitrogen environments [94]. The results of their study recommended that Z application had a strong influence on the tillers and tillers buds growth regulation, therefore Z application supports the strong soil anchorage of plants and ultimately creates a more lodging tolerance environment for plants. Nevertheless, the relationship between phytohormones (auxin) and lignin in the maize-soybean intercropping system is still uncertain yet (Figure 2).

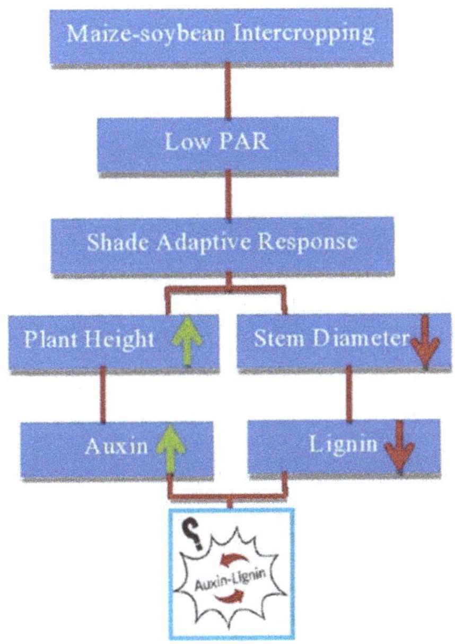

Figure 2. Shade adaptive response of soybean to low light intensity and relationship between phytohormone (auxin) and lignin.

3. Agro-Techniques for the Management of Lodging Stress

3.1. Genetic Manipulation for Increasing Lodging Resistance

The poor resistance to lodging could reduce the soybean yield potential. Previously, independent studies have indicated a significant number of observations of quantitative trait loci (QTL) for lodging resistance [103]. A recent investigation on the integration of QTLs in the lodging resistance of soybean indicated the four QTLs which resulting in the two considerable QTL integrations on chromosomes 6 and 19. Their finding could be useful to increase the lodging resistance in soybean. Their results find a strong and pleiotropic relationship between the lodging resistance and QTL integration on chromosome 6 [104]. Several genes and their QTLs revealed the resistance to lodging and its related traits have been reported in rice (4CL gene family), wheat and barley [105–107]. Along with conventional breeding, we have to focus on the identification of lodging-resistance genes especially for cereal crops [108]. Now, identification and transformation of desired genes are much easier because of recent advances in breeding, genomics and biotechnology, which eventually help to increase crop productivity [109]. So, the transformation of lodging susceptible genes to lodging resistance genotypes has the potential to increase the grain yield of cereal crops under lodging prone zones [110]. The population based studies have been reported for QTL mapping in cereals and exhibited lodging resistance especially in wheat [111]. The QTLs study of phenotypic traits which are directly associated with lodging resistance in cereals had already been reported like plant stalk strength, pith diameter, culm diameter and culm wall thickness in wheat [112].

These QTLs have significant effects on lodging but further validation is needed by fine mapping or other advanced techniques. Currently, some of them have already been validated, for instance, in winter wheat a dominant Rht5 gene related to dwarfism in plant height, has been marked on chromosome 3BS linked with molecular marker Xbarc102 [113]. The recent studies revealed that this region on chromosome 6A showed higher phenotypic variations for plant height and could be used further for QTL mapping studies [114,115]. *Rht3* is another dwarfing gene in wheat which shortened plant height up to 59% but has not been used in commercial varieties [116]. *Rht5*, a plant hormone (GA)-responsive gene for dwarfism, which significantly shortened the plant height up to 55% without having negative effects over coleoptile length and seedling vigor [117,118]. On the other hand, some researchers also reported that *Rht5* has negative effect over flowering time and delayed by 4.8–14% in a thermal environment [119,120]. In addition, there is still an attention required from molecular actors to identify the sequence of candidate genes through QTLs intervals, map-based cloning to enhance the lodging resistance in soybean.

3.2. Proper Sowing Time and Planting Density

Sowing time and planting density are both key factors in the maize-soybean intercropping system that affect the lodging resistance and yield of soybean crop [121]. Sowing time is a crucial factor that enhanced the competition within the species which ultimately reduced the crop yield [122]. Therefore, the selection of an optimal sowing time is vital perspective to enhance the lodging resistance and yield under the maize-soybean intercropping system [123]. It has been observed that delay sowing significantly decreased the risk of lodging by shortening the internodal length, plant height and center of gravity point and by increasing the cell wall thickness, diameter and grain filling period [124]. For instance, only two weeks late sowing could reduce up to 30% threat of lodging in wheat [125]. Under the maize-soybean intercropping the tall stature crops adversely affected the short stature crops at the both vegetative and reproductive stages as compared to relay intercropping in which the adverse effect could only be observed at vegetative stage [7]. However, in intercropping systems, the competition between the intercropped species could be decreased by fluctuating the planting time [126]. Most importantly, under relay intercropping systems, the crop sown first have more competition than the second crop [127]. On the other hand, under the relay intercropping system, selecting an optimal planting time could reduce the co-growth duration among the intercropped species and adverse effect of first sown crop could also be minimized on the second grown crop [128]. Furthermore, it is shown that fluctuating the planting time, 50 days of co-growth period of maize and soybean in relay intercropping system enhanced the crop growth rate 17–64% as compared to 70–90 days of co-growth period. In maize-soybean relay intercropping the soybean production was recorded maximum 2.11 t/ha at 50 days co-growth duration of maize and soybean [126].

High plant populations have been used extensively to increase crop production [77]. Under high planting densities, mutual shading of plants disturbs the light environment of crop which reduced the photosynthetic activities and carbohydrates accumulation in the stem that leads to lodging easily [25,129]. It has been witnessed that planting densities were negatively correlated with the lodging resistance of stem; however, high plant population had promoted the lodging and lower grain production [129–132]. In addition, optimum planting densities could expand the structure of the plant population and provide a better light environment, enhanced photosynthetic rate and ultimately increased the lodging resistance of stem [74,133]. Furthermore, it was found that in strip intercropping optimum planting densities that is, 17 and 20 plants/m^2 had significantly enhanced the stem diameter by 4.3% and 6%, respectively as compared to 25 plants/m^2 planting density. Most specifically, the plant height was decreased by 6.2% and 9.4% at 17 and 20 plants/m^2, respectively than that of 25 plants/m^2 planting density. In addition, decreasing the planting densities such as 17 and 20 plants/m^2 could decrease the lodging percentage by 50.3% and 19.3%, respectively as compared to 25 plants/m^2 planting density [134]. However, further field experiments to optimize the planting densities in the maize-soybean intercropping system Table 2.

Table 2. Comparative analysis of maize-soybean intercropping system (MSIS) in different countries.

	Planting Density in Monocropping ($\times10^3$ Plants ha^{-1})	Planting Density in Intercropping ($\times10^3$ Plants ha^{-1})	
Country	Maize Soybean	Maize Soybean	Reference (s)
China (MSIS)	59 117	59 117	[124,134,135]
Egypt	71 143	24 95	[136]
Ethiopia	44 500	44 375	[137]
Ghana	56 111	56 222	[138]
India	83 333	42 250	[139]
Nepal	40 200	40 200	[140]
Nigeria	33 200	33 200	[141]
Iran	36 400	36 400	[142]

3.3. *Efficient Use of Fertilizers*

Nitrogen (N), phosphorus (P) and potassium (K) are the important macro-elements, vital for crop growth and development [143]. N fertilizer application rate and time of application had also a significant effect on the lodging resistance of crops [144,145]. It has been described that high nitrogen rate had reduced the diameter of basal or lower internode that cause lodging [146]. On the other hand, the low application of nitrogen had enhanced the concentrations of water soluble carbohydrates as in the middle internodes 21% and in the basal internodes 42% than that of high nitrogen rate [147]. Several researchers suggested that a high rate of nitrogen application promoted vegetative growth and decreased the root anchorage in the soil and stem secondary cell wall composition (lignin content) which resulted in lodging of the crops [25,133,148]. A higher rate of nitrogen application had reduced the activities of lignin related enzymes (PAL, POD, CAD and 4CL) and lignin deposition in the cell wall which decreased the lodging resistance [24]. Berry et al. [149] also observed that a low rate of N had reduced the height of plant and also proliferated the stem diameter and cell wall width which turns in high stem strength. Another study had found that increasing the amount of N in wheat crop had progressively increased the cell wall constituents (lignin and cellulose content) and afterward it was decreased gradually [148]. Many recent studies showed that lodging resistance of crop could be improved at the price of yield sacrificing by minimizing the N application rate and rescheduling the application of N fertilizers [149,150]. Furthermore, the relation between N and K has a fundamental role in improving crop grain production and quality [146,150]. Increasing the level of K^+ along with elevated $NH4^+$ could decrease the stem cell wall thickness. For an instance, the application of a high level of N and P fertilizers in the absence of K decreased the 30%–35% grain production in rice due to lodging [151]. However, with the application of K, the stem mechanical strength could be increased [143]. An equal application rate of N and K had significantly promoted the root growth and enhanced the root anchorage which resulted in lodging resistance [152].

The optimum level of K^+ nutrition to plants was positively correlated with lignin accumulation into the vascular bundles and cells of sclerenchyma of plant cell wall which ultimately enhanced the stem diameter and lodging resistance [153]. In the same way, it has been noticed that K^+ considerably inhibited the adverse effects of a higher level of NH_4^+ which in turns amplified the 20–27% of stem mechanical strength in wheat [146]. However, K^+ has a pivotal role in the process of photosynthesis and metabolic activities of carbohydrates production in plants [154,155]. However, appropriate fertilization of nitrogen could enhance the lignin content in the basal internode and improved the stem lodging resistance [24,79,156]. Conclusively, there is a still gap in nitrogen fertilization of soybean under maize-soybean intercropping; henceforth a more attention is required from agronomists and plant breeders for rescheduling the rate and application of nitrogen fertilizer which play a major role in lodging resistance and crop final production.

Previous experiments have revealed that the efficiency of intercrops to absorb nitrogen (N) is more than the sole-cropping, however, the total uptake of N by intercropped soybean and wheat is greater than the total of the sole crops [157,158]. According to an estimation, soybean can fix nitrogen about 39 to 182 kg N ha^{-1} [159]. It has been noticed that under high nitrogen conditions legumes are usually shaded by the maize which results in shade avoidance response and low grain production [160]. These adverse effects of cereal crops on legumes can be alleviated by fluctuating the sowing date [161,162]. A recent meta-analysis explained the land and fertilizer nitrogen use efficiency of intercropped maize and soybean [163]. They concluded that maize-soybean intercropping system has greater potential to attain high land equivalent ratio (LER) and fertilizer nitrogen equivalent ratio (FNER) by utilizing the optimum levels of nitrogen inputs. Whereas further studies are needed to pinpoint the optimum combinations of sowing configuration, planting dates and fertilizer rate and time to attain the high yields by reducing the lodging stress.

3.4. Development of Lodging Resistant Cultivars

Agronomists characterized the soybean cultivars into three sets depending on the genotypes response towards lodging resistance: highly, moderately and susceptible cultivars [164]. In some previously conducted experiments, different cultivars of soybean were grown under the maize-soybean intercropping system to distinguish the more suitable and lodging resistance cultivars. In a previous experiment, four recombinant inbred lines (B3, B15, B23 and B24) of Nandou-12 (that is shade tolerant and widely grown in maize-soybean intercropping system of China) and Nan 032-4 (that is shade susceptible cultivar) in were used [99], the lignin content in stem and lodging resistance index of B23 and B24 was significantly higher than that of B3 and B15 under both monocropping and intercropping systems. Furthermore, another experiment was conducted in which three cultivars were selected on the base of their response to lodging and shade stress [19]. A shade susceptible cultivar (Nan 032-4), a moderate cultivar to shade and lodging tolerance (Jiuyuehang) and shade tolerance and lodging resistance cultivar (Nandou-12) [19]. Their findings revealed that Nandou-12 had more accumulation of lignin in stem and high enzymatic activities of lignin related enzymes (PAL, 4CL, CAD and POD), hence more lodging resistance as compared to Nan 032-4 and Jiuyuehang cultivars under both monocropping and intercropping systems (Figure 3). Along with these research outcomes, there is a dire need to work out on the soybean cultivars which are well suited for intercropping systems with greater lodging resistance.

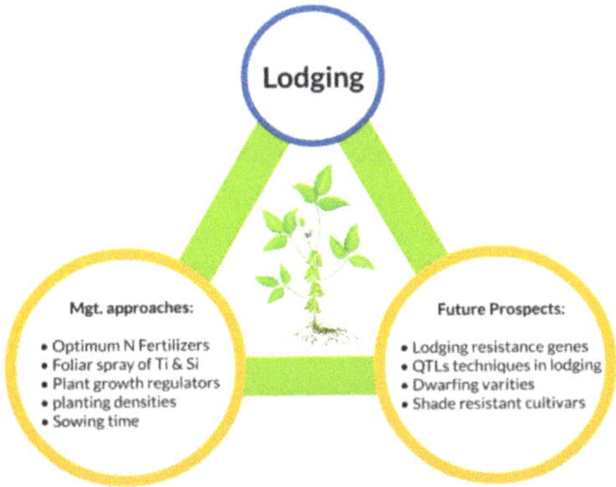

Figure 3. An overview of the most suitable approaches including molecular breeding techniques, agronomical managements and chemical controls to mitigate the lodging stress in maize-soybean intercropping.

3.5. Role of Silicon and Titanium

Si plays a pivotal role in the growth and development of plants as a beneficial micronutrient. A significant impact of silicon has been observed on the plant height, internodal length and stem strength and lodging tolerance [165,166]. It was found that the application of Si could be dispersed from third to the fourth internode, which result in enhancement of lodging resistance [167]. A previous study revealed that Si had enhanced the thickness of the cell wall and the vascular bundles size in the rice stem [168]. With increase in the amount of Si application, the cellulose and hemicellulose content are increased which contribute in the cell wall formation of rice stem [169]. However, it was also identified that Si acts as ligands binding with hemicellulose in the rice cell walls [54,170]. A recent research showed that appropriate concentrations of Si could improve the enzymatic activity of lignin related enzymes and also prompted the gene expression of related enzymes [75]. It also revealed that Si content also increased the lignin content in the stem cell wall and promoted the lodging resistance of soybean under low light conditions. Moreover, Si also accelerated the photosynthetic activities, stomatal conductance and increased the chlorophyll content of tobacco under cadmium stress [171]. Wang et al. [172] observed that Si element could amplify the photosynthetic rate by altering the leaf structure and the content of chlorophyll in rice plants. In addition, Si element can change the leaf anatomy to capture more light and enhance the light interception in the plant's leaves and to improve the vascular bundle sheath and sclerenchyma tissues, which help in lodging resistance [173,174]. Another research has concluded that Si could be used as fuel to ignite the process of lignification and silicification for the cell wall and collenchyma cells thickness that increases the development of keratinocyte and cellulose contents resulting in lodging tolerance [28].

On the other hand, the biological importance and role of titanium (Ti) in growth and development has been studied for decades but still, it is not recognized as an essential nutrient for the plants. However, recent research revealed that optimum concentration of Ti improves the leaf chloroplast structure, total biomass, chlorophyll fluorescence, root architecture, RuBisCO enzyme activity and total chlorophyll content of soybean plants under low light conditions [175]. In addition, how Ti affects the lignin content is not elaborated yet under low light conditions. Therefore, in future RNAseq transcriptional studies, proteomics and genomic profiling should be done to gain deeper insights into the effects and benefits of Si and Ti on soybean stem strength, lodging resistance under the maize-soybean intercropping system.

3.6. Chemical Approaches

Foliar application of plant growth regulators at the suitable growth stage of a crop can enhance the stem mechanical strength, reduce the plant height and inhibit lodging [176,177]. Plant growth hormones that stop the biosynthesis of GA are being widely used in high input cropping systems to decrease straw content and also increase the lodging resistance [176]. Many plant growth regulators were extensively utilized to minimize lodging index through shortening the plant height and to obtain a stable grain production [177,178]. The most common growth regulators that inhibit the GA production have been used to decrease the growth of stem are the onium type elements, which have Chlormequate chloride (2-chloroethyl-*N*, *N*, *N*-trimethyl-ammonium chloride, CCC) and Mepiquat-Cl. Some other plant growth regulators which comprise N heterocycles for example, triazoles and imidazoles could also be used to minimize the lodging risk [179]. The application of paclobutrazol had considerably increased the lignin content in the stem cell wall and its function is closely related to enzymes present in the basal second internode [180]. It had also enhanced the stem diameter, internode filling degree and cell wall thickness with increasing the lodging resistance [71]. The Paclobutrazol chemical (PP333) could prompt the enzymatic activities of lignin related enzymes that is, tyrosine ammonia-lyase (TAL), phenylalanine ammonia-lyase (PAL) and cinnamyl alcohol dehydrogenase (CAD) and ultimately proliferated the lignin content and lodging resistance [181]. A plant growth regulator (Trinexapac-ethyl) has reduced the height of the plant in wheat [182,183]. In addition, some other chemicals are needed to be introduced which are capable of enhancing the lodging resistance in intercropped soybean to get better results.

4. Future Prospects

In the maize-soybean intercropping system, shade has a drastic effect on the normal growth and development of soybean at both vegetative and reproductive stages. Shading of maize disturbs the microclimate of soybean which results in shade avoidance response of soybean (stem elongation) and finally causes lodging. The influence of lodging concerning the reduction of grain yield depends on the types of cultivars and planting geometries. This review mainly focused on the multiple approaches and genetic techniques which would be helpful to control the lodging under intercropping systems (Figure 3). The damages induced by lodging in maize-soybean intercropping could be actively reduced with more advanced crop breeding techniques. We further explored that, lignin and cellulose are the main constituents of plant cell wall which play a vital role in plant vigor against biotic and abiotic factors such as lodging. However, the molecular mechanism of lignin and cellulose formation and their relation with hormones (indole-acetic acid, IAA) that how they affect each other has not been explored thoroughly yet. Therefore, it is necessary to explore the biochemical association of hormones and lignin biosynthesis pathways in the maize-soybean intercropping system.

In addition, researchers have to develop new techniques and tools to modify the lignin content in soybean stem without altering its normal functions. In consequences of natural calamities like high-speed winds and rainfall could damage the crops catastrophically through lodging. Therefore, to escape from these devastating effects of lodging following approaches could be adopted: (i) breeding of soybean cultivars with stronger and harder roots without disturbing the existing root numbers per plant, (ii) suitable agronomic management's that is, use of lodging resistant cultivars, rescheduling the planting time and density and use of fertilizers. Furthermore, plant growth regulators can also manipulate the height of the plant which helps in lowering the risk of lodging. Optimum levels of N, P, K and Si fertilizers could play a significant role in the maize-soybean intercropping system. Additionally, further studies are required to alter the plant canopy area which is a vital part of modern agronomy techniques and it is usually obtained through the application of balanced nitrogen fertilizers.

Author Contributions: Conceptualization, A.R., L.W., W.Y.; Supervision, L.W., W.Y.; Visualization, C.B., M.A.A., B.A., M.I.H, I.S., W.L., Z.Y.; Writing—original draft, A.R.,M.Y., M.A.A.; Writing—review and editing, A.R., L.W., M.Y., T.I., I.A. and M.A.A. B.A., M.I.H. All authors have read and agreed to submit the manuscript.

Funding: This work was funded by the National Key Research and Development Program of China (2018YFD1000905) and Sichuan Innovation Team Project of National Modern Agricultural Industry Technology System, SCCXTD−2020−20 and The APC was funded by above mentioned funding agencies.

Acknowledgments: We thank the native English speaker colleagues for editing the English language and grammar correction.

Conflicts of Interest: The authors declare no conflict of interest.

References

1. Oelbermann, M.; Echarte, L. Evaluating soil carbon and nitrogen dynamics in recently established maize-soyabean inter-cropping systems. *Eur. J. Soil Sci.* **2011**, *62*, 35–41. [CrossRef]
2. Tscharntke, T.; Clough, Y.; Wanger, T.C.; Jackson, L.; Motzke, I.; Perfecto, I.; Vandermeer, J.; Whitbread, A. Global food security, biodiversity conservation and the future of agricultural intensification. *Biol. Conserv.* **2012**, *151*, 53–59. [CrossRef]
3. Rosenberg, M. Current World Population. 2011. Available online: www.about.com (accessed on 1 February 2011).
4. Vogel, E.; Meyer, R. Climate Change, Climate Extremes, and Global Food Production—Adaptation in the Agricultural Sector. In *Resilience*; Elsevier: Amsterdam, The Netherlands, 2018; pp. 31–49.
5. Yang, F.; Liao, D.; Fan, Y.; Gao, R.; Wu, X.; Rahman, T.; Yong, T.; Liu, W.; Liu, J.; Du, J.; et al. Effect of narrow-row planting patterns on crop competitive and economic advantage in maize–soybean relay strip intercropping system. *Plant Prod. Sci.* **2016**, *20*, 1–11. [CrossRef]
6. Kermah, M.; Franke, A.C.; Adjei-Nsiah, S.; Ahiabor, B.D.K.; Abaidoo, R.C.; Giller, K.E. Maize-grain legume intercropping for enhanced resource use efficiency and crop productivity in the Guinea savanna of northern Ghana. *Field Crops Res.* **2017**, *213*, 38–50. [CrossRef] [PubMed]
7. Iqbal, N.; Hussain, S.; Ahmed, Z.; Yang, F.; Wang, X.; Liu, W.; Yong, T.; Du, J.; Shu, K.; Yang, W.; et al. Comparative analysis of maize–soybean strip intercropping systems: A review. *Plant Prod. Sci.* **2019**, *22*, 131–142. [CrossRef]
8. Knörzer, H.; Graeff-Hönninger, S.; Guo, B.; Wang, P.; Claupein, W. The rediscovery of intercropping in China: A traditional cropping system for future Chinese agriculture—A review. In *Climate Change, Intercropping, Pest Control and Beneficial Microorganisms*; Springer: Berlin/Heidelberg, Germany, 2009; pp. 13–44.
9. Hong, Y.; Heerink, N.; Zhao, M.; van der Werf, W. Intercropping contributes to a higher technical efficiency in smallholder farming: Evidence from a case study in Gaotai County, China. *Agric. Syst.* **2019**, *173*, 317–324. [CrossRef]
10. Li, L.; Li, S.-M.; Sun, J.-H.; Zhou, L.-L.; Bao, X.-G.; Zhang, H.-G.; Zhang, F.-S. Diversity enhances agricultural productivity via rhizosphere phosphorus facilitation on phosphorus-deficient soils. *Proc. Natl. Acad. Sci. USA* **2007**, *104*, 11192–11196. [CrossRef]
11. Du, J.-B.; Han, T.-F.; Gai, J.-Y.; Yong, T.-W.; Xin, S.; Wang, X.-C.; Feng, Y.; Jiang, L.; Kai, S.; Liu, W.-G. Maize-soybean strip intercropping: Achieved a balance between high productivity and sustainability. *J. Integr. Agric.* **2018**, *17*, 747–754. [CrossRef]
12. Fan, Y.; Chen, J.; Cheng, Y.; Raza, M.A.; Wu, X.; Wang, Z.; Liu, Q.; Wang, R.; Wang, X.; Yong, T. Effect of shading and light recovery on the growth, leaf structure, and photosynthetic performance of soybean in a maize-soybean relay-strip intercropping system. *PLoS ONE* **2018**, *13*, e0198159. [CrossRef]
13. Ahmed, S.; Raza, M.A.; Zhou, T.; Hussain, S.; Khalid, M.H.B.; Feng, L.; Wasaya, A.; Iqbal, N.; Ahmed, A.; Liu, W. Responses of soybean dry matter production, phosphorus accumulation, and seed yield to sowing time under relay intercropping with maize. *Agronomy* **2018**, *8*, 282. [CrossRef]
14. Feng, L.; Raza, M.A.; Li, Z.; Chen, Y.; Khalid, M.H.B.; Du, J.; Liu, W.; Wu, X.; Song, C.; Yu, L. The influence of light intensity and leaf movement on photosynthesis characteristics and carbon balance of soybean. *Front. Plant Sci.* **2019**, *9*, 1952. [CrossRef] [PubMed]

15. Yang, F.; Liao, D.; Wu, X.; Gao, R.; Fan, Y.; Raza, M.A.; Wang, X.; Yong, T.; Liu, W.; Liu, J. Effect of aboveground and belowground interactions on the intercrop yields in maize-soybean relay intercropping systems. *Field Crops Res.* **2017**, *203*, 16–23. [CrossRef]
16. Raza, M.A.; Feng, L.Y.; van Der Werf, W.; Iqbal, N.; Khalid, M.H.B.; Chen, Y.K.; Wasaya, A.; Ahmed, S.; Din, A.M.U.; Khan, A. Maize leaf-removal: A new agronomic approach to increase dry matter, flower number and seed-yield of soybean in maize soybean relay intercropping system. *Sci. Rep.* **2019**, *9*, 1–13. [CrossRef] [PubMed]
17. Gardiner, B.; Berry, P.; Moulia, B. Wind impacts on plant growth, mechanics and damage. *Plant Sci.* **2016**, *245*, 94–118. [CrossRef]
18. Wassmann, R.; Jagadish, S.; Sumfleth, K.; Pathak, H.; Howell, G.; Ismail, A.; Serraj, R.; Redona, E.; Singh, R.; Heuer, S. Regional vulnerability of climate change impacts on Asian rice production and scope for adaptation. *Adv. Agron.* **2009**, *102*, 91–133.
19. Liu, W.; Deng, Y.; Hussain, S.; Zou, J.; Yuan, J.; Luo, L.; Yang, C.; Yuan, X.; Yang, W. Relationship between cellulose accumulation and lodging resistance in the stem of relay intercropped soybean [*Glycine max* (L.) Merr.]. *Field Crops Res.* **2016**, *196*, 261–267. [CrossRef]
20. Yamaguchi, N.; Sayama, T.; Yamazaki, H.; Miyoshi, T.; Ishimoto, M.; Funatsuki, H. Quantitative trait loci associated with lodging tolerance in soybean cultivar 'Toyoharuka'. *Breed. Sci.* **2014**, *64*, 300–308. [CrossRef]
21. Mizuta, K.; Araki, H.; Takahashi, T. Shifting timing of intensive nitrogen topdressing later to the stem-elongation phase reduced lower internodes length and lodging risk of wheat. *Plant Prod. Sci.* **2020**, 1–9. [CrossRef]
22. Nagashima, H.; Hikosaka, K. Plants in a crowded stand regulate their height growth so as to maintain similar heights to neighbours even when they have potential advantages in height growth. *Ann. Bot.* **2011**, *108*, 207–214. [CrossRef]
23. Hubbart, S.; Smillie, I.R.; Heatley, M.; Swarup, R.; Foo, C.C.; Zhao, L.; Murchie, E.H. Enhanced thylakoid photoprotection can increase yield and canopy radiation use efficiency in rice. *Commun. Biol.* **2018**, *1*, 1–12. [CrossRef] [PubMed]
24. Chen, X.; Wang, J.; Wang, Z.; Li, W.; Wang, C.; Yan, S.; Li, H.; Zhang, A.; Tang, Z.; Wei, M. Optimized nitrogen fertilizer application mode increased culms lignin accumulation and lodging resistance in culms of winter wheat. *Field Crops Res.* **2018**, *228*, 31–38. [CrossRef]
25. Zheng, M.; Chen, J.; Shi, Y.; Li, Y.; Yin, Y.; Yang, D.; Luo, Y.; Pang, D.; Xu, X.; Li, W. Manipulation of lignin metabolism by plant densities and its relationship with lodging resistance in wheat. *Sci. Rep.* **2017**, *7*, 1–12. [CrossRef] [PubMed]
26. Weng, F.; Zhang, W.; Wu, X.; Xu, X.; Ding, Y.; Li, G.; Liu, Z.; Wang, S. Impact of low-temperature, overcast and rainy weather during the reproductive growth stage on lodging resistance of rice. *Sci. Rep.* **2017**, *7*, 46596. [CrossRef] [PubMed]
27. Ma, J.F. Role of silicon in enhancing the resistance of plants to biotic and abiotic stresses. *Soil Sci. Plant Nutr.* **2004**, *50*, 11–18. [CrossRef]
28. Dorairaj, D.; Ismail, M.R.; Sinniah, U.R.; Kar Ban, T. Influence of silicon on growth, yield, and lodging resistance of MR219, a lowland rice of Malaysia. *J. Plant Nutr.* **2017**, *40*, 1111–1124. [CrossRef]
29. Abdel Latef, A.A.; Tran, L.-S.P. Impacts of priming with silicon on the growth and tolerance of maize plants to alkaline stress. *Front. Plant Sci.* **2016**, *7*, 243. [CrossRef]
30. Berry, P.M.; Spink, J. Predicting yield losses caused by lodging in wheat. *Field Crops Res.* **2012**, *137*, 19–26. [CrossRef]
31. Sun, Z.; Zhang, Z.; Cai, Z.; Zeng, W.; Lai, Z.; Chen, H.; Yang, S.; Tang, X.; Yanzhu, S.; Gai, J. Establishment of an evaluation system of shade tolerance in soybean and its variation in southern China germplasm population. *Sci. Agric. Sin.* **2017**, *50*, 729–801.
32. Haghighi, M.; Heidarian, S.; Da Silva, J.A.T. The effect of titanium amendment in N-withholding nutrient solution on physiological and photosynthesis attributes and micronutrient uptake of tomato. *Biol. Trace Elem. Res.* **2012**, *150*, 381–390. [CrossRef]
33. Boykov, I.N.; Shuford, E.; Zhang, B. Nanoparticle titanium dioxide affects the growth and microRNA expression of switchgrass (*Panicum virgatum*). *Genomics* **2019**, *111*, 450–456. [CrossRef] [PubMed]

34. Samadi, N.; Yahyaabadi, S.; Rezayatmand, Z. Effect of TiO2 and TiO2 nanoparticle on germination, root and shoot Length and photosynthetic pigments of Mentha piperita. *Int. J. Plant Soil Sci.* **2014**, *3*, 408–418. [CrossRef]
35. Assefa, T.; Otyama, P.I.; Brown, A.V.; Kalberer, S.R.; Kulkarni, R.S.; Cannon, S.B. Genome-wide associations and epistatic interactions for internode number, plant height, seed weight and seed yield in soybean. *BMC Genom.* **2019**, *20*, 527. [CrossRef] [PubMed]
36. Longin, C.F.H.; Würschum, T. Genetic variability, heritability and correlation among agronomic and disease resistance traits in a diversity panel and elite breeding material of spelt wheat. *Plant Breed.* **2014**, *133*, 459–464. [CrossRef]
37. Specht, J.; Chase, K.; Macrander, M.; Graef, G.; Chung, J.; Markwell, J.; Germann, M.; Orf, J.H.; Lark, K. Soybean response to water: A QTL analysis of drought tolerance. *Crop Sci.* **2001**, *41*, 493–509. [CrossRef]
38. Piñera-Chavez, F.; Berry, P.; Foulkes, M.; Jesson, M.; Reynolds, M. Avoiding lodging in irrigated spring wheat. I. Stem and root structural requirements. *Field Crops Res.* **2016**, *196*, 325–336. [CrossRef]
39. Shah, A.N.; Tanveer, M.; ur Rehman, A.; Anjum, S.A.; Iqbal, J.; Ahmad, R. Lodging stress in cereal—Effects and management: An overview. *Environ. Sci. Pollut. Res.* **2017**, *24*, 5222–5237. [CrossRef]
40. Su, B.; Song, Y.; Song, C.; Cui, L.; Yong, T.; Yang, W. Growth and photosynthetic responses of soybean seedlings to maize shading in relay intercropping system in Southwest China. *Photosynthetica* **2014**, *52*, 332–340. [CrossRef]
41. Xu, Y.; Zhang, R.; Hou, Z.; Yan, C.; Xia, X.; Ma, C.; Dong, S.; Gong, Z. Mechanical properties of soybean plants under various plant densities. *Crop Pasture Sci.* **2020**, *71*, 249–259. [CrossRef]
42. Gommers, C.M.; Visser, E.J.; St Onge, K.R.; Voesenek, L.A.; Pierik, R. Shade tolerance: When growing tall is not an option. *Trends Plant Sci.* **2013**, *18*, 65–71. [CrossRef]
43. Park, Y.; Runkle, E.S. Far-red radiation promotes growth of seedlings by increasing leaf expansion and whole-plant net assimilation. *Environ. Exp. Bot.* **2017**, *136*, 41–49. [CrossRef]
44. Green-Tracewicz, E.; Page, E.R.; Swanton, C.J. Shade avoidance in soybean reduces branching and increases plant-to-plant variability in biomass and yield per plant. *Weed Sci.* **2011**, *59*, 43–49. [CrossRef]
45. Casal, J.J. Shade avoidance. *Arab. Book/Am. Soc. Plant Biol.* **2012**, *10*. [CrossRef] [PubMed]
46. Zhang, W.-J.; Wu, L.-M.; Ding, Y.-F.; Fei, W.; Wu, X.-R.; Li, G.-H.; Liu, Z.-H.; She, T.; Ding, C.-Q.; Wang, S.-H. Top-dressing nitrogen fertilizer rate contributes to decrease culm physical strength by reducing structural carbohydrate content in japonica rice. *J. Integr. Agric.* **2016**, *15*, 992–1004. [CrossRef]
47. Liu, X.; Rahman, T.; Song, C.; Su, B.; Yang, F.; Yong, T.; Wu, Y.; Zhang, C.; Yang, W. Changes in light environment, morphology, growth and yield of soybean in maize-soybean intercropping systems. *Field Crops Res.* **2017**, *200*, 38–46. [CrossRef]
48. Liu, W.-G.; Hussain, S.; Ting, L.; Zou, J.-L.; Ren, M.-L.; Tao, Z.; Jiang, L.; Feng, Y.; Yang, W.-Y. Shade stress decreases stem strength of soybean through restraining lignin biosynthesis. *J. Integr. Agric.* **2019**, *18*, 43–53. [CrossRef]
49. Cui, H.; Jin, L.; Li, B.; Zhang, J.; Zhao, B.; Dong, S.; Liu, P. Effects of shading on stalks morphology, structure and lodging of summer maize in field. *Sci. Agric. Sin.* **2012**, *45*, 3497–3505.
50. Weng, J.-K.; Li, X.; Stout, J.; Chapple, C. Independent origins of syringyl lignin in vascular plants. *Proc. Natl. Acad. Sci. USA* **2008**, *105*, 7887–7892. [CrossRef]
51. He, C.; Ma, J.; Wang, L. A hemicellulose-bound form of silicon with potential to improve the mechanical properties and regeneration of the cell wall of rice. *New Phytol.* **2015**, *206*, 1051–1062. [CrossRef]
52. Koyama, K.; Ikeda, H.; Poudel, P.R.; Goto-Yamamoto, N. Light quality affects flavonoid biosynthesis in young berries of Cabernet Sauvignon grape. *Phytochemistry* **2012**, *78*, 54–64. [CrossRef]
53. Wang, Y.; Gao, L.; Shan, Y.; Liu, Y.; Tian, Y.; Xia, T. Influence of shade on flavonoid biosynthesis in tea (*Camellia sinensis* (L.) O. Kuntze). *Sci. Horticult.* **2012**, *141*, 7–16. [CrossRef]
54. Yu, J.-K.; Graznak, E.; Breseghello, F.; Tefera, H.; Sorrells, M.E. QTL mapping of agronomic traits in tef [*Eragrostis tef* (Zucc) Trotter]. *BMC Plant Biol.* **2007**, *7*, 30. [CrossRef] [PubMed]
55. Hussain, S.; Iqbal, N.; Rahman, T.; Liu, T.; Brestic, M.; Safdar, M.E.; Asghar, M.A.; Farooq, M.U.; Shafiq, I.; Ali, A. Shade effect on carbohydrates dynamics and stem strength of soybean genotypes. *Environ. Exp. Bot.* **2019**, *162*, 374–382. [CrossRef]
56. Zhang, J.; Li, G.; Song, Y.; Liu, Z.; Yang, C.; Tang, S.; Zheng, C.; Wang, S.; Ding, Y. Lodging resistance characteristics of high-yielding rice populations. *Field Crops Res.* **2014**, *161*, 64–74. [CrossRef]

57. Acreche, M.M.; Slafer, G.A. Lodging yield penalties as affected by breeding in Mediterranean wheats. *Field Crops Res.* **2011**, *122*, 40–48. [CrossRef]
58. Arai-Sanoh, Y.; Ida, M.; Zhao, R.; Yoshinaga, S.; Takai, T.; Ishimaru, T.; Maeda, H.; Nishitani, K.; Terashima, Y.; Gau, M. Genotypic variations in non-structural carbohydrate and cell-wall components of the stem in rice, sorghum, and sugar vane. *Biosci. Biotechnol. Biochem.* **2011**, 1105072478. [CrossRef]
59. Wu, L.; Zhang, W.; Ding, Y.; Zhang, J.; Cambula, E.D.; Weng, F.; Liu, Z.; Ding, C.; Tang, S.; Chen, L. Shading contributes to the reduction of stem mechanical strength by decreasing cell wall synthesis in japonica rice (*Oryza sativa* L.). *Front. Plant Sci.* **2017**, *8*, 881. [CrossRef]
60. Loades, C.J. *Manipulation of Defence Related Lignification in Wheat*; University of Southampton: Southampton, UK, 2003.
61. Boudet, A.M.; Kajita, S.; Grima-Pettenati, J.; Goffner, D. Lignins and lignocellulosics: A better control of synthesis for new and improved uses. *Trends Plant Sci.* **2003**, *8*, 576–581. [CrossRef]
62. Wang, C.; Hu, D.; Liu, X.; She, H.; Ruan, R.; Yang, H.; Yi, Z.; Wu, D. Effects of uniconazole on the lignin metabolism and lodging resistance of culm in common buckwheat (*Fagopyrum esculentum* M.). *Field Crops Res.* **2015**, *180*, 46–53. [CrossRef]
63. Ahmad, I.; Kamran, M.; Guo, Z.; Meng, X.; Ali, S.; Zhang, P.; Liu, T.; Cai, T.; Han, Q. Effects of uniconazole or ethephon foliar application on culm mechanical strength and lignin metabolism, and their relationship with lodging resistance in winter wheat. *Crop Pasture Sci.* **2020**, *71*, 12–22. [CrossRef]
64. Liu, W.-G.; Ren, M.-L.; Ting, L.; Du, Y.-L.; Tao, Z.; Liu, X.-M.; Jiang, L.; Hussain, S.; Yang, W.-Y. Effect of shade stress on lignin biosynthesis in soybean stems. *J. Integr. Agric.* **2018**, *17*, 1594–1604. [CrossRef]
65. Moura, J.C.M.S.; Bonine, C.A.V.; de Oliveira Fernandes Viana, J.; Dornelas, M.C.; Mazzafera, P. Abiotic and biotic stresses and changes in the lignin content and composition in plants. *J. Integr. Plant Biol.* **2010**, *52*, 360–376. [CrossRef] [PubMed]
66. Boerjan, W.; Ralph, J.; Baucher, M. Lignin biosynthesis. *Annu. Rev. Plant Biol.* **2003**, *54*, 519–546. [CrossRef] [PubMed]
67. Yao, X.; Li, C.; Li, S.; Zhu, Q.; Zhang, H.; Wang, H.; Yu, C.; Martin, S.K.S.; Xie, F. Effect of shade on leaf photosynthetic capacity, light-intercepting, electron transfer and energy distribution of soybeans. *Plant Growth Regul.* **2017**, *83*, 409–416. [CrossRef]
68. Peng, D.; Chen, X.; Yin, Y.; Lu, K.; Yang, W.; Tang, Y.; Wang, Z. Lodging resistance of winter wheat (*Triticum aestivum* L.): Lignin accumulation and its related enzymes activities due to the application of paclobutrazol or gibberellin acid. *Field Crops Res.* **2014**, *157*, 1–7. [CrossRef]
69. Huber, H.; de Brouwer, J.; von Wettberg, E.J.; During, H.J.; Anten, N.P. More cells, bigger cells or simply reorganization? Alternative mechanisms leading to changed internode architecture under contrasting stress regimes. *New Phytol.* **2014**, *201*, 193–204. [CrossRef]
70. Xue, J.; Gou, L.; Zhao, Y.; Yao, M.; Yao, H.; Tian, J.; Zhang, W. Effects of light intensity within the canopy on maize lodging. *Field Crops Res.* **2016**, *188*, 133–141. [CrossRef]
71. Okuno, A.; Hirano, K.; Asano, K.; Takase, W.; Masuda, R.; Morinaka, Y.; Ueguchi-Tanaka, M.; Kitano, H.; Matsuoka, M. New approach to increasing rice lodging resistance and biomass yield through the use of high gibberellin producing varieties. *PLoS ONE* **2014**, *9*, e86870. [CrossRef]
72. Hirano, K.; Ordonio, R.L.; Matsuoka, M. Engineering the lodging resistance mechanism of post-Green Revolution rice to meet future demands. *Proc. Jpn. Acad. Ser. B* **2017**, *93*, 220–233. [CrossRef]
73. Spielmeyer, W.; Ellis, M.H.; Chandler, P.M. Semidwarf (sd-1), "green revolution" rice, contains a defective gibberellin 20-oxidase gene. *Proc. Natl. Acad. Sci. USA* **2002**, *99*, 9043–9048. [CrossRef]
74. Xiang, D.B.; Zhao, G.; Wan, Y.; Tan, M.L.; Song, C.; Song, Y. Effect of planting density on lodging-related morphology, lodging rate, and yield of tartary buckwheat (*Fagopyrum tataricum*). *Plant Prod. Sci.* **2016**, *19*, 479–488. [CrossRef]
75. Hussain, S.; Shuxian, L.; Mumtaz, M.; Shafiq, I.; Iqbal, N.; Brestic, M.; Shoaib, M.; Sisi, Q.; Li, W.; Mei, X.; et al. Foliar application of silicon improves stem strength under low light stress by regulating lignin biosynthesis genes in soybean (*Glycine max* (L.) Merr.). *J. Hazard. Mater.* **2020**, *401*, 123256. [CrossRef] [PubMed]
76. Wen, B.; Zhang, Y.; Hussain, S.; Wang, S.; Zhang, X.; Yang, J.; Xu, M.; Qin, S.; Yang, W.; Liu, W. Slight Shading Stress at Seedling Stage Does not Reduce Lignin Biosynthesis or Affect Lodging Resistance of Soybean Stems. *Agronomy* **2020**, *10*, 544. [CrossRef]

77. Raza, M.A.; Feng, L.Y.; Iqbal, N.; Ahmed, M.; Chen, Y.K.; Khalid, M.H.B.; Din, A.M.U.; Khan, A.; Ijaz, W.; Hussain, A. Growth and development of soybean under changing light environments in relay intercropping system. *PeerJ* **2019**, *7*, e7262. [CrossRef] [PubMed]
78. Wu, Y.-S.; Feng, Y.; Gong, W.-Z.; Ahmed, S.; Fan, Y.-F.; Wu, X.-L.; Yong, T.-W.; Liu, W.-G.; Kai, S.; Jiang, L. Shade adaptive response and yield analysis of different soybean genotypes in relay intercropping systems. *J. Integr. Agric.* **2017**, *16*, 1331–1340. [CrossRef]
79. Luo, Y.; Ni, J.; Pang, D.; Jin, M.; Chen, J.; Kong, X.; Li, W.; Chang, Y.; Li, Y.; Wang, Z. Regulation of lignin composition by nitrogen rate and density and its relationship with stem mechanical strength of wheat. *Field Crops Res.* **2019**, *241*, 107572. [CrossRef]
80. Chen, X.; Shi, C.; Yin, Y.; Wang, Z.; Shi, Y.; Peng, D.; Ni, Y.; Cai, T. Relationship between lignin metabolism and lodging resistance in wheat. *Acta Agron.Sin.* **2011**, *37*, 1616–1622. [CrossRef]
81. Del Río, J.C.; Rencoret, J.; Prinsen, P.; Martínez, A.n.T.; Ralph, J.; Gutiérrez, A. Structural characterization of wheat straw lignin as revealed by analytical pyrolysis, 2D-NMR, and reductive cleavage methods. *J. Agric. Food Chem.* **2012**, *60*, 5922–5935. [CrossRef]
82. Hyles, J.; Vautrin, S.; Pettolino, F.; MacMillan, C.; Stachurski, Z.; Breen, J.; Berges, H.; Wicker, T.; Spielmeyer, W. Repeat-length variation in a wheat cellulose synthase-like gene is associated with altered tiller number and stem cell wall composition. *J. Exp. Bot.* **2017**, *68*, 1519–1529. [CrossRef]
83. Xu, Z.; Zhang, D.; Hu, J.; Zhou, X.; Ye, X.; Reichel, K.L.; Stewart, N.R.; Syrenne, R.D.; Yang, X.; Gao, P. Comparative genome analysis of lignin biosynthesis gene families across the plant kingdom. *BMC Bioinform.* **2009**, *10*, S3. [CrossRef]
84. Wang, C.; Ruan, R.; Yuan, X.; Hu, D.; Yang, H.; Li, Y.; Yi, Z. Relationship between lignin metabolism and lodging resistance of culm in buckwheat. *J. Agric. Sci.* **2014**, *6*, 29. [CrossRef]
85. Dranski, J.A.L.; Malavasi, U.C.; Malavasi, M.D.M. Relationship between lignin content and quality of Pinus taeda seedlings1. *Rev. Árvore* **2015**, *39*, 905–913. [CrossRef]
86. Lu, H.; Zhao, Y.-l.; Jiang, X.-N. Stable and specific expression of 4-coumarate: Coenzyme A ligase gene (4CL1) driven by the xylem-specific Pto4CL1 promoter in the transgenic tobacco. *Biotechnol. Lett.* **2004**, *26*, 1147–1152. [CrossRef] [PubMed]
87. Maeda, H.A. Lignin biosynthesis: Tyrosine shortcut in grasses. *Nat.Plants* **2016**, *2*, 1–2. [CrossRef] [PubMed]
88. Fagerstedt, K.V.; Kukkola, E.M.; Koistinen, V.V.; Takahashi, J.; Marjamaa, K. Cell wall lignin is polymerised by class III secretable plant peroxidases in Norway spruce. *J. Integr. Plant Biol.* **2010**, *52*, 186–194. [CrossRef]
89. Berthet, S.; Thevenin, J.; Baratiny, D.; Demont-Caulet, N.; Debeaujon, I.; Bidzinski, P.; Leple, J.-C.; Huis, R.; Hawkins, S.; Gomez, L.-D. Role of plant laccases in lignin polymerization. In *Advances in Botanical Research*; Elsevier: Amsterdam, The Netherlands, 2012; Volume 61, pp. 145–172.
90. Ma, Q.-H. Characterization of a cinnamoyl-CoA reductase that is associated with stem development in wheat. *J. Exp. Bot.* **2007**, *58*, 2011–2021. [CrossRef]
91. Nguyen, T.-N.; Son, S.; Jordan, M.C.; Levin, D.B.; Ayele, B.T. Lignin biosynthesis in wheat (*Triticum aestivum* L.): Its response to waterlogging and association with hormonal levels. *BMC Plant Biol.* **2016**, *16*, 28. [CrossRef]
92. Tamaoki, D.; Karahara, I.; Nishiuchi, T.; Wakasugi, T.; Yamada, K.; Kamisaka, S. Involvement of auxin dynamics in hypergravity-induced promotion of lignin-related gene expression in Arabidopsis inflorescence stems. *J. Exp. Bot.* **2011**, *62*, 5463–5469. [CrossRef]
93. Gutiérrez, J.; Núñez-Flores, M.J.L.; Gómez-Ros, L.V.; Uzal, E.N.; Carrasco, A.E.; Díaz, J.; Sottomayor, M.; Cuello, J.; Barceló, A.R. Hormonal regulation of the basic peroxidase isoenzyme from Zinnia elegans. *Planta* **2009**, *230*, 767–778. [CrossRef]
94. Cai, T.; Meng, X.; Liu, X.; Liu, T.; Wang, H.; Jia, Z.; Yang, D.; Ren, X. Exogenous hormonal application regulates the occurrence of wheat tillers by changing endogenous hormones. *Front. Plant Sci.* **2018**, *9*, 1886. [CrossRef]
95. Liu, Y.; Gu, D.; Ding, Y.; Wang, Q.; Li, G.; Wang, S. The relationship between nitrogen, auxin and cytokinin in the growth regulation of rice (*Oryza sativa* L.) Tiller buds. *Aust. J. Crop Sci.* **2011**, *5*, 1019.
96. Chatfield, S.P.; Stirnberg, P.; Forde, B.G.; Leyser, O. The hormonal regulation of axillary bud growth in Arabidopsis. *Plant J.* **2000**, *24*, 159–169. [CrossRef] [PubMed]
97. Franklin, K.A. Shade avoidance. *New Phytol.* **2008**, *179*, 930–944. [CrossRef] [PubMed]

98. Keuskamp, D.H.; Sasidharan, R.; Pierik, R. Physiological regulation and functional significance of shade avoidance responses to neighbors. *Plant Signal. Behav.* **2010**, *5*, 655–662. [CrossRef] [PubMed]
99. Hussain, S.; Liu, T.; Iqbal, N.; Brestic, M.; Pang, T.; Mumtaz, M.; Shafiq, I.; Li, S.; Wang, L.; Gao, Y.; et al. Effects of lignin, cellulose, hemicellulose, sucrose and monosaccharide carbohydrates on soybean physical stem strength and yield in intercropping. *Photochem. Photobiol. Sci.* **2020**, *19*, 462–472. [CrossRef]
100. Boomsma, C.R.; Santini, J.B.; Tollenaar, M.; Vyn, T.J. Maize morphophysiological responses to intense crowding and low nitrogen availability: An analysis and review. *Agron. J.* **2009**, *101*, 1426–1452. [CrossRef]
101. Carriedo, L.G.; Maloof, J.N.; Brady, S.M. Molecular control of crop shade avoidance. *Curr. Opin. Plant Biol.* **2016**, *30*, 151–158. [CrossRef] [PubMed]
102. Cui, W.; Song, Q.; Zuo, B.; Han, Q.; Jia, Z. Effects of Gibberellin (GA4+7) in Grain Filling, Hormonal Behavior, and Antioxidants in High-Density Maize (*Zea mays* L.). *Plants* **2020**, *9*, 978. [CrossRef]
103. Wang, D.; Graef, G.L.; Procopiuk, A.M.; Diers, B.W. Identification of putative QTL that underlie yield in interspecific soybean backcross populations. *Theor. Appl. Genet.* **2004**, *108*, 458–467. [CrossRef]
104. Hwang, S.; Lee, T.G. Integration of lodging resistance QTL in soybean. *Sci. Rep.* **2019**, *9*, 6540. [CrossRef]
105. Berry, P.; Berry, S. Understanding the genetic control of lodging-associated plant characters in winter wheat (*Triticum aestivum* L.). *Euphytica* **2015**, *205*, 671–689. [CrossRef]
106. Yadav, S.; Singh, U.M.; Naik, S.M.; Venkateshwarlu, C.; Ramayya, P.J.; Raman, K.A.; Sandhu, N.; Kumar, A. Molecular mapping of QTLs associated with lodging resistance in dry direct-seeded rice (*Oryza sativa* L.). *Front. Plant Sci.* **2017**, *8*, 1431. [CrossRef] [PubMed]
107. Yano, K.; Ookawa, T.; Aya, K.; Ochiai, Y.; Hirasawa, T.; Ebitani, T.; Takarada, T.; Yano, M.; Yamamoto, T.; Fukuoka, S. Isolation of a novel lodging resistance QTL gene involved in strigolactone signaling and its pyramiding with a QTL gene involved in another mechanism. *Mol. Plant* **2015**, *8*, 303–314. [CrossRef] [PubMed]
108. Liu, F.; Wang, P.; Zhang, X.; Li, X.; Yan, X.; Fu, D.; Wu, G. The genetic and molecular basis of crop height based on a rice model. *Planta* **2018**, *247*, 1–26. [CrossRef] [PubMed]
109. Varshney, R.K.; Bansal, K.C.; Aggarwal, P.K.; Datta, S.K.; Craufurd, P.Q. Agricultural biotechnology for crop improvement in a variable climate: Hope or hype? *Trends Plant Sci.* **2011**, *16*, 363–371. [CrossRef]
110. Wu, W.; Ma, B.L. A new method for assessing plant lodging and the impact of management options on lodging in canola crop production. *Sci. Rep.* **2016**, *6*, 31890. [CrossRef]
111. Cui, F.; Li, J.; Ding, A.; Zhao, C.; Wang, L.; Wang, X.; Li, S.; Bao, Y.; Li, X.; Feng, D. Conditional QTL mapping for plant height with respect to the length of the spike and internode in two mapping populations of wheat. *Theor. Appl. Genet.* **2011**, *122*, 1517–1536. [CrossRef]
112. Hai, L.; Guo, H.; Xiao, S.; Jiang, G.; Zhang, X.; Yan, C.; Xin, Z.; Jia, J. Quantitative trait loci (QTL) of stem strength and related traits in a doubled-haploid population of wheat (*Triticum aestivum* L.). *Euphytica* **2005**, *141*, 1–9. [CrossRef]
113. Ellis, M.; Rebetzke, G.; Azanza, F.; Richards, R.; Spielmeyer, W. Molecular mapping of gibberellin-responsive dwarfing genes in bread wheat. *Theor. Appl. Genet.* **2005**, *111*, 423–430. [CrossRef]
114. Griffiths, S.; Simmonds, J.; Leverington, M.; Wang, Y.; Fish, L.; Sayers, L.; Alibert, L.; Orford, S.; Wingen, L.; Snape, J. Meta-QTL analysis of the genetic control of crop height in elite European winter wheat germplasm. *Mol. Breed.* **2012**, *29*, 159–171. [CrossRef]
115. Spielmeyer, W.; Hyles, J.; Joaquim, P.; Azanza, F.; Bonnett, D.; Ellis, M.; Moore, C.; Richards, R. A QTL on chromosome 6A in bread wheat (*Triticum aestivum*) is associated with longer coleoptiles, greater seedling vigour and final plant height. *Theor. Appl. Genet.* **2007**, *115*, 59–66. [CrossRef] [PubMed]
116. Shearman, V.J. *Changes in the Yield Limiting Processes Associated with the Genetic Improvement of Wheat*; University of Nottingham: Nottingham, UK, 2001.
117. Rebetzke, G.; Ellis, M.; Bonnett, D.; Mickelson, B.; Condon, A.; Richards, R. Height reduction and agronomic performance for selected gibberellin-responsive dwarfing genes in bread wheat (*Triticum aestivum* L.). *Field Crops Res.* **2012**, *126*, 87–96. [CrossRef]
118. Daoura, B.G.; Chen, L.; Du, Y.; Hu, Y.-G. Genetic effects of dwarfing gene Rht-5 on agronomic traits in common wheat (*Triticum aestivum* L.) and QTL analysis on its linked traits. *Field Crops Res.* **2014**, *156*, 22–29. [CrossRef]
119. Daoura, B.G.; Chen, L.; Hu, Y.-G. Agronomic traits affected by dwarfing gene'Rht-5'in common wheat (*'Triticum aestivum'* L.). *Aust. J. Crop Sci.* **2013**, *7*, 1270.

120. Chen, L.; Yang, Y.; Cui, C.; Lu, S.; Lu, Q.; Du, Y.; Su, R.; Chai, Y.; Li, H.; Chen, F. Effects of Vrn-B1 and Ppd-D1 on developmental and agronomic traits in Rht5 dwarf plants of bread wheat. *Field Crops Res.* **2018**, *219*, 24–32. [CrossRef]
121. Huang, C.; Liu, Q.; Li, H.; Li, X.; Zhang, C.; Zhang, F. Optimised sowing date enhances crop resilience towards size-asymmetric competition and reduces the yield difference between intercropped and sole maize. *Field Crops Res.* **2018**, *217*, 125–133. [CrossRef]
122. Egbe, O. Effects of plant density of intercropped soybean with tall sorghum on competitive ability of soybean and economic yield at Otobi, Benue State, Nigeria. *J. Cereals Oilseeds* **2010**, *1*, 1–10.
123. Zhou, T.; Du, Y.; Ahmed, S.; Liu, T.; Ren, M.; Liu, W.; Yang, W. Genotypic differences in phosphorus efficiency and the performance of physiological characteristics in response to low phosphorus stress of soybean in southwest of China. *Front. Plant Sci.* **2016**, *7*, 1776. [CrossRef]
124. Dai, X.; Wang, Y.; Dong, X.; Qian, T.; Yin, L.; Dong, S.; Chu, J.; He, M. Delayed sowing can increase lodging resistance while maintaining grain yield and nitrogen use efficiency in winter wheat. *Crop J.* **2017**, *5*, 541–552. [CrossRef]
125. Spink, J.; Semere, T.; Sparkes, D.; Whaley, J.; Foulkes, M.; Clare, R.; Scott, R. Effect of sowing date on the optimum plant density of winter wheat. *Ann. Appl. Biol.* **2000**, *137*, 179–188. [CrossRef]
126. Ahmed, S.; Raza, M.A.; Yuan, X.; Du, Y.; Iqbal, N.; Chachar, Q.; Soomro, A.A.; Ibrahim, F.; Hussain, S.; Wang, X.; et al. Optimized planting time and co-growth duration reduce the yield difference between intercropped and sole soybean by enhancing soybean resilience toward size-asymmetric competition. *Food Energy Secur.* **2020**, *9*. [CrossRef]
127. Yu, Y.; Stomph, T.-J.; Makowski, D.; van der Werf, W. Temporal niche differentiation increases the land equivalent ratio of annual intercrops: A meta-analysis. *Field Crops Res.* **2015**, *184*, 133–144. [CrossRef]
128. Zhang, F.; Li, L. Using competitive and facilitative interactions in intercropping systems enhances crop productivity and nutrient-use efficiency. *Plant Soil* **2003**, *248*, 305–312. [CrossRef]
129. Song, D.; Tariq, A.; Pan, K.; Khan, S.U.; Saleh, T.A.; Gong, S.; Zhang, A.; Wu, X. Influence of planting distance and density on the yield and photosynthetic traits of sweet potato (*Ipomoea balatas* L.) under an intercropping system with walnut (*Juglans regia*) saplings. *Soil Tillage Res.* **2020**, *196*, 104484. [CrossRef]
130. Wang, C.; Wu Ruan, R.; Hui Yuan, X.; Hu, D.; Yang, H.; Li, Y.; Lin Yi, Z. Effects of nitrogen fertilizer and planting density on the lignin synthesis in the culm in relation to lodging resistance of buckwheat. *Plant Prod. Sci.* **2015**, *18*, 218–227. [CrossRef]
131. Sher, A.; Khan, A.; Ashraf, U.; Liu, H.H.; Li, J.C. Characterization of the effect of increased plant density on canopy morphology and stalk lodging risk. *Front. Plant Sci.* **2018**, *9*, 1047. [CrossRef] [PubMed]
132. Fan, Y.; Wang, Z.; Liao, D.; Raza, M.A.; Wang, B.; Zhang, J.; Chen, J.; Feng, L.; Wu, X.; Liu, C. Uptake and utilization of nitrogen, phosphorus and potassium as related to yield advantage in maize-soybean intercropping under different row configurations. *Sci. Rep.* **2020**, *10*, 1–10. [CrossRef]
133. Khan, S.; Anwar, S.; Kuai, J.; Noman, A.; Shahid, M.; Din, M.; Ali, A.; Zhou, G. Alteration in yield and oil quality traits of winter rapeseed by lodging at different planting density and nitrogen rates. *Sci. Rep.* **2018**, *8*, 1–12. [CrossRef]
134. Cheng, B.; Raza, A.; Wang, L.; Xu, M.; Lu, J.; Gao, Y.; Qin, S.; Zhang, Y.; Ahmad, I.; Zhou, T.; et al. Effects of Multiple Planting Densities on Lignin Metabolism and Lodging Resistance of the Strip Intercropped Soybean Stem. *Agronomy* **2020**, *10*, 1177. [CrossRef]
135. Chen, P.; Du, Q.; Liu, X.; Zhou, L.; Hussain, S.; Lei, L.; Song, C.; Wang, X.; Liu, W.; Yang, F. Effects of reduced nitrogen inputs on crop yield and nitrogen use efficiency in a long-term maize-soybean relay strip intercropping system. *PLoS ONE* **2017**, *12*, e0184503. [CrossRef]
136. Said, M.; Hamd-Alla, W. Impact of foliar spraying with antioxidant and intercropping pattern of maize and soybean on yields and its attributes. *J. Plant Prod.* **2018**, *9*, 1069–1073. [CrossRef]
137. Kebebew, S.; Belete, K.; Tana, T. Productivity evaluation of maize-soybean intercropping system under rainfed condition at Bench-Maji Zone, Ethiopia. *Eur. Res.* **2014**, *79*, 1301–1309.
138. Addo-Quaye, A.; Darkwa, A.; Ocloo, G. Yield and productivity of component crops in a maize-soybean intercropping system as affected by time of planting and spatial arrangement. *J. Agric. Biol. Sci.* **2011**, *6*, 50–57.

139. Yogesh, S.; Halikatti, S.; Hiremath, S.; Potdar, M.; Harlapur, S.; Venkatesh, H. Light use efficiency, productivity and profitability of maize and soybean intercropping as influenced by planting geometry and row proportion. *Karnataka J. Agric. Sci.* **2014**, *27*, 1–4.
140. Khatri, N.; Dahal, K.; Amgain, L.; Karki, T. Productivity and economic assessment of maize and soybean intercropping under various tillage and residue levels in Chitwan, Nepal. *World J. Agric. Res.* **2014**, *2*, 6–12. [CrossRef]
141. Ijoyah, M.; Fanen, F. Effects of different cropping pattern on performance of maize-soybean mixture in Makurdi, Nigeria. *Sci. J. Crop Sci.* **2012**, *1*, 39–47.
142. Mahmoudi, R.; Jamshidi, K.; Pouryousef, M. Evaluation of grain yield of maize (*Zea mays* L.) and soybean (*Glycine max* L.) instrip intercropping. *Int. J. Agron. Plant Prod.* **2013**, *4*, 2388–2392.
143. Zaman, U.; Ahmad, Z.; Farooq, M.; Saeed, S.; Ahmad, M.; Wakeel, A. Potassium fertilization may improve stem strength and yield of Basmati rice grown on nitrogen-fertilized soils. *Pak. J. Agric. Sci.* **2015**, *52*, 437–443.
144. Zhang, M.; Wang, H.; Yi, Y.; Ding, J.; Zhu, M.; Li, C.; Guo, W.; Feng, C.; Zhu, X. Effect of nitrogen levels and nitrogen ratios on lodging resistance and yield potential of winter wheat (*Triticum aestivum* L.). *PLoS ONE* **2017**, *12*, e0187543. [CrossRef]
145. Mariani, L.; Ferrante, A. Agronomic management for enhancing plant tolerance to abiotic stresses—Drought, salinity, hypoxia, and lodging. *Horticulturae* **2017**, *3*, 52. [CrossRef]
146. Kong, L.; Sun, M.; Wang, F.; Liu, J.; Feng, B.; Si, J.; Zhang, B.; Li, S.; Li, H. Effects of high NH+ 4 on K+ uptake, culm mechanical strength and grain filling in wheat. *Front. Plant Sci.* **2014**, *5*, 703. [CrossRef] [PubMed]
147. Pan, J.; Zhao, J.; Liu, Y.; Huang, N.; Tian, K.; Shah, F.; Liang, K.; Zhong, X.; Liu, B. Optimized nitrogen management enhances lodging resistance of rice and its morpho-anatomical, mechanical, and molecular mechanisms. *Sci. Rep.* **2019**, *9*, 1–13. [CrossRef] [PubMed]
148. Wei, F.-Z.; Li, J.-C.; Wang, C.-Y.; Qu, H.-J.; Shen, X.-S. Effects of nitrogenous fertilizer application model on culm lodging resistance in winter wheat. *Acta Agron. Sin.* **2008**, *34*, 1080–1085. [CrossRef]
149. Berry, P.; Griffin, J.; Sylvester-Bradley, R.; Scott, R.; Spink, J.; Baker, C.; Clare, R. Controlling plant form through husbandry to minimise lodging in wheat. *Field Crops Res.* **2000**, *67*, 59–81. [CrossRef]
150. Peake, A.; Bell, K.; Carberry, P.; Poole, N.; Raine, S. Vegetative nitrogen stress decreases lodging risk and increases yield of irrigated spring wheat in the subtropics. *Crop Pasture Sci.* **2016**, *67*, 907–920. [CrossRef]
151. Yousaf, M.; Li, J.; Lu, J.; Ren, T.; Cong, R.; Fahad, S.; Li, X. Effects of fertilization on crop production and nutrient-supplying capacity under rice-oilseed rape rotation system. *Sci. Rep.* **2017**, *7*, 1–9. [CrossRef] [PubMed]
152. Bhiah, K.; Guppy, C.; Lockwood, P.; Jessop, R. Effect of potassium on rice lodging under high nitrogen nutrition. In Proceedings of the Brisbane (AU): 19th World Congress of Soil Science "Soil Solution for a Changing World, Brisbane, Australia, 1–6 August 2010; pp. 1–6.
153. Kong, E.; Liu, D.; Guo, X.; Yang, W.; Sun, J.; Li, X.; Zhan, K.; Cui, D.; Lin, J.; Zhang, A. Anatomical and chemical characteristics associated with lodging resistance in wheat. *Crop J.* **2013**, *1*, 43–49. [CrossRef]
154. Wang, M.; Zheng, Q.; Shen, Q.; Guo, S. The critical role of potassium in plant stress response. *Int. J. Mol. Sci.* **2013**, *14*, 7370–7390. [CrossRef]
155. Hafsi, C.; Debez, A.; Abdelly, C. Potassium deficiency in plants: Effects and signaling cascades. *Acta Physiol. Plant.* **2014**, *36*, 1055–1070. [CrossRef]
156. Pitre, F.E.; Cooke, J.E.; Mackay, J.J. Short-term effects of nitrogen availability on wood formation and fibre properties in hybrid poplar. *Trees* **2007**, *21*, 249–259. [CrossRef]
157. Yong, T.; Liu, X.; Yang, F.; Song, C.; Wang, X.; Liu, W.; Su, B.; Zhou, L.; Yang, W. Characteristics of nitrogen uptake, use and transfer in a wheat-maize-soybean relay intercropping system. *Plant Prod. Sci.* **2015**, *18*, 388–397. [CrossRef]
158. Li, L.; Sun, J.; Zhang, F.; Li, X.; Yang, S.; Rengel, Z. Wheat/maize or wheat/soybean strip intercropping: I. Yield advantage and interspecific interactions on nutrients. *Field Crops Res.* **2001**, *71*, 123–137. [CrossRef]
159. Peoples, M.; Brockwell, J.; Herridge, D.; Rochester, I.; Alves, B.; Urquiaga, S.; Boddey, R.; Dakora, F.; Bhattarai, S.; Maskey, S. The contributions of nitrogen-fixing crop legumes to the productivity of agricultural systems. *Symbiosis* **2009**, *48*, 1–17. [CrossRef]
160. Yang, F.; Huang, S.; Gao, R.; Liu, W.; Yong, T.; Wang, X.; Wu, X.; Yang, W. Growth of soybean seedlings in relay strip intercropping systems in relation to light quantity and red: Far-red ratio. *Field Crops Res.* **2014**, *155*, 245–253. [CrossRef]

161. Yu, Y.; Makowski, D.; Stomph, T.J.; van der Werf, W. Robust Increases of Land Equivalent Ratio with Temporal Niche Differentiation: A Meta-Quantile Regression. *Agron. J.* **2016**, *108*, 2269–2279. [CrossRef]
162. Yu, Y.; Stomph, T.-J.; Makowski, D.; Zhang, L.; van der Werf, W. A meta-analysis of relative crop yields in cereal/legume mixtures suggests options for management. *Field Crops Res.* **2016**, *198*, 269–279. [CrossRef]
163. Xu, Z.; Li, C.; Zhang, C.; Yu, Y.; van der Werf, W.; Zhang, F. Intercropping maize and soybean increases efficiency of land and fertilizer nitrogen use; A meta-analysis. *Field Crops Res.* **2020**, *246*, 107661. [CrossRef]
164. Upadhya, D.; Dhakal, R.; Khadka, K.; Rana, S.; Acharya, P.; Rana, R.; Chaudhary, P. Local knowledge on climate-induced traits in rice for improving crop yield, food security and climate resilience. *Int. Agric. Innov. Res. J.* **2016**, *5*, 385–396.
165. Waseem, M.; Ahmad, R.; Randhawa, M.A.; Aziz, T.; Mahmood, N. Influence of silicon application on blast incidence and lodging resistance in rice. *J. Agric. Res.* **2016**, *54*, 435–443.
166. Fallah, A. Studies effect of silicon on lodging parameters in rice plant under hydroponics culture in a greenhouse experiment. In Proceedings of the Silicon in Agriculture Conference, Wild Coast Sun, South Africa, 26–31 October 2008.
167. Salman, D.; Morteza, S.; Dariush, Z.; Abbas, G.M.; Reza, Y.; Ehsan, G.D.; Reza, N.A. Application of nitrogen and silicon rates on morphological and chemical lodging related characteristics in rice (*Oryza sativa* L.) at North of Iran. *J. Agric. Sci.* **2012**, *4*, 12. [CrossRef]
168. Kim, S.G.; Kim, K.W.; Park, E.W.; Choi, D. Silicon-induced cell wall fortification of rice leaves: A possible cellular mechanism of enhanced host resistance to blast. *Phytopathology* **2002**, *92*, 1095–1103. [CrossRef]
169. Zhang, J.; Zou, W.; Li, Y.; Feng, Y.; Zhang, H.; Wu, Z.; Tu, Y.; Wang, Y.; Cai, X.; Peng, L. Silica distinctively affects cell wall features and lignocellulosic saccharification with large enhancement on biomass production in rice. *Plant Sci.* **2015**, *239*, 84–91. [CrossRef] [PubMed]
170. He, C.; Wang, L.; Liu, J.; Liu, X.; Li, X.; Ma, J.; Lin, Y.; Xu, F. Evidence for 'silicon'within the cell walls of suspension-cultured rice cells. *New Phytol.* **2013**, *200*, 700–709. [CrossRef] [PubMed]
171. Yingang, L.; Jun, M.; Ying, T.; Junyu, H.; Christie, P.; Lingjia, Z.; Wenjie, R.; Zhang, M.; Shiping, D. Effect of silicon on growth, physiology, and cadmium translocation of tobacco (*Nicotiana tabacum* L.) in cadmium-contaminated soil. *Pedosphere* **2018**, *28*, 680–689.
172. Wang, Y.; Zhang, B.; Jiang, D.; Chen, G. Silicon improves photosynthetic performance by optimizing thylakoid membrane protein components in rice under drought stress. *Environ. Exp. Bot.* **2019**, *158*, 117–124. [CrossRef]
173. Rehman, B. Silicon elicited varied physiological and biochemical responses in Indian mustard (*Brassica juncea*): A concentration dependent study. *Israel J. Plant Sci.* **2016**, *63*, 158–166. [CrossRef]
174. Hernandez-Apaolaza, L. Can silicon partially alleviate micronutrient deficiency in plants? A review. *Planta* **2014**, *240*, 447–458. [CrossRef]
175. Hussain, S.; Iqbal, N.; Brestic, M.; Raza, M.A.; Pang, T.; Langham, D.R.; Safdar, M.E.; Ahmed, S.; Wen, B.; Gao, Y.; et al. Changes in morphology, chlorophyll fluorescence performance and Rubisco activity of soybean in response to foliar application of ionic titanium under normal light and shade environment. *Sci. Total Environ.* **2019**, *658*, 626–637. [CrossRef]
176. Xu, C.; Gao, Y.; Tian, B.; Ren, J.; Meng, Q.; Wang, P. Effects of EDAH, a novel plant growth regulator, on mechanical strength, stalk vascular bundles and grain yield of summer maize at high densities. *Field Crops Res.* **2017**, *200*, 71–79. [CrossRef]
177. Berry, P.; Sterling, M.; Spink, J.; Baker, C.; Sylvester-Bradley, R.; Mooney, S.; Tams, A.; Ennos, A. Understanding and reducing lodging in cereals. *Adv. Agron.* **2004**, *84*, 215–269.
178. Rajala, A. Plant Growth Regulators to Manipulate Cereal Growth in Northern Growing Conditions. 2003. Available online: https://helda.helsinki.fi/bitstream/handle/10138/20698/plantgro.pdf?sequence=1 (accessed on 2 June 2012).
179. Bhaskaruni, S.V.; Maddila, S.; Gangu, K.K.; Jonnalagadda, S.B. A Review on multi-component green synthesis of N-containing heterocycles using mixed oxides as heterogeneous catalysts. *Arabian J. Chem.* **2020**, *13*, 1142–1178. [CrossRef]
180. Kamran, M.; Ahmad, I.; Wu, X.; Liu, T.; Ding, R.; Han, Q. Application of paclobutrazol: A strategy for inducing lodging resistance of wheat through mediation of plant height, stem physical strength, and lignin biosynthesis. *Environ. Sci. Pollut. Res.* **2018**, *25*, 29366–29378. [CrossRef] [PubMed]

181. Shah, L.; Yahya, M.; Shah, S.M.A.; Nadeem, M.; Ali, A.; Ali, A.; Wang, J.; Riaz, M.W.; Rehman, S.; Wu, W. Improving lodging resistance: Using wheat and rice as classical examples. *Int. J. Mol. Sci.* **2019**, *20*, 4211. [CrossRef] [PubMed]
182. Leśniowska-Nowak, J.; Nowak, M.; Zapalska, M.; Dudziak, K.; Kowalczyk, K. Influence of CCC and trinexapac-ethyl on the expression of genes involved in gibberellic biosynthesis and metabolism pathway in isogenic line with Rht12 dwarfing gene. *Acta Sci. Polonorum. Hortorum Cultus* **2017**, *16*, 141–151. [CrossRef]
183. Espindula, M.; Rocha, V.; Grossi, J.; Souza, M.; Souza, L.; Favarato, L. Use of growth retardants in wheat. *Planta Daninha* **2009**, *27*, 379–387. [CrossRef]

plants

MDPI

Article

Effect of Cadmium Toxicity on Growth, Oxidative Damage, Antioxidant Defense System and Cadmium Accumulation in Two Sorghum Cultivars

Muhammad Jawad Hassan [1,†], Muhammad Ali Raza [2,†], Sana Ur Rehman [3,†], Muhammad Ansar [3], Harun Gitari [4], Imran Khan [1], Muhammad Wajid [5], Mukhtar Ahmed [3,6], Ghulam Abbas Shah [3], Yan Peng [1] and Zhou Li [1,*]

1 Department of Grassland Science, Animal Science and Technology College, Sichuan Agricultural University, Chengdu 611130, China; jawadhassan3146@gmail.com (M.J.H.); Imran.62k@gmail.com (I.K.); pengyanlee@163.com (Y.P.)
2 College of Agronomy, Sichuan Agricultural University, Chengdu 611130, China; Razaali0784@yahoo.com
3 Department of Agronomy, PMAS-Arid Agriculture University, Rawalpindi 46000, Pakistan; Sana07139@gmail.com (S.U.R.); muhammad.ansar@uaar.edu.pk (M.A.); ahmadmukhtar@uaar.edu.pk (M.A.); shahga@uaar.edu.pk (G.A.S.)
4 Department of Agricultural Science and Technology, School of Agriculture and Enterprise Development, Kenyatta University, Nairobi P. O. Box 43844-00100, Kenya; harun.gitari@ku.ac.ke
5 College of Plant Science and Technology, Huazhong Agricultural University, Wuhan 430070, China; hafizwajid.agri@gmail.com
6 Department of Agricultural Research for Northern Sweden, Swedish University of Agricultural Sciences, 90183 Umeå, Sweden
* Correspondence: lizhou1986814@163.com; Tel.: +86-028-86291010
† These authors contributed equally to this work.

Received: 30 October 2020; Accepted: 11 November 2020; Published: 13 November 2020

Abstract: Heavy metal stress is a leading environmental issue reducing crop growth and productivity, particularly in arid and semi-arid agro-ecological zones. Cadmium (Cd), a non-redox heavy metal, can indirectly increase the production of reactive oxygen species (ROS), inducing cell death. A pot experiment was conducted to investigate the effects of different concentrations of Cd (0, 5, 25, 50, 100 μM) on physiological and biochemical parameters in two sorghum (*Sorghum bicolor* L.) cultivars: JS-2002 and Chakwal Sorghum. The results showed that various concentrations of Cd significantly increased the Cd uptake in both cultivars; however, the uptake was higher in JS-2002 compared to Chakwal Sorghum in leaf, stem and root. Regardless of the cultivars, there was a higher accumulation of the Cd in roots than in shoots. The Cd stress significantly reduced the growth and increased the electrolyte leakage (EL), hydrogen peroxide (H_2O_2) concentration and malondialdehyde (MDA) content in both cultivars, but the Chakwal Sorghum showed more pronounced oxidative damage than the JS-2002, as reflected by higher H_2O_2, MDA and EL. Moreover, Cd stress, particularly 50 μM and 100 μM, decreased the activity of different antioxidant enzymes, including superoxide dismutase (SOD), peroxidase (POD), and catalase (CAT). However, the JS-2002 exhibited higher SOD, POD and CAT activities than the Chakwal Sorghum under different Cd-levels. These findings revealed that JS-2002 had a stronger Cd enrichment capacity and also exhibited a better tolerance to Cd stress due to its efficient antioxidant defense system than Chakwal Sorghum. The present study provides the available information about Cd enrichment and tolerance in *S. bicolor*, which is used as an important agricultural crop for livestock feed in arid and semi-arid regions.

Keywords: Cadmium; heavy metal; food security; oxidative damage; antioxidants

1. Introduction

Plants are adversely affected by various environmental factors that hinder their growth and development. Heavy metal stress is one of the most critical abiotic factors that has gained enormous attention over the last 30 years [1,2]. A heavy metal is defined as any element exhibiting high density and that exerts its lethal effects even when available in trace amounts. In short, heavy metals belong to a class of metals with an atomic density of more than 4 g cm^{-3} [1]. Among all elements discovered so far, 53 elements have been identified as heavy metals. However, most of them have no beneficial function in plant metabolism. Among them, chromium (Cr), lead (Pb), cadmium (Cd), silver (Ag), cobalt (Co), platinum (Pt), arsenic (As) and nickel (Ni) play the most devastating role in plant physiology [3]. Heavy metals exist in soils naturally in amounts that exert no apparent harm. However, due to an increase in anthropogenic and geological activities in the past few decades, the concentration of such metals in the environment has increased greatly, which is extremely harmful to plant and animal species [3]. Heavy metals restrict plant growth by lowering the performance of different cell components, such as the thylakoid membranes of chloroplast, lipids and proteins [4]. Besides, heavy metals get incorporated into the food chain from plants to animals, thus leading to a potential risk of different disorders in human beings [5].

Among all the heavy metals, Cd has gained extreme importance due to its massive involvement in food chain contamination, as it is easily absorbed by the cells of various plant species [6–8]. Cd is readily soluble in water, thus entering the semi-permeable membrane of root cells via an active transport mechanism, following an analogous pathway used for zinc transport [9,10]. Cd ions' movement across the semi-permeable membrane takes place with the help of metal transporters (ZIP family), which are a special type of transmembrane protein used for the transport of metals across biological membranes [11,12]. Due to advancements in agricultural and industrial sectors, serious attention has been given to issues related to Cd toxicity in the past few decades [13]. Cd presence in excessive amounts ($\geq$1 mM) in the soil is detrimental for plants, resulting in hazardous effects on many physiological processes, including growth inhibition, root browning, chlorophyll degradation and chlorosis of leaves [14–18]. Cd toxicity results from the increased concentration of Cd ions in both roots and shoots, thus decreasing the crop biomass [19–22]. Agricultural crops absorb Cd via their roots from water present in the soil, so that its uptake correlates with the total Cd available in soil solution [23–26]. Heavy metal accumulation, especially Cd, has been thoroughly investigated, but the exact mechanism involved in Cd stress injury to plants is still not clearly understood.

Cd toxicity is the leading cause of oxidative damage, resulting in the reduced growth of plants by inducing changes in membrane permeability and the later production of reactive oxygen species at the organelle level. The different types of reactive oxygen species (ROS) formed due to Cd stress include hydrogen peroxides (H_2O_2), hydroxyl radical (OH^-) and superoxide anion (O^{2-}), and these are responsible for membrane damage [27,28]. These ROS are the primary causes of membranous proteins and lipids oxidation, which are associated with cell death [29,30]. However, plants have a natural defense system consisting of enzymatic and non-enzymatic antioxidants to protect them against oxidative damage. Enzymatic antioxidants include superoxide dismutase (SOD), peroxidase (POD), catalase (CAT) and ascorbate peroxidase (APX), while non-enzymatic antioxidants consist of glutathione (GSH), ascorbate (ASA), carotenoids and α-tocopherol, which are an effective defensive mechanism to safeguard plants against stress conditions [27,31]. At the cellular level, the first step in the antioxidant defense mechanism is the production of superoxide dismutase for scavenging O^{2-}. The CAT and APX play an important role in H_2O_2 quenching [32–34]. Many studies have been published concerning the defensive role of antioxidants against Cd stress in plants [35,36].

Sorghum (*Sorghum bicolor*) stands at the fifth position among the economically important crops worldwide, while its number is third in the USA. It is a multipurpose crop that can be used not only for food, fodder, and feed, but also as an important ingredient in the beverage industry and in biofuels [35]. Due to its high climatic adaptability, greater fodder yield potential, better quality, palatability, digestibility and bioactive compounds content, sorghum has been widely cultivated in south-Asia, Africa and Central America as an excellent animal feed [35]. It is also a rich source of various phenolic compounds (luteolinidin, apigeninidin, naringenin, etc.) and essential vitamins such as vitamin E [36]. Furthermore, sorghum is tolerant to various abiotic stresses, including high temperature, drought, and heavy metal toxicity. Studies have shown that sorghum plants accumulate large quantities of heavy metals in their roots and shoots with higher biomass production compared to other summer crops [37,38]. The 'JS-2002' and 'Chakwal Sorghum' are two important sorghum cultivars that are widely used as forages to feed livestock under the arid and semiarid regions of Pakistan. The objectives of this study were therefore to evaluate the impact of Cd toxicity on physiological and biochemical parameters, such as growth, Cd accumulation, oxidative damage and antioxidative enzyme activities, in two sorghum cultivars during the stress period. These findings will provide available information about Cd enrichment and tolerance in sorghum, which is used as an important agricultural crop for livestock feed in arid and semi-arid regions. The results will also help to estimate forage safety when the sorghum is cultivated in Cd contaminated soil.

2. Materials and Methods

2.1. Planting Material and Treatments

The present experiment was conducted during the summer season of 2017 in a research area of the Agronomy Department Pir Mehr Ali Shah Arid Agriculture University, Rawalpindi, Pakistan. The study area was located at 33.649° N latitude and 73.082° E longitude. Seeds of two sorghum cultivars, viz. JS-2002 and Chakwal Sorghum, were obtained from National Agriculture Research Centre Islamabad. Before sowing, seeds (50 g for each cultivar) were surface sterilized with 75% ethanol for 10 min and thoroughly washed with distilled water. Both JS-2002 and Chakwal Sorghum are diploids. Seeds were sown in plastic pots (18 cm diameter and 22 cm depth) containing 5 kg of sandy loam soil that was collected from the research fields of the Department of Soil Sciences, PMAS Arid Agriculture University Rawalpindi, Pakistan. Eight seeds were sown in each pot and then thinned to three at the four-leaf stage. The pH of the soil was 7.3 with total Cd (0.21 mg/Kg), and pots were arranged under a completely randomized design with three replicates. Plants were watered daily to maintain field capacity and avoid drought stress. Four different Cd concentrations (5, 25, 50, and 100 μM) were fertigated in the form of $CdCl_2$ before sowing, while a control (without Cd) was kept for comparison. A 400 mL per kg of Cd solution was irrigated evenly in each plastic pot. At 55 days after sowing (DAS), fully expanded and mature leaves were collected for the measurement of morphological parameters, oxidative damage associated with the reactive oxygen species and activities of different antioxidant enzymes. Leaf samples were dissected and immediately immersed in liquid nitrogen then stored at −80 °C in the freezer for further analysis. Similarly, leaf, stem and root samples were collected at 55 DAS for the measurement of Cd accumulation.

2.2. Measurement of Growth Parameters

The height and the leaf number of each sorghum plant were measured at the 55th day after sowing to obtain a clear view of Cd toxicity.

2.3. Estimation of Cd Accumulation

Fresh root, stem and leaf samples at the heading stage were thoroughly washed with deionized water. Later, samples were dried in an oven for 24 h at 70 °C, then converted to powder form and placed in a muffle furnace for 4 h at 550 °C for dry-ashing. After this, the obtained ash residues were added with 1M nitric acid to make a standard volume. The amount of Cd accumulated was measured by using a flame atomic absorption spectrophotometer (Unicam, 929 AAS).

2.4. Determination of Cell Membrane Stability and Oxidative Damage

For each treatment, three plants were taken to measure the electrolyte leakage (EL) of leaves [39]. The leaf samples (1 g) were washed with distilled water to clean the contamination from the surface. These were kept in closed vials containing 10 mL of distilled water, incubated in a rotary shaker at 25 °C for 24 h, and the electrical conductivity (EC) of the solution ($L1$) was estimated. Later, samples were autoclaved for 20 min at 120 °C, and final electrical conductivity ($L2$) was achieved after equilibration at 25 °C. The EL (%) was measured as shown in Equation (1).

$$\text{EL}\ (\%) = \frac{L1}{L2} \times 100 \tag{1}$$

Malondialdehyde (MDA), as the final product of lipid peroxidation, was estimated following the procedure of [40], with minor changes. The reaction mixture comprised 500 μL of enzyme extract, 0.65% (w/v) thiobarbituric acid (TBA) in 20% trichloroacetic acid (TCA), which was heated for thirty minutes, then quickly cooled to stop the reaction. After this, the reaction mixture was centrifuged at 10,000× g for ten minutes. The absorbance value of the reaction mixture was recorded spectrophotometrically at 532 nm, while the non-specific absorption value at a wavelength of 600 nm was deducted from absorbance value gradually.

MDA content was calculated using Equation (2).

$$\text{MDA content} = C \times \frac{V}{W} \tag{2}$$

where W represents 'sample fresh weight', V represents 'extract volume', while C is the 'variation in non-specific absorption between two wavelengths'. It is given as mmol g^{-1} FW.

Hydrogen peroxide (H_2O_2) content was estimated by following the method described by [41]. Exactly 0.3 g of fresh leaf samples were taken and ground mechanically using 4 mL ice-cold acetone (CH_3COCH_3). The homogenate was centrifuged at 12,000× g for 20 min at 4 °C. Then, the supernatant was collected and mixed in 5% (w/v) ammonium hydroxide and titanyl sulphate solution. The mixture was centrifuged again at 3000× g for 10 min at 4 °C. After this, the obtained supernatant was thoroughly mixed in 2 M sulphuric acid, and the absorbance was recorded at 415 nm. H_2O_2 content was measured using a standard curve.

2.5. Estimation of Antioxidant Enzyme Activity

Fresh leaf samples (0.4 g) were homogenized with 2 mL of ice-cooled 100 mM Na_2PO_4 buffer (pH 7.8) comprising 1.0% polyvinylpyrrolidone (PVPP) and 0.1 mM EDTA. The mixture was later centrifuged at 12,000× g for 20 min at 4 °C. After centrifugation, the supernatant was collected and utilized for various antioxidant enzymes analysis. The activity of SOD was estimated spectrophotometrically at 560 nm by measuring the potential of an enzyme to inhibit the photochemical reduction of nitroblue tetrazolium (NBT) by O^{-2} radicals liberated by light-induced chemical reactions [42]. The POD was determined spectrophotometrically by recording the fluctuation in absorbance value at 470 nm resulting from guaiacol oxidation [43]. The CAT activity was measured by observing the decomposition rate of hydrogen peroxide at 240 nm [43].

2.6. Statistical Analysis

Statistical analysis was carried out using statistix 8.1 (version 8.1. Statistix, Tallahassee, FL, USA). Significant differences among all treatments were measured by using ANOVA (one way) in combination with LSD test. The significance of differences was assessed at the 5% probability level ($p < 0.05$).

3. Results

3.1. Effects of Exogenous Cadmium on Growth Parameters

Different concentrations of Cd significantly affected the plant height and leaf number in both sorghum cultivars (Figure 1). Among the treatments, the highest plant heights (68 or 64 cm) and leaf numbers (9 or 8) were reported in two controls, whereas the respective lowest plant heights (21 or 17 cm) and leaf numbers (5 or 4) were recorded in JS-2002 or Chakwal Sorghum under 100 μM Cd treatment. The results showed that JS-2002 exhibited significantly higher plant height and number of leaves, particularly under the 50 μM treatment, than Chakwal Sorghum, as shown in Figure 1. When compared with the respective control, Cd stress reduced the plant height of the JS-2002 cultivar by 11%, 47%, 53% or 69% in treatments receiving Cd concentrations of 5, 25, 50 and 100 μM, respectively, whereas the respective reductions in the height of the Chakwal Sorghum cultivar were 12%, 53%, 59% and 73%. Moreover, the interactive effect of Cd and plant growth was antagonistic, since plant height and leaf number reduced greatly with a gradual increase in Cd concentration (Figure 1).

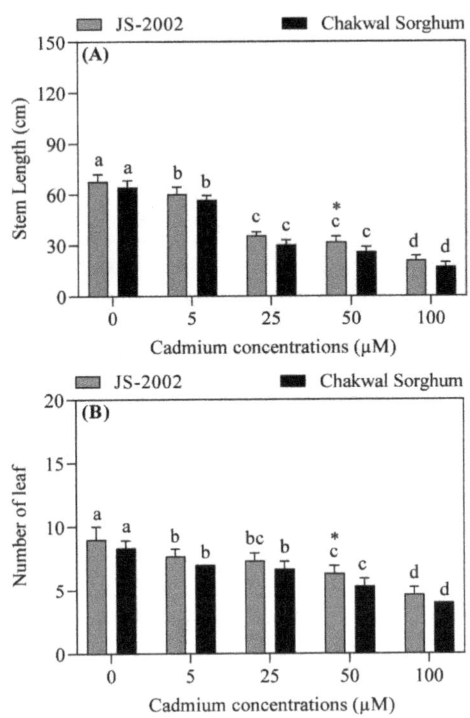

Figure 1. Effect of cadmium toxicity on (**A**) stem length and (**B**) number of leaves in two sorghum cultivars. Values are mean ± standard error ($n = 5$). Different letters in vertical column show significant differences for a cultivar under different cadmium concentrations, whereas "*" shows a significant difference between two sorghum cultivars under a particular cadmium concentration. Comparison of mean was confirmed by LSD at $p < 0.05$.

3.2. Effects of Exogenous Cadmium on Cadmium Accumulation

The highest (42.44 or 37.9 μg Cd g^{-1} DW) Cd accumulation was found in roots, while the lowest value (6.83 or 6.31 μg Cd g^{-1} DW) was observed in leaves of JS-2002 or Chakwal Sorghum (Figure 2). In contrast to the control, different Cd stress levels significantly increased Cd accumulation in various plant organs (leaf, stem and root) of both cultivars, however such an increment was more pronounced when Cd was applied at 50 or 100 μM, respectively. Moreover, JS-2002 showed a significant difference in Cd accumulation compared to Chakwal Sorghum in all plant organs under different levels of Cd toxicity (Figure 2).

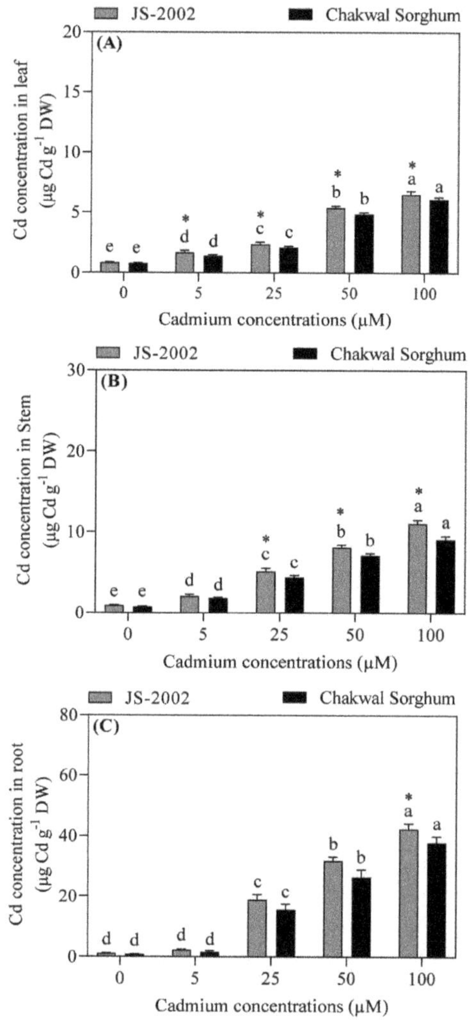

Figure 2. Effect of cadmium toxicity on cadmium accumulation in (**A**) leaf, (**B**) stem, and (**C**) root of two sorghum cultivars. Values are mean ± standard error (n = 5). Different letters in the vertical column show significant differences for a cultivar under different cadmium concentrations, whereas "*" shows a significant difference between two sorghum cultivars under a particular cadmium concentration. Comparison of mean was confirmed by LSD at $p < 0.05$.

3.3. *Effects of Exogenous Cadmium on Cell Membrane Stability and Oxidative Damage*

The Cd application significantly enhanced membrane damage, which intensified with the increase in Cd dosage (Figure 3A). The JS-2002 exhibited considerably lower EL as compared to the Chakwal Sorghum under 50 and 100 μM Cd levels, while no significant difference was noticed in all other treatments, as shown in (Figure 3A). The Cd toxicity induced the massive production of ROS resulting in oxidative stress. The results showed that Cd application significantly elevated the H_2O_2 content in the leaves of two sorghum cultivars. However, the JS-2002 showed a significantly lower H_2O_2 concentration as compared with Chakwal Sorghum (50 and 100 μM Cd levels). Maximum values of 466 μmol g^{-1} FW and 482 μmol g^{-1} FW were observed in the 100 μM treatment, while the corresponding lowest values of 309 and 317 μmol g^{-1} FW were recorded in the control of JS-2002 and Chakwal Sorghum (Figure 3B). Based on the results of MDA contents in both cultivars grown under 5, 25, 50, or 100 μM Cd stress, a 27%, 54%, 47%, or 39% increase in the JS-2002 and a 31%, 58%, 51%, or 43% increase in the Chakwal Sorghum was found in contrast to their respective control treatment (Figure 3C). However, JS-2002 exhibited significantly lower MDA content as compared to Chakwal Sorghum under all Cd treatments.

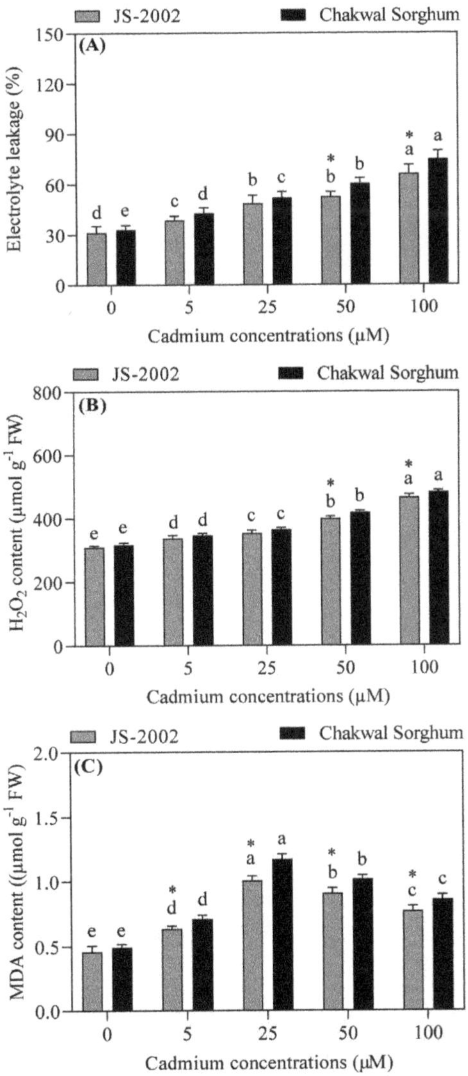

Figure 3. Effect of cadmium toxicity on (**A**) electrolyte leakage (EL), and (**B**) hydrogen peroxide (H_2O_2)

or (**C**) malondialdehyde (MDA) content in two sorghum cultivars. Values are mean ± standard error ($n = 5$). Different letters in vertical column show significant differences for a cultivar under different cadmium concentrations, whereas "*" shows a significant difference between two sorghum cultivars under a particular cadmium concentration. Comparison of mean was confirmed by LSD at $p < 0.05$.

3.4. *Effect of Exogenous Cadmium on Antioxidant Enzymes Activity*

Significant differences were observed in the antioxidant enzyme activities in the leaves of two sorghum cultivars grown under different levels of Cd stress (Figure 4). The JS-2002 exhibited significantly higher SOD activity under all Cd treatments, while POD and CAT showed significant difference under the 5 or 25 μM treatments when compared to Chakwal Sorghum. Except for controls, the maximum values of POD and CAT activities were recorded in the 25 μM Cd treatments in both sorghum cultivars, however for SOD activity, the maximum values were observed in the 5 μM Cd treatments (Figure 4A–C). The SOD activity increased by 12%, 22%, 28% or 9% in JS-2002 compared to that in the Chakwal Sorghum under 5, 25, 50 or 100 μM Cd stress, respectively (Figure 4A). Moreover, the JS-2002 showed a 12%, 11%, 15%, or 19% increase in POD and a 7%, 8%, 12% or 13% increase in CAT in contrast to the Chakwal Sorghum under 5, 25, 50, or 100 μM Cd stress, respectively (Figure 4B,C).

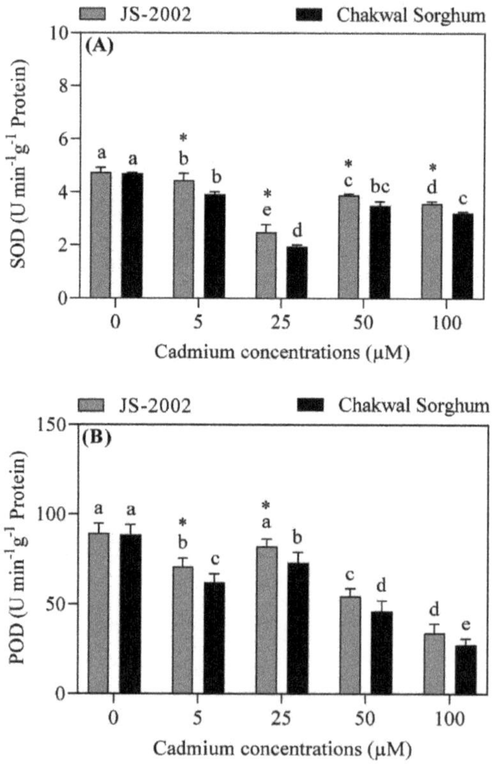

Figure 4. *Cont.*

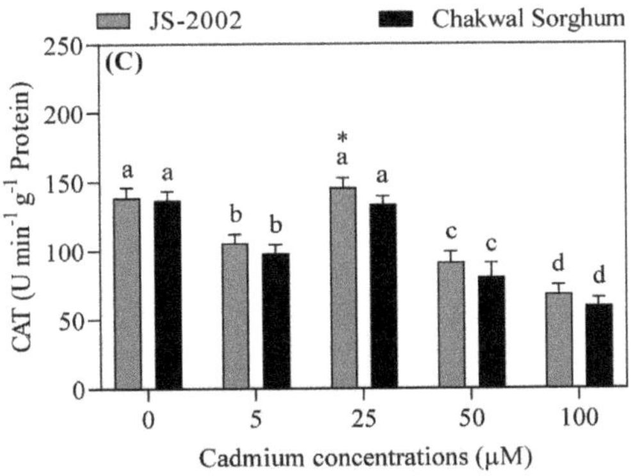

Figure 4. Effect of cadmium toxicity on (**A**) superoxide dismutase (SOD), (**B**) peroxidase (POD), and (**C**) catalase (CAT) activities in leaves of two sorghum cultivars. Values are mean ± standard error ($n = 5$). Different letters in vertical column show significant differences for a cultivar under different cadmium concentrations, whereas "*" shows a significant difference between two sorghum cultivars under a particular cadmium concentration. Comparison of mean was confirmed by LSD at $p < 0.05$.

4. Discussion

Cd is a heavy metal with no apparent beneficial role in plant metabolism [14]. Stunting is the most usual and visible response when plants are exposed to Cd stress, which could be correlated with the Cd ion toxicity in plant metabolism, or a Cd induced interference in the uptake of several bio-elements that are essential for growth and development [19]. It has been reported that phosphorus deficiency was the main reason for stunted growth in plants under Cd stress, because Cd and phosphorus form insoluble complexes [19]. A lack or shortage of photosynthetic machinery (leaf) results in the inhibition of organic metabolites, thus leading toward stunted growth under heavy metal stress [44]. Our results showed that Cd application reduced the growth of two sorghum cultivars, JS-2002 and Chakwal Sorghum, as evident from a significant decrease in plant height and leaf number, with pronounced hazardous effects at higher levels of stress. Our results are in line with the study of Daud et al. [45] who observed a linear decrease in plant height and leaf width in cotton cultivars under heavy metal stress.

Plant species vary significantly in their potential to absorb Cd from the soil and transport Cd towards different organs [46]. The amount of Cd absorbed by the roots and Cd translocated towards the shoot depend on its bonding with the extracellular matrix, root efflux, complexion within cells and transport efficiency [47,48]. In the present study, it was observed that the roots accumulated more Cd than the stems or leaves in JS-2002 and Chakwal Sorghum. This possibly could be connected to the fact that the root is the first organ coming into contact with the Cd ions in soil. In addition, the JS-2002 accumulated a higher concentration of Cd in the root, stem and leaves compared to Chakwal Sorghum under Cd stress, which is in accordance with the findings of Wang et al. [26]. At the root level, the first resistance against Cd stress may be provided by the cell wall and extracellular carbohydrates that play an important role in reducing Cd uptake and transport [49]. The Cd accumulation in the roots often hinders the uptake and translocation of other bio-elements, which aggravates Cd toxicity to plants [50].

Abiotic stresses, such as heavy metals, drought, high temperature and salt stress, initially disrupt cell membrane integrity, resulting in increased membrane damage [51–54]. EL is an imperative index in cell stress physiology and is used to evaluate the leakage of cell components. The results from this study showed that Cd toxicity to both sorghum cultivars enhanced the ROS production, leading to a noticeable increase in membrane damage. Our results are similar to the previous findings in pea and barley [46,55], respectively. Significant increases in membrane damage of Chakwal Sorghum might be connected to the strong imbalance between ROS production and the activity of antioxidative enzymes for ROS scavenging. Cd is considered a non- redox heavy metal that lacks the potential to take part in

Fenton-type reactions. Much clear evidence has shown that Cd can indirectly increase the production of ROS through disturbance of electron transport, which is one of the main events taking place in photosystem II [56]. In our study, the application of different Cd levels significantly elevated the H_2O_2 concentration in both sorghum cultivars compared to control, however the MDA content first increased and then decreased with the gradual increase in Cd concentration (Figure 3B,C). Our results regarding the increase in MDA and H_2O_2 concentration were in accordance with a previous study on wheat [28] and rice in response to Cd stress [57]. In addition, the degree of oxidative damage was lower in JS-2002 in contrast to that in Chakwal Sorghum under all Cd treatments (Figure 3), which could be associated with the more efficient ROS scavenging defense system in JS-2002.

Plants are provided with a natural defense system consisting of enzymatic and non-enzymatic antioxidants to protect them against the oxidative damage induced by various environmental stresses. SOD plays the most important role in the antioxidant defense mechanism, because it is the most effective enzyme in stress resistance, involved in the dismutation of O^{2-} into H_2O_2 and molecular oxygen in plants under stress conditions [58]. In the present study, it was observed that the SOD activity in both cultivars was initially decreased by Cd application. However, the JS-2002 exhibited significantly higher SOD activity than the Chakwal Sorghum under all Cd concentrations. Interestingly, the SOD activity firstly declined and then increased in response to 5, 25, 50 and 100 μM Cd concentrations, which is in line with the findings of Liu et al. [59] in Sorghum. The possible reason could be associated with the increased production of O^{2-} radicals, leading to the activation of existing enzyme stock [58]. In addition, a large amount of H_2O_2 accumulation is extremely harmful for cell metabolism. The CAT and POD enzymes are responsible for the conversion of H_2O_2 to water and oxygen by dissociation of H_2O_2, and thus play necessary roles in providing tolerance to unfavorable conditions in plants [53,60,61]. Our findings showed that the JS-2002 maintained higher POD and CAT activities than the Chakwal Sorghum under Cd toxicity (25 μM), indicating the better antioxidant capacity of JS-2002. Interestingly, the CAT and POD activities increased significantly in the leaves under 25 μM Cd stress in both cultivars, as compared to other Cd stress concentration (5, 50 and 100 μM). Considerable increases in POD and CAT activities could be induced by excessive production of H_2O_2 (a by-product of superoxide dismutation by SOD) under 25 μM Cd stress. However, these responses were limited due to the severe oxidative damage under higher Cd concentrations (50 and 100 μM) in sorghums. Previous studies on other plant species have also shown that changes in antioxidant enzyme activities were associated with the severity of Cd stress [27,57,58].

5. Conclusions

In the present study, different Cd doses were used to investigate the effects on growth, Cd accumulation, oxidative damage, and antioxidative defense system in two sorghum cultivars. The results found out that the majority of Cd was enriched in the roots, and the JS-2002 had a greater Cd enrichment capacity than the Chakwal Sorghum in all plant organs (leaf, stem, and root). In addition, the JS-2002 also showed a better tolerance to Cd stress in comparison with the Chakwal Sorghum, which could be associated with higher antioxidative enzyme activities to cope with the Cd-induced oxidative damage. The current study therefore provides a better understanding of the concentration-dependent role of Cd in sorghum plants; however, intensive work is still required to explain the interaction of Cd with various physiological and genetic functions in sorghum or other plant species.

Author Contributions: Conceptualization, M.A. (Muhammad Ansar), M.J.H., M.A. (Mukhtar Ahmed) and G.A.S.; methodology, I.K. and M.J.H.; formal analysis, M.W., I.K., M.A.R. and S.U.R.; investigation, S.U.R. and M.J.H.; writing—original draft preparation, M.J.H.; writing—review and editing, Z.L., Y.P., M.A.R., and H.G.; funding acquisition, Z.L. All authors have read and agreed to the published version of the manuscript.

Funding: This research was supported by the Chunhui Program of the Ministry of Education (Z2017095).

Acknowledgments: Muhammad Jawad Hassan's thanks, to loving parents (Muhammad Akram and Salima Akram), Muhammad Shahzad Hassan, Muhammad Zeeshan Hassan, Arooj Fatima, Anum Fatima and Muhammad Nauman Hassan for their prayers, love and support, to Muhammad Naveed Aslam and Syed Ahmad Shah for their encouragement and continuous help throughout the research, to Zhou Li and Muhammad Ansar for their love, sincere advice, expert guidance and developing my knowledge, and interest towards scientific research.

Conflicts of Interest: All authors declare no conflict of interest.

References

1. Hawkes, S.J. What is a "heavy metal"? *J. Chem. Educ.* **1997**, *74*, 1374. [CrossRef]
2. De Caroli, M.; Furini, A.; Dalcorso, G.; Rojas, M.; Sansebastiano, G.P.D. Endomembrane Reorganization Induced by Heavy Metals. *Plants* **2020**, *9*, 482. [CrossRef] [PubMed]
3. Chibuike, G.U.; Obiora, S.C. Heavy metal polluted soils: Effect on plants and bioremediation methods. *Appl. Environ. Soil Sci.* **2014**, *2014*, 752708. [CrossRef]
4. Kim, Y.-H.; Khan, A.L.; Kim, D.-H.; Lee, S.-Y.; Kim, K.-M.; Waqas, M.; Jung, H.-Y.; Shin, J.-H.; Kim, J.-G.; Lee, I.-J. Silicon mitigates heavy metal stress by regulating P-type heavy metal ATPases, Oryza sativa low silicon genes, and endogenous phytohormones. *BMC Plant Biol.* **2014**, *14*, 1–13. [CrossRef] [PubMed]
5. Nagajyoti, P.C.; Lee, K.D.; Sreekanth, T. Heavy metals, occurrence and toxicity for plants: A review. *Environ. Chem. Lett.* **2010**, *8*, 199–216. [CrossRef]
6. Gomes-Junior, R.A.; Moldes, C.A.; Delite, F.S.; Pompeu, G.B.; Gratao, P.L.; Mazzafera, P.; Lea, P.J.; Azevedo, R.A. Antioxidant metabolism of coffee cell suspension cultures in response to cadmium. *Chemosphere* **2006**, *65*, 1330–1337. [CrossRef]
7. Liu, J.; Qian, M.; Cai, G.; Yang, J.; Zhu, Q. Uptake and translocation of Cd in different rice cultivars and the relation with Cd accumulation in rice grain. *J. Hazard. Mater.* **2007**, *143*, 443–447. [CrossRef]
8. Vergine, M.; Aprile, A.; Sabella, E.; Genga, A.; Bellis, L.D. Cadmium Concentration in Grains of Durum Wheat (*Triticum turgidum* L. subsp. durum). *J. Agric. Food Chem.* **2017**, *65*, 6240–6246. [CrossRef]
9. Grant, C.; Buckley, W.; Bailey, L.; Selles, F. Cadmium accumulation in crops. *Can. J. Plant Sci.* **1998**, *78*, 1–17. [CrossRef]
10. Hart, J.J.; Welch, R.M.; Norvell, W.A.; Sullivan, L.A.; Kochian, L.V. Characterization of cadmium binding, uptake, and translocation in intact seedlings of bread and durum wheat cultivars. *Plant Physiol.* **1998**, *116*, 1413–1420. [CrossRef]
11. Hall, J.Á.; Williams, L.E. Transition metal transporters in plants. *J. Exp. Bot.* **2003**, *54*, 2601–2613. [CrossRef] [PubMed]
12. Milner, M.J.; Seamon, J.; Craft, E.; Kochian, L.V. Transport properties of members of the ZIP family in plants and their role in Zn and Mn homeostasis. *J. Exp. Bot.* **2013**, *64*, 369–381. [CrossRef] [PubMed]
13. Ashraf, M.Y.; Roohi, M.; Iqbal, Z.; Ashraf, M.; Öztürk, M.; Gücel, S. Cadmium (Cd) and lead (Pb) induced changes in growth, some biochemical attributes, and mineral accumulation in two cultivars of mung bean [*Vigna radiata* (L.) Wilczek]. *Commun. Soil Sci. Plant Anal.* **2016**, *47*, 405–413.
14. Doganlar, Z.B.P.; Yurekli, F. Interactions between cadmium and phytochelatin accumulation in two different sunflower cultivars. *Fresenius Environ. Bull.* **2009**, *18*, 304–310.
15. Michel-López, C.Y.; Espadas y Gil, F.; Fuentes Ortíz, G.; Santamaría, J.M.; González-Mendoza, D.; Ceceña-Duran, C.; Grimaldo Juarez, O. Bioaccumulation and effect of cadmium in the photosynthetic apparatus of Prosopis juliflora. *Chem. Speciat. Bioavailab.* **2016**, *28*, 1–6. [CrossRef]
16. Okem, A.; Moyo, M.; Stirk, W.; Finnie, J.; Van Staden, J. Investigating the effect of cadmium and aluminium on growth and stress-induced responses in the micropropagated medicinal plant Hypoxis hemerocallidea. *Plant Biol.* **2016**, *18*, 805–815. [CrossRef]
17. Schat, H.; Sharma, S.S.; Vooijs, R. Heavy metal-induced accumulation of free proline in a metal-tolerant and a nontolerant ecotype of Silene vulgaris. *Physiol. Plant.* **1997**, *101*, 477–482. [CrossRef]
18. Vijendra, P.D.; Huchappa, K.M.; Lingappa, R.; Basappa, G.; Jayanna, S.G.; Kumar, V. Physiological and Biochemical Changes in Moth Bean (*Vigna aconitifolia* L.) under Cadmium Stress. *J. Bot.* **2016**, *2016*. [CrossRef]

19. Gonçalves, J.F.; Antes, F.G.; Maldaner, J.; Pereira, L.B.; Tabaldi, L.A.; Rauber, R.; Rossato, L.V.; Bisognin, D.A.; Dressler, V.L.; de Moraes Flores, E.M. Cadmium and mineral nutrient accumulation in potato plantlets grown under cadmium stress in two different experimental culture conditions. *Plant Physiol. Biochem.* **2009**, *47*, 814–821. [CrossRef]
20. Jiang, W.; Liu, D.; Hou, W. Hyperaccumulation of cadmium by roots, bulbs and shoots of garlic (*Allium sativum* L.). *Bioresour. Technol.* **2001**, *76*, 9–13. [CrossRef]
21. Liu, J.; Cai, G.; Qian, M.; Wang, D.; Xu, J.; Yang, J.; Zhu, Q. Effect of Cd on the growth, dry matter accumulation and grain yield of different rice cultivars. *J. Sci. Food Agric.* **2007**, *87*, 1088–1095. [CrossRef]
22. Ozdener, Y.; Kutbay, H.G. Toxicity of copper, cadmium, nickel, lead and zinc on seed germination and seedling growth in Eruca sativa. *Fresenius Environ. Bull.* **2009**, *18*, 26–31.
23. Gounden, D.; Kisten, K.; Moodley, R.; Shaik, S.; Jonnalagadda, S.B. Impact of spiked concentrations of Cd, Pb, As and Zn in growth medium on elemental uptake of Nasturtium officinale (Watercress). *J. Environ. Sci. Health Part B* **2016**, *51*, 1–7. [CrossRef] [PubMed]
24. Kutrowska, A.; Małecka, A.; Piechalak, A.; Masiakowski, W.; Hanć, A.; Barałkiewicz, D.; Andrzejewska, B.; Zbierska, J.; Tomaszewska, B. Effects of binary metal combinations on zinc, copper, cadmium and lead uptake and distribution in Brassica juncea. *J. Trace Elem. Med. Biol.* **2017**, *44*, 32–39. [CrossRef] [PubMed]
25. Rahman, M.; Haq, N.; Williams, I. Phytoaccumulation of arsenic, cadmium and lead by Brassica juncea parents and their F1 hybrids. *J. Environ. Prot.* **2016**, *7*, 613–622. [CrossRef]
26. Wang, M.; Zou, J.; Duan, X.; Jiang, W.; Liu, D. Cadmium accumulation and its effects on metal uptake in maize (*Zea mays* L.). *Bioresour. Technol.* **2007**, *98*, 82–88. [CrossRef] [PubMed]
27. Anjum, S.A.; Tanveer, M.; Hussain, S.; Bao, M.; Wang, L.; Khan, I.; Ullah, E.; Tung, S.A.; Samad, R.A.; Shahzad, B. Cadmium toxicity in Maize (*Zea mays* L.): Consequences on antioxidative systems, reactive oxygen species and cadmium accumulation. *Environ. Sci. Pollut. Res.* **2015**, *22*, 17022–17030. [CrossRef]
28. Khan, N.; Samiullah; Singh, S.; Nazar, R. Activities of antioxidative enzymes, sulphur assimilation, photosynthetic activity and growth of wheat (*Triticum aestivum*) cultivars differing in yield potential under cadmium stress. *J. Agron. Crop Sci.* **2007**, *193*, 435–444. [CrossRef]
29. Cho, U.-H.; Seo, N.-H. Oxidative stress in Arabidopsis thaliana exposed to cadmium is due to hydrogen peroxide accumulation. *Plant Sci.* **2005**, *168*, 113–120. [CrossRef]
30. Khaliq, A.; Aslam, F.; Matloob, A.; Hussain, S.; Geng, M.; Wahid, A.; Ur Rehman, H. Seed priming with selenium: Consequences for emergence, seedling growth, and biochemical attributes of rice. *Biol. Trace Elem. Res.* **2015**, *166*, 236–244. [CrossRef]
31. Noctor, G.; Foyer, C.H. Ascorbate and glutathione: Keeping active oxygen under control. *Annu. Rev. Plant Biol.* **1998**, *49*, 249–279. [CrossRef] [PubMed]
32. Mobin, M.; Khan, N.A. Photosynthetic activity, pigment composition and antioxidative response of two mustard (*Brassica juncea*) cultivars differing in photosynthetic capacity subjected to cadmium stress. *J. Plant Physiol.* **2007**, *164*, 601–610. [CrossRef] [PubMed]
33. Khaliq, A.; Zia-ul-Haq, M.; Ali, F.; Aslam, F.; Matloob, A.; Navab, A.; Hussain, S. Salinity tolerance in wheat cultivars is related to enhanced activities of enzymatic antioxidants and reduced lipid peroxidation. *Clean Soil Air Water* **2015**, *43*, 1248–1258. [CrossRef]
34. Ekmekçi, Y.; Tanyolac, D.; Ayhan, B. Effects of cadmium on antioxidant enzyme and photosynthetic activities in leaves of two maize cultivars. *J. Plant Physiol.* **2008**, *165*, 600–611. [CrossRef] [PubMed]
35. De Morais Cardoso, L.; Pinheiro, S.S.; Martino, H.S.D.; Pinheiro-Sant'Ana, H.M. Sorghum (*Sorghum bicolor* L.): Nutrients, bioactive compounds, and potential impact on human health. *Crit. Rev. Food Sci. Nutr.* **2017**, *57*, 372–390. [CrossRef] [PubMed]
36. De Morais Cardoso, L.; Montini, T.A.; Pinheiro, S.S.; Pinheiro-Sant'Ana, H.M.; Martino, H.S.D.; Moreira, A.V.B. Effects of processing with dry heat and wet heat on the antioxidant profile of sorghum. *Food Chem.* **2014**, *152*, 210–217. [CrossRef]
37. Epelde, L.; Mijangos, I.; Becerril, J.M.; Garbisu, C. Soil microbial community as bioindicator of the recovery of soil functioning derived from metal phytoextraction with sorghum. *Soil Biol. Biochem.* **2009**, *41*, 1788–1794. [CrossRef]
38. Zhuang, P.; Wensheng, S.; Zhian, L.; Bin, L.; Jintian, L.; Jingsong, S. Removal of metals by sorghum plants from contaminated land. *J. Environ. Sci.* **2009**, *21*, 1432–1437. [CrossRef]

39. Lutts, S.; Kinet, J.; Bouharmont, J. NaCl-induced senescence in leaves of rice (Oryza sativaL.) cultivars differing in salinity resistance. *Ann. Bot.* **1996**, *78*, 389–398. [CrossRef]
40. Cakmak, I.; Marschner, H. Magnesium deficiency and high light intensity enhance activities of superoxide dismutase, ascorbate peroxidase, and glutathione reductase in bean leaves. *Plant Physiol.* **1992**, *98*, 1222–1227. [CrossRef]
41. Mukherjee, S.; Choudhuri, M. Implications of water stress-induced changes in the levels of endogenous ascorbic acid and hydrogen peroxide in Vigna seedlings. *Physiol. Plant.* **1983**, *58*, 166–170. [CrossRef]
42. Dhindsa, R.S.; Plumb-Dhindsa, P.; Thorpe, T.A. Leaf senescence: Correlated with increased levels of membrane permeability and lipid peroxidation, and decreased levels of superoxide dismutase and catalase. *J. Exp. Bot.* **1981**, *32*, 93–101. [CrossRef]
43. Cakmak, I.; Strbac, D.; Marschner, H. Activities of hydrogen peroxide-scavenging enzymes in germinating wheat seeds. *J. Exp. Bot.* **1993**, *44*, 127–132. [CrossRef]
44. Rodríguez-Serrano, M.; Romero-Puertas, M.C.; Pazmino, D.M.; Testillano, P.S.; Risueño, M.C.; Luis, A.; Sandalio, L.M. Cellular response of pea plants to cadmium toxicity: Cross talk between reactive oxygen species, nitric oxide, and calcium. *Plant Physiol.* **2009**, *150*, 229–243. [CrossRef] [PubMed]
45. Daud, M.; Variath, M.; Ali, S.; Najeeb, U.; Jamil, M.; Hayat, Y.; Dawood, M.; Khan, M.I.; Zaffar, M.; Cheema, S.A. Cadmium-induced ultramorphological and physiological changes in leaves of two transgenic cotton cultivars and their wild relative. *J. Hazard. Mater.* **2009**, *168*, 614–625. [CrossRef] [PubMed]
46. Metwally, A.; Safronova, V.I.; Belimov, A.A.; Dietz, K.-J. Genotypic variation of the response to cadmium toxicity in *Pisum sativum* L. *J. Exp. Bot.* **2005**, *56*, 167–178. [CrossRef]
47. Marchiol, L.; Leita, L.; Martin, M.; Peressotti, A.; Zerbi, G. Physiological responses of two soybean cultivars to cadmium. *J. Environ. Qual.* **1996**, *25*, 562–566. [CrossRef]
48. Akhter, M.; Macfie, S. Species-specific relationship between transpiration and cadmium translocation in lettuce, barley and radish. *J. Plant Stud.* **2012**, *1*, 2.
49. Wagner, G.J. Accumulation of cadmium in crop plants and its consequences to human health. *Adv. Agron.* **1993**, *51*, 173–212.
50. Gomes, M.P.; Soares, A.M. Cadmium effects on mineral nutrition of the Cd-hyperaccumulator Pfaffia glomerata. *Biologia* **2013**, *68*, 223–230. [CrossRef]
51. Jia, G.-X.; Zhu, Z.-Q.; Chang, F.-Q.; Li, Y.-X. Transformation of tomato with the BADH gene from Atriplex improves salt tolerance. *Plant Cell Rep.* **2002**, *21*, 141–146.
52. Jawad Hassan, M.; Ali Raza, M.; Khan, I.; Ahmad Meraj, T.; Ahmed, M.; Abbas Shah, G.; Ansar, M.; Awan, S.A.; Khan, N.; Iqbal, N. Selenium and Salt Interactions in Black Gram (*Vigna mungo* L): Ion Uptake, Antioxidant Defense System, and Photochemistry Efficiency. *Plants* **2020**, *9*, 467. [CrossRef] [PubMed]
53. Li, Z.; Peng, Y.; Zhang, X.-Q.; Ma, X.; Huang, L.-K.; Yan, Y.-H. Exogenous spermidine improves seed germination of white clover under water stress via involvement in starch metabolism, antioxidant defenses and relevant gene expression. *Molecules* **2014**, *19*, 18003–18024. [CrossRef] [PubMed]
54. Djanaguiraman, M.; Prasad, P.V.; Seppanen, M. Selenium protects sorghum leaves from oxidative damage under high temperature stress by enhancing antioxidant defense system. *Plant Physiol. Biochem.* **2010**, *48*, 999–1007. [CrossRef] [PubMed]
55. Wu, F.; Zhang, G.; Dominy, P.; Wu, H.; Bachir, D.M. Differences in yield components and kernel Cd accumulation in response to Cd toxicity in four barley genotypes. *Chemosphere* **2007**, *70*, 83–92. [CrossRef]
56. Sandalio, L.; Dalurzo, H.; Gomez, M.; Romero-Puertas, M.; Del Rio, L. Cadmium-induced changes in the growth and oxidative metabolism of pea plants. *J. Exp. Bot.* **2001**, *52*, 2115–2126. [CrossRef]
57. Chien, H.-F.; Wang, J.-W.; Lin, C.C.; Kao, C.H. Cadmium toxicity of rice leaves is mediated through lipid peroxidation. *Plant Growth Regul.* **2001**, *33*, 205–213. [CrossRef]
58. Zhang, S.; Zhang, H.; Qin, R.; Jiang, W.; Liu, D. Cadmium induction of lipid peroxidation and effects on root tip cells and antioxidant enzyme activities in *Vicia faba* L. *Ecotoxicology* **2009**, *18*, 814–823. [CrossRef]
59. Liu, D.-L.; Zhang, S.-P.; Zheng, C.; Qiu, W.-W. Soil cadmium regulates antioxidases in sorghum. *Agric. Sci. China* **2010**, *9*, 1475–1480. [CrossRef]
60. Liu, J.-N.; Zhou, Q.-X.; Sun, T.; Ma, L.Q.; Wang, S. Growth responses of three ornamental plants to Cd and Cd–Pb stress and their metal accumulation characteristics. *J. Hazard. Mater.* **2008**, *151*, 261–267. [CrossRef]

61. Zornoza, P.; Sánchez-Pardo, B.; Carpena, R.O. Interaction and accumulation of manganese and cadmium in the manganese accumulator *Lupinus albus*. *J. Plant Physiol.* **2010**, *167*, 1027–1032. [CrossRef] [PubMed]

Publisher's Note: MDPI stays neutral with regard to jurisdictional claims in published maps and institutional affiliations.

plants MDPI

Article

Allelopathic Potential of Aqueous Extract from *Acacia melanoxylon* R. Br. on *Lactuca sativa*

M. Iftikhar Hussain [1,2,*], Mohamed A. El-Sheikh [3] and Manuel J. Reigosa [1]

1 Department of Plant Biology and Soil Science, Faculty of Biology, University of Vigo, Campus Lagoas-Marcosende, E-36310 Vigo, Spain; mreigosa@uvigo.es
2 CITACA, Agri-Food Research and Transfer Cluster, Campus da Auga, University of Vigo, 32004 Ourense, Spain
3 Botany & Microbiology Department, College of Science, King Saud University, P.O. Box 2455, Riyadh 11451, Saudi Arabia; melsheikh@ksu.edu.sa
* Correspondence: mih786@gmail.com

Received: 13 August 2020; Accepted: 14 September 2020; Published: 18 September 2020

Abstract: We studied the polyphenol (phenolic compounds and flavonoids) composition and allelopathic effects of *Acacia melanoxylon* R. Br. aerial foliage aqueous extract (0%, 25%, 50%, 75% and 100%) on the seedling growth and plant biomass of the general biotest species, lettuce (*Lactuca sativa*). Mean leaf fresh weight, leaf dry weight, root fresh weight and root dry weight were decreased following exposure to *Acacia* aerial foliage, flowers aqueous extract (AFE) and phyllodes aqueous extract (APE) after 6 days. The reduction in plant dry biomass was more than 50% following treatment with AFE. The decrease in mean root length was approximately 37.7% and 29.20% following treatment with *Acacia* flowers extract (AFE) at 75% and 100% concentration, respectively. Root dry weight of *L. sativa* was reduced by both flowers and phyllodes extract. The reduction of protein contents in lettuce leaves following *Acacia* foliage extract proved that both AFE and APE exhibit polyphenols that causes the toxicity which led to decrease in leaf protein contents. High-Performance Liquid Chromatography (HPLC) was employed to analyze the *A. melanoxylon* flowers and phyllodes. A total of 13 compounds (accounting for most abundant compounds in flowers and phyllodes) include different flavonoids and phenolic compounds. The phytochemical compounds detected were: Gallic acid, protocatechuic acid, *p*-hydroxybenzoic acid, *p*-hydroxybenzaldehyde, vanillic acid, syringic acid, *p*-coumaric acid, and ferulic acid. The major flavonoid compounds identified include rutin, luteolin, apigenin, and catechin. Allelopathic effects of flower and phyllodes extracts from *A. melanoxylon* may be due to the presence of above compounds identified by HPLC analysis.

Keywords: allelopathic potential; chemical composition; phenolics; *Acacia melanoxylon*; *Lactuca sativa*; HPLC seedling growth Flavonoides

1. Introduction

Allelopathy can be referred as "any process that involves secondary metabolites produced by plants, algae, bacteria and fungi that influence the growth and development of biological systems" [1]. For the last decades, the study of the allelopathic phenomenon has reached a specialized level of knowledge, but it is still necessary to introduce it into the modern ecophysiological concept to give answers to several unanswered questions [2,3]. The intensity of the allelopathic effect in the field will depend, between others, on the different transformations that the organic compounds will suffer after the release to the environment. Exotic plants are causing a serious threat to the native plants and ecosystem through interfering in growth and ecophysiological attributes of neighbours [4–7].

Nonindigenous plants present a serious risk to their neighboring plant [8]. Invasive plants possess several phytotoxic compounds, when released into the environment impede the germination and seedling growth of surrounding plant species at both ecosystem and species level [9,10]. Secondary metabolites are active allelopathic compounds released in the natural plant-soil-environment ecosystem from allelopathic crops, weeds, halophytes, shrubs and trees and their natural leachates might interfere with growth and physiological attributes of neighbouring plant species [3,10]. These toxic metabolites can be stored in the vacuole, polymerised or directly liberated, but anyway, they will be finally released to the environment where they can act as allelopathic agents on the metabolism of neighbouring plants, and giving usually an advantage to the producer [11].

Acacia melanoxylon R. Br. commonly known as Blackwood belonging to the family Fabaceae, subfamily Mimosoidae is a perennial tree, is native to Australia and Tasmania. It has now spread to various parts of the world including Asia, Africa and Europe [12]. The tree has wider ecological amplitude and has potential to grow in a wide range of soil types, and currently invading coastal habitats of North Western Iberian Peninsula, (Spain and Portugal). It has been listed as one of the most dangerous invasive species and currently invading the agriculture open fields, along the water channels, and rivers [13]. The allelopathic effects of *A. melanoxylon* on neighboring plants has been implied [14,15], but not thoroughly investigated. The phyllodes of *A. melanoxylon* have also been previously shown to contain allelochemicals [14] and affecting the natural vegetation [16]. The search for plant protection measures as an alternative to chemical weed control, which are widely used in agronomic and field and horticulture crops is of paramount importance [17,18].

Alternative control methods can take advantage of ecological processes such as using natural water extract of allelopathic plants, natural leachates and mixing the plant residues in the soil or applying allelopathic substances through foliar spray [3,19]. Several authors evaluated the role of allelopathy for weed management through allelopathic plant tissue aqueous extract. Brassica, mulberry and sorghum aqueous extracts have demonstrated their allelopathic potential in the laboratory and field experiments mainly through plant density reduction and biomass inhibition of *Cyperus rotundus* L. and *Trianthema portulacastrum* L. [20,21]. Allelopathic water extracts are water-soluble allelochemicals extracted from plants. Allelopathic water extract is used as natural herbicide because most of allelochemical compounds are water-soluble and are easy to apply without additional wetting agent, and they are more environmentally friendly than synthetic herbicide. According to Bajwa et al. [18], water extracts of mature fine grains greatly control the population and biomass of a weed competitor, and fine grain water extract reduces weed growth and density and increases wheat yield. However, allelochemicals from different plants and the concentration of the extract has different effects on the target plants, respectively [18].

Although, the role of *A. melanoxylon* allelopathy in controlling weeds and its phytotoxic impact is well known; nonetheless, influence of aerial foliage aqueous extract on the horticulture crops like lettuce has not been explored extensively. Meanwhile, as one of the most common cultivated vegetables in the open agriculture fields that are colonized by *Acacia melanoxylon* R. Br., lettuce' seedlings are very sensitive to the changes in the environmental factorss and often used as a general biotest species in allelopathic bioassays [19]. The findings from this study will elaborate a solid foundation to better understand the allelopathic mechanism that drives the successful colonization of *A. melanoxylon* in the North Western Iberian Peninsula. These findings will also help policy makers to prepare guidelines and measures for prevention and possible control of invasive alien species. This study elucidated the three hypothesis: (I) *A. melanoxylon* foliage water extract recruit adverse influences on root/shoot length, fresh and dry weights of leaves and roots of lettuce and the adverse influences notably upsurge with increasing concentration of *Acacia* foliage extract; (II) flower and phyllodes aqueous extract intensify the allelopathy of *A. melanoxylon* on seedling growth of lettuce; (III) the allelopathic potential of *A. melanoxylon* foliage might be due to the presence of different polyphenols (phenolic compounds and flavonoids) and hence, these polyphenols can be used as lead compounds in herbicide discovery program.

2. Results

2.1. *Effect of A. melanoxylon Foliage on Lettuce Growth and Biomass Accumulation*

The phytotoxicity of *A. melanoxylon* (flowers and phyllodes) was evaluated by measuring the leaf and root length and biomass accumulation of *L. sativa* seedlings grown in the presence or absence of the *Acacia* aqueous extract. Tables 1 and 2 show the mean leaf fresh weight (LFW), leaf dry weight (LDW), root fresh weight (RFW) and root dry weight (RDW) following exposure to *Acacia* aqueous extract (AAE) after 6 days. It can be seen that *Acacia* flower extract (AFE) (100%) significantly decreased the LFW (Table 1) while lettuce LFW was also reduced following treatment with *A. melanoxylon* phyllodes extract (Table 1). *Acacia* flower extract (AFE) exhibits a slight but significant phytotoxic effect on *L. sativa*, as revealed by a decrease in mean root fresh and root dry weight of approximately 51% and 42.5%, respectively, compared to the control (Table 2). Treatments, including AFE, are much more effective in reducing the leaf/root biomass, growth in *L. sativa*, with an inhibition percentage reaching more than 60% after treatment with *Acacia* aqueous extract aqueous extract (50%). There were non-significant differences in leaf dry/fresh weight ratio, root dry/fresh weight ratios, of the lettuce grown with different levels of *Acacia* flowers/phyllodes aqueous extract.

Table 1. Effect of aqueous extract of aerial foliage of *Acacia melanoxylon* R. Br. on leaf fresh weight (g) and leaf dry weight (g) of one month old general biotest species, *Lactuca sativa* L. LDFW ratio (Leaf dry weight fresh weight ratio).

Growth Characteristics	Treatments	100%	75%	50%	25%
Leaf fresh weight (g)	Control	10.58 ± 0.60	10.58 ± 0.60	10.58 ± 0.60	10.58 ± 0.60
	Acacia phyllodes	7.73 ± 0.3 *	7.93 ± 0.5 *	8.63 ± 0.2 *	8.29 ± 1.0 *
	Acacia flowers	5.10 ± 0.4 *	5.62 ± 0.3 *	7.25 ± 0.34 *	6.85 ± 1.0 *
Leaf dry weight (g)	Control	1.63 ± 0.15	1.63 ± 0.15	1.63 ± 0.15	1.63 ± 0.15
	Acacia phyllodes	1.78 ± 0.04 *	1.77 ± 0.1 *	1.90 ± 0.4 *	1.98 ± 0.3 *
	Acacia flowers	1.23 ± 0.2 *	1.21 ± 0.4 *	1.45 ± 0.8 *	1.64 ± 0.2
LDFW ratio	Control	0.154 ± 0.07	0.154 ± 0.07	0.154 ± 0.07	0.154 ± 0.07
	Acacia phyllodes	0.230 ± 0.0 *	0.223 ± 0.00 *	0.221 ± 0.0 *	0.238 ± 0.05 *
	Acacia flowers	0.272 ± 0.1 *	0.215 ± 0.0 *	0.200 ± 0.0 *	0.240 ± 0.01 *

Each value represents the mean (± S.E.) of three replicates. * Asterik indicates significant differences as compared to control for $p < 0.05$ according to LSD test.

Table 2. Effect of aqueous extract of aerial foliage of *Acacia melanoxylon* R. Br. on on root fresh weight (g) and root dry weight (g) of one month old general biotest species, *Lactuca sativa* L. RDFW ratio (Root dry weight fresh weight ratio).

Growth Characteristics	Treatments	100%	75%	50%	25%
Root fresh weight (g)	Control	6.31 ± 1.1	6.31 ± 1.0	6.31 ± 1.1	6.31 ± 1.2
	Acacia phyllodes	4.57 ± 0.3 *	3.9 ± 0.6 *	4.11 ± 0.13 *	4.5 ± 0.4 *
	Acacia flowers	3.82 ± 0.1 *	3.44 ± 1.3 *	3.79 ± 1.3 *	3.07 ± 0.6 *
Rood dry weight (g)	Control	0.40 ± 0.02	0.40 ± 0.03	0.40 ± 0.04	0.40 ± 0.05
	Acacia phyllodes	0.27 ± 0.04 *	0.32 ± 0.01 *	0.29 ± 0.09 *	0.37 ± 0.04 *
	Acacia flowers	0.28 ± 0.2 *	0.28 ± 0.04 *	0.23 ± 0.001 *	0.27 ± 0.09 *
RDFW ratio	Control	0.064 ± 0.02	0.064 ± 0.02	0.064 ± 0.02	0.064 ± 0.02
	Acacia phyllodes	0.059 ± 0.01	0.082 ± 0.01 *	0.071 ± 0.01 *	0.082 ± 0.01 *
	Acacia flowers	0.073 ± 0.14 *	0.081 ± 0.01 *	0.061 ± 0.01 *	0.088 ± 0.01 *

Each value represents the mean (± S.E.) of three replicates. * Asterik indicates significant differences as compared to control for $p < 0.05$ according to LSD test.

The shoot length of *L. sativa* significantly inhibited by *Acacia* phyllodes extract (APE) at each concentration tested, while 100% proved to be very lethal because it caused 50.78% phytotoxic effect in

L. sativa shoot length (Table 3). The decrease in mean root length of was approximately 37.7% and 29.20% following treatment with *Acacia* flowers extract (AFE) (75% and 100%), respectively, compared to the control (Table 3). Treatments, including *Acacia* flowers extract (AFE), are much more effective in reducing the root growth in *L. sativa*, with an inhibition percentage reaching more than 50%.

Table 3. Effect of aqueous extract of aerial foliage of *Acacia melanoxylon* R. Br. on shoot length (cm) and root length (cm) of one month old general biotest species, *Lactuca sativa* L. RS ratio (Root shoot ratio).

Growth Characteristics	Treatments	100%	75%	50%	25%
Shoot length (cm)	Control	15.30 ± 0.23	15.30 ± 0.23	15.30 ± 0.23	15.30 ± 0.23
	Acacia phyllodes	7.53 ± 0.5 *	8.96 ± 1.5 *	7.66 ± 0.5 *	9.16 ± 1.6 *
	Acacia flowers	8.96 ± 2.2 *	8.0 ± 0.5 *	8.16 ± 2.0 *	9.66 ± 0.5 *
Root length (cm)	Control	27.77 ± 0.7	27.77 ± 0.7	27.77 ± 0.7	27.77 ± 0.7
	Acacia phyllodes	21.33 ± 2.3 *	19.0 ± 1.5 *	22.66 ± 2.5 *	21.5 ± 2.7 *
	Acacia flowers	19.66 ± 2.0 *	17.3 ± 2.0 *	22.0 ± 1.7 *	21.3 ± 2.0 *
RS ratio	Control	1.82 ± 0.03	1.82 ± 0.03	1.82 ± 0.03	1.82 ± 0.03
	Acacia phyllodes	2.83 ± 0.02 *	2.12 ± 0.02 *	2.95 ± 0.02 *	2.34 ± 0.02 *
	Acacia flowers	2.19 ± 0.01 *	2.16 ± 0.02 *	2.69 ± 0.01 *	2.20 ± 0.02 *

Each value represents the mean (± S.E.) of three replicates. * Asterisk indicate significant differences as compared to control for $p < 0.05$ according to LSD test.

The concentration-induced differences in root fresh weight that was significantly affected following treated with *Acacia* flowers extract (AFE) at each concentration tested are given in Table 4. Root dry weight of *L. sativa* was also reduced by both flowers and phyllodes extract (Table 4). However, AFE decreased the root dry weight upto 40% and 39% after treatment with 75% and 50%, respectively.

Table 4. Effect of aqueous extract of aerial foliage of *Acacia melanoxylon* R. Br. on root fresh weight (g) and root dry weight (g) of one month old general biotest species, *Lactuca sativa* L.

Growth Characteristics	Treatments	100%	75%	50%	25%
Root fresh weight (g)	Control	6.31 ± 0.34	6.31 ± 0.34	6.31 ± 0.34	6.31 ± 0.34
	Acacia phyllodes	4.57 ± 0.7 *	3.90 ± 0.07 *	4.12 ± 0.3 *	4.49 ± 0.4 *
	Acacia flowers	3.82 ± 0.1 *	3.44 ± 0.29 *	3.79 ± 0.6 *	3.07 ± 0.6 *
Root dry weight (g)	Control	0.45 ± 0.042	0.45 ± 0.042	0.45 ± 0.042	0.45 ± 0.042
	Acacia phyllodes	0.27 ± 0.02 *	0.323 ± 0.04 *	0.290 ± 0.09 *	0.373 ± 0.04 *
	Acacia flowers	0.28 ± 0.06 *	0.286 ± 0.04 *	0.273 ± 0.01 *	0.270 ± 0.09 *
RDWFW ratio	Control	0.071 ± 0.02	0.071 ± 0.02	0.071 ± 0.02	0.071 ± 0.02
	Acacia phyllodes	0.059 ± 0.01 *	0.083 ± 0.01 *	0.070 ± 0.01ns	0.083 ± 0.01 *
	Acacia flowers	0.073 ± 0.04ns	0.083 ± 0.04 *	0.072 ± 0.01ns	0.088 ± 0.01 *

Each value represents the mean (± S.E.) of three replicates. * Asterisk indicate significant differences as compared to control for $p < 0.05$ according to LSD test.

2.2. *Effect of A. melanoxylon Foliage on Lettuce Leaf Protein Contents*

Both AFE and APE decreased the protein contents of leaves, (0.78 mg g^{-1}) and (0.83 mg g^{-1}), respectively, following exposure at 100% as compared to control (1.03 mg g^{-1}) (Figure 1). This reduction of leaf protein contents in lettuce leaves following *Acacia* aerial part extract proved that both AFE and APE exhibit polyphenols that causes the toxicity which led to decrease in leaf protein contents.

2.3. *Chemical Composition of the Acacia melanoxylon Aerial Foliage*

Table 5 demonstrtae the chemical composition of *Acacia melanoxylon* aerial foliage (flowers and phyllodes). Several phytochemical constituents were identified from foliage extracts but we concentrated mainly on phenolics and flavonoids. Among the phenolic compounds, eight (8) were most important and include Gallic acid, protocatechuic acid, *p*-hydroxybenzoic acid, *p*-hydroxybenzaldehyde, vanillic acid, syringic acid, *p*-coumaric acid, ferulic acid constituents.

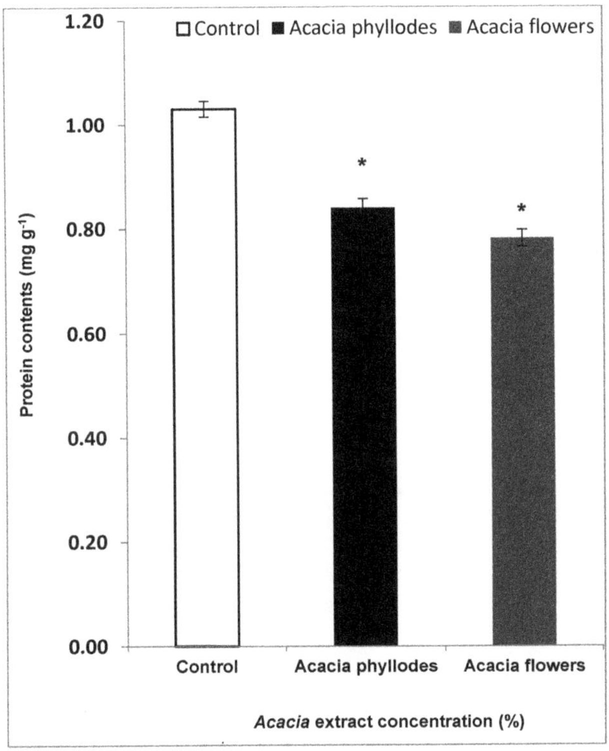

Figure 1. Inhibitory effects of *A. melanoxylon* flowers and phyllodes extract (0% and 100%) on the leaf protein contents (mg/g) of *Lactuca sativa*. Each bar represents the mean (± S.E.) of three replicates. * Asterisks indicate significant differences at level 0.05 with respect to the control.

Table 5. Phenolics and flavonoids found in flower and phyllode extracts of *Acacia melanoxylon* R. Br. RT: retention time (minutes) of compounds in column.

Sr. No.	Common Name	Scientific Name	Flowers mg/L	Phyllodes mg/L	RT
	PHENOLICS				
1	Gallic acid	3,4,5-trihydroxy benzoic acid	4.4	3.41	6.5
2	Protocatechuic acid	3,4-dihydroxybenzoic acid	5.06		12.9
3	*p*-Hydroxybenzoic acid	4-hydroxybenzoic acid	12.33	1.46	21.9
4	*p*-Hydroxybenzaldehyde		0.91	0.16	28.6
5	Vanillic acid	4-hydroxy-3-methoxybenzoic acid	9.7	1.64	32.2
6	Syringic acid	4-hydroxy-3,5-dimethoxybenzoic acid		2.64	40.1
7	*p*-Coumaric acid	4-hydroxycinnamic acid	3.19	3.6	49.7
8	Ferulic acid	4-hydroxy-3-methoxycinnamic acid	3.87		57.9
	FLAVONOIDS				
9	Rutin		5342.39	5032.87	20.4
10	Quercetin	3,3′,4′,5,7-Pentahydroxyflavone	326.4		25.2
11	Luteolin	3′,4′,5,7-Tetrahydroxyflavone	388.89	706.89	26.2
12	Apigenin	4′,5,7-Trihydroxyflavone		85.55	28.5
13	Catechin	(±)-3,3′,4′,5,7-Flavanpentol		765.44	7.9

Table 5 showed that *p*-hydroxybenzoic acid was in highest content (12.33 mg/L), followed by vanillic acid (9.7 mg/L) and were the principal compounds in the flowers and phyllodes extract.

In the phyllodes methanol extract, the major flavonoid compounds were rutin, luteolin, apigenin, and catechin.

3. Discussion

Allelochemicals are released to the environment in different ways (by volatilisation, leaching, exudation or decomposition) and can act in a direct or an indirect way on the receptor plants [16,19,22,23]. Their chemical nature is complex and diverse (organic acids, aldehydes, coumarins, quinones, flavonoids, alkaloids, terpenoids, etc.), but the most part of them comes from three principal biosynthetic routes, the route of shikimic acid (benzoic and cinnamic acids and their derivatives, coumarins, glycosides, alkaloids, etc.), and the routes of acetic and mevalonic acids (terpenoids, steroids, complex quinones, etc.) [3,9,10,16]. Allelopathic compounds are present in almost all plants and they can be found in many parts of the plant like in roots, seeds, leave, fruits, stems, etc.

The leaf and root growth of lettuce was decreased following treatment with *Acacia melanoxylon* R.Br. aerial foliage aqueous extract at all concentration tested. The phytotoxicity of *A. melanoxylon* R.Br. was previously studied by Souto et al. [14], who reported that soil bioassays showed clear inhibitory effects on growth and germination of understory plants, particularly soils from *Eucalyptus* and *Acacia* stands. In this study, the effects of water soluble allelochemicals appear to alter a variety of morpho-physiological functions and significantly reduced the leaf fresh and dry biomass. Zhang et al. [24], showed that allelochemical (isoliquiritigenin) in the concentrations ranged between 0.2–1.0 mM exhibited toxic impacts that decreased the lettuce seedling growth. However, the results were dose dependent. They reported that isoliquiritigenin is an important phytochemical that caused >40% inhibition of radicle elongation in lettuce seedlings following exposure to 0.8 mM concentration. In another study, Sánchez-Moreiras et al. [25] documented that natural product 2-(3*H*)-benzoxazolinone caused a 50% reduction in the radicle length of lettuce at a concentration of 0.9 mM. Several authors indicated that it is extremely difficult to separate the growth and physiological secondary effects afrom the primary impacts caused by allelochemicals. However, the secondary impacts led to the disruption of cell differentiation, plant water status, water uptake, respiration, signal transduction, photosynthesis and enzyme function [3,6,7,10,17,26].

The germination and seedling growth (root, shoot length, plant biomass) of plant seeds constitutes a primary step in the growth and development of many plant species and demonstrate an importance to highlight the allelopathic activity. Our seedling growth bioassays showed that AFE and APE exhibit significant amount of water soluble allelochemicals and had phytotoxic impact ($p > 0.05$) on lettuce growth. These water soluble polyphenols (phenolic compounds and flavonoids) have allelopathic effect on seedling growth, biomass and biochemical traits (leaf protein contents) of lettuce. In some trials, low dose of allelochemicals showed stimulation while higher concentration was lethal. Similarly, Al-Wakeel et al. [27] documented that *A. nilotica* leaf extract enhanced the plant growth in peas, while a higher concentration was toxic that significantly decrease plant growth in the pea.

Several authors documented that, apart from environmental stresses, different invasive plant species showed phytotoxicity on the physiological and biochemical attributes of lettuce. In our results, *Acacia melanoxylon* flowers was highly deleterious at (100%) concentration tested and significantly inhibited protein contents in *L. sativa* as compared to control. However, Wang et al. [28] evaluated the leaf extract of alien invasive plant species, *Solidago canadensis* L. alone and joint application of nitrogen and cadmium stress. The *S. canadensis* employee distinctly allelopathy on germination and growth characteristics of lettuce and remarkably increased with increasing concentration of leaf extracts. However, N application relieved the allelopathy, while Cd treatment proved to be lethal, inhibiting the physiological traits of the lettuce.

In another study, BOA caused a significant reduction in leaf proteins due to protein denaturation, following exposure to allelochemicals. They reported that protein synthesis was significantly reduced following treatment with BOA [25]. However, contrary, there was an increase in leaf proteins in maize and decrease in kidney beans following treatment with *Acacia nilotica* extract [29]. Another study conducted by Lu et al. [30] showed that leaf extracts from three invasive plant species [(*Solidago canadensis* L., *Erigeron annuus* (L.) Pers., and *Conyza canadensis* L. Cronq)] inhibited the seed germination of lettuce. Meanwhile, heavy metals (Cu, Pb) promoted the invasion of all three plant species and allelopathic phenomena were more severe in the presence of heavy metals. Allelopathyic impact was significantly varied from one to other invasive species and *S. Canadensis* demonstrated moe phytotix than others. The allelopathic activity depended on a number of factors such as evaluating vegetation, extract types and concentration, and environmental attributes. Our results showed that aqueous extract of flowers and phyllodes from *A. melanoxylon* proved phytotoxic in nature (due to the presence of phytochemicals) and hence decreased the growth and morpho-physiological parameters of *L. sativa* in a dose-dependent manner (Tables 1 and 2). These results are in conformity with the earlier reports of allelopathy in *A. melanoxylon* [13–15]. According to Chou et al. [31], foliage extract of *Acacia confusa* Merr showed significant inhibitory effects on target plant species. Similarly, different concentrations of *Acacia auriculiformis* A. Cunn. ex Benth. also showed phytotoxicity against a number of vegetations [32]. Similar results were reported for *Acacia nilotica* [29], *Acacia auriculiformis* [33], *Acacia nilotica* [27]. In the present study, we found that flowers of *A. melanoxylon* were more phytotoxic as compared to phyllodes. It might be due to the difference in phytotoxicity of different secondary metabolites present in two plant organs and their chemical structure. Several other allelopathic crops, such as sorghum root extract, significantly reduced growth and development of several weeds in the wheat fields [34]. The allelopathic crop residue also showed inhibition of germination of different weeds seeds [35].

The potential effect of the allelochemical will be higher or lower depending on the concentration, the soil transformations, and the target species, but also on the environmental factors and the simultaneous occurrence of other biotic or abiotic stresses. Most of the polyphenols, identified in the present study (Table 5) are assumed to be water soluble and, mostly, phenolics have good potential as templates for new herbicide classes [36,37]. Different polyphenols (phenolic compounds, flavonoids, derivatives of hydoxybenzoic acids, derivatives of cinnamic acids) were identified and seem to be water soluble; previously, it was elaborated that most of the allelochemicals, when in contact with plant cell walls, caused phytotoxicity and interfered in the ecophysiological parameters of the target plants [36,38,39]. These phytochemicals possess certain properties and can be used them as lead compounds for new herbicide discovery program [3,19,23,40,41].

In terms of the allelopathy bioassay employed in this study, it has been widely used in evaluating the allelopathic impact and is also short, simple and highly sensitive [42,43]. Germination, growth and seedling development conditions of the pot experiment are similar to those in the natural environment than that of the laboratory [30]. Therefore, it was suggested that growth and ecophysiological study of the target plant in the natural environment needs further investigation to study the allelopathic stress of the donor plant. Growth, fresh weight, dry weight, shoot length and root length attributes are important biological attributes that were selected as to evaluate the *A. melanoxylon* allelopathy, whose extract concentration will affect the abundance and competitiveness of recipient plants in natural habitat. However, at present, we have not studied the allelopathic effect of *A. melanoxylon* extracts on *L. sativa* seedlings at the molecular level, so the effects on cell division and elongation, growth regulation system and respiration need to be determined in the future context.

The polyphenol composition of *Acacia melanoxylon* aerial foliage is shown in Table 5. Several phytochemical constituents were identified from foliage extracts but we concentrated mainly on phenolics and flavonoids. Among the phenolic compounds, eight were the most important and include Gallic acid, protocatechuic acid, *p*-hydroxybenzoic acid, *p*-hydroxybenzaldehyde, vanillic acid, siringic acid, *p*-coumaric acid, and ferulic acid constituents.

The *p*-hydroxybenzoic acid was highest in content (12.33 mg/L), followed by vanillic acid (9.7 mg/L) and these were the principal compounds in the flower and phyllode extracts as shown in Table 5. In the phyllode methanol extract, major flavonoid compounds were rutin, luteolin, apigenin, and catechin. According to studies of Souto et al. (14), allelopathy was the main process involved in the inhibition of soil microbial activity and retard the germination attributes of *Lactuca sativa* L., *Dactylis glomerata* L.

and *Trifolium repens* L. following *Acacia* and *Eucalyptus* population as compared to the autochthonous *Quercus robur* L. The phytotoxic effects were more prominent during *Acacia* blooming stages. In another investigation carried out through gas chromatography and mass spectroscopy, Freire et al. [44] documented the presence of several polyphenols and phytotoxic molecules such as cinnamic acid derivatives, spinasteryl glucoside and dihydrospinasteryl glucosides. Physiological and biochemical parameters (seedling growth, sugar contents and chlorophyll content) in *Zea mays* L. (maize) and *Phaseolus vulgaris* L. were decreased by *Acacia nilotica* and *Eucalyptus rostrata* leaf leachates (29). In a pot experiment, *Acacia nilotica* leaves extract showed a significant reduction in the growth and metabolic activities of 45-day-old pea (*Pisum sativum* L.) plants. Meanwhile, higher concentration was inhibitory while lower concentration was stimularory. Qualitative and quantitative HPLC analysis of water extract of *Acacia nilotica* leaves revealed that protocatechuic and caffeic acids were the principal phenolic compounds accompanied by ferulic, cinnamic acids and apigenin in higher quantity, whereas pyrogallic, *p*-coumaric, syringic acids and coumarin were found in trace amounts [27].

4. Methods and Materials

4.1. Plant Materials

The fresh aerial parts (shoots exhibiting flowers and phyllodes) of *Acacia melanoxylon* were gathered from natural population from the mountainous area, Lagoas Marcosende campus, University of Vigo, during blooming stage (Pontecedra province, Spain, 42.2406° N, 8.7207° W). The seeds of Lettuce (*Lactuca sativa* L. cv. Great Lakes California (Asteraceae, Asterales), a food and cash crop, were purchased from Semillas Fito (Barcelona, Spain) and used for seedling growth bioassays.

4.2. Extraction of Polyphenols and HPLC Analysis

The fresh plant tissue was cut into small pieces with scissors and placed in the laboratory at room temperature (25 °C) for air drying because phenolics are not stable to drying in an oven at high temperature. The dry flowers or leaves (1 g), were mixed with 50 mL of methanol/HCl (1000:1, *v*/*v*) in a 100-mL Erlenmeyer flask and kept in darkness for 12 h with hand shaking every three hours. The mixture was filtered by using a filter paper and a funnel. The filtrate was saved in refrigerator and the process was repeated with residual plant material for another 12 h by adding 50 mL of methanol/HCl. The mixture was filtered again and two methanolic fractions were combined (approx. 100 mL). This mixture was dried in rotary evaporator under vacuum. The temperature of rotary evaporator was maintained below 35 °C. The remaining material was dissolved in 40 mL of ethanol/water (2:8, *v*/*v*) and filtered. Afterwards, three sequential extractions were carried out with 20 mL of diethyl ether. The mixture was extracted with an extraction funnel by shaking vigorously for one minute each time, waiting until complete separation into two phases: the aqueous one in the lower part and an organic one in the upper part of funnel. The organic phases were removed and saved. We collected three ethereal phases in an Erlenmeyer flask. After that, three sequential extractions were carried out with 20 mL ethyl acetate on aqueous phase, obtaining three new organic phases that were collected and combined with ethereal ones. The total organic fractions obtained in this way were dehydrated with anhydrous sodium sulphate for 30 min to remove minimum residual water. Subsequently, it was filtered to withdraw sodium sulphate and evaporated to dryness in rotary evaporator. The final residue containing phenolics were re-dissolved in 1 mL methanol/water (1:1, *v*/*v*) and filtered through a 0.45-μm-pore-size nylon membrane filter and saved in refrigerator at 4 °C until HPLC analysis. The henolics were extracted by the method as proposed by Macheix et al. [45] and was optimized by Bolaño et al. [46].

4.3. Extraction of Flavonoids from Flowers and Leaves of *A. melanoxylon*

The flavonoids were extracted by the method of Markham 1989 with certain modifications. The fresh plant tissue was cut into small pieces with scissors and placed in the laboratory at room temperature (25 °C) for air drying. The dried and powdered flowers or leaves (10 g) were macerated with 300 mL of methanol/water (80:20) for 24 h at room temperature with continuous shaking. The extract was obtained through filtration by using a Buchner funnel with a filter paper. This mixture was dried in rotary evaporator under vacuum until all methanol was removed. The remaining material was re-dissolved in 40 mL of ethanol/water (2:8, *v*/*v*) and filtered. The resultant aqueous material was extracted with petroleum ether in an extraction funnel to remove fats, terpens, chlorophylls and xanthophylls. This extraction was repeated three times. Afterwards the mixture was extracted as per described in Phenolics acids, i.e., three sequential extractions with diethyl ether followed by three sequential extractions with ethyl acetate. The total organic fractions obtained in this way were dehydrated with anhydrous sodium sulphate for 30 min to remove minimum residual water. Subsequently, it was filtered to withdraw sodium sulphate and evaporated to dryness in rotary evaporator. The final residue was re-dissolved in 2.5 mL methanol and filtered through a 0.45 μm pore size nylon membrane filter and saved in refrigerator at −20 °C until HPLC analysis.

4.4. UV-DIODE ARRAY Chemical Analyses

Analysis was performed using a Shimadzu chromatograph equipped with a UV-DIODE ARRAY detector to identify phenolic and flavonoids. Identification of the compounds was made by using a reverse-phase Waters Nova-Pak C-18 (4.6 × 250 mm) column with a 4μm particle size. For flavonoids, the extracts were analyzed using two mobile phases: (A) methanol:phosphoric acid 999:l and (B) water: phosphoric acid 999:1. Linear gradients starting with 20% (A) and ending with 100% (A) were used over the first 40 min with an additional 5 min at 100% (A). The flow rate of the mobile phase was 1 mL/min and the eluate was analyzed at 250–400 nm. For the phenolic compounds, extracts were analyzed using two mobile phases: (A) water: acetic acid 98:2 and (B) water: methanol: acetic acid 68:30:2. Linear gradients starting with 100% (A) and ending with 20% (A) were used over the first 59 min, with an additional 6 min at 20% (A). The flow rate of the mobile phase was 0.8 mL/min and the eluate was analysed at 210–400 nm.

4.5. Acacia melanoxylon Flowers and Phyllodes Aqueous Extract Preparation for Bioassays

Fresh shoots of *Acacia melanoxylon* were collected from natural population in the surrounding area of Lagoas Marcosende campus of University of Vigo, during flowering period. The flowers and phyllodes were separated from branches and soaked in distilled water in the ratio of 1:1 (*w*/*v*) at room temperature and left in the laboratory for 24 h. Similar extraction techniques were used by Molina et al. [47], in their studies of *Eucalyptus* spp. and Lorenzo et al. [48], in their studies of allelopathic effects of *Acacia dealbata* L. The extract was collected, filtered through filter paper and described as 100%. Distilled water was added in this solution to make different dilution (75%, 50% and 25%). We examined effects of both phyllodes and flowers extracts made by mixing plant material in water, as opposed to the often criticized method of tissue disruption and organic solvent extraction that may yield artificially higher levels of specific compounds [37].

4.6. Plant Material and Growth Conditions

Lettuce (*Lactuca sativa* L. cv. Great Lakes California (Asteraceae, Asterales)), a food and cash crop and the most widely used model species in allelopathic studies [49], was selected because of its fast germination and homogeneity. The seeds of test species were purchased commercially from Semillas Fito (Barcelona, Spain). The seeds were placed in plastic trays (32 × 20 × 6 cm) with a 5-cm-deep layer of perlite (500 g/tray). The trays were irrigated on alternate days with tap water until germination of seeds and thereafter with 500 mL 1:1 Hoagland solution/tray, twice in a week. Seedlings were germinated in darkened at 20 °C temperatures in environmentally controlled growth chamber. For seedling growth, the environmental conditions were as follows; temperature: 18/8 °C (day/night) and 12/12 h (light/darkness) photoperiod, 80% relative humidity and 200 μmol m^{-2} s^{-1} irradiance. One month old seedlings (when plants have three fully expanded leaves), were transferred to pots (10 cm) containing perlite (70g) to stimulate the development of root system and shifted to the glass house with same growing condition and nutrient solution (100 mL/pot). Every second day the pots were well watered with tap water through an automatic irrigation system. One week later, treatment solutions (100 mL/pot) were applied three times (day one, three and five) and measurements were taken on each day. The temperature in the glass house was maintained at 21 ± 2 °C with a relative humidity of 75%. The glasshouse was ventilated with outside air to ensure steady CO_2. Aqueous extract (100%, 75%, 50%, 25%) of *Acacia melanoxylon* (flowers and phyllodes) and control were watered three times (days 1, 3 and 5) with 100 mL solution. The experiment was laid out in Randomized Complete Block Design (RCBD) and replicated thrice.

4.7. Plant Growth Measurements

Information about plant height/root length was obtained with a ruler and values were expressed in cm. The fresh and oven dry plant leaves and roots weight were obtained by first weighting independently fresh leaves and roots then after drying these samples in a circulatory air oven at 70 °C for 72 h. The samples were weighed again to get dry weight of plant.

4.8. Statistical Analysis

Data was analyzed with the statistical package SPSS® (version 21.00) for Windows® (SPSS Inc., Chicago, IL, USA). Data were analysed by using one-way analysis of variance (ANOVA) (Sokal and Rohlf 1995) (when variance were homogeneous) or Kruskal-Wallis test (when heterogeneous). The LSD test as post hoc test was used to determine main differences between treatment means. Significant differences between means of treatment were compared at 5% probability level.

5. Conclusions

The exotic invasive tree in North Western Iberian Peninsula (Galicia, Spain), *Acacia melanoxylon*, their aerial foliage (flowers and phyllodes) extracts have negative effects on the growth and biomass of a general biotest species *Lactuca sativa* seedlings. Allelopathic effect of *A. melanoxylon* extracts at higher concentration was stronger than that of lower concentration and flowers extract showed stronger inhibition in different attributes than phyllodes extract. According to HPLC data of flowers and phyllodes extracts, several phyto-chemicals were identified and they are represented as derivatives of benzoic acid and cinnamic acid. Apart from these phenolic acids, a significant amount of flavonoids such as rutin, quercetin, luteolin, apigenin, and catechin were identified that enhanced the phytotoxicity of the extract. The presence of all these polyphenols thus affects the growth, biomass and root development rate of *L. sativa*. What's more, the inhibitory effect of flower extract is stronger than that of phyllodes extract, which is one of the reasons that the allelopathic molecules are in much higher amounts in flowers than phyllodes in *A. melanoxylon*.

Author Contributions: M.I.H. conducted the experiment, collected data, and drafted the manuscript. M.J.R. designed the experiment and followed upon data collection, helped in article draft, correction, and revision. M.A.E.-S. provided support through the statistical analysis, English language and grammar correction, Ms revision, and edited the final manuscript. All authors have read and agreed to the published version of the manuscript.

Funding: The research was supported by the Project 10PXIB310261PR from the Galician Government and Project AGL2010-17885 (sub-programa AGR) "Estudio del modo de acción del citral en larespuesta temprana de plantas superior" funded by Ministerio de Ciencia e Innovación, Subdirección General de Proyectos de Investigación y Proyectos de Investigación Fundamental no orientada. Author M.A.El-Sheikh extends his appreciation to the Researchers Supporting Project Number (RSP-2020/182), King Saud University, Riyadh, Saudi Arabia.

Acknowledgments: We thank Aldo Barreiro, Carlos Bolano and Paula Lorenzo with field and laboratory assistance. Author M.A.El-Sheikh extends his appreciation to the Researchers Supporting Project Number (RSP-2020/182), King Saud University, Riyadh, Saudi Arabia.

Conflicts of Interest: The authors declare no conflict of interest.

References

1. Anaya, A.L.; Pelayo-Benavides, H.P. Allelopathic potential of *Mirabilis jalapa* L. (*Nyctaginaceae*): Effects on germination, growth and cell division of some plants. *Allelop. J.* **1997**, *4*, 57–68.
2. Reigosa, M.J.; Sanchez-Moreiras, A.M.; Gonzalez, L. Ecophysiological approaches in allelopathy. *Crit. Rev. Plant Sci.* **1999**, *18*, 83–88. [CrossRef]
3. Hussain, M.I.; Reigosa, M.J. Allelochemical stress inhibits growth, leaf water relations, PSII photochemistry, non-photochemical fluorescence quenching and heat energy dissipation in three C_3 perennial species. *J. Exp. Bot.* **2011**, *62*, 4533–4545. [CrossRef]
4. Lockwood, J.L.; Simberloff, D.; Mckinney, M.L.; Von Holle, B. How many, and which, plants will invade natural areas. *Biol. Invasions* **2001**, *3*, 1–8. [CrossRef]
5. Hussain, M.I.; Reigosa, M.J. Characterization of xanthophyll pigments, photosynthetic performance, photon energy dissipation, reactive oxygen species generation and carbon isotope discrimination during artemisinin-induced stress in *Arabidopsis thaliana*. *PLoS ONE* **2015**, *10*, e0114826. [CrossRef]
6. Hussain, M.I.; El-Keblawy, A.; Tsombou, F.M. Leaf age, canopy position, and habitat affect the carbon isotope discrimination and water-use efficiency in three C_3 leguminous *Prosopis* species from a hyper-arid climate. *Plants* **2019**, *8*, 402. [CrossRef]
7. Hussain, M.I.; Tsombou, F.M.; El-Keblawy, A. Surface canopy position determines the photosystem II photochemistry in invasive and native *Prosopis* Congeners at Sharjah Desert, UAE. *Forests* **2020**, *11*, 740. [CrossRef]
8. Ma, H.; Chen, Y.; Chen, J.; Zhang, Y.; Zhang, T.; He, H. Comparison of allelopathic effects of two typical invasive plants: *Mikania micrantha* and *Ipomoea cairica* in Hainan island. *Sci. Rep.* **2020**, *10*, 1–10. [CrossRef]
9. Chen, B.M.; Liao, H.X.; Chen, W.B.; Wei, H.J.; Peng, S.L. Role of allelopathy in plant invasion and control of invasive plants. *Allelop. J.* **2017**, *41*, 155–166. [CrossRef]
10. Kato-Noguchi, H. Involvement of allelopathy in the invasive potential of *Tithonia diversifolia*. *Plants* **2020**, *9*, 766. [CrossRef] [PubMed]
11. Duke, S.O.; Blair, A.C.; Dayan, F.E.; Johnson, R.D.; Meepagala, K.M.; Cook, D.; Bajsa, J. Is (−)-Catechin a novel weapon of spotted knapweed (*Centaurea stoebe*)? *J. Chem. Ecol.* **2009**, *35*, 141–153. [CrossRef] [PubMed]
12. Knapic, S.; Tavares, F.; Pereira, H. Heartwood and sapwood variation in *Acacia melanoxylon* R. Br. trees in Portugal. *Forest* **2006**, *79*, 371–380. [CrossRef]
13. Souto, X.C.; González, L.; Reigosa, M.J. Comparative analysis of the allelopathic effects produced by four forestry species during the decomposition process in their soils in Galicia (NW Spain). *J. Chem. Ecol.* **1994**, *20*, 3005–3015. [CrossRef] [PubMed]
14. Souto, X.C.; Bolano, J.C.; González, L.; Reigosa, M.J. Allelopathic effects of tree species on some soil microbial populations and herbaceous plants. *Biolog. Plant.* **2001**, *44*, 269–275. [CrossRef]
15. González, L.; Souto, X.C.; Reigosa, M.J. Allelopathic effects of *Acacia melanoxylon* R. Br. phyllodes during their decomposition. *For. Ecol. Manag.* **1995**, *77*, 53–63. [CrossRef]
16. Hussain, M.I.; González, L.; Souto, C.; Reigosa, M.J. Ecophysiological responses of native plants to phytotoxic effect of *Acacia melanoxylon* R. Br. *Agrofor. Syst.* **2011**, *83*, 149–166. [CrossRef]
17. Farooq, M.; Khan, I.; Nawaz, A.; Cheema, M.A.; Siddique, K.H. Using sorghum to suppress weeds in autumn planted maize. *Crop Prot.* **2020**, *133*, 105162. [CrossRef]
18. Bajwa, A.A.; Nawaz, A.; Farooq, M. Allelopathic crop water extracts application improves the wheat productivity under low and high fertilizer inputs in a semi-arid environment. *Int. J. Plant Prod.* **2020**, *14*, 23–35. [CrossRef]
19. Hussain, M.I.; González, L.; Reigosa, M.J. Phytotoxic effect of allelochemicals and herbicides on photosynthesis, growth and carbon isotope discrimination in *Lactuca sativa*. *Allelop. J.* **2010**, *26*, 157–174.
20. Cheema, Z.A.; Khaliq, A.; Farooq, M. Sorghum allelopathy for weed management in wheat. In *Allelopathy in Sustainable Agriculture and Forestry*; Zeng, R.S., Mallik, A.U., Luo, S.M., Eds.; Springer: New York, NY, USA, 2008; pp. 255–270.
21. Cheema, Z.A.; Khaliq, A.; Saeed, S. Weed control in maize (*Zea mays* L.) through sorghum allelopathy. *J. Sustain. Agric.* **2004**, *23*, 73–86. [CrossRef]
22. Duke, S.O.; Dayan, F.E.; Romagni, J.G.; Rimando, A.M. Natural products as sources of herbicides: Current status and future trends. *Weed Res.* **2000**, *40*, 99–111. [CrossRef]

23. Hussain, M.I.; Shackleton, R.T.; El-Keblawy, A.; Del Mar Trigo Pérez, M.; González, L. Invasive Mesquite (*Prosopis juliflora*), an allergy and health challenge. *Plants* **2020**, *9*, 141. [CrossRef] [PubMed]
24. Zhang, S.; Sun, S.W.; Shi, H.L.; Zhao, K.; Wang, J.; Liu, Y.; Liu, X.H.; Wang, W. Physiological and biochemical mechanisms mediated by allelochemical isoliquiritigenin on the growth of lettuce seedlings. *Plants* **2020**, *9*, 245. [CrossRef]
25. Sánchez-Moreiras, M.A.; Reigosa, M.J. Whole plant response of lettuce after root exposure to BOA (2(3*H*)-benzoxazolinone). *J. Chem. Ecol.* **2005**, *31*, 2689–2703. [CrossRef] [PubMed]
26. Farooq, N.; Abbas, T.; Tanveer, A.; Jabran, K. Allelopathy for weed management. In *Co-Evolution of Secondary Metabolites*; Mérillon, J.M., Ramawat, K., Eds.; Reference Series in Phytochemistry; Springer: Cham, Switzerland, 2020; pp. 505–519. [CrossRef]
27. AL-Wakeel, S.A.M.; Gabr, M.A.; Hamid, A.A.; Abu-EL-Soud, W.M. Allelopathic effects of *Acacia nilotica* leaf residue on *Pisum sativun* L. *Allelop. J.* **2007**, *19*, 23–34.
28. Wang, S.; Wei, M.; Wu, B.; Cheng, H.; Wang, C. Combined nitrogen deposition and Cd stress antagonistically affect the allelopathy of invasive alien species Canada goldenrod on the cultivated crop lettuce. *Sci. Hortic.* **2020**, *261*, 108955. [CrossRef]
29. EL-Khawas, S.; Shehata, M.M. The allelopathic potential of *Acacia nilotica* and *Eucalyptus rostrata* on monocot (*Zea mays* L.) and dicot (*Phaseolus Vulgaris* L.) plants. *Biotechnology* **2005**, *4*, 23–34.
30. Lu, Y.; Wang, Y.; Wu, B.; Wang, S.; Wei, M.; Du, D.; Wang, C. Allelopathy of three Compositae invasive alien species on indigenous *Lactuca sativa* L. enhanced under Cu and Pb pollution. *Sci. Hortic.* **2020**, *267*, 109323. [CrossRef]
31. Chou, C.H.; Fu, C.Y.; Li, S.Y.; Wang, Y.F. Allelopathic potential of *Acacia confusa* and related species in Taiwan. *J. Chem. Ecol.* **1998**, *24*, 2131–2150. [CrossRef]
32. Hoque, A.T.M.R.; Ahmad, R.; Uddin, M.B.; Hossain, M.K. Allelopathic effect of different concentration of water extracts of *Acacia auriculiformis* leaf on some initial growth parameters of five common agricultural crops. *Pak. J. Agron.* **2003**, *2*, 92–100.
33. Oyun, M.B. Allelopathic potentialities of *Gliricidia sepium* and *Acacia auriculiformis* on the germination and seedling vigour of Maize (*Zea maya* L.). Amer. *J. Agric. Biol. Sci.* **2006**, *1*, 44–47. [CrossRef]
34. Roth, C.M.; Shroyer, J.P.; Paulsenice, G.M. Allelopathy of sorghum on wheat under several tillage systems. *Agron. J.* **2000**, *92*, 855–860. [CrossRef]
35. Khaliq, A.; Matloob, A.; Irshad, M.S.; Tanveer, A.; Zamir, M.S.I. Organic weed management in maize (*Zea mays* L.) through integration of allelopathic crop residues. *Pak. J. Weed Sci. Res.* **2010**, *16*, 409–420.
36. Djurdjevic, L.; Dinic, A.; Pavlovic, P.; Mitrovic, M.; Karadzic, B.; Tesevic, V. Allelopathic potential of *Allium ursinum* L. *Biochem. Syst. Ecol.* **2004**, *32*, 533–544. [CrossRef]
37. Inderjit; Duke, S.O. Ecophysiological aspects of allelopathy. *Planta* **2003**, *217*, 529–539. [CrossRef] [PubMed]
38. Thiébaut, G.; Tarayre, M.; Rodríguez-Pérez, H. Allelopathic effects of native versus invasive plants on one major invader. *Front. Plant Sci.* **2019**, *10*, 10. [CrossRef] [PubMed]
39. Rob, M.; Hossen, K.; Iwasaki, A.; Suenaga, K.; Kato-Noguchi, H. Phytotoxic activity and identification of phytotoxic substances from *Schumannianthus dichotomus*. *Plants* **2020**, *9*, 102. [CrossRef]
40. Hussain, M.I.; Reigosa, M.J. Higher peroxidase activity, leaf nutrient contents and carbon isotope composition changes in *Arabidopsis thaliana* are related to rutin stress. *J. Plant Physiol.* **2014**, *171*, 1325–1333. [CrossRef]
41. Irimia, R.E.; Lopes, S.M.; Sotes, G.; Cavieres, L.A.; Eren, Ö.; Lortie, C.J.; French, K.; Hierro, J.L.; Rosche, C.; Callaway, R.M.; et al. Biogeographic differences in the allelopathy of leaf surface extracts of an invasive weed. *Biol. Invasions* **2019**, *21*, 3151–3168. [CrossRef]
42. Zhang, F.J.; Guo, J.Y.; Chen, F.X.; Liu, W.X.; Wan, F.H. Identification of volatile compounds released by leaves of the invasive plant croftonweed (*Ageratina adenophora*, compositae), and their inhibition of rice seedling growth. *Weed Sci.* **2012**, *60*, 205–211. [CrossRef]
43. Shen, S.; Xu, G.; Li, D.; Clements, D.R.; Jin, G.; Liu, S.; Yang, Y.; Chen, A.; Zhang, F.; Kato-Noguchi, H. Allelopathic potential of sweet potato (*Ipomoea batatas*) germplasm resources of Yunnan Province in southwest China. *Acta Ecol. Sin.* **2018**, *38*, 444–449. [CrossRef]
44. Freire, C.S.; Silvestre, A.J.; Neto, C.P. Demonstration of long-chain n-alkyl caffeates and Δ^7-steryl glucosides in the bark of *Acacia* species by gas chromatography–mass spectrometry. *Phytochem. Anal. Int. J. Plant Chem. Biochem. Tech.* **2007**, *18*, 151–156. [CrossRef] [PubMed]
45. Macheix, J.J.; Fleuriet, A.; Billot, Y. *Fruit Phenolics*; C.K.C. Press Inc.: Boca Ratón, FL, USA, 1990.

46. Bolaño, J.C.; González, L.; Souto, X.C. Análisis de compuestos fenólicos de bajo peso molecular inducidos por condiciones de estrés, mediante HPLC. In *Manual de Técnicas en Ecofisiología Vegetal*; Pedrol, N., Reigosa, M.J., Eds.; Gamesal: Vigo, Spain, 1997; pp. 31–38.
47. Molina, A.; Reigosa, M.J.; Carballeira, A. Release of allelochemical agents from litter, throughfall, and topsoil in plantations of *Eucalyptus globulus* Labill in Spain. *J. Chem. Ecol.* **1991**, *17*, 147–160. [CrossRef] [PubMed]
48. Lorenzo, P.; Pazos-Malvido, E.; Gonzalez, L.; Reigosa, M.J. Allelopathic interference of invasive *Acacia dealbata*: Physiological effects. *Allelop. J.* **2008**, *22*, 452–462.
49. Macías, F.A.; Castellano, D.; Molinillo, J.M.G. Search for a standard phytotoxic bioassay for allelochemicals. Selection of standard target species. *J. Agric. Food Chem.* **2000**, *48*, 2512–2521. [CrossRef]

MDPI
St. Alban-Anlage 66
4052 Basel
Switzerland
Tel. +41 61 683 77 34
Fax +41 61 302 89 18
www.mdpi.com

Plants Editorial Office
E-mail: plants@mdpi.com
www.mdpi.com/journal/plants

www.ingramcontent.com/pod-product-compliance
Lightning Source LLC
Chambersburg PA
CBHW061601220926
51850CB00005B/388